Intelligent Systems and Sustainable Computational Models

The fields of intelligent systems and sustainability have been gaining momentum in the research community. They have drawn interest in such research fields as computer science, information technology, electrical engineering, and other associated engineering disciplines. The possibility of applying intelligent systems to sustainability is becoming a reality due to the recent advancements in the Internet of Things (IoT), artificial intelligence, big data, blockchain, deep learning, and machine learning. The emergence of intelligent systems has given rise to a wide range of techniques and algorithms using an ensemble approach to implement novel solutions for complex problems associated with sustainability.

Intelligent Systems and Sustainable Computational Models: Concepts, Architecture, and Practical Applications explores this ensemble approach towards building a sustainable future. It explores novel solutions for such pressing problems as smart healthcare ecosystems, energy efficient distributed computing, affordable renewable resources, mitigating financial risks, monitoring environmental degradation, and balancing climate conditions. The book helps researchers to apply intelligent systems to computational sustainability models to propose efficient methods, techniques, and tools.

The book covers areas such as the following:

- Intelligent and adaptive computing for sustainable energy, water, and transportation networks
- Blockchain for decentralized systems for sustainable applications, systems, and infrastructure
- IoT for sustainable critical infrastructure
- Explainable AI (XAI) and decision-making models for computational sustainability
- Sustainable development using edge computing, fog computing, and cloud computing
- Cognitive intelligent systems for e-learning
- Artificial intelligence and machine learning for large scale data
- Green computing and cyber-physical systems
- Real-time applications in healthcare, agriculture, smart cities, and smart governance.

By examining how intelligent systems can build a sustainable society, the book presents systems solutions that can benefit researchers and professionals in fields such as information technology, health, energy, agriculture, manufacturing, and environmental protection.

Intelligent Systems and Sustainable Computational Models

Concepts, Architecture, and Practical Applications

Edited by
Rajganesh Nagarajan, Senthil Kumar Narayanasamy,
Ramkumar Thirunavukarasu, and Pethuru Raj

CRC Press
Taylor & Francis Group
Boca Raton London New York

CRC Press is an imprint of the
Taylor & Francis Group, an **informa** business
AN AUERBACH BOOK

First edition published 2024
by CRC Press
2385 NW Executive Center Drive, Suite 320, Boca Raton FL 33431

and by CRC Press
4 Park Square, Milton Park, Abingdon, Oxon, OX14 4RN

CRC Press is an imprint of Taylor & Francis Group, LLC

ISBN: 978-1-032-52703-1 (hbk)
ISBN: 978-1-032-52705-5 (pbk)
ISBN: 978-1-003-40795-9 (ebk)

DOI: 10.1201/9781003407959

Typeset in Times
by Newgen Publishing UK

Contents

Preface..ix
About the Editors xi

Chapter 1 Smart Power Management in Data Centers Using Machine-Learning Techniques........ 1

D.V. Ashoka, P.M. Rekha, and P.R. Sudha

Chapter 2 Exploring the Power of Deep Learning and Big Data in Flood Forecasting: State-of-the-Art Techniques and Insights.. 14

G. Selva Jeba and P. Chitra

Chapter 3 Storage Management Techniques for Medical Internet of Things (MIoT) 34

S.U. Muthunagai, M.S. Girija, B. Praveen Kumar, and R. Anitha

Chapter 4 A Study on Trending Technologies for IoT Use Cases Aspires to Build Sustainable Smart Cities...48

Mangayarkarasi Ramaiah, R. Mohemmed Yousuf, R. Vishnukumar, and Adla Padma

Chapter 5 Hydro-Meteorological Disaster Prediction Using Deep Learning Techniques..........61

P. Kaviya and P. Chitra

Chapter 6 Assessment of ICT for Sustainable Developments with Reference to Fog and Cloud Computing 82

H.K. Shilpa, D.K. Girija, M. Rashmi, and N. Yogeesh

Chapter 7 Explainable Artificial Intelligence (XAI) for Computational Sustainability: Concepts, Opportunities, Challenges, and Future Directions 103

B. Prabadevi, M. Pradeepa, and S. Kumaraperumal

Chapter 8 Edge Computing-Based Intrusion Detection Systems: A Review of Applications, Challenges, and Opportunities 117

Posham Uppamma and Sweta Bhattacharya

Chapter 9 Recent Advancements in IoT Security-Based Challenges: A Brief Review 136

Suranjeet Chowdhury Avik, Abdullahi Chowdhury, Ranesh Naha, Shahriar Kaisar, Arunkumar Arulappan, and Aniket Mahanti

Chapter 10 An Approach to Smart Targeted Advertising Using Deep Convolutional
Neural Networks...150

A. Gayathri, D. Ruby, N. Manikandan, and T. Gopalakrishnan

Chapter 11 Text Classification of Customer and Salesperson Conversations to
Predict Sales Using Ensemble Models...162

T. Chellatamilan and Neel Rakesh Choksi

Chapter 12 Sentimental Analysis on Amazon Book Reviews: A Deep Learning Approach 188

A. Vijayalakshmi, Koesha Sinha, and Debopriya Bose

Chapter 13 A Deep LSTM Recurrent Learning Approach for Sentiment Analysis on
Movie Reviews 202

*G.R. Khanaghavalle, V. Rajalakshmi, R. Jayabhaduri, A. Kala,
and P. Sharon Femi*

Chapter 14 Cognitive Intelligent Personal Learning Assistants for Enriching
Personalized Learning 212

D. Ramalingam and Mahalakshmi Dharmalingam

Chapter 15 Natural Language Processing for Fake News Detection Using Hybrid Deep
Learning Techniques 224

*B. Valarmathi, Aditya Kocherlakota, Yuvraj Das, Aritam,
N. Srinivasa Gupta, and V. Mohanraj*

Chapter 16 A Comparative Analysis of Deep Learning Models for Fake News Detection
and Popularity Prediction of Articles 246

Jayanthi Devaraj

Chapter 17 Internet of Things (IoT)-Based Smart Maternity Healthcare Services 266

P. Vinothiyalakshmi, V. Pallavi, N. Rajganesh, and V. Adityavignesh

Chapter 18 A Real-Time Automated Face Recognition and Detection System for
Competitive Examination 275

Rajalakshmi Gurusamy and B. Ben Sujitha

Chapter 19 Medical Image Analysis with Vision Transformers for Downstream Tasks
and Clinical Report Generation...288

Evans Kotei and Ramkumar Thirunavukarasu

Chapter 20 Ensemble Embedding and Convolutional Neural Network-Based Big Data
Framework for Structure Prediction of Proteins 308

*Leo Dencelin Xavier, Ramkumar Thirunavukarasu, Rajganesh Nagarajan,
and Mohamed Uvaze Ahamed Ayoobkhan*

Chapter 21 Deep Learning-Based Automated Diagnosis and Prescription of
Plant Diseases...321

R.K. Kapila Vani, P. Geetha, D. Abhishek, K. Gokul Krishna, and V. Akaash

Chapter 22 Intelligent Farming Through Weather Forecasting Using Deep Learning
Techniques for Enhancing Crop Productivity 339

V. Ezhilarasi, S. Selvamuthukumaran, and N. Srinivasan

Chapter 23 Plant Disease Detection and Classification Using a Deep Learning Approach
for Image-Based Data...352

D. Tamil Priya and A. Vijayarani

Chapter 24 Deep Learning-Based Object Detection in Real-Time Video 369

T. Sukumar

Chapter 25 Prediction of COVID Stages Using Data Analysis and Machine Learning 383

Rajalakshmi Gurusamy, S. Siva Ranjani, and G. Susan Shiny

Chapter 26 A Statistical Analysis of Suitable Drugs for Major Drug Resistant
Mutations in the HIV-1 Group M Virus 397

*N. Durga Shree, D.A. Steve Mathew, Ramkumar Thirunavukarasu,
and J. Arun Pandian*

Index...415

Preface

In the realm of technological advancements, the domain of intelligent systems (IS) has witnessed an extraordinary surge, casting a profound impact on numerous real-time applications that confer a competitive advantage over traditional approaches. This surge has not only unleashed a plethora of techniques and algorithms to tackle complex challenges but has also endowed us with the flexibility to navigate the intricate landscape of sustainability. The nexus emerged for intelligent systems with sustainable paradigms has proven to be a dynamic force, capable of unraveling anomalies, mitigating uncertainties, and achieving balance in the intricate real-world problems.

As we stand amidst this confluence of intelligence and sustainability, it has become evident that the intelligent systems have demonstrated a profound potential to redefine its approach to sustainability across diverse sectors. The journey of sustainable computing has made an increasing dominance with threads of a multidisciplinary approach in collaborating with emerging technologies and with some of the state-of-the-art techniques. In the context of dwindling natural resources, the importance of sustainability magnifies, pressing us to cultivate intelligent strategies for complex challenges. The advent of the Internet of Things (IoT) and the proliferation of deep learning forge the path for innovative solutions and resilient systems. Similarly, there has been tremendous improvements witnessed in recent years through the lens of machine learning, natural language processing, cloud computers, artificial intelligence, and many more sub-fields still emerging gradually.

Within the chapters of this book, the readers can embark on a journey through the advancement of intelligent systems and sustainable computing. As we are contemplating the power of technology to steer our world toward sustainability, this book covers the topics ranging from handling health crises, fake news detection, optimizing smart systems for real-world scenarios, sentiment analysis, and financial stock predictions. The compilation of this book is a testament to the collective wisdom of experts, innovators, and scholars who have convened to demystify the intricacies of intelligent systems and sustainable computing. The insights shared within these pages stand as a testament to the potential of harmonizing technological prowess with environmental stewardship. As we continue to push the boundaries of knowledge, this book would certainly serve as a guiding light towards a future defined by intelligence, compassion, and sustainability.

Dr. Rajganesh Nagarajan
Dr. Senthil Kumar Narayanasamy
Dr. Ramkumar Thirunavukarasu
Dr. Pethuru Raj

About the Editors

Rajganesh Nagarajan is presently working as Associate Professor in the Department of Computer Science and Engineering, Sri Venkateswara College of Engineering, Sriperumbudur, Tamilnadu, India. He obtained a Ph.D degree from Anna University during the year 2018 for his thesis entitled "Fuzzy based Intelligent Semantic Cloud Service Discovery for Effective Utilization of Services". He has 18 years of experience in teaching and has contributed research findings in various reputed international journals. During his career, he has attended more than 20 Faculty Development Programs/Workshops/Seminars, which are sponsored by AICTE, UGC, ISTE, and Anna University. He has functioned as a resource person in more than 10 Faculty Development Programmes and organized seminars and workshops. He is functioning as an active reviewer for top-notch journals from IEEE, Springer, Elsevier, and other publishers.

Senthilkumar Narayanasamy is Assistant Professor (Senior) in the Department of Computer Applications, School of Computer Science Engineering and Information Systems, Vellore Institute of Technology (VIT), Vellore. He has been working at VIT for more than 15+ years and, in total, he has 18+ years of teaching experience. Currently, he is the RAAC coordinator for the school. Prior to that, he was the Proctor Coordinator and Project Coordinator of the school. He has delivered guest lectures, special talks, and webinars at various engineering colleges on the topics of natural language processing, big data analytics, data science and cyber security. His research interests are NLP, machine learning, and semantic web. In this connection, he has published more research articles in SCOPUS indexed journals and also published research papers at various conferences as well. He is an avid reader and always encourages others to read a lot on the subject that they are really inclined.

Ramkumar Thirunavukarasu is presently working as Professor in the School of Computer Science Engineering and Information Systems, Vellore Institute of Technology, Vellore, India. He obtained a Ph.D degree from Anna University, Chennai during the year 2010 for his thesis entitled "Synthesizing Global Association Rules in Multi-Database Mining". Having 22 years of experience in higher education and research, regularly he contributes papers in various reputed international journals. Some of his research works have been published in journals from Springer, Elsevier, Wiley, World-Scientific, and other publishers. He has functioned as a resource person in more than 20 Faculty Development Programmes and organized seminars and workshops, which are funded by ISTE, AICTE and others. He has successfully guided three research scholars to achieve Ph.D degrees.

Pethuru Raj is working as Vice President and Chief Architect at Reliance Jio Platforms Ltd. (JPL) Bangalore. Previously, Dr. Raj has worked in IBM Global Cloud Centre of Excellence (CoE), Wipro consulting services (WCS), and Robert Bosch Corporate Research (CR). He has gained over 22 years of IT industry experience and 9 years of research experience. He finished the CSIR-sponsored Ph.D degree at Anna University, Chennai and continued with the UGC-sponsored postdoctoral research in the Department of Computer Science and Automation, Indian Institute of Science (IISc), Bangalore. After that, he was granted two international research fellowships (JSPS and JST) to work as a research scientist for 3.5 years in two leading Japanese universities. He is focusing on some of the emerging technologies such as the Internet of Things (IoT), artificial intelligence (AI) model optimization techniques, prompt engineering for large language models (LLMs), efficient, explainable, and edge AI, blockchain, digital twins, cloud-native computing, edge and serverless computing, site reliability engineering (SRE), platform engineering, 5G, etc.

1 Smart Power Management in Data Centers Using Machine-Learning Techniques

D.V. Ashoka, P.M. Rekha, and P.R. Sudha
Department of Information Science and Engineering, JSS Academy of
Technical Education, Bengaluru, Karnataka, India

1.1 INTRODUCTION

Energy and power consumption are the two critical challenges that are faced by mankind. To reduce the impact of climate variation, energy efficiency or maintenance of power are extremely important. It involves choosing an appropriate set of sources to produce energy by minimizing costs and losses. Data centers and the information and communication (DIC) sector produces more than 2% of the global (CO_2) emissions because of the developments in internet services and cloud computing technology. 60% of the network traffic is because of the usage of mobile devices, live video streaming, and online gaming, by the year 2025 it is predicted to become 80%. Data centers are computer warehouses that hold huge data to handle day-to-day transactions. They run uninterruptedly and servers are installed to use and store a huge amount of data to ensure efficiency. This contributes to heavy energy consumption because it needs cooling equipment, lighting, and power requirements. Around 40% of the energy is consumed by cooling systems, followed by water chillers, which supply frozen water to coils to remove excess heat discharged by the servers.

To ensure optimization of power, renewable resources that are freely available need to be utilized to full extent. Hence minimizing the use of conventional generation plants obtained from the grid. Due to the increase in consumption of energy, leading to an alarming increase in climate change and CO_2 emissions, various policies and measures have been present in the conventional or existing power systems.

A smart power system is equipped with sensors, which can self-heal and self-monitor, whereas existing power systems involving very few sensors do not monitor themselves and need manual restoration. A smart power system accomplishes remote tests, extensive control, and provides users with plenty of options, as compared to the existing system, which performs manual checks, exhibit limited control, and provides only a few choices to customers.

This system makes use of a melange of different parameters, such as 60% CPU usage, 30% memory usage, 5% disk-IO, and 5% network-IO, and combines it into one single parameter called total CPU utilization to give an overall output of the power consumed [1].

1.2 OBJECTIVES OF THE PROPOSED WORK

A smart energy system has to ensure the following parameters:

- That a large amount of inexhaustible energy resources are integrated as an alternative for existing exhaustible energy resources.
- That energy demanded by the customer is satisfied without compromising the safety of the system and its supplies.

DOI: 10.1201/9781003407959-1

Several VMs can be consolidated into one single physical machine using cloud computing techniques. These support various virtualization technologies like kernel-based virtual machine (KVM), XEN (open source hypervisor), etc. The source host has a state of guest OS, which can be replicated on the destination host. This requires the state of the processor, memory, local storage, and the network. This is called live migration where the VM can only stop for a very small duration.

1.3   RELATED WORK

Several works have been carried out by many authors to maximize the utilization of power in data centers. Some of them are stated here.

To manage power resources efficiently, an energy management system (EMS) which is primarily combined with power networks, is needed [2]. It monitors, regulates, and improves the systems for generating power and transmitting it. It permits the utilities to analyze and store the data. It performs network modeling and also determines the power consumption and approaches to lessen power consumption. An energy management system is executed for both utilities and customers to examine the electricity usage and to conserve the same. Supervisory control and data acquisition (SCADA) refers to the monitoring and control processes. A nanoelectromechanical system (NEMS) is a system intended to regulate electromechanical facilities in the building that lead to enormous energy consumption, like heating and lighting facilities. Energy management systems are used to provide functions to gas, electricity, and water. The methodology used in this chapter is a SCADA architecture. The components in this architecture are as follows:

Customer information systems (CIS), geographical information systems (GIS), historical archival system, operation support, distribution network analysis, remote terminal unit (RTUs), billing, relay, meters and functions like network topology, short circuit analysis, and optimal switching.

Larik et al. [3] discussed the need to create data-center efficiency and furnish some approaches for getting GREEN operations at the data centers. Due to uninterrupted availability and its cooling system, energy consumption at data centers is raising at an unbelievable rate; there is an increased need to maintain power consumption. Hence these components need to be preserved at normal temperatures to make sure of achieving reliability on data centers, longevity, and maximum return on the investments.

Poor [4] emphasizes systems with multi-user agents that control a set of power sources organized as a micro-grid to gain optimal solution. The micro-grid was made autonomous to interact with other agents and take necessary steps to optimize power. The system is made up of different types of energy producers called loads, sources, energy consumers. If the produced energy is insufficient to provide the energy demanded by the loads, then extra energy in the grid has to be supplied. In order to make the solution global and available to other agents the source features are blended with those of other agents. A micro-grid contains three parameters.

1. **Loads**: Power demanded by the load, whether it is reversible or not and what type of energy resources are used.
2. **Sources**: Power, capacity, condition of the power, its efficiency, and total cost incurred.
3. **Storage systems**: Charging power, discharging power, capacity, and efficiency of the system.

Smart grids [5] are applied for effective control and sensing of power consumption. It has been proved that smart grid technology can become a powerful technology for customer satisfaction, stability of the power system, distribution of load, as well as any grid functions. Smart grids provide the following: control of resources, security with cost-effective operations, grid observations, improved performance, system planning and maintenance. Reconfiguration, self-correctiveness, and restoration capabilities of a new grid is because of smart grid technology. It provides pervasive computing abilities and two-way communications to get better control, reliability, and efficiency of the consumption of power. Providing all the options for storage and power generation and ensuring that it is efficient and operationally inexpensive are the objectives of the smart grid. The major disadvantage is security as smart grid systems store critical and private data across digital communication networks. An appropriate mechanism should be implemented to ensure privacy and security in systems. Cyber security is the key challenge to a recent smart grid system. Smart grids have been implemented in developed countries whereas in less developed countries it is lagging behind.

Avgerinou et al. [6] intended to analyze, evaluate, and outline the most recent improvements in efficiencies and energy consumption of data centers. It uses the data provided by many companies to reduce the amount of energy consumption. It implements the methods to bring the efficiency of data-center components to maintain eco-friendly and green design. This analysis, identified as power usage effectiveness (PUE) of the deftness to reduce consumption, is declining year by year and has reached a prime value of 1.64. The organization implementing these conceptualizations can reduce the variations in climate issues and efficient usage of electricity and power.

Sorrell [7] discussed the challenges and issues in reducing energy demand. This gives a clear idea of the deterioration and power consumption problem. The ideas from natural science, psychology, economics, sociology, and innovation studies are integrated in order to provide different solutions to this problem. It describes the link that exists between the scale of physical and human systems and energy demand. It also emphasizes energy consumption and economic growth. Large energy flow is required by complex systems.

Gizli et al. [8] describe the recent trends in energy consumption at data centers. Efficiency of data centers has to be improved. An evolutionary prototype for the mentioned requirement is done in this chapter. This helped companies to address their personalized energy efficiency needs. An interactive system was built, which supported the computation, assessment, and presentation of the data-center energy efficiency. It helps in finding the efficiency of data centers. A prototype was implemented that was found to be flexible, wherein new features to increase the efficiency could also be added. It could record the values that are measured and the other data in order to perform calculations of data-center efficiency. Challenges in data centers are depicted in Table 1.1.

1.4 PROPOSED METHODOLOGY

In this section, the two-fold proposed methodology is focused on the key concerns of power saving and energy monitoring of real-time power consumption in data centers using appropriate techniques. The detailed description of the two-fold methodology is stated below:

- Firstly, in the pre-processing module the real-time data-set is used for simulating the values of a real data center and is generated by considering several parameters such as CPU utilization, memory usage, network-IO, and disk-IO.
- Second is the implementation of a support vector algorithm and the virtual machine migration.

TABLE 1.1
Review of the Literature and Challenges in Data-Centre-Related Issues

Challenges	References	Description
Reliability	[9]; [10]; [11]	Requires new framework for unified integration with reliability techniques, protocols and standards.
Security	[12]; [13]	Addresses the issues related to cyber security for power system communication infrastructure in the data center.
Complexity	[14]; [15], [16]; [17]; [18]; [19]; [20]; [21]	System is complex due to data-center infrastructure.

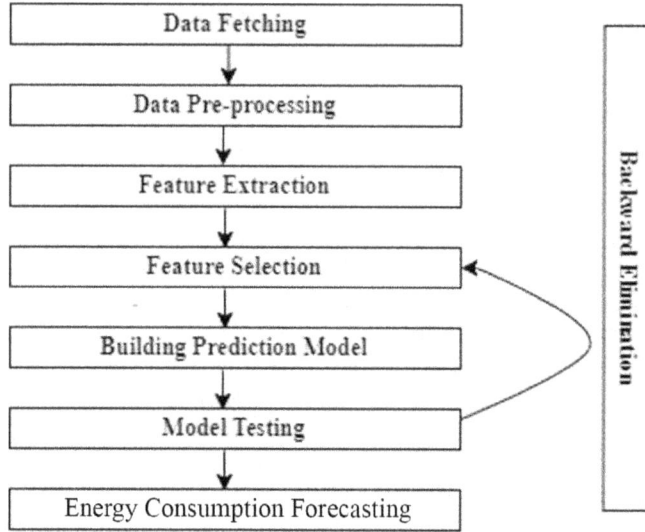

FIGURE 1.1   Proposed model.

1.4.1   EXPERIMENTAL SETUP

A prediction model is determined, which can predict effective and responsive power consumption. This is accomplished by associating energy usage with atmospheric conditions close to the data center. The prediction model can be used to forecast data-center power prerequisites for forthcoming atmospheric conditions. In order to forecast, the prediction system make use of data-center data on weather and thermodynamic efficiency.

The different phases in testing the proposed model and its thorough execution are shown in Figure 1.1.

1.4.2   DATA COLLECTION AND PRE-PROCESSING

The initial step in is to gather weather-related data from the data-center testbed sensors. Past weather data was used for the construction of a robust forecasting model, which utilized the data gathered in the process in a sequential fashion. The retrieval of data was being done through python scripts and SPARQL queries [20]. The SPARQL protocol is used to direct queries across various data sources, whether the data is stored intrinsic or noticed through the middleware. The data was collected in CSV files where python scripts are incorporated for data processing and for acquisition requests. These python scripts were used to optimize the request-retrieval process. Later, pre-processing data is used to determine the most important weather conditions associated with power consumption. The features consisted of different weather conditions used to build a forecast model. Dependent and independent variables are illustrated in Figure 1.2, where the variables were fetched from the testbed.

The primary objective of pre-processing is to ensure that entire information is coordinated by generating a timestamp. A timestamp is defined by effective data structures with control variables for each of power sensor used. The observed power measurements in comparison with the timestamp of weather measurements were not matched. Few timestamps failed to synchronize due to delay and undesired data and this was also filtered out. The resulting csv files consisted of weather data and the values from the sensor indicating whether it is active or reactive.

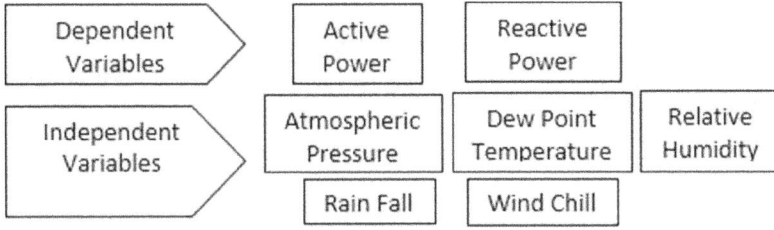

FIGURE 1.2 Independent and dependent variables.

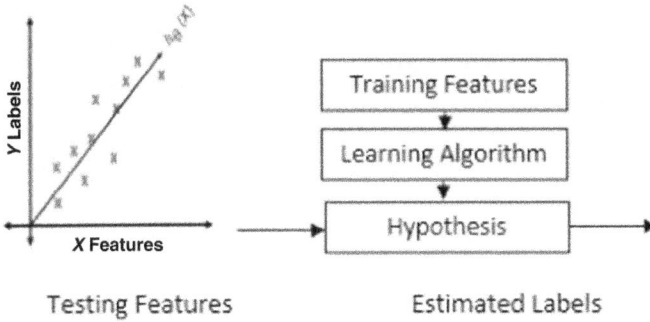

FIGURE 1.3 Single variable linear regression example.

1.4.3 FEATURE EXTRACTION AND SELECTION

Once the data is pre-processed, the feature is extracted to create a regression model in the next step [22]. This uses atmospheric conditions collected in the earlier step. Predominancy of features used for forecasting power consumption was ensured by collecting features in a recurrent manner. The study reveals that the measured energy consumption is suitable with independent weather variables. To achieve this, correlation is implemented using python programming using its library tools. The weather sensor data and active/reactive power values were evaluated. This resulted in identification of the right sensor that has greater correlation to the weather conditions.

1.4.4 PREDICTION MODEL

A linear regression approach is used to implement the prediction model. This regression machine-learning model was used to predict, as it has a wide range of applications that require continuous output. There are two groups of data: training and testing. This ensured successful building of the model. Finding the regression model's necessary parameters will serve as the foundation for the training model, which will map input data into output data. The single variable linear regression model is represented by a graph in Figure 1.3 where hypothesis is the mapping function.

The feature matrix is represented as

$$X = \{X_1, X_2, \ldots\ldots\ldots, X_n\}$$

The label vector is represented by

$$Y = \{Y_1, Y_2, \ldots\ldots\ldots, Y_n\}$$

$n \rightarrow$ number of input samples considered.

Matrix $X \rightarrow i^{th}$ training example that could contain multiple features.

$H_\theta (X_i) \rightarrow$ contains the parameters $\theta = \{\theta_0, \theta_1, ..., \theta_m\}$ corresponds to the ith training example with m features.

The hypothesis is

$$H_\theta (X_i) = \theta_0 + \theta_1 X_i 1 + \ldots\ldots\ldots\ldots + \theta_m X_i m$$

Cost function was minimized using the root mean squared error given by

$$\underset{\theta}{\text{minimize}} \, J\left\{\theta_0, \theta_1, ..., \theta_m\right\} = \frac{1}{2n} \sum_{i=1}^{n} \left(h_\theta\left(X^i\right) - y^i\right)^2 \tag{1.1}$$

The main objective is to reduce the cost function. The model performs better when fewer features are used.

1.4.5 BACKWARD ELIMINATION

There are a wide range of independent variables in linear egression application. Less significant variables had to be eliminated. The most crucial variables can be identified through backward elimination, which then assist in classifying the data that is required. Thus, it can minimize the volume of data that needs to be analyzed in upcoming forecasting procedures. The algorithm represented below elucidates this elimination of the variables. The higher the P-value, the lower the significance of the variable, which is therefore eliminated.

Algorithm 1
Backward Elimination

1: Calculate P-values for all features.
2: Select a significant level (SL) to stay in the model.
3: Fit the regression model with all features.
4: **while** $P \geq$ SL **do**
5: Remove the feature with highest P-value from feature set.
6: Fit the regression model with the updated feature set.
7: Recalculate the P-values.
8: **end while**
9: Set of independent variables for the forecast model is ready.

1.4.6 DEVELOPING THE MODEL AND TESTING

The backward elimination was first performed, followed by the sensors that have implemented linear regression using linear regression libraries. This resulted in the creation of regressor in python. The independent variables in the system matched with the classifier we built that could generate the energy consumption prediction model. The linear regression model was built by importing the csv files that were obtained previously. Once the system construction is complete, the information flow is examined to assess the forecasted amount of active/reactive power and to evaluate the performance of the regression model's effectiveness. Due to the existence of observed values we establish the correlation between the weather and the corresponding potential passive and active energy values for the total. The cooling system's time delay response was computed.

This established a correlation between energy measurements and the present conditions of the atmosphere.

1.4.7 PREDICTING THE ENERGY CONSUMPTION

The developed model is fed with training data. This was useful in estimating how potential energy would be used based on weather details. Consequently, testing data was used in the prediction model, the forecasted usage and the actual energy consumed was compared (test series). The proportion of training data was 80% of the complete set of sensor data and the remainder is utilized as test series to equate the prediction with both actual values.

There are some popular ways and existing time series methods, which employ standard time series techniques for prediction. The same is confirmed to be infeasible when applied to specific patterns in applications [3]. To resolve regression issues the data is nonlinearly mapped onto high-dimensional feature space 'Φ' SVM uses an insensitive linear function. Transformation function is denoted by 'Φ', employed to forecast the original datapoints from the initial inputspace to a typically higher-dimensional feature space (F). The primary goal of employing the 'ξ' insensitive loss function is to identify a function that best fits the most recent training data with a deviation is less than/equal to 'ξ'. Standard w is minimized in an effort to find a compact weight vector. In terms of the accuracy of solving non-linear regression, SVR performs excellently. SVR reduces structural risks to determine the function f and also effectively addresses the issues of over fitting.

$$\text{Min} \frac{1}{2}\|w^2\| + c\sum_{i-1}^{n}\left(\xi_i + \xi_i\right) \tag{1.2}$$

Here N-dimensional space is considered. SVM determines the hyperplane to classify the datapoints precisely. There are numerous hyperplanes that are used to split the two groups of datapoints. Datapoints with a higher proximity to hyperplanes are known as support vectors. The entire data center is optimized as a result of these algorithms' ability to estimate the threshold value at which the complete transfer will occur.

Since each server's architecture may be different, migration of virtual machines is dependent not only on each server's accessibility but also on each server's architecture. After obtaining the threshold value through the SVR algorithm, the next is to think about choosing the virtual machine ready for migration. Most systems have a limited number of resources. So the existing virtual machine chosen to migrate will create an impact on the whole data center. There exist few methods to select VMs for migration.

Random migration: From a list of VMs waiting, one VM is selected at random. This strategy has an advantage of faster selection and equal chances for all VMs. A disadvantage is a greater number of virtual machine migrations occur.

Minimum memory quota choice migration: The application's memory value prior to assigning it to the virtual machine is known as memory quota. The VM that has the lowest memory quota is chosen as the one that is ready for migration.

Dynamic prediction migration: The approach is based on the dynamic use of virtual machine memory. This method is employed to choose VM with a blend of 30% of memory usage, 60% of CPU usage, 5% of disk-IO, and 5% of network-IO apt for migration.

$$\text{Power consumption total} = \{(60\% \text{ of CPU}) + (30\% \text{ of memory}) + (5\% \text{ of network}) + (5\% \text{ of disk})\} \tag{1.3}$$

First select VM that needs to be migrated and then select a target host.

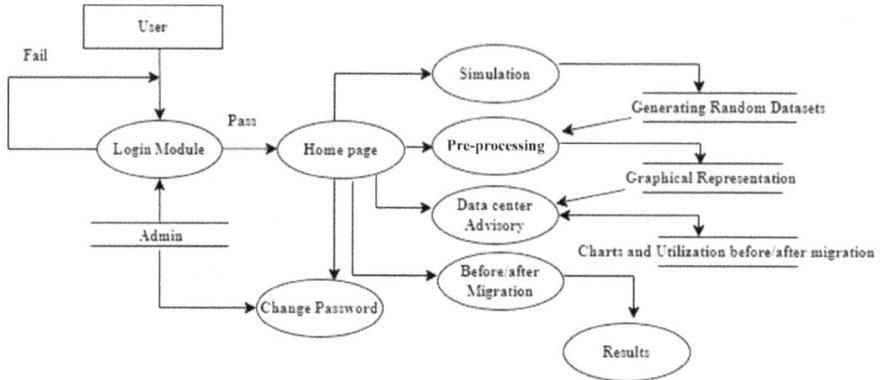

FIGURE 1.4 Data flow diagram for a power management system in data centers.

The following are the categories:

1. Minimum load priority.
2. Load balance.
3. Optimum based on prediction

The VM_DM algorithm target host for VM deployment is chosen. Figure 1.4 shows the working of the system once the user logs in and various pages are controlled by the admin and user. The data flow diagram for a power management system in data centers consists of three sections: admin section, data processing section, result prediction section.

1.5 RESULTS AND DISCUSSIONS

The analysis of the suggested approach has focused on a novel form of power management system, which is completely distinct from other technologies like [8, 22–25] contributing to a novel form of green computing, without affecting the server performance. It has been demonstrated to consistently save up to 9.5% on energy costs and to monitor power usage in real time. To make maximum benefit, the SVR algorithm predicts well in allocating the under-utilized VM's of a host to the available VM's. Comparison of original power consumption and predicted power consumption is shown in the graphical form in Figure 1.5.

One best practice to optimize power consumption is to carry out the assessment periodically utilizing software designs to identify the servers demanding high energy. The simplest and fastest way to cool the servers quickly and in tight budget is to add cold aisle containment devices to the current systems. When power supply to the servers is based on the priority, then a limited amount of energy is used by the less useful servers, which is also one way to optimize power. Using sustainable power management reduces the carbon emissions and also increases energy efficiency.

There is rise in efficiency of the grid, which results in declining pressure on the cost of electricity. Creating a smart grid requires a lot of manpower, thus creating new employment opportunities and parallelly can achieve a country's economic growth. In order to provide power to sophisticated systems supported by computer control, the power supply sector prefers quick and reliable solutions. As a result, high-performance computations can profitably optimize their use of power.

Through various experiments and observations, it is clear that increased predictability and correlation of the dependent variables are caused by combination of a multitude of independent variables. Figures 1.6 and 1.7 depict the significant variation in power measurement versus time graph. This is

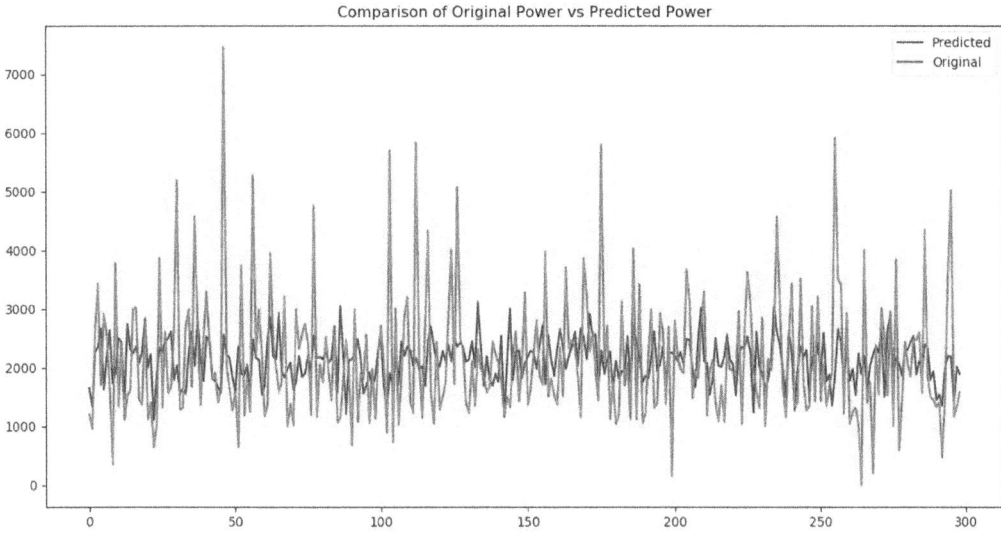

FIGURE 1.5 Graph of predicted power consumption using SVMs versus the original power consumption.

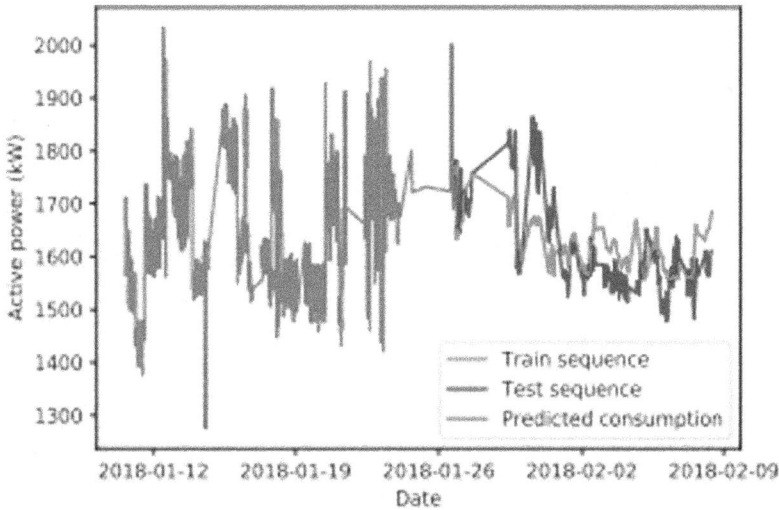

FIGURE 1.6 Active power forecasting in wind chill.

due to merging. Also, the cleansing of data creates a difference between the measurements of active / reactive power.

The red color of the graph represents the forecast results, which use a single variable regression model. This is on a par with how the actual values are represented. The forecast of the atmospheric conditions can be used to represent how power is consumed in data centers. This is so that the cooling of data centers can be affected by atmospheric conditions. The backward elimination and equations are utilized to calculate the parameters used in the model. These parameters are obtained using the historical values of atmospheric conditions that we get from the regression algorithm. The results are verified to obtain reliable forecast values.

Figures 1.6 and 1.7 show how the power consumed is related with wind chill for active and reactive power. The data center is selected by deploying sensors at various locations. The greater

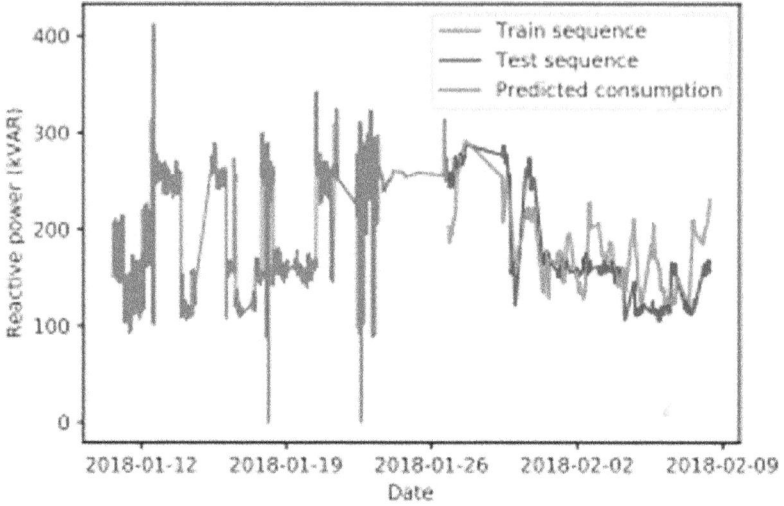

FIGURE 1.7   Reactive power forecasting in wind chill.

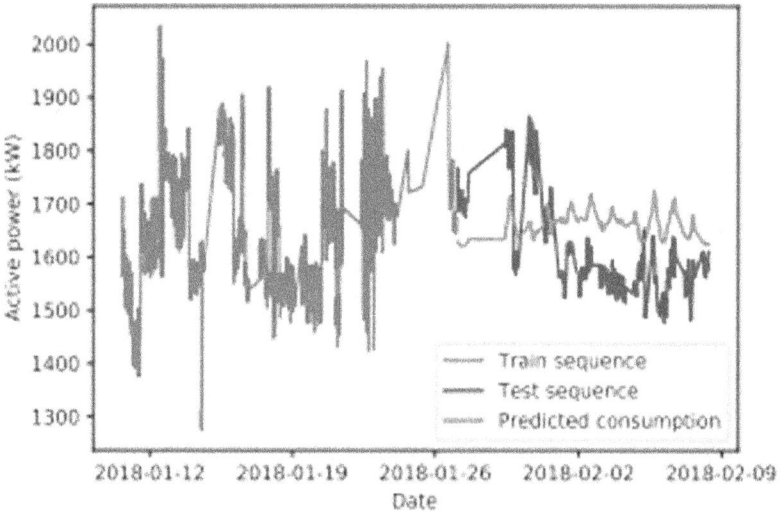

FIGURE 1.8   Active-power forecasting based on relative humidity.

active or reactive power correlation results in different atmospheric circumstances. This proves that the data centers located in different regions can give different values. The comparison can be done by studying the features that are less correlated, as seen in Figures 1.8 and 1.9.

Figure 1.10 presents the power consumed in active and reactive power by considering more than one feature of the atmospheric conditions. The input features that have high correlation are considered in the experiment. The results are not improved by multiple features when they are less correlated with power consumption.

1.6   SIGNIFICANCE OF STUDY

Supervised learning and reinforced learning like powerful energy analytics for data centers contributes persistent energy savings. It helps reduce cost, manage capacity, and increase environmental

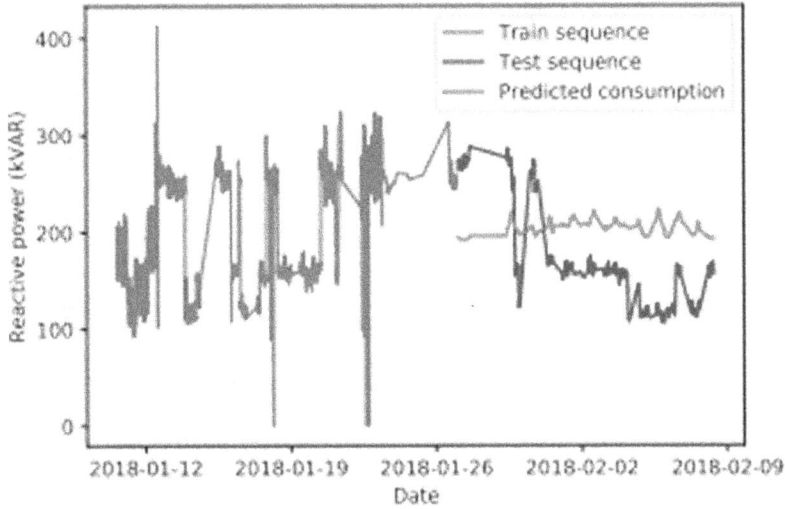

FIGURE 1.9 Reactive power forecasting based on relative humidity.

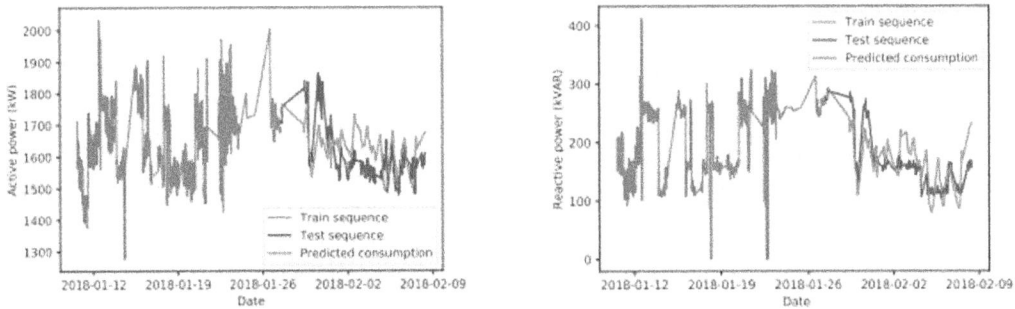

FIGURE 1.10 Active (left) and reactive (right) forecast with various features of atmospheric conditions.

responsibility. The demand for power in data centers has been increasing rapidly. It also gives a solid understanding of energy consumption by the data center.

1.7 CONCLUSION AND FUTURE ENHANCEMENT

In a world with 509.147 data centers, power management is a key job. Server under-utilization affects organizations' monetary worth and raises atmospheric carbon dioxide emissions and contributes to the environment degradation, leading to global warming. In this chapter, an optimized SVM model is built for accurate prediction of short term and medium term. The experiment conducted showed the desirable results and could be able to achieve significant results. Future work will be focused on carrying out the empirical studies to determine the optimal strategies for every class of prediction.

REFERENCES

[1] Molnár, F. (2023). Smart Solutions for Securing the Power Supply of Smart Cities. *Interdisciplinary Description of Complex Systems: INDECS*, 21(2), 161–167.

[2]   Amaral, J., Reis, C., & Brandao, R. F. M. (2013). Energy Management Systems. 48th International Universities' Power Engineering Conference (UPEC). doi:10.1109/upec.2013.6715015

[3]   Larik, R. M., Mustafa, M. W., & Qazi, S. H. (2015). *Smart Grid Technologies in Power Systems: An Overview. Research Journal of Applied Sciences, Engineering and Technology*, 11, 633–638. doi: 10.19026/rjaset.11.2024

[4]   Poor, H. (1985). *An Introduction to Signal Detection and Estimation*. New York: Springer-Verlag, https://doi.org/10.1007/978-1-4757-3863-6

[5]   Mittal, S. (2014). Power Management Techniques for Data Centers: A Survey. https://doi.org/10.48550/arXiv.1404.6681

[6]   Avgerinou, M., Bertoldi, P., & Castellazzi, L. (2017). Trends in Data Centre Energy Consumption under the European Code of Conduct for Data Centre Energy Efficiency. *European Journal*. https://data.europa.eu/doi/10.2760/358256

[7]   Sorrell, S. (2015). Reducing Energy Demand: A Review of Issues, Challenges and Approaches. ELSEVIER, *Renewable and Sustainable Energy Reviews*, 47, 74–82.

[8]   Gizli, V., & Gomez, J. M. (2018). A Framework to Optimize Energy Efficiency in Data Centers Based on Certified KPIs. *MDPI Journal, Technologies*. https://doi.org/10.1007/978-3-319-65687-8_24

[9]   Mets, K., Verschueren, T., Haerick, W., Develder, C., & De Turck, F. (2010, April). Optimizing smart energy control strategies for plug-in hybrid electric vehicle charging. In 2010 IEEE/IFIP network operations and management symposium workshops (pp. 293–299). IEEE.

[10] Moslehi, K., & Kumar, R. (2010a). A reliability perspective of the smart grid. IEEE transactions on smart grid, 1(1), 57–64.

[11] Moslehi, K., & Kumar, R. (2010b, January). Smart grid-a reliability perspective. In 2010 Innovative Smart Grid Technologies (ISGT) (pp. 1–8). IEEE.

[12] Ericsson, G. N. (2007). Toward a framework for managing information security for an electric power utility—CIGRÉ experiences. IEEE transactions on power delivery, 22(3), 1461–1469.

[13] Ericsson, G. N. (2010). Cyber security and power system communication—Essential parts of a smart grid infrastructure, IEEE Trans. Power Del, 25(3), pp. 1501–1507.

[14] Choi, S. & Chan, A. (2004). A Virtual Prototyping System for Rapid Product Development. *Computer-Aided Design*, 36(5), 401–412.

[15] Cristaldi, L., Ferrero, A., Lazzaroni, M., Monti, A., & Ponci, F. (2002). Multiresolution Modeling: An Experimental Validation. Proceeding of the 19th Instrumentation and Measurement Technology Conference (IMTC, 2002), 893–898.

[16] Dougal, R. A., & Monti, A. (2007). The Virtual Test Bed as a Tool for Rapid System Engineering. Proceeding of the 1st Annual IEEE Systems Conference, 1–6.

[17] Dougal, R., A. Monti, A., & Ponci, F. (2006). The Incremental Design Process for Power Electronic Building Blocks. Proceeding of IEEE Power Engineering Society General Meeting, p. 8.

[18] Wen, J., Arons, P., & Liu, W. H. (2010). The Role of Remedial Action Schemes in Renewable Generation Integrations. Proceeding IEEE of Innovative Smart Grid Technologies (ISGT), pp. 1–6. doi: 10.1109/ISGT.2010.5434770

[19] Krebs, R., Buchholz, B., Styczynski, Z., Rudion, K., Heyde, C., & Sassnick, Y. (2008). Vision 2020-Security of the Network Operation Today and in the Future: German Experiences. Proceeding of the IEEE Power and Energy Society General Meeting-Conversion and Delivery of Electrical Energy in the 21st Century, 1–6.

[20] Dollen, V. (2008). Utility Experience with Developing a Smart Grid Roadmap. Proceeding of the IEEE Power and Energy Society General Meeting-Conversion and Delivery of Electrical Energy in the 21st Century, pp. 1–5. doi: 10.1109/PES.2008.4596927

[21] Ponci, F., Santi, E., & Monti, A. (2009). Discrete-Time Multi-Resolution Modeling of Switching Power Converters Using Wavelets. *Simulation*, 85(2), 69–88.

[22] Petropoulos, F., Apiletti, D., Assimakopoulos, V., Babai, M. Z., Barrow, D. K., Taieb, S. B., & Ziel, F. (2022). Forecasting: Theory and Practice. *International Journal of Forecasting*, 38(3), 705–871. https://doi.org/10.1016/j.ijforecast.2021.11.001

[23] Bertoldi, P. A. (17–22 August 2014). Market Transformation Programme for Improving Energy Efficiency in Data Centers. In Proceedings of the MELS: Taming the Beast, ACEEE Summer Study on Energy Efficiency in Buildings, Pacific Grove, CA, USA, 9–14.

[24] Koomey, J. (2011). Growth in Data Centre Electricity Use 2005–2010. A Report by Analytical Press, Completed at the Request of the New York Times, 9, 161.

[25] Tatchell-Evans, M., Kapur, N., Summers, J., Thompson, H., & Oldham, D. (2017). On Power Consumption. *Applied Energy*, 186, 457–469, doi: 10.1016/j.apenergy.2016.03.076.

2 Exploring the Power of Deep Learning and Big Data in Flood Forecasting

State-of-the-Art Techniques and Insights

G. Selva Jeba[1]* and P. Chitra[2]
[1] Department of Computer Science and Engineering, Thiagarajar College of Engineering, Madurai, India
[2] Department of Computer Applications, Thiagarajar College of Engineering, Madurai, India

2.1 INTRODUCTION

Deep learning is a powerful technique in the field of artificial intelligence that has revolutionized many areas, including flood prediction. It involves training neural networks on vast amounts of data to analyze complex interactions between inputs and outputs.

Figures 2.1 and 2.2 represent the general framework of a deep learning model and internal representation of a basic deep learning network, respectively. Some of the basic terminologies used in deep learning network are the following.

Neural network layers

In a network, the input layer is the first layer, and the output layer is the last. The output of one layer serves as the input for the following layer as data is transferred through many layers. The term "hidden layer" refers to every layer that exists between the input and the output.

Weights

The weight matrix in a neural network represents the connection weights between two layers, with its dimensions determined by the sizes of the connected layers. The number of rows represent the neurons in the originating layer and the number of columns represent the neurons in the target layer. These weights, which define the connections between artificial neurons, are typically initialized with random values initially.

Activation function

The activation function plays a crucial role in normalizing the output of a neuron in a neural network. It encompasses the mathematical equations used to calculate the output of the network, which can be linear or non-linear in nature. Selecting an appropriate activation function is essential as it significantly impacts the accuracy of the results produced by the neural network [1].

Bias

Bias is a parameter used to shift the activation function by adding a constant value to the input. By adjusting the bias, the activation function can be moved to the left or right, allowing for better fitting

DOI: 10.1201/9781003407959-2

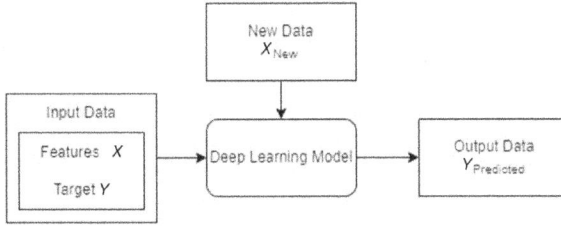

FIGURE 2.1   General framework of a deep learning model.

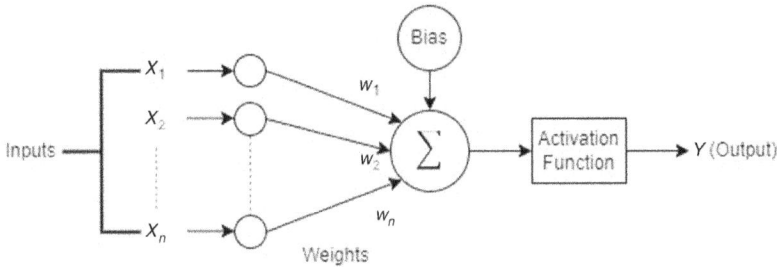

FIGURE 2.2   Internal representation of a basic deep learning network.

of the data. However, it's important to note that the bias only affects the output values and does not directly impact the weights or connections within the neural network.

Cost function

The cost function measures the disparity between the actual and predicted values in a neural network. A lower cost function indicates a closer approximation to the desired target value. Minimizing the cost function is a common objective in training neural networks to improve the accuracy of predictions.

In flood prediction, deep learning algorithms can effectively process and analyze large datasets, such as historical weather patterns, precipitation data, river flow data, and satellite imagery, to predict the likelihood of floods in specific areas. One of the main advantages of using deep learning for flood prediction is its ability to handle and process massive amounts of data in real time. Traditional flood prediction models often struggle to analyze and extract meaningful patterns from the immense volumes of data involved in understanding complex weather systems and hydrological processes. Deep learning models, on the other hand, excel at automatically learning intricate representations and relationships within the data, allowing them to capture subtle patterns that might be missed by traditional methods. The ability of deep learning models to continuously update and improve over time is another significant advantage. Flood prediction relies on understanding and predicting complex weather patterns, which can change rapidly and be challenging to model accurately. Deep learning models can be trained on updated data as it becomes available, enabling them to adapt and improve their predictions as they receive new information. This adaptability makes deep learning particularly useful in predicting floods with unpredictable weather patterns, where the ability to incorporate real-time data can significantly enhance the accuracy of predictions.

Additionally, deep learning models can leverage the power of convolutional neural networks (CNN), recurrent neural networks (RNN), long-short term memory (LSTM), gated recurrent unit (GRU), generative adversarial networks (GAN), transformers to process spatial and temporal information, generate synthetic data, compress data, parallel processing of data in flood-related datasets. CNNs are well-suited for analyzing satellite imagery and other geospatial data, allowing the models to extract relevant features and spatial patterns related to flood-prone areas. RNNs including LSTM and GRU, on the other hand, are effective in capturing temporal dependencies in time series data, such as river flow and weather patterns, enabling the models to understand and predict how these factors contribute to the likelihood of floods. GANs are used to generate synthetic data to increase the amount of training data for flood prediction. Autoencoders are used for compressing input data and then train the model with the compressed data. Transformers can be used to handle sequential data in parallel and learn long-term dependencies in time-series data.

The application of deep learning in flood prediction holds significant promise for authorities and communities dealing with flood risks. By analyzing diverse datasets and learning complex relationships, deep learning models can provide timely and accurate flood predictions. These predictions can assist in making informed decisions about emergency response, evacuation plans, and resource allocation, ultimately reducing the impact of floods on communities and potentially saving lives.

However, it's worth noting that deep learning models require large amounts of labeled training data [2], which can be a challenge in some cases. Data quality, data availability, and data labeling efforts are crucial factors in training accurate and reliable deep learning models for flood prediction. Additionally, the interpretability of deep learning models can be a concern, as their decision-making processes can be seen as black boxes. Researchers and practitioners are actively working on developing methods to interpret and explain the predictions made by deep learning models in flood prediction, enhancing their transparency and trustworthiness.

2.1.1 Definition of Flood Prediction and Its Significance

Flood prediction refers to the process of taking into account various meteorological, hydrological, and topographical factors for forecasting the likelihood of a flood event and it's potential severity. The purpose of flood prediction is to issue advanced warning to communities and emergency management officials, enabling them to prepare and respond effectively.

The prediction of floods has been an important aspect of human society since ancient times, as it saves lives and reduces property damage and economic losses. Early warning can help people evacuate, emergency responders can be deployed to the affected areas, and necessary preparations can be made to mitigate the impact of the flood. Accurate flood predictions also enable local authorities to develop and implement effective emergency response plans, reducing the risk of injury, loss of life, and damage to property. Flood prediction is essential for the safety and well-being of communities and is crucial in reducing the impact of floods on society.

2.1.2 Overview of Deep Learning and Big Data Technologies

Deep learning and big data are two buzzwords that have taken the world of technology by storm. Deep learning is a subset of machine learning that is a subset of artificial intelligence (AI), which trains neural networks to learn from large amounts of data. It is a self-learning process where the algorithms get better as they process more information. Deep learning aims to create algorithms that can learn from data and forecast outcomes without explicit programming. Big data refers to the massive amounts of information generated by businesses, governments, and individuals every day. This data can come from a variety of sources such as social media, Internet of Things (IoT) devices, customer interactions, etc. It is difficult to process and analyze this data using traditional methods due to its sheer volume, variety, and velocity.

Big data and deep learning have the potential to completely transform a variety of sectors, including marketing, finance, and healthcare. Deep learning algorithms can use big data to train themselves, allowing them to make accurate predictions, and identify patterns in the data. Deep learning algorithms, for instance, can be used to analyze vast amounts of weather data to forecast weather, floods, landslides, and other disasters. In the healthcare industry, deep learning algorithms can be used to analyze vast amounts of medical data to identify disease patterns, forecast patient outcomes, and create individualized treatments. Deep learning algorithms can be applied in finance to analyze stock market data and forecast future market movements. Large volumes of image data can be used to train deep learning algorithms to find patterns and anticipate what can be found in an image. This prompted the creation of computer vision systems that are capable of carrying out operations like object detection and image classification.

Big data and deep learning are complementary technologies with the potential to significantly progress a variety of businesses. These technologies work together to enable organizations to handle, examine, and interpret massive amounts of data, resulting in better and more accurate decision-making. The future of technology will continue to be greatly influenced by the combination of these technologies.

Deep learning and big data are two interconnected concepts that have had a profound impact on the field of technology. Deep learning, a subset of AI, focuses on training neural networks to learn and make predictions from large volumes of data. On the other hand, big data refers to the massive amounts of data generated by various sources, such as social media, IoT devices, and customer interactions, which is challenging to process and analyze using traditional methods.

Deep learning algorithms excel at processing and extracting valuable insights from big data. By leveraging the power of neural networks with multiple layers, these algorithms can automatically learn intricate patterns and relationships within the data, allowing them to make accurate predictions and classifications without explicit programming. As more data is processed, deep learning models continuously improve their performance, making them adept at handling complex and dynamic datasets.

The combination of deep learning and big data has the potential to transform numerous industries. In marketing, organizations can leverage deep learning algorithms to analyze vast amounts of customer data, such as purchasing behavior and preferences, to gain insights into customer segmentation, personalize marketing campaigns, and improve customer experience. By understanding patterns in customer data, companies can target their advertising efforts more effectively and make data-driven decisions to optimize their marketing strategies.

In finance, deep learning algorithms can analyze massive amounts of stock market data, including historical prices, trading volumes, and news sentiment, to forecast market movements, identify trading patterns, and make investment recommendations. These algorithms can uncover hidden patterns and correlations that might not be apparent to human analysts, enabling them to generate insights and predictions that can inform trading decisions and risk management strategies.

In the healthcare industry, deep learning algorithms can analyze large and diverse medical datasets, including electronic health records, medical images, genomics data, and clinical notes. By processing this data, deep learning models can assist in diagnosing diseases, predicting patient outcomes, identifying treatment options, and even contributing to the discovery of new therapies. The ability to leverage big data and deep learning in healthcare has the potential to revolutionize patient care by enabling more precise and personalized medicine.

Furthermore, computer vision, a subfield of deep learning, has witnessed significant advancements with the availability of big data. By training deep learning models on massive image datasets, computer vision algorithms can accurately detect objects, classify images, and even generate realistic visual content. This has led to advancements in autonomous vehicles, facial recognition systems, quality control in manufacturing, and many other applications that rely on image and video analysis.

The combination of deep learning and big data offers organizations the ability to extract valuable insights, discover patterns, and make data-driven decisions at an unprecedented scale. However, it is important to note that there are challenges associated with the use of big data and deep learning, including data quality, data privacy, computational requirements, and interpretability of the models.

Organizations must address these challenges while ensuring ethical and responsible use of data to fully harness the potential of deep learning and big data in various industries.

Deep learning and big data are intertwined technologies that have the power to revolutionize industries by enabling organizations to process, analyze, and interpret vast amounts of data. The combination of deep learning algorithms' ability to learn from data and big data's abundance of information opens up new possibilities for innovation, decision-making, and problem-solving across multiple domains. As these technologies continue to advance, we can expect to see further transformative applications and advancements in fields ranging from healthcare and finance to marketing and beyond.

2.2 BACKGROUND

2.2.1 HISTORICAL EVOLUTION OF FLOOD PREDICTION

Floods are one of the dangerous natural disasters, which can cause serious loss in the world [3,4]. Floods possess intricate characteristics, including their expeditious occurrence, uncertainty, higher frequency, and varying spatial distribution. Consequently, accurately predicting floods has always presented a challenge in hydrological research [5]. Flood prediction models play a crucial role in assessing hazards and managing extreme events, making them highly significant. Accurate and reliable predictions are vital for effective water resource management, policy recommendations, analysis, and evacuation planning [6].

Consequently, there is a strong emphasis on developing advanced systems for both short-term and long-term flood prediction to mitigate damages [7]. Despite ongoing research in this field for several decades, flood prediction remains a challenging problem that requires continued attention [8]. Flood prediction has come a long way from relying on basic observations and local knowledge to using intricate computer models that incorporate a range of data sources for precise predictions. In the past, flood forecasting relied on natural indicators like shifts in river levels, animal behavior, or the emergence of specific plants, which communities were informed about to prepare for impending floods. With the emergence of science and governmental meteorological agencies during the Middle Ages, data was gathered and forecasts were offered, but the accuracy was limited due to the lack of modern technology and scarcity of data.

Scientific techniques to forecast floods were established in the 19th century with the emergence of modern hydrology and meteorological stations. The first flood forecasting models were developed using mathematical equations to mimic the movement of water in rivers and streams. Traditionally, physically based models have been extensively used for flood forecasting [9,10]. With the emergence of computers and satellite data in the 20th century, flood forecasting was revolutionized with modern computer models that could combine information from radar, satellite imaging, and other meteorological observations for quicker responses and more precise forecasts. The availability of abundant monitoring data and advancements in computing technologies have led to an increasing adoption of statistical models [11]. These models require fewer input parameters compared to numerical models and have demonstrated higher forecasting accuracy in various scenarios [12]. In recent studies, researchers have begun embracing more sophisticated models to enhance the capabilities of neural networks and capture intricate relationships with greater precision [13]. These advancements allow for a deeper understanding of complex patterns and phenomena within the flood forecasting domain.

Today, flood prediction systems are used globally to mitigate the impact of floods and protect lives and property. Big data, which refers to enormous datasets generated from numerous sources, including social media, satellite imagery, and weather sensors, has emerged as a crucial tool in flood prediction. Real-time analysis and flood early warning systems are made possible by big data algorithms and deep learning techniques, which merge data from remote sensing techniques, satellite imagery, historical records, and social media to provide up-to-date information about the conditions on the ground.

2.2.2 Overview of Big Data and Its Role in Flood Prediction

Big data has recently emerged as a crucial tool in the arena of flood prediction. Real-time analysis and flood early warning systems are made possible by the use of big data algorithms and deep learning techniques. Remote sensing techniques and satellite imagery are used to observe changes in water levels and to assess the condition of the land. To forecast the possibility of a flood event, the data from these sources is merged with historical records and other pertinent data. By providing up-to-date information about the conditions on the ground, social media sites are also helping to predict floods. It is feasible to obtain a more complete picture of the situation and make a more precise flood prediction by combining different data sources. The ability of authorities to respond to flood events has been significantly enhanced by the use of big data in flood prediction, which is a significant development. Early warnings and efficient crisis management are made possible, which helps to lessen the harm that flooding causes to communities and the environment.

2.2.3 Overview of Traditional Methods of Flood Prediction

Before the invention of modern technology, flood prediction was dependent on historical records and local observations, such as changes in river levels, weather, and soil moisture. The preciseness of the forecasts was moderate, and the technique was subjective and time. Since the beginning of the 20th century, there has been active research and advancement in the field of flood prediction. Empirical models, conceptual models, and physically based models are some of the approaches that were used to construct flood prediction models. The traditional methods of flood prediction are depicted in Figure 2.3.

During the early 20th century, the primary approach for flood prediction was the use of empirical models, which were based on statistical relationships between observed hydrological variables and past flood events. The first empirical flood prediction models were based on the observation that floods tend to follow a regular pattern. These models used correlation-based or regression-based analysis, to develop relationships between flood characteristics and historical data. The methods attempt to link a series of inputs to a sequence of outcomes, to learn from the data about how the system functions. Empirical models are sometimes referred to as black-box modeling [14]. The development of conceptual models marked a shift towards an approach that represented the hydrological processes

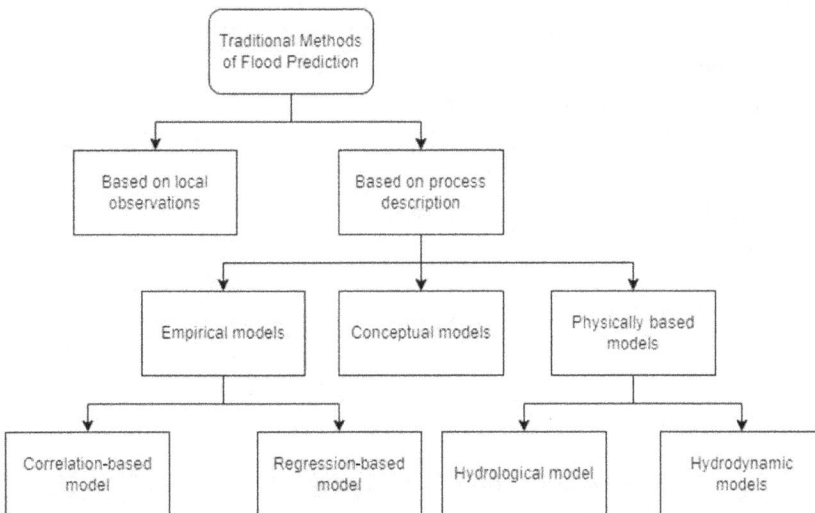

FIGURE 2.3 Traditional methods of flood prediction.

involved in a flood event in a simplified form and were used to simulate the effects of different rainfall conditions on the water balance in a catchment. Physically based models use detailed representations of the hydrological processes involved in a flood event, including precipitation, evaporation, infiltration, and runoff. They are based on mathematical representations and are of two types, namely, hydrological and hydrodynamic models. Hydrological models are used to predict floods by simulating the flow of water in rivers, lakes, and groundwater systems. Hydrodynamic models are used to predict floods by simulating the flow of water and the interactions between water and other components of the hydrological system. The development of flood prediction models has continued to evolve, with a focus on improving the accuracy and reliability of predictions. The use of remote sensing and satellite data, improved data assimilation techniques, and the development of machine learning algorithms have all contributed to the continued improvement of flood prediction models.

2.2.4 INTRODUCTION TO DEEP LEARNING IN FLOOD PREDICTION

Deep learning, a branch of machine learning, is the process of training neural networks with numerous layers to identify patterns in data. Deep learning has been used in a number of sectors, including flood prediction where forecasting the possibility and severity of floods in a specific region can aid in preventing casualties and property damage. Large datasets of historical flood data can be used to train deep learning algorithms. These datasets include meteorological and hydrological variables, including precipitation, river discharge, and water levels. On the basis of current and anticipated weather and hydrological conditions, these algorithms can then be utilized to forecast future floods. Overall, deep learning has shown promise in improving the accuracy and speed of flood prediction, allowing for more effective response to flood events and potentially saving lives and property.

Deep learning has garnered significant interest due to its ability to enhance prediction systems by offering superior performance and cost-effective solutions [15]. Deep learning is an effective technique for flood forecasting because it can examine vast volumes of data and spot intricate patterns that conventional statistical models would overlook. Analysis of a range of data sources, including rainfall patterns, river flow rates, and topographical data, is required for flood prediction. Hybrid models offer a wide range of applications, covering diverse forecast units such as hourly or seasonal mean forecasts, varying lead times from short-term to long-term predictions [16] with their efficacy in the accurate forecasting of floods [17]. The employment of deep learning models in real-time flood forecasting and warning systems is also possible to warn of forthcoming floods and assist citizens in taking the appropriate preparations by continuously assessing data from many sources. By giving more precise and timely information to decision-makers and the general public, the application of deep learning in flood prediction can help lessen the damage caused by floods and save lives. Currently, a major obstacle in employing deep learning methods for forecasting systems is the requirement to handle time-sensitive data and provide timely and accurate assistance [18].

2.3 OVERVIEW OF BIG DATA IN FLOOD PREDICTION

Big data has emerged as an essential means in the flood prediction process due to its ability to facilitate the collecting, processing, and analysis of enormous volumes of data from several sources. The first step is to identify the kind of data required for the study. This could include information from satellites, the weather, river gauges, and other pertinent sources. Following the identification of data sources, data is acquired from a variety of sources. Here are some examples of potential data types required for flood prediction:

Rainfall data: Data on rainfall should be obtained through weather stations or satellite pictures. Given that torrential rains can cause flash floods or river floods, this information is essential for estimating the likelihood of flooding.

Topographic data: Acquire details on the terrain, such as elevation and slope, which may have an effect on how water moves and accumulates during a flood.

Hydrological data: Water levels, discharge rates, and flow rates from rivers, streams, and other water bodies that can cause floods should be gathered.

Land use data: Data on land use should be gathered to help identify places that are more vulnerable to floods as a result of human activity, such as those that are urban, agricultural, or forested.

Historical data: Compile details about previous floods in the area, such as their severity, duration, and effects, in order to detect patterns and trends in their behavior.

Remote sensing data: To identify changes in land cover or identify potential flood hazards, remote sensing data can be gathered via satellites or other remote sensing technology.

Geographic information systems (GIS) data: Data from GIS data provide details on the physical environment, such as terrain, land use, and other elements that influence the occurrence of flooding. The development of flood models and the estimation of the scope and severity of flood rely on this data.

Social media data: During a flood disaster, social media sites like Twitter, Facebook, and Instagram can offer useful information. On social media, people frequently post details on flood conditions, water levels, and other pertinent information. This information can be gathered and evaluated to offer real-time flood conditions.

Sensor data: It is becoming more common to employ sensors to gather information on flood conditions. To measure water levels, flow rates, and other factors, these sensors can be placed in various locations. This information can be utilized to develop real-time flood models and offer flood early warning systems.

Predicting flood-related situations involves analyzing various factors and utilizing different methods to assess the likelihood and severity of flooding. Here are some key nuances followed in flood prediction:

Rainfall analysis: Monitoring and analyzing rainfall patterns is crucial for flood prediction. Meteorological data from weather stations and radar systems are collected to assess the intensity, duration, and spatial distribution of rainfall.

Watershed characteristics: Understanding the characteristics of the watershed or river basin is essential. Factors such as topography, soil type, land cover, and vegetation affect how water is absorbed, flows, and accumulates. Detailed knowledge of these features helps in modeling water runoff and estimating flood potential.

Hydrological modeling: Hydrological models simulate the behavior of water within a watershed. They consider factors such as rainfall, evaporation, infiltration, and river flow to predict how water will move through the system. These models help estimate river levels, floodplains, and potential flood extents.

River gauges: River gauges are instruments that measure water levels and flow rates in rivers and streams. Real-time data from these gauges provide critical information about rising water levels and the current state of rivers. By comparing current readings with historical data, experts can identify increasing trends and assess flood risks.

2.3.1 DATA PROCESSING AND CLEANING TECHNIQUES

Data processing and cleaning are critical steps in the data analysis pipeline, as they help ensure the accuracy and reliability of the results obtained from the data. This involves identifying and handling missing values, duplicates, inconsistencies, and outliers. Missing values entail substituting estimated values for missing data based on the information that is currently available. Data

FIGURE 2.4 Steps in pre-processing of Big Data.

normalization entails scaling the data to a standard range, such as 0 to 1, to enable comparison of different variables. By using these methods, the quality of the data can be enhanced, which may yield forecasts and insights that are more precise and reliable.

The various data pre-processing steps are given in Figure 2.4 and are as follows:

Data acquisition: Obtaining of raw flood-related datasets from various sources, such as remote sensing platforms, weather stations, river gauges, and social media. It also involves the collection of ancillary data, including topographic information, land use/land cover maps, and hydrological models.

Data cleaning: It comprises the process of handling missing values and removal of outliers.

Handling missing values: To identify missing data points and apply techniques like interpolation or data imputation to fill the gaps.

Removal of outliers: To detect and remove data points that significantly deviate from the expected range or show abnormal behavior.

Data integration: Combining multiple datasets (e.g., precipitation, water level, soil moisture) into a unified representation for a comprehensive analysis.

Data augmentation: In some cases, data augmentation techniques are employed to artificially increase the diversity and size of the dataset. Augmentation methods can include rotation, scaling, flipping the data. This helps to improve the model's ability to generalize and handle variations in the input data.

Data normalization: Normalization is a process of transforming data into a standard range or scale to remove variations and bring the values within a specific range.

2.3.1.1 Data Normalization

Data normalization is an important pre-processing step for training deep learning models for flood prediction. Normalization helps to improve the performance of the model by reducing the impact of differences in the scale and distribution of input features. The first step is to identify the input features that will be used to train the deep learning model. This could include features such as precipitation, temperature, soil moisture, and river discharge. Once the input features are identified, the range and distribution of each feature is to be determined. Then the method of normalization has to be chosen. There are different ways to perform data normalization using deep learning, namely, the following:

Batch normalization: Batch normalization is a process used to normalize the activations of the previous layer at each batch. It helps in the reduction of internal covariate shift and speed up training process.

Layer normalization: Instead of normalizing the activations over the batch dimension, layer normalization normalizes over the feature dimension.

Instance normalization: Instead of normalizing the activations over the batch dimension, instance normalization normalizes over the spatial dimensions.

Group normalization: Group normalization is a variation of batch normalization that splits the channels into small groups and normalizes the activations within each group.

Min-max scaling normalization: Min-max scaling is the process of scaling the data to a fixed range, typically between the range 0 and 1. It is a simple normalization method that can be easily implemented in deep learning models.

Z-score normalization: Z-score normalization scales the data to have zero mean and unit variance. It is a common normalization method used in statistics and can also be used in deep learning.

The normalization technique is chosen depending on the nature of the data and the requisites of the deep learning model. In practice, a combination of different normalization techniques may be used to achieve optimal performance [19].

Once the normalization method is chosen, it is applied to the input data. It's important to apply the same normalization method used for the training set to the testing set to ensure consistency in the data. It can help the model converge faster and avoid numerical instabilities, ultimately leading to better predictions.

2.4   DEEP LEARNING MODELS FOR FLOOD PREDICTION

In 1943, McCulloch and Pitts in [20] created a computer model based on the neural networks of the human brain, which is regarded as the inception of deep learning. Deep learning is a branch of machine learning and artificial intelligence that layers neural networks to process massive amounts of data. The millions of neurons that the human brain uses to process information are the inspiration for deep learning algorithms. A broad term of statistical and deep learning methods that forecast future events using both historical and real-time data is known as predictive analytics. Predictive analytics examines past and present data to assist us in understanding potential future events. For applications involving disasters, accurate forecasting is crucial. Deep learning approaches are appropriate for time-series forecasting problems, aiming to reduce the maximum forecasting error, as demonstrated in [21]. Data science, which includes statistics and predictive modeling, is fundamentally dependent on deep learning. Deep learning accelerates and streamlines the process of analyzing and interpreting vast volumes of data. The key steps in building a deep learning model are as follows:

(i) Feature extraction

It is the process of eliciting all of the features needed to solve the problem. The various feature extraction techniques in flood prediction using deep learning are the below:

Time-series analysis: Flood prediction often involves analyzing time-series data, such as water levels, precipitation, or streamflow. Time-series analysis helps to extract trend, seasonality, and residual components from the time series. These components can provide valuable information about the historical patterns and dynamics of flood-related variables.

Spatial feature analysis: Spatial information can be informative for flood prediction, especially when considering the characteristics of the area surrounding the target location. Spatial features can include land cover type, slope, elevation, proximity to water bodies or infrastructure, and hydrological characteristics. GIS techniques can be applied to extract and derive these spatial features.

Satellite imagery analysis: Deep learning models can leverage satellite imagery for flood prediction. CNNs can be employed to extract features automatically from satellite images, capturing relevant patterns, land cover changes, or water extent.

Recurrent pattern analysis: RNNs and their variants, such as LSTM or GRU, are effective for capturing temporal dependencies and recurrent patterns in time-series data. These models can extract features that capture long-term dependencies and memory of past flood events, which can aid in flood prediction.

(ii) Feature selection

It involves choosing significant features that increase the deep learning model's performance. The criteria for feature selection in flood prediction using deep learning are as below:

Relevance to the prediction task: The selected features should be directly relevant to the flood prediction task. They should capture the key factors influencing flooding, such as rainfall patterns, historical water levels, or hydrological characteristics. Domain knowledge and expert consultation can help identify the most relevant features.

Predictive correlation: Features with high predictive power, i.e., those that strongly correlate with flood occurrences or can effectively discriminate between flood and non-flood conditions, should be selected. Statistical measures like correlation analysis or information gain can be used to evaluate the relationship between features and the target variable.

Removal of redundant features: Redundant features that provide similar information may increase model complexity without adding significant predictive power. It is important to identify and remove redundant features to avoid overfitting and enhance model interpretability. Techniques like correlation analysis or principal component analysis (PCA) can help identify redundant features.

Generalizability: The selected features should be applicable across different flood prediction scenarios and generalize well to unseen data. Features that are stable and consistent in their predictive power across various time periods, locations, and flood events are preferred. Cross-validation or evaluating feature stability on different datasets can help assess the generalizability of selected features.

Manually extracting features necessitates a thorough understanding of both the subject and the domain. Although feature engineering is a time-consuming process, it can now be automated with deep learning [22,23].

2.4.1 OVERVIEW OF POPULAR DEEP LEARNING ARCHITECTURES

Deep learning models can be effective in predicting floods by leveraging the large amounts of data generated by various sources such as weather stations, satellite imagery, river gauges, and social media. Some deep learning models that can be applied to flood prediction are listed in Figure 2.5.

2.4.1.1 Convolutional Neural Network (CNN)

Convolutional neural networks (CNN) are among the most often used deep neural networks. According to Albawi et al., it gets its name from the linear mathematical convolution of two matrices [24]. The layers that make up the CNN architecture are convolutional, pooling, and fully connected layers. A feature extractor and a classification neural network are combined in convolutional neural networks. Pairs of convolutional and pooling layers are placed on top of one another in the feature extraction neural network. The convolution layer applies the convolution technique to modify the image. The convolution layer emphasizes specific features of the original image using feature maps. This layer generates as many output images as there are convolution filters in the network. The convolution filter is merely a two-dimensional matrix in fact. In the pooling layer, adjacent pixels are combined to create a single pixel. The pooling layer downsamples the feature maps thus reducing the image's dimension. The fully connected layer is responsible for accepting input from all of the different feature extraction stages and performing a global analysis of all of the previous layers' output. The fully connected layers process the retrieved features and map them to the output classes, such as flood or non-flood.

In order to overcome the typical artificial neural network (ANN)'s constraint on evaluating complicated networks in terms of learning speed [25], in [26] a deep CNN method is used to build a

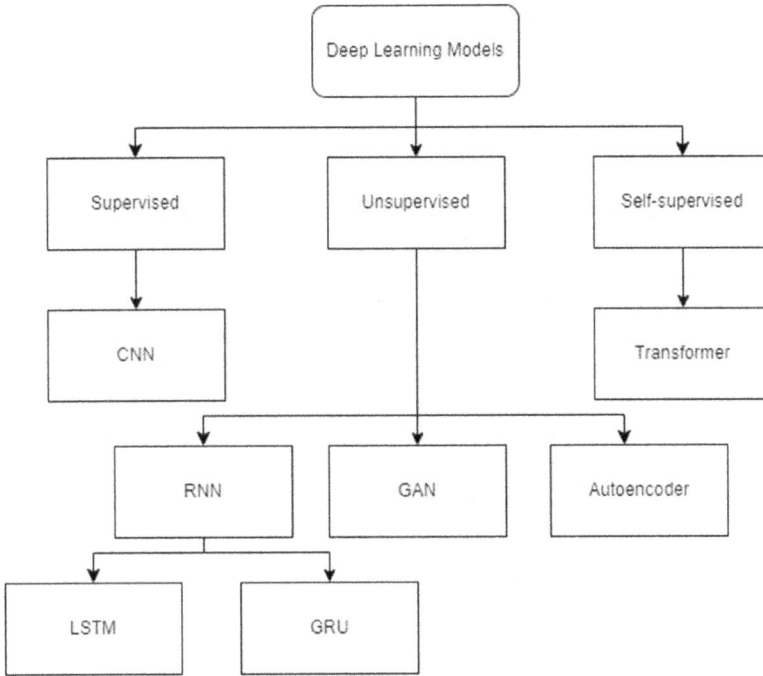

FIGURE 2.5 Deep learning models for flood prediction.

flood susceptibility map for Iran. Ten flood conditioning elements (slope, altitude, aspect, plan curvature, profile curvature, rainfall, geology, land use, distance from roads, and distance from rivers) were identified and utilized to build a map of Iran's flood vulnerability using data from 2769 historical flood locations. The slope, height, aspect, plan curvature, and profile curvature were calculated using topographic maps. ArcGIS 10.2's multiple ring buffer capabilities were used with data from Iranian rivers and motorways to estimate distances from rivers and roadways. The generated map was quantitatively analyzed to identify the areas with the highest and lowest flood risk, as well as the major factors influencing this risk. This analysis confirmed CNN's classification of flood-prone areas as acceptable and of good quality. With the aid of a CNN-based technique, the authors calculated the regionally dispersed water depths for two large flood events that occurred in Carlisle, England [27]. For the creation and operation of LISFLOOD-FP, which forecasts flood inundation, hydrometric and topography data were necessary. A digital elevation model and upstream boundary conditions, such as flow hydrographs, were essential to replicate the Carlisle flood episodes. Comparison of the CNN model's performance to an SVR approach provides additional evidence for its effectiveness. The outcomes demonstrated that the CNN model performed significantly better than SVR. The CNN performs remarkably well in deep learning tasks due to its simplicity, superior performance, and computing economy [28]. CNNs have developed into a hotspot for research and the de facto benchmark for many deep learning and computer vision applications since its inception in 1990.

2.4.1.2 Recurrent Neural Network

Sequential data is utilized by RNNs to make predictions on forthcoming events. RNNs are called RNNs because they perform the same process for each element of a sequence and base their output on previous computations. A combination of the input at times $x(t)$ and $x(t-1)$ makes up the input at any given time t. The output at any given time is fetched back to the network in order to improve the output. Vanishing gradients are a challenge for RNNs. The parameter updates become insignificant when the gradient diminishes too much since the gradients include information that is used by the RNN. Long data sequences are therefore challenging to learn.

2.4.1.3 Long-Short Term Memory Network

Recurrent neural network models can effectively employ historical data for prediction, but they have a limited short-term dependency handling capacity and are susceptible to the vanishing gradient problem. To address these issues, Hochreiter and Schmidhuber developed the LSTM in 1997 [29]. It is currently extensively employed in a variety of applications with spectacular outcomes.

For time-series data processing and prediction, especially the forecasting of flood disasters, the LSTM variation of recurrent neural network architecture is often utilized. The LSTM design is a suitable option for flood prediction since it can find long-term dependencies in the time-series data. The cornerstone of an LSTM architecture is the LSTM cell. Each cell contains a number of gates, including an input gate, an output gate, and a forget gate, which control the flow of information within the network. The input gate determines the amount of new data that should be added to the cell, the forget gate determines the amount of data that should be forgotten, and the output gate determines the amount of data that should be sent to the network's next tier. In order to train an LSTM model for flood prediction, historical data on water levels, precipitation, and river flow rates is fed into the network. The network learns to forecast these variables' future values based on past observations using historical data and complex interactions between numerous elements. After training the model, real-time data can be utilized to predict future flood instances.

The parameters rainfall, temperature, and humidity were collected weekly in the proposed LSTM for weekly rainfall forecasting in a study made by Prasetya et al. [30]. The past 10 year's climate data were generalized using LSTM for the training phase. For weighting renewal, stochastic gradient descent (SGD) and adaptive moment estimation (Adam) were utilized. The results showed that dataset size and learning rates are the two main factors that have a significant impact on accuracy, with data learning over the past 10 years producing over 96% precision for new data and more than 98% precision for training data.

2.4.1.4 Gated Recurrent Unit

Another variation of the RNN architecture called the GRU has been successfully used to predict floods. In [31] Cho et al. developed GRU, a novel deep learning architecture with just two gates: update and reset, by simplifying the LSTM structure. When compared to LSTM, GRU minimizes and simplifies time complexity. The reset gate assists in determining how much historical data should be eliminated, while the update gate assists in determining how much past data ought to be carried on to the future. The hidden state is forced to discard the prior hidden state and is reset with the current input when the reset gate is almost zero. GRUs are faster at training than the LSTM model and are built to leverage past data information from predictions in an efficient manner.

GRU models account for the temporal correlations between various weather parameters and their influence on the occurrence of floods when making flood predictions. This is especially helpful when working with time-series data because a prediction at one time point depends on the historical data and the data points are highly dependent on one another. A hybrid deep learning-based flood prediction model utilizing meteorological data was proposed in [32]. The hybrid model combining RNN and GRU, demonstrated superior prediction accuracy over the LSTM architecture. Advanced variants of the RNN architecture, such as LSTM and GRU, were developed to provide a better representation of chronological sequences and their long-range interactions than conventional RNNs.

2.4.1.5 Generative Adversarial Networks

A deep learning neural network called a GAN developed in 2014 by Goodfellow confronts two neural networks against one another [33]. Underpinned on a given dataset, the first network, referred to as the generator, produces new data. The created data is assessed and classified as real or fake by the discriminator network, which is the second network. A technique known as adversarial training is used to train both the discriminator and the generator simultaneously. In order to trick the discriminator during training, the generator attempts to generate new data. Consequently, the discriminator works to precisely distinguish real data from fake data. Through this rivalry, the discriminator improves its capacity to distinguish between real and fake data, while the generator learns to produce increasingly realistic data. GANs are still a topic of current research because they have demonstrated considerable potential in many disciplines. However, they also pose difficulties like instability during training and the possibility to produce inaccurate or biased data.

The pluvial flooding spurred on by nonlinear spatial heterogeneity rainfall events was predicted using a deep convolutional generative adversarial network known as floodGAN. The floodGAN model is up to 106 times faster than hydrodynamic models because it uses an image-to-image translation methodology. The accuracy and generalizability of the model were examined using both simulated events and a real-world rainfall event, showing promise for real-time applications like early warning systems [34]. A real-time predictive deep convolutional generative adversarial network (DCGAN) for flood forecasting was proposed by Meiling et al. Dynamic flow learning and real-time forecasting are the two stages of the technique. The model is tested against an extreme flood occurrence in Greve, Denmark, with a 100-year return period, and the results demonstrate good agreement with a high fidelity model for a lead of 900 time step forecast [35].

2.4.1.6 Autoencoders

A type of neural network known as an autoencoder learns to compress and then reconstruct data, such as text, audio, or images. It has two major components: an encoder and a decoder. After receiving the input data, the encoder creates a compressed representation, also known as a bottleneck, of the data. The decoder creates a reconstruction of the original data using this compressed representation. Minimizing the difference between the input and the reconstructed output is the objective of an autoencoder. This is accomplished by utilizing an optimization approach like stochastic gradient descent to train the network using a dataset of instances. In order to reduce the discrepancy between the input and output for each application during training, the network modifies the weights of its connections. Autoencoders have several applications, including image and audio compression, anomaly detection, and denoising. They can also be used for dimensionality reduction, by training on a high-dimensional dataset and then using the compressed representation as input to another machine learning algorithm.

To aid in emergency management during floods, Feng et al. introduced the SAE-RNN model, for precise and timely multistep-ahead flood forecasts. The method utilizes a combination of deep learning techniques to address flood prediction, including a stacked autoencoder for compressing high-dimensional flood data into a lower-dimensional representation, a recurrent neural network for forecasting flood features based on local rainfall patterns, and another stacked autoencoder for reconstructing flood data in a region. The approach involves extracting the nonlinear dependence structure of the data throughout the process and nonlinear conversion of rainfall sequences are credited with the accuracy of the model [36].

2.4.1.7 Transformers

The transformer architecture is a powerful neural network architecture, introduced by Google in 2017 [37]. It has shown great success in various natural language processing tasks, however, it can be adapted for other applications such as spatio-temporal prediction. The temporal and spatial linkages between the different environmental factors that influence floods can be discovered using the transformer architecture. A sequence of encoder-decoder layers may be used in the transformer architecture for flood prediction. The input sequence of environmental parameters would be fed into the encoder layers, which would then encode them into a number of hidden states. Additionally, the transformer can include spatial attention techniques that let the model focus on various areas of the input data that are crucial for forecasting flooding in particular areas. This can be especially helpful in regions where local weather patterns or geographic elements contribute to flooding. By utilizing the temporal and spatial linkages between many environmental parameters that cause flooding, the transformer model has the potential to substantially boost flood forecast accuracy.

Xu et al. evaluated the transformer neural network model's effectiveness in predicting lake water levels using Poyang Lake as a case study. Furthermore, they investigated into the Yangtze River's influence on lake water level variations and evaluated the effect of lead time on the preciseness of the model. As a consequence, it was found that the transformer model does exceptionally well of modeling variation in the lake water level and can accurately capture Poyang Lake's characteristics of temporal water level variation [38].

2.5 EVALUATION METRICS AND RESULTS

2.5.1 OVERVIEW OF EVALUATION METRICS FOR DEEP LEARNING MODELS

Performance metrics are the measures used to evaluate prediction accuracy for flood prediction.

Mean absolute error (MAE)

MAE is the metric used in regressive prediction problems. It is defined as the average of the sum of the absolute differences between the forecasted and the observed values. The formula to calculate MAE is as follows in equation (2.1).

$$MAE = \frac{\sum_{i=1}^{n} Y_i - X_i}{n} \tag{2.1}$$

Here, Y = observed values and X = forecasted values.

Root mean squared error (RMSE)

RMSE is defined as the square root of the average of the squared difference between the observed and the forecasted values. RMSE is calculated using the formula as in equation (2.2).

$$RMSE = \sqrt{\frac{\sum_{i=1}^{n} \left(Y_i - X_i\right)^2}{n}} \tag{2.2}$$

Here, Y = observed values and X = forecasted values.

R-squared (R^2)

The coefficient of determination, also known as R-squared, is a metric that assesses how well a model fits the data. The dependent variable is said to be predicted more accurately by the model when R-squared value is closer to 1. The formula to calculate R^2 is as follows in equation (2.3).

$$R^2 = \frac{n(\Sigma Y_i . X_i) - (\Sigma X_i).(\Sigma Y_i)}{\sqrt{[n\Sigma Y_i^2 - (\Sigma Y_i)^2].[n\Sigma X_i^2 - (\Sigma X_i)^2]}} \tag{2.3}$$

Here, Y = observed values and X = forecasted values.

Nash-Sutcliffe efficiency (NSE)

The NSE metric is commonly used in hydrology and other environmental sciences to assess the performance of hydrological models. The NSE is a performance metric that measures the relative magnitude of errors in the predicted and observed values. It ranges from $-\infty$ to 1, with a value closer to 1 indicating a perfect fit. The formula to calculate NSE is as follows in equation (2.4):

$$\mathrm{NSE} = 1 - \frac{\sum_{i=1}^{n}\left(Y_i - X_i\right)^2}{\sum_{i=1}^{n}\left(Y_i - \bar{Y}\right)^2} \tag{2.4}$$

Here, Y = observed values and X = forecasted values.

2.5.2 Comparison of Deep Learning Models

The choice of architecture depends on the specific problem at hand, the available data, and the computational resources available. A few of the deep learning architectures are compared with their advantages, disadvantages, and applicability to flood prediction in Table 2.1.

2.6 CONCLUSION

2.6.1 Summary of Key Findings

Deep learning-based artificial intelligence has demonstrated great potential in flood prediction using big data. The following are some of the key findings:

TABLE 2.1
Comparison of Deep Learning Models for Flood Prediction

Architecture	Advantages	Disadvantages	Applicability to Flood Prediction
CNN	Good at learning spatial features	Not good at handling sequential data	Can be used for analyzing satellite imagery of flood-prone areas, but not suitable for time-series prediction
RNN	Can handle sequential data well	Prone to vanishing/exploding gradient problem	Suitable for predicting time-series data, such as water levels and precipitation
LSTM	Can handle long-term dependencies in sequential data	Can be computationally expensive	Useful for flood prediction as it can capture complex patterns and dependencies in time-series data
GRU	Similar to LSTM but has few parameters	May not perform as well as LSTM in some cases	Can be a more efficient alternative to LSTM for flood prediction
GAN	Can generate synthetic data to augment training data	May be difficult to train and require large amounts of data	Can be used to generate synthetic data to increase the amount of training data available for flood prediction
Autoencoder	Can be used for unsupervised feature learning and data compression	May not perform as well as supervised methods when labeled data is available	Can be useful for compressing data before training a supervised model for flood prediction
Transformer	Can handle sequential data in parallel and learn long-term dependencies	Can be computationally expensive	Can be a powerful tool for flood prediction, especially when dealing with large datasets and long time series

- Deep learning algorithms can be used to extract beneficial features from different types of big data sources, such as satellite imagery, weather data, and social media data.
- These algorithms can be trained to forecast flood events with high accuracy by utilizing both historical data and real-time observations.
- Deep learning algorithms can also be used to identify flood-prone areas.
- Hybrid methods that combine multiple deep learning models can significantly improve the accuracy of flood predictions.
- However, the accuracy of deep learning-based flood prediction models largely depends on the quality and availability of the data used in training and testing.
- Integrating multiple data sources, including remote sensing and crowdsourcing, can further improve the accuracy of flood predictions.
- Deep learning-based models for predicting floods have the prospective to transform flood management by giving decision-makers and emergency personnel the access to timely and precise information that will help them better prepare for and respond to flood disasters.

2.6.2 DISCUSSION OF LIMITATIONS AND CHALLENGES

Although deep learning-based AI has demonstrated promising outcomes in predicting floods, there are several challenges that must be addressed. One of the main challenges is the absence of standardization and integration of various data sources. Furthermore, deep learning algorithms necessitate significant computational resources, which may impede real-time flood prediction. Nevertheless, with advancements in technology and the availability of more comprehensive data, deep learning-based AI is positioned to play a significant role in predicting and managing floods. The current need for forecasting systems to deal with time-sensitive data and provide the most accurate assistance on time presents a significant obstacle for the use of deep learning techniques [39]. The complexity of predicting systems remains tangled to produce accurate systems with respect to identification of relevant features [40].

2.6.3 CONCLUSION AND RECOMMENDATIONS FOR FUTURE WORK

The utilization of deep learning-based AI in flood prediction with big data presents a significant potential to enhance flood forecasting and minimize the impact of natural disasters. Deep learning algorithms can analyze large amounts of data from numerous sources, such as weather sensors, satellite imagery, and social media, to provide accurate flood predictions. As technology advances, we can expect even more advanced models and systems to be developed in the future to better predict and manage floods.

Uncertainty quantification: A crucial aspect of flood prediction is uncertainty quantification, which can provide insights into the reliability of predictions. Researchers can investigate Bayesian deep learning methods that can provide probabilistic predictions together with measures of uncertainty.

Real-time forecasting: Developing deep learning models that can provide real-time flood forecasting is another scope of future work. This necessitates the development of algorithms that can process vast quantities of data quickly and efficiently and provide accurate real-time predictions. The use of edge computing and cloud-based services can facilitate this development.

Integration with other technologies: Integrating deep learning models with other technologies such as IoT sensors and drones can provide a more comprehensive and precise understanding of flood events. This can result in better predictions and mitigation measures.

Data management and processing: Another essential area of future work is to develop effective methods for managing and processing vast volumes of flood-related data. This can entail utilizing

big data technologies such as Hadoop, Spark, and NoSQL databases, as well as developing efficient algorithms for data pre-processing, feature extraction, and data augmentation.

Deploying models in the real world: Lastly, it is crucial to consider the deployment of deep learning
models in the real world. This involves addressing challenges such as data privacy, security, and
ethical considerations, as well as ensuring the scalability and reliability of the models.

In conclusion, the future of deep learning-based artificial intelligence in flood prediction with big
data is promising, and there are several potential directions for future research and development. By
addressing these challenges and exploring new methods and techniques, we can improve the preciseness of flood prediction and mitigation strategies.

REFERENCES

[1] Nwankpa, C. E., Ijomah, W., Gachagan, A., & Marshall, S. (2021). "Activation functions: Comparison
of trends in practice and research for deep learning", 2nd International Conference on Computational
Sciences and Technology, Jamshoro, Pakistan, 124–133.

[2] Davis, J. et al. (2022). "Spatial relationship-driven computer vision image data set annotation",
2022 International Joint Conference on Neural Networks (IJCNN), Padua, Italy, 1–8, doi: 10.1109/
IJCNN55064.2022.9892975

[3] Ahmadlou, M., Karimi, M., Alizadeh, S., Shirzadi, A., Parvinnejhad D., Shahabi, H., & Panahi, M.
(2019). "Flood susceptibility assessment using integration of adaptive network-based fuzzy inference
system (ANFIS) and biogeography-based optimization (BBO) and BAT algorithms (BA)", *Geocarto
International*, 34(11), 1252–1272, DOI: 10.1080/10106049.2018.1474276

[4] Li, B., Hou, J., & Ma, Y. (2022). "A coupled high-resolution hydrodynamic and cellular automata-
based evacuation route planning model for pedestrians in flooding scenarios", *Nat Hazards*, 110, 607–
628. https://doi.org/10.1007/s11069-021-04960-x

[5] Linda, J. S., Cranston, M. D., White, C. J., & Kelly, L. (2021). "Operational and emerging capabilities
for surface water flood forecasting", *WIREs Water*, 8(3), e1517. doi:10.1002/wat2.1517

[6] Xie, K., Ozbay, K., Zhu, Y., & Yang, H. (2017). "Evacuation zone modeling under climate change: A
data-driven method", *Journal of Infrastructure Systems*, 23, 04017013.

[7] Pitt, M. (2008). "Learning Lessons from the 2007 Floods", Cabinet Office: London, UK.

[8] Maspo, N.-A., Bin Harun, A. N., Goto, M., Cheros, F., Haron, N. A., & Mohd Nawi, M. N. (2020).
"Evaluation of machine learning approach in flood prediction scenarios and its input parameters: A systematic review", *IOP Conference Series: Earth and Environmental Science,* 479, 012038. doi:10.1088/
1755-1315/479/1/012038

[9] Devia, G. K., Ganasri, B. P., & Dwarakish, G. S. (2015). "A review on hydrological models", Aquatic
Procedia, 4, 1001–1007, ISSN 2214-241X, https://doi.org/10.1016/j.aqpro.2015.02.126

[10] Castangia, M., Medina Grajales, L. M., Aliberti, A., Rossi, C., Macii, A., Macii, E., & Patti, E. (2023).
"Transformer neural networks for interpretable flood forecasting", *Environmental Modelling &
Software*, 160, 105581, ISSN 1364-8152, https://doi.org/10.1016/j.envsoft.2022.105581

[11] Li, Y., & Hong H. (2023). "Modelling flood susceptibility based on deep learning coupling with
ensemble learning models", *Journal of Environmental Management*, 325, 116450, ISSN 0301-4797,
https://doi.org/10.1016/j.jenvman.2022.116450

[12] Aziz, K., Rahman, A., & Fang, G. (2014). "Application of artificial neural networks in regional
flood frequency analysis: A case study for Australia", *Stochastic Environmental Research and Risk
Assessment*, 28, 541–554. https://doi.org/10.1007/s00477-013-0771-5

[13] Sit, M., Demiray, B. Z., Xiang, Z., Ewing, G. J., Sermet, Y., & Demir, I. (2020). "A comprehensive
review of deep learning applications in hydrology and water resources", *Water Science and Technology*,
82, 2635–2670, https://doi.org/10.2166/wst.2020.369

[14] Kirkby, M. J. (1975). "Hydrograph modelling strategies", *Processes in Physical and Human
Geography*, chapter 3, 69–90. Heinemann, London.

[15] Hinton, G. E., & Sejnowski, T. J. (1983). "Optimal perceptual inference", in Proceedings of the IEEE
Conference on Computer Vision and Pattern Recognition (Citeseer), 448–453.

[16] Ravuri, S., Lenc, K., Willson, M., Kangin, D., Lam, R., Mirowski, P., Fitzsimons, M., Athanassiadou, M., Kashem, S., Madge, S., Prudden, R., Mandhane, A., Clark, A., Brock, A., Simonyan, K., Hadsell, R., Robinson, N., Clancy, E., Arribas, A., & Mohamed, S. (2021). "Skilful precipitation nowcasting using deep generative models of radar", *Nature*, 597, 672–677, https://doi.org/10.1038/s41586-021-03854-z

[17] Slater, L. J. & Villarini, G. (2018). "Enhancing the predictability of seasonal streamflow with a statistical-dynamical approach", *Geophysical Research Letters,* 45, 6504–6513.

[18] Tian, H. & Chen, S.-C. (2017). "A video-aided semantic analytics system for disaster information integration", *2017 IEEE Third International Conference on Multimedia Big Data (BigMM)*, Laguna Hills, CA, 2017, 242–243. doi: 10.1109/BigMM.2017.31

[19] Zhou, X., Zhang, R., Yang, K., Yang, C., & Huang, T. (2020). "Using hybrid normalization technique and state transition algorithm to VIKOR method for influence maximization problem", *Neurocomputing*, 410, 41–50, ISSN 0925-2312. https://doi.org/10.1016/j.neucom.2020.05.084

[20] McCulloch, W. S., & Pitts, W. (1943). "A logical calculus of the ideas immanent in nervous activity", *Bulletin of Mathematical Biophysics,* 5, 115–133. https://doi.org/10.1007/BF02478259

[21] Cecaj, A., Lippi, M., Mamei, M., & Zambonelli, F. (2020). "Comparing deep learning and statistical methods in forecasting crowd distribution from aggregated mobile phone data", *Applied Sciences*, 10(18), 6580. https://doi.org/10.3390/app10186580

[22] Deng, L., & Yu, D. (2014). "Deep learning: Methods and applications", *Found Trends Signal Process*, 7(3–4), 197–387.

[23] Sarker, I. H. (2021). "Deep learning: A comprehensive overview on techniques, taxonomy, applications and research directions", *SN Computer Science*, 2, 420. https://doi.org/10.1007/s42 979-021-00815-1

[24] Albawi, S., Mohammed, T. A., & Al-Zawi, S. (2017). "Understanding of a convolutional neural network", *2017 International Conference on Engineering and Technology (ICET)*, Antalya, Turkey, 2017, 1–6, doi: 10.1109/ICEngTechnol.2017.8308186

[25] Lu, H. et al. (2017). "Cultivated land information extraction in UAV imagery based on deep convolutional neural network and transfer learning", *Journal of Mountain Science*, 14(4), 731–741.

[26] Khosravi, K., Panahi, M., Golkarian, A., Keesstra, S. D., Saco, P. M., Bui, D. T., & Lee, S. (2020). "Convolutional neural network approach for spatial prediction of flood hazard at national scale of Iran", *Journal of Hydrology*, 591, 125552. doi:10.1016/j.jhydrol.2020.125552

[27] Kabir, S., Patidar, S., Xia, X., Liang, Q., Neal, J., & Pender, G. (2020). "A deep convolutional neural network model for rapid prediction of fluvial flood inundation" *Journal of Hydrology*, 590, 125481. doi:10.1016/j.jhydrol.2020.125481

[28] Kim, P. (2017). "Convolutional neural network", In: MATLAB Deep Learning. Apress, Berkeley, CA. https://doi.org/10.1007/978-1-4842-2845-6_6

[29] Hochreiter, S., & Schmidhuber, J. (1997). "Long short-term memory", *Neural Computation*, 9(8), 1735–1780. https://doi.org/10.1162/neco.1997.9.8.1735

[30] Prasetya, E. P., & Djamal, E. C. (2019). "Rainfall forecasting for the natural disasters preparation using recurrent neural networks", *2019 International Conference on Electrical Engineering and Informatics (ICEEI)*, Bandung, Indonesia, 2019, 52–57, doi: 10.1109/ICEEI47359.2019.8988838

[31] Cho, K., Van Merrienboer, B., Gulcehre, C., Bougares, F., Schwenk, H., & Bengio, Y. (2017). "Learning phrase representations using rnn encoder-decoder for statistical machine translation", arXiv preprint arXiv:1406.1078.

[32] Jeba, G. S., Chitra, P. and Rajasekaran, U. M. 2022, "Time-series analysis and flood prediction using a deep learning approach", *2022 International Conference on Wireless Communications Signal Processing and Networking (WiSPNET)*, Chennai, India, 139–142, doi: 10.1109/WiSPNET54241.2022.9767102

[33] Goodfellow, I., Pouget-Abadie, J., Mirza, M., Xu, B., Warde-Farley, D., Ozair, S., Courville, A., & Bengio, Y. (2014). "Generative adversarial networks", *Advances in Neural Information Processing Systems*, 63(11), 139–144. doi: 10.1145/3422622

[34] Hofmann, J., & Schuttrumpf, H. (2021). "floodGAN: Using deep adversarial learning to predict pluvial flooding in real time" *Water*, 13(16), 2255. https://doi.org/10.3390/w13162255

[35] Cheng, M., Fang, F., Navon, I. M., & Pain, C. C. (2021). "A real-time flow forecasting with deep convolutional generative adversarial network: Application to flooding event in Denmark", *Physics of Fluids*, 33(5), 056602. https://doi.org/10.1063/5.0051213

[36] Kao, I.-F., Liou, J.-Y., Lee, M.-H., & Chang, F.-J. (2021). "Fusing stacked autoencoder and long short-term memory for regional multistep-ahead flood inundation forecasts", *Journal of Hydrology*, 598, 126371, ISSN 0022-1694, https://doi.org/10.1016/j.jhydrol.2021.126371

[37] Vaswani, A., Shazeer, N., Parmar, N., Uszkoreit, J., Jones, L., Gomez, A., Kaiser, L., & Polosukhin, I. (2017). "Attention is all you need". 31st Conference on Neural Information Processing Systems (NIPS 2017), Long Beach, CA, 5998–6008.

[38] Xu, J., Fan, H., Luo, M., Li, P., Jeong T., & Xu, L. (2023). "Transformer based water level prediction in Poyang Lake, China", *Water* 15(3), 576. https://doi.org/10.3390/w15030576

[39] Tian, H. & Chen, S.-C. (2017). "A video-aided semantic analytics system for disaster", *2017 IEEE Third International Conference on Multimedia Big Data (BigMM)*, Laguna Hills, CA, USA, 2017, 242–243. doi: 10.1109/ BigMM.2017.31

[40] Pouyanfar, S., Sadiq, S., Yan, Y., Tian, H., Tao, Y., Reyes, M. P., Shyu, M. L., Chen, S. C., and Iyengar, S. S. (2018). "A survey on deep learning: Algorithms, techniques, and applications", *ACM Computing Surveys*, 51(5).

3 Storage Management Techniques for Medical Internet of Things (MIoT)

S.U. Muthunagai[1,2], M.S. Girija[1,2], B. Praveen Kumar[3], and R. Anitha[1,2]

[1] School of Computer Science and Engineering, VIT, Vellore, Tamilnadu, India
[2] Department of Computer Science and Engineering, Sri Venkateswara College of Engineering, Sriperumbudur, Chennai, Tamilnadu, India
[3] Department of Computer Science and Engineering, GITAM (Deemed to be University) School of Technology, Bengaluru, Karnataka, India

3.1 INTRODUCTION

Sensors have been incorporated in a variety of medical products in recent decades as a result of advances in computer technology, having a significant impact on the medical and healthcare industries [1]. Due to the implantation of different sensors at various levels in the healthcare environment, the data collected from the sensors is getting increased. Hence, the majority of IoT healthcare data is redundant and is only reserved for a certain time span in order to forecast the future. It is challenging to aggregate data for important insights and analysis due to the non-uniformity of data and communication protocol [2]. IoT collects data in bulk, and in order to perform proper data analysis, the data must be segmented into chunks without being overloaded with precision and accuracy. An overabundance of data may have a long-term impact on the hospitality industry's decision-making process. In response to this issue, the proposed work makes an effort to offer an optimal data storage mechanism at the edge that reduces both storage space and processing time [3–6]. A faster decision-making process is made possible by the proposed architecture, which details the storage methods used for IoT healthcare at the edge in conjunction with the analysis of variance approach to compute time-series forecasting on IoT healthcare data. In this chapter, the data management lifecycle from the standpoint of IoT architecture is highlighted and demonstrates how it should differ from traditional data management systems.

The data that is created is what powers all of these wearable sensors, and a number of parties are vying for control by combining the data streams. Microsoft created the HealthVault, an electronic medical record that serves as a safe. HereIsMyData is a database that patients can use to save personal health information and control who has access to it. It was first offered by the Radboud University Medical Centre in Holland [7]. Over the past few years, the media and the general public have used the term "big data" a lot. The "three V's"—volume (huge amounts of data), variety (heterogeneity in the types of data available), and velocity (speed at which a data scientist or user may access and analyze the data)—seem to be the common denominators among the various definitions that have been presented [8]. As a result, healthcare has emerged as one of the major new adopters of big data. For instance, Fitbit and Apple's Research Kit can give academics access to enormous repositories of biometric information on individuals, which can subsequently be used to test theories about diet, exercise, illness development, treatment effectiveness, and other topics. The ability to continuously

DOI: 10.1201/9781003407959-3

update the annotations based on new information while maintaining the data's placement is one of the things that the data analysis should offer. Billing data from the Centres for Medicare and Medicaid Services (CMS) is available in enormous quantities and can be mined. A typical scenario engaged an elderly patient healing from a health issue at home associated to a variety of connected services that transmit data to various parties, such as family members and doctors. Hospitals have tried to lower re-admission rates by focusing on patients where predictive algorithms indicate people who may be at highest risk based on an analysis of the data collected from their current patients. However, there are a number of technological, legal, and ethical issues that are related to, among other things and informed permission, as well as research and therapeutic ethics that underlie these and many more potential uses [9].

The rest of the chapter is organized as follows: Section 3.2 elaborates on various existing data management frameworks on the basis of MIoT. Section 3.3 elaborates about the MIoT applications and challenges. Section 3.4 illustrates a data management framework in MIoT and Section 5 concludes the chapter.

3.2 RELATED WORKS

This chapter provides the recent works that are carried out on the impact of IoT in healthcare, electronic health records, cyber-physical systems and benefits of digitizing healthcare institutions. Also, it gives a detailed survey on data management techniques in MIoT.

3.2.1 IMPACT OF IoT IN HEALTHCARE

The Internet of Things (IoT) is a network of physical objects and other products that are connected to the internet and that are equipped with electronics, software, sensors, and network connectivity [1]. The most significant and intimate will likely be its effect on medicine. Health-related IoT technology will account for 40% of all IoT-related technology by 2025, accounting for a $117 billion market [2]. By cutting costs, eliminating waste, and saving lives, the fusion of medicine and information technologies, such as medical informatics, will revolutionize healthcare. Figure 3.1 depicts how this medical revolution will actually play out in a typical IoT hospital.

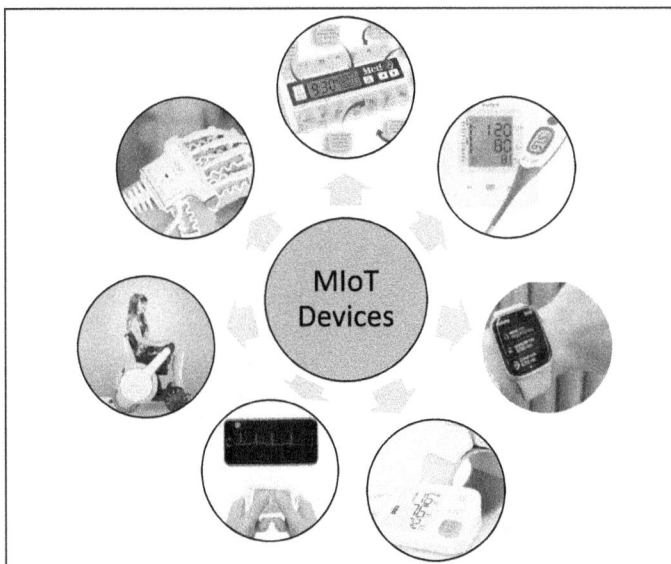

FIGURE 3.1 MIoT application devices.

3.2.2 ELECTRONIC HEALTH RECORDS

A patient with diabetes will have an ID card that, when scanned, connects to a secure cloud where their electronic health records, including vital signs and lab results as well as medical and prescription history, are kept. This record can be easily accessed on a tablet or desktop computer by doctors and nurses. The implementation of electronic health records (EHRs) is a major changer, despite the fact that it seems quite simple. A paper-and-ink method of managing records that dates back thousands of years will be replaced by a digital one in less than 10 years [3]. The benefits are numerous and clear. Researchers and other healthcare professionals may not be able to access paper documents that are filed away in filing cabinets because they are frequently written in sloppy handwriting. Instead, EHRs will eliminate many inefficiencies and save lives by putting all the crucial information in one location and making it simple to share.

Communication is one of the biggest obstacles to IoT implementation because even though many devices have sensors to collect data today, they frequently communicate with the server in their own language. Since every manufacturer has a unique set of proprietary protocols, sensors from several manufacturers may not be able to communicate with one another. The IoT as a whole is undermined by the fragmented software environment, privacy worries, and the tendency of bureaucracies to hoard all obtained data. This leads to important data being regularly marooned on data islands. In the upcoming years, the term "precision medicine" will be used frequently [4]. With multi-scale data for analysis and interpretation, it starts with genomics and moves through the remaining omics platforms [5]. The Collaborative Cancer Cloud, a high-performance analytics platform that gathers and securely maintains confidential medical data that may be utilized for cancer research, was introduced in 2015 as a joint venture between Intel and the Oregon Health & Science University. Although the technology was initially developed for the treatment of cancer, Intel plans to make the federated cloud network available to organizations researching treatments for disorders including Parkinson's. Through the use of engineering simulation technologies and the Medical Internet of Things (MIoT), medicine is becoming participatory, personalized, predictive, and preventative (P4 medicine) [6]. Pharma IoT will make it possible for patients and medical staff to use medications with cutting-edge sensor gear and create personalized care services and procedures (Product 2.0). The linked sensor wearables for people with Parkinson's disease and multiple sclerosis are good examples of pharmaceutical IoT solutions because they provide medication management, which enhances patient outcomes and quality of life [10]. Additionally, existing medical device goods like insulin pens and inhalers can be integrated with sensor and connectivity technologies to gather data for personalized medication and other care analytics [11]. Because patient care data offers new sources of innovation and competitiveness, all of this will significantly improve individual medication and care processes.

3.2.3 CYBER-PHYSICAL SYSTEMS

A number of academic research studies and the business world have focused on the IoT, which has emerged as one of the most promising technologies in the last 10 years [12]. The IoT system has a complex structure, with many parts communicating with one another to offer the user a range of options. This multi-tiered system enables real-time data collecting, communication between devices, data transfers, and analytics to manage end-user applications [13]. Because they enable data-driven decision-making and mix human contact with computer-based structures, cyber-physical systems are essential parts of any IoT architecture. A connected ecosystem made up of cyber-physical systems is created by the IoT, which blends human interaction with computer-based systems and enables data-driven decision-making. IoT technologies that are supported by sensors, actuators, and communication protocol networks include smart grids, smart homes, intelligent logistics, and smart communities. The IoT offers numerous real-time solutions by fusing data analytics with in-machine sensors.

3.2.4 Benefits of Digitizing Healthcare Institutions

The IoT [14] is crucial in a number of applications in the sphere of healthcare. This criterion's three steps are clinical treatment, remote monitoring, and situational awareness. With the rapid advancement of wearable/implantable sensors and wireless communication, researchers are growing more interested in enhancing the health sector in response to human needs by [15] digitizing and decentralizing healthcare institutions and providing continuous and remote medical monitoring. The MIoT, a healthcare application of IoT technology, is a network of connected devices that detect critical data in real-time [16]. The MIoT, which is essential for lowering medical costs, delivering quick medical assistance, and enhancing the standard of medical care, is becoming more and more significant in the healthcare industry to build the safety and health system in human civilization [17].

Small wearable devices or implantable sensors are used by MIoT, a cutting-edge e-healthcare technology, to gather vital body metrics and track patients' pathological conditions. Numerous healthcare organizations [18] are deploying MIoT apps to enhance patient care, manage drugs, manage diseases, reduce errors, and improve patient satisfaction [19]. Compliance with healthcare service standards will also be improved by greater patient involvement in [20] decision-making. The MIoT, machine learning, artificial intelligence (AI) [21], and big data [22] analytics have all contributed to the development of the digital healthcare industry [23]. As a result, technology adoption will quicken in the upcoming years, driving the market's worth to $254.2 billion by 2026 on a global scale. The IoT is a dynamic, worldwide network architecture in which "Things" (subsystems and distinct physical and virtual entities) may be recognized, are autonomous and can be configured on their own. The expectation is that "Things" will exchange data produced by sense, respond to events, and initiate activities in order to interact with the environment and manage the physical world [24]. The goal of the IoT should be to create a standardized platform for the creation of cooperative services and applications that harness the combined power of the resources made accessible by individual "Things" and any associated support systems.

3.2.5 Data Management Techniques in MIoT

The amount of knowledge that may be made available through the merging of data created in real-time and data kept in long-term repositories is at the heart of these resources. A useful source for trend analysis and strategic prospects, this information can enable the realization of novel and unconventional applications and value-added services. To do this, a thorough management structure for the data created and stored by IoT objects is required. Data management is a broad term that refers to the designs, practices, and methods used to effectively manage a system's data lifecycle requirements. Data management should operate as a layer in the IoT context between the objects and devices that generate the data and the applications that access the data for services and analysis. The actual devices themselves can be grouped into autonomously governed subsystems or subspaces with internal hierarchical management [25]. Depending on the level of privacy requested by the subsystem owners, the functionality and data provided by these subsystems are to be made available to the IoT network.

The unique properties of IoT data render the use of conventional relational-based database management systems outdated. By periodically delivering observations on certain monitored phenomena or reporting the presence of specific or abnormal occurrences of interest, millions of varied devices will generate a tremendous amount of heterogeneous, streaming, and geographically scattered real-time data [26]. Due to its streaming and continuous nature, periodic observations are the most demanding in terms of communication overhead and storage, whereas events present time-strain with end-to-end reaction timings depending on the urgency of the response required for the event. In addition to the data generated by "Things," there is metadata that describes "Things"; examples of this data include item identity, location, operations, and services offered. IoT data will transmit the network from dynamic and mobile objects to concentration storage places and be stored statically in databases with a fixed- or flexible-schema.

Up until centralized data repositories, this will go on. So, while designing data management solutions for the Medical Internet of Things, communication, storage, and procedure will be key considerations. Recent predictions suggest that there is a resurgence in interest in database systems research that concentrates on alternatives to the conventional relational paradigm. IoT can benefit greatly from the move away from conventional database models in a number of ways, including the use of remote storage at the Things layer, non-structural data support, relaxation of the atomicity, consistency, isolation, and durability (ACID) properties to trade-off consistency and availability, and incorporation of energy efficiency as a data management design primitive [27].

3.3 MEDICAL INTERNET OF THINGS

3.3.1 MIoT Applications

The quickly developing technologies can offer affordable, accessible healthcare services, but they cannot completely eradicate chronic diseases. Because routine healthcare services are costly, technology has changed the focus of routine health checks from being hospital-centric to patient-centric (at home), which lessens the need for hospitalization. The widespread use of IoT in the healthcare paradigm makes it possible for doctors to work more effectively and give patients better care. Applications are broadly divided into the two categories of dispersed applications and congregate applications in this section. This section includes applications of wearable application types.

- Blood pressure monitoring
- Body temperature monitoring
- Blood glucose monitoring
- ECG monitoring
- Galvanic skin response (GSR) monitoring
- Oxygen saturation monitoring
- Activity monitoring
- Rehabilitation system
- Smart wheelchair system
- Medication control system

3.3.2 Challenges in Healthcare

The challenges can be divided into two groups: policy and technology. In policy matters, healthcare professionals in a fee-for-service system can only be paid when they interact directly with patients. This strongly biases against promoting non-face-to-face contact-enhancing technologies. Moving away from that model and more towards value-based care, wherein global risk-based payments are made to delivery organizations, there will be more incentive to use new technologies that cut down on pointless in-office encounters. Face-to-face interactions are actually a cost center rather than a profit center in such a setting, and population health improvements are rewarded.

In technology, the condition of health data is the main technical obstacle to reaching this aim. Health information has been mainly dispersed into institution-specific silos by outdated EHR systems. Even though the silos can be quite huge, they are still silos. Much of the current work is focused on the exchange of individual records between silos using more standardized vocabularies (code sets) and message formats. That, however, does not address the issue of data fragmentation. The next generation of health technology will revolve around aggregating data rather than just sharing copies of individual records (the conventional query-response strategy), as more and more people in the health information exchange space are beginning to realize.

The only way to make the data truly meaningful is to gather it from a variety of sources, normalize it into a recognizable structure, and resolve it around distinct patient and provider IDs. Two further benefits of aggregated data. (1) The compatibility issue is resolved. (2) It is also adaptable enough to support machine learning and AI to function in a real-time fashion. The data gathered from IoT devices need to be organized in the storage in order to perform intelligent actions and hence this chapter aims to provide storage management techniques suitable for MIoT. Improving the storage efficiency in a cloud by removing redundant data to develop single instance storage system with elimination of redundant of data especially from IoT devices has become the main challenges to be resolved. The next section provides the data management framework suitable for MIoT.

3.4 DATA MANAGEMENT FRAMEWORK IN MIOT

As shown in Figure 3.1 [28], the MIoT data management system is divided into an online, real-time frontend and an offline backend. The online frontend application interacts with the connected IoT items and sensors directly, while the offline backend manages bulk storage and in-depth analysis of IoT data. The data management frontend involves extensive connectivity with sensors and intelligent objects to propagate query requests and results. The backend uses a lot of storage because it stores generated data in large quantities for further processing, analysis, and in-depth queries. The storage components are located on the backend, but they frequently communicate with the frontend through ongoing updates, making them considered online.

In conventional database systems, the majority of the data is gathered from predefined, limited sources and stored as scalar data relations in accordance with rigid normalization criteria. Specific summary views of the system can be retrieved via queries, as can specific database elements. When necessary, fresh data is also added to the database via insertion queries. Query activities are typically local, with processing and intermediate storage acting as execution costs. To enforce overall data integrity, transaction management systems must ensure the ACID properties [19]. Query processing and distributed transaction management are required even if the database is spread across several locations [21].

With a vast and expanding number of data sources, including sensors, RFIDs, embedded systems, and mobile devices, the situation is drastically different with IoT systems. When compared to the infrequent updates and queries made to traditional DBMSs, data is continually pouring from a wide variety of Things to IoT data stores, and queries are more frequent and have more varied needs. It may be possible to modify the rigid relational database design and relational normalization practice in favor of more open-ended and adaptable forms that can accommodate complex queries and a variety of data kinds. Although query optimization in distributed database management systems (DBMS) is based on communication considerations, optimizers still make choices based on fixed, well-defined schemas.

The following section provides details of some data management solutions for IoT that integrate all or part of the above mechanisms. These proposals are then analyzed in light of a set of design primitives that we deem necessary for a comprehensive data management solution for IoT.

3.4.1 DATA MANAGEMENT FRAMEWORK FOR IOT

Since wireless sensor network (WSN) are only a small portion of the overall IoT landscape, the majority of the current data management options do not specifically address the more complex architectural aspects of IoT. Most sensors are immobile and resource-constrained, which makes it difficult to provide complex analysis and services. With limited permanent storage capacity for long-term use, the primary goal of MIoT data management systems is to quickly gather real-time data for speedy decision-making. This is merely a portion of the more flexible IoT system, which attempts to collect data from a range of sources, including those that are fixed and mobile, intelligent and embedded, resource-constrained and resource-rich, real-time and archival. Therefore, the

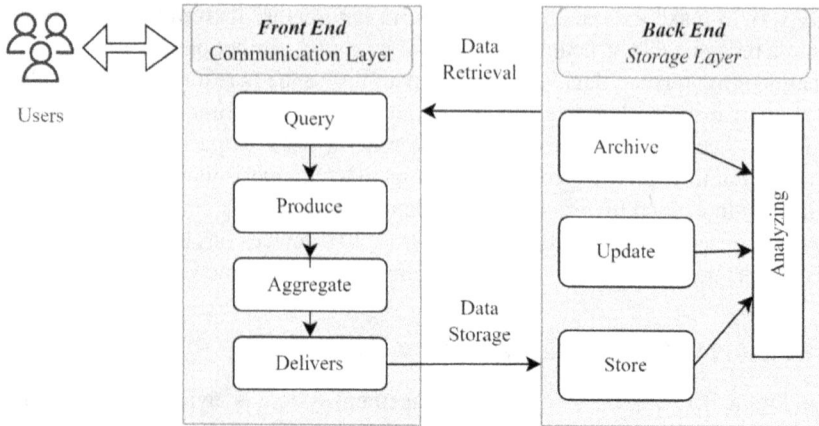

FIGURE 3.2   IoT data lifecycle and data management.

primary goal of IoT-based data management goes beyond the wireless network's original design to include provisions for a seamless approach to access the amounts of heterogeneous data in order to uncover intriguing global trends and business prospects.

In order to assist the integration of data from diverse networks, several of the existing ideas offer abstractions, opening the door for the adaption and seamless integration of additional IoT subsystems. There is still a need for reliable and thorough data management solutions that facilitate communication between various subsystems and incorporate the full lifecycle of data management with the existence of mobile objects and context-awareness needs. This chapter focuses on a framework for managing IoT data that is more in line with the lifespan of IoT data and takes care of the design primitives covered previously. To ensure the independence of distinct IoT subspaces as well as flexible join/leave architecture, the proposed framework uses a layered approach that focuses on data- and source-centric middleware and concepts from federated database management systems.

3.4.1.1   Framework Description

Four stacked layers make up the proposed IoT data management structure. The framework layers closely correspond to the stages of the IoT data lifecycle shown in Figure 3.2, with lookup/orchestration being viewed as an additional operation that is not strictly a part of the data lifecycle. IoT sensors, smart objects (data production items), and modules for in-network processing and data collection/real-time aggregation (processing, aggregation) are all included in the "Things" layer. The data layers handle organized and unstructured data in turn. This layer is where data sources are found, catalogued, and stored, as well as where collected data is stored and indexed. For local, autonomous data repository sites, the data layer additionally manages data and query processing. In collaboration with data sources, the query layer manages the specifics of query processing and optimization. The aggregation sublayer of the query layer manages aggregation and fusion queries using a variety of data sources and locations. Both the requester of data or analysis needs and the consumer of data and analysis outputs are members of the application/analysis layer. The layers of the suggested MIoT data management architecture is shown in Figure 3.3 along with the functional modules that go with each layer. The layers are then explained in the subsections that follow.

3.4.1.2   Things Layer

The entities that produce data for the IoT system and the modules that carry out in-network, real-time operations whose outcomes are to be transmitted further up the system are both included in the Things layer.

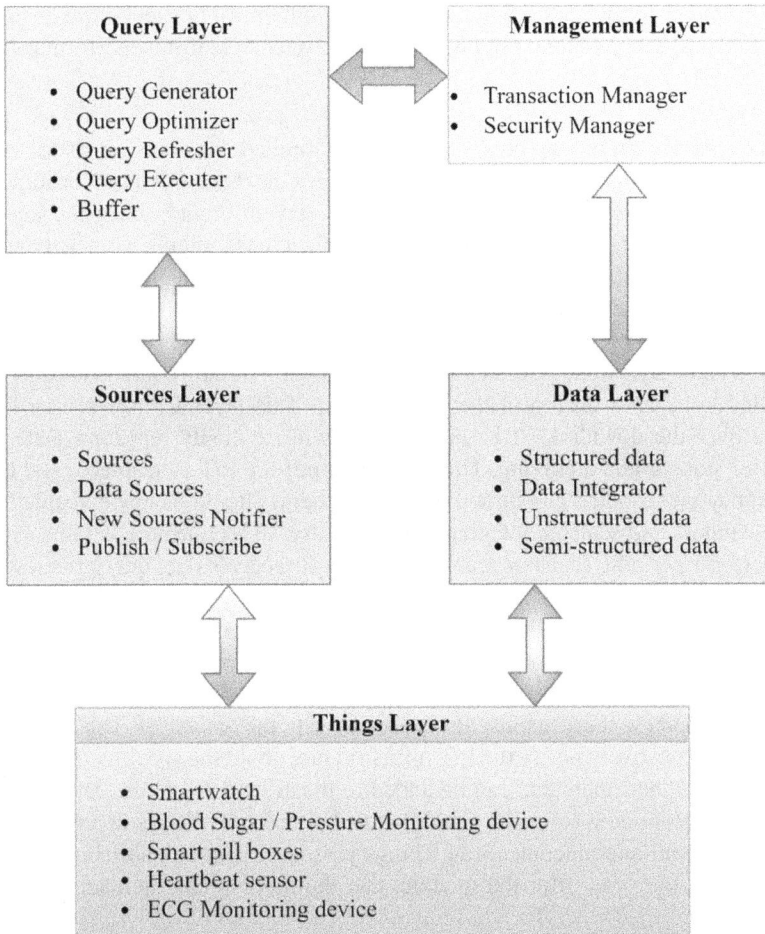

FIGURE 3.3   MIoT data management framework.

A device that has embedded intelligent devices with communication capabilities are referred to as an entity. Entities can be sensors, RFID tags, mobile phones, laptops, wearable sensors, accelerometers, gyroscopes, magnetic field sensors, and other similar devices. A body area network connected to a patient, a vehicle with a variety of intelligent devices, or an environmental sensor network are examples of entities that can be viewed as collections or as what is known as virtual entities. Each of these things is regarded as a data source for the IoT network, regardless of whether it functions independently or as a component of a larger network system.

There must be a method to specifically identify the data sources in order to have access to the data at the Things layer. Platform-dependent identification, which can be cross-referenced with location-based identification, is used to identify things that belong to a certain platform or entity and are connected to that platform or entity for a general purpose. An example of this would be body area networks that are connected to a particular patient. Because of this, the IoT objects' unique identification is closely linked to them and is quickly accessible at the Things layer by both in-network processes and crawling agents at higher layers of the framework that need to identify specific objects that might fulfil particular requests.

The aggregation point assists in gathering and condensing information from numerous objects and Things, which may or may not be homogeneous or part of a single system. The processing and efficiency requirements of the underlying subsystem or entity must be met before using such aggregation points, and any delays should not undermine the system's real-time performance.

3.4.1.3 Source Layer

Database fragments are stored in predetermined and limited locations in distributed database systems. The system's design requires that metadata stores database fragment locations in advance for querying and updating. For IoT data, the situation is very different because the sources that will generate and deliver the data are dispersed over various locations, ranging from dedicated objects to implants on objects, are not finite, and there is no unified schema or definition of metadata. Due to these properties, it can be difficult to execute real-time queries sent to the layers of the thing if it is impossible to determine which sources or subsystems can respond to the requests.

The seamless and transparent identification and unification of data sources for query processing must be handled by a layer on top of the Things layer. This layer focuses primarily on serving requests made over the downlink. It is unconcerned with routinely reporting generated data to higher layers for storage and archiving. The architecture of the IoT is adaptable and can take new subsystems as they are installed and turned on. Such systems should be made visible to the system as a whole if seamless and scalable integration into the overall architecture is to be supported. The crawler can perform periodic scans or scans only when it receives data query requests to perform reactive discovery of new sources.

There needs to be a notification system in place, a new source notifier, so that real-time, continuously updated queries can be made aware of new sources as they are discovered by the sources crawler. In order to be integrated with the objects catalog and metadata store, the data specifications of these new sources must also be reported to the data layer. The new source notifier can also be used to proactively notify the IoT infrastructure about the specifications of new devices or subsystems so that their metadata can be added to the system for future reference and access. The publish/subscribe module serves as a middleman between the Things layer and the data layer, taking both queries and announcements of Things presence. Things publish descriptions of the information/services they can offer the module, and the module can then actively involve them in queries.

3.4.1.4 Data Layer

The foundational component of data management is the data layer. For later updates, queries, and access to the data, it is crucial to comprehend where and how the data is stored. Regarding IoT data management, there are two primary concerns that need to be addressed: transaction manager, applications layer, security manager, health monitoring medical care, environmental monitoring transportation, unstructured data store, structured data store metadata store, local query analysis module, local data integrator, data layer, new source notifier sources, source crawler, publish/subscribe module data sources. In this framework, a hybrid approach to data storage is suggested, with temporal, real-time data stored nearby the objects that are generating this data and persistent, long-term data that will be used for analysis catalogued and stored at dedicated facilities. This results in a beneficial trade-off between the costs of storage space and data transmission, on the one hand, and the accessibility of data for in-depth analysis and queries, on the other.

The functionalities of the modules in the data layer are comparable to those of their counterparts in the upper layers. On the level of a local IoT site, the local query/analysis module performs the same tasks as the query processing and optimization modules that will be covered later in the query layer. Information needed to define sources and access data is stored in the data layer catalogs. It has been noted that medical devices have generated a significant amount of data. Massive amounts of data generated in the healthcare industry are transferred to the cloud for further processing because the data providers do not provide adequate storage capacity. This is done for computational intelligence and remote accessibility.

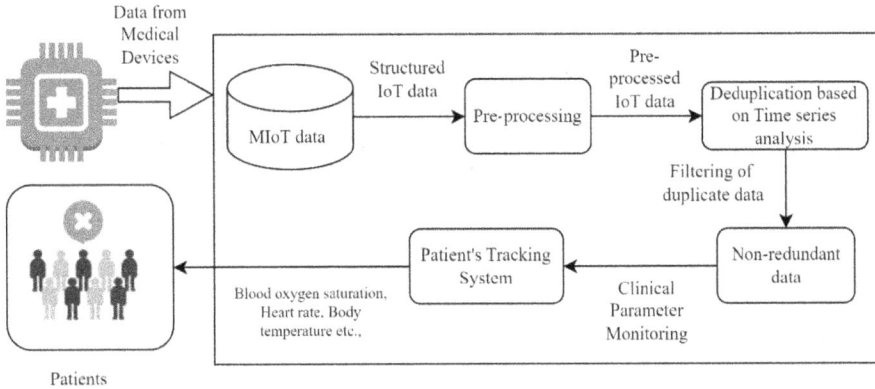

FIGURE 3.4 Proposed framework.

When redundant data generated by MIoT sensors is loaded into cloud storage, it results in an inefficient use of storage space and makes it difficult to access the data there. Deduplication is a popular data reduction technique used to significantly improve storage network access. This technique removes duplicate data by retaining one instance of the redundant data and connecting the other instances to it. The geographically distributed cloud data centers have a lot of redundant data. As a result, managing data, storage space, and networks in a cloud environment is getting more difficult. A potential fix for the aforementioned issue is to use a data deduplication approach for MIoT at cloud servers.

Time-series analysis is a method for examining a collection of data points acquired over a period of time. In contrast to occasional or arbitrary data collection, time-series analysis gathers data points at regular intervals over a predetermined timeframe. As presented in Figure 3.4, the proposed strategy uses observed sensor value to conduct deduplication prior to data accumulating in the cloud server. After the data has been received from the healthcare system, noisy and irrelevant data are eliminated from it using a preprocessing procedure. Before the deduplication operation is finished, time-based partitions are created from the preprocessed data.

The deduplication process greatly lowers the redundancy rate of the enormous volume of data, which also reduces the amount of storage needed. If the redundancy rate of MIoT data is decreased, a large portion of the cloud storage capacity will be used for other computations. Therefore, deduplication techniques are required to remove the superfluous data. The optimum storage method is deduplication, which eliminates redundant copies of data from storage. For MIoT users, deduplication would also be strongly advised if any computational performance were incorporated [29–30].

3.4.1.5 Query Layer

The components required for generating, optimizing, and executing queries on the IoT database are contained in the query layer. The query plan generator converts the specifications for the desired output into a query that is written in a standard query language format using input from the application or directly from the user. The query optimizer then receives the plans for a query and determines which plan should be executed. The query optimizer receives a set of query plans for a query and chooses the one that will execute the query in the quickest and most effective manner. Each plan is given a cost that is estimated using evaluation criteria that are either provided at runtime by the user submitting the query or are predefined in a data dictionary.

For time-series, location-aware, and source-specific data, evaluation criteria include the quantity of input/output operations, processing time, delay constraints, and temporal/spatial/modal constraints. Every time new sources become available or existing sources stop providing results for the query, optimization is restarted to include or exclude sources. The query refresher is turned on for queries whose sources dynamically join/leave the network or whose results are streaming data that is periodically generated or updated. The allocated storage space for the query results must be taken into account in dynamic real-time updates. This can be accomplished in one of three ways: by condensing results into synopses, gracefully aging results in accordance with relevance criteria, or by providing approximations of query results with assurances regarding the approximation accuracy in representing the real data. The query's incremental intermediate results as they come in can be stored in a temporary storage buffer for results.

3.4.1.6 Management Layer

The mechanisms required to provide access and security to the various data stores in the framework's data layer are the focus of the management layer. The transaction manager is in charge of handling transactions that are more closely related to business services and processes. The manager can use either a traditional single-source execution mechanism or global or distributed execution strategies, depending on the kind of transaction submitted to it.

It is possible to relax the strict ACID requirements for successful transactions in favor of eventual consistency guarantees, which are more in line with current trends [28]. After an electrical failure, a crash, damaged files, etc., the recovery manager is responsible for bringing the data repositories back to their most recent consistent state. This is typically accomplished by undoing all transactions or operations that were ongoing within the data management system but had not yet been successfully committed. The mechanisms required to provide access and security to the various data stores in the framework's data layer are the focus of the management layer. The execution of transactions that are more related is handled by the transaction manager. The transaction manager is in charge of handling transactions that are more closely related to business services and processes. The manager can use either a traditional single-source execution mechanism or global or distributed execution strategies, depending on the kind of transaction that has been submitted to it.

For sensitive systems with stringent data availability requirements, replicas of data repositories that are updated concurrently with the primary repositories can be used. Archiving is one method for recovering lost or damaged data in primary storage space. Due to the redo/undo recovery mechanisms already in place for database management systems, the recovery manager may only be concerned with independent repository maintenance.

3.4.1.7 Applications/Analysis Layer

Utilizing the wealth of information that can be generated from the correlation of data and extracting potentially interesting patterns/trends is the main justification for implementing a massive data storage strategy in MIoT. The system's smooth operation, upcoming improvements, and novel or unusual business opportunities can all benefit from this information. By providing a range of services and analysis capabilities to the IoT system's end users, the applications/analysis layer will be concerned with utilizing the information generated by IoT.

With the help of this layer, the healthcare systems are being improved. Conventional medical equipment is being replaced with more advanced medical equipment that can remotely monitor biological parameters. Innovations open up new opportunities for everyone to actively participate in remote clinical parameter monitoring in non-clinical settings. Numerous studies have shown that routine treatment of acute and chronic illnesses improves the patient's quality of life. Through interfaces and dashboards, sensors allow healthcare professionals to monitor, track, and assess physiological factors.

For predicting the disease, these medical sensors are becoming more accurate and trustworthy [29]. The wearable sensor primarily provides patients with comfort and flexibility. Depending on the clinical applications, wearable sensors can be placed anywhere on the body, including the wrist, ankle, waist, chest, arm, and fingers. The newly developed Medical Internet of Things keeps track of routine activities like standing, walking, and postures as well as blood oxygen saturation, heart rate, body temperature, galvanic skin responses, and hand postures while moving.

Additionally, some inertial sensors based on micro-electro-mechanical systems (MEMS), such as accelerometers, gyroscopes, and magnetic field sensors, are frequently used for assessing activity-related events. Gyroscopes have been used to perform gait analysis in order to present a patient's motion tracking system in the healthcare industry because an accelerometer alone is unable to provide accurate data regarding motion. This has led to the development of an appropriate accelerometer-based system for monitoring breathing and heart rates as well as postural changes.

The majority of applications falling under the healthcare domain are data-intensive, with different data/analysis requirements depending on the services they offer to businesses and end users. Moreover, depending on the real-time needs and complexity level of the analysis, different applications might require data access and processing at the lower Things layer or the higher data layer, data and pattern recognition numerous IoT applications can make use of sensor mining techniques.

They must, however, take into account the three elements that make IoT data unique: the growing volume of data, the highly unstructured and heterogeneous nature of the data itself, and the geographically dispersed storage facilities.

3.5   CONCLUSION AND FUTURE WORK

In this chapter, a few data management solutions for the Medical Internet of Things are put forth with an emphasis on the necessary design considerations that must be made in order to offer a complete solution. Insofar as they address the data management requirements of IoT subsystems like WSNs, the current solutions are only partial. We described the elements of a thorough IoT data management framework with core data and source layers to make up for this deficiency. The framework emphasizes the necessity of a two-way, multi-layered design strategy that can handle both real-time and archival query, analysis, and service requirements. Future work entails more precise mapping of the proposed framework's specifics to the MIoT reference model, in-depth research and creation of a data management solution that builds on the framework, and incorporation of data security and privacy considerations into the framework design in accordance with the requirements for the MIoT's dynamic and heterogeneous environment. The integration of heterogeneous data sources and systems within the IoT is a further area that the authors wish to investigate. Heterogeneity here goes beyond the traditional idea of various data types and formats to include various data sources, time and geotags, and geographically dispersed locations.

REFERENCES

1.   Zanella A., Bui N., Castellani A., Vangelista L., Zorzi M. Internet of things for smart cities. *IEEE Transaction on Internet Things*, Vol. 1, No.1, pp. 22–32, 2014.
2.   Bauer H., Patel M., Veira J. *The Internet of Things: Sizing Up the Opportunity [Internet]*. New York (NY): McKinsey & Company, 2016 Available from: www.mckinsey.com/industries/hightech/our-insights/the-internet-of-things-sizing-up-the opportunity.
3.   Kruse C.S., Kothman K., Anerobi K., Abanaka L. Adoption factors of the electronic health record: a systematic review. *JMIR Med Inform*, Vol. 4, No. 2, 2016.
4.   Scheen A.J. Precision medicine: the future in diabetes care. *Diabetes Res Clin Pract*, Vol. 117, pp. 12–21, 2016.
5.   van Leeuwen N., Swen J.J., Guchelaar H.J., Hart L.M. The role of pharmacogenetics in drug disposition and response of oral glucose-lowering drugs. *Clin Pharmacokinet*, Vol. 52, No. 10, pp. 833–852, 2013.

6.	Flores M., Glusman G., Brogaard K., Price N.D., Hood L. P4 medicine: how systems medicine will transform the healthcare sector and society. *Per Med*, Vol. 10, No. 6, pp. 565–576, 2013.

7.	HereIsMyData [Internet]. *Nijmegen: Radboud University Medical Center*; c2015 [cited at 2016 July 1]. Available from: www.hereismydata.com/.

8.	Auffray C., Balling R., Barroso I., Bencze L., Benson M., Bergeron J., et al. Making sense of big data in health research: towards an EU action plan. *Genome Med*, Vol. 8, No. 1, p. 71, 2016.

9.	Filkins B.L., Kim J.Y., Roberts B., Armstrong W., Miller M.A., Hultner M.L., et al. Privacy and security in the era of digital health: what should translational researchers know and do about it? *Am J Transl Res*, Vol. 8, No. 3, pp. 1560–1580, 2016.

10.	van Uem J.M., Maier K.S., Hucker S., Scheck O., Hobert M.A., Santos A.T., et al. Twelve-week sensor assessment in Parkinson's disease: impact on quality of life. Mov Disord 2016 May 31 [Epub]. http://dx.doi.org/10.1002/ mds.26676

11.	Dzubur E., Li M., Kawabata K., Sun Y., McConnell R., Intille S., et al. Design of a smartphone application to monitor stress, asthma symptoms, and asthma inhaler use. *Ann Allergy Asthma Immunol*, Vol. 114, No. 4, pp. 341–2.e2, 2015.

12.	Gubbi J., Buyya, R. Marusic, S. Palaniswami, M. Internet of things (IoT): A vision, architectural elements and future directions. *Future Gener. Comput. Syst.*, Vol. 29, pp. 1645–1660, 2013.

13.	Perwej Y., AbouGhaly M.A., Kerim B., Mahmoud Harb H.A. *An extended review on internet of things (IoT) and its promising applications.* Communications on Applied Electronics (CAE), New York, USA, Vol. 9, No. 26, pp. 8–22, February 2019.

14.	Islam, S.M.R., Kwak, D., Kabir, M.H., Hossain, M., Kwak, K. The Internet of Things for health care: a comprehensive survey. *IEEE Access,* Vol. 3, pp. 678–708, 2015.

15.	Perwej Y., Omer M.K., Sheta O.E., Harb H.A.M., Adrees M.S. The future of internet of things (IoT) and its empowering technology. *International Journal of Engineering Science and Computing (IJESC)*, Vol. 9, No. 3, pp. 20192–20203, March 2019.

16.	Stellios I., Kotzanikolaou P., Psarakis M., Alcaraz, C., Lopez, J. A survey of iot-enabled cyberattacks: assessing attack paths to critical infrastructures and services. *IEEE Commun. Surv. Tutor*, 20, 3453–3495, 2018.

17.	Parwej F., Akhtar N., Perwej Y. An empirical analysis of web of things (WoT). *International Journal of Advanced Research in Computer Science*, Vol. 10, No. 3, pp. 32–40, 2019. DOI: 10.26483/ijarcs. v10i3.6434

18.	Perwej Y., Ahamad F., Khan M.Z., Akhtar N. An empirical study on the current state of internet of multimedia things (IoMT). *International Journal of Engineering Research in Computer Science and Engineering (IJERCSE), ISSN (Online) 2394–2320*, Vol. 8, No. 3, pp. 25–42, March 2021. DOI: 10.1617/vol8/iss3/pid85026

19.	Pantelopoulos A., Bourbakis N.G. A survey on wearable sensor-based systems for health monitoring and prognosis. *IEEE Trans. Syst. Man Cybern. Part C (Appl. Rev.)*, Vol. 40, No. 1, pp. 1–12, 2010.

20.	Akhtar N., Rahman S., Sadia H., Perwej Y. A holistic analysis of medical internet of things (MIoT). *Journal of Information and Computational Science (JOICS)*, ISSN: 1548–7741, Vol. 11, No. 4, pp. 209–222, April 2021. DOI:10.12733/JICS.2021/V11I3.535569.31023

21.	Perwej Y., Parwej F. A neuroplasticity (brain plasticity) approach to use in artificial neural network. *International Journal of Scientific & Engineering Research (IJSER)*, France, ISSN 2229–5518, Vol. 3, No. 6, pp. 1–9, June 2012. DOI: 10.13140/2.1.1693.2808

22.	Perwej Y. An experiential study of the big data, International Transaction of Electrical and Computer Engineers System (ITECES), *USA, Science and Education Publishing*, Vol. 4, No. 1, pp. 14–25, March 2017. DOI: 10.12691/iteces-4-1-3

23.	Y. Jin, Low-cost and active control of radiation of wearable medical health device for wireless body area network. *J. Med. Syst.*, Vol. 43, No. 5, p. 137, 2019.

24.	Vermesan O., Harrison M., Vogt H., Kalaboukas K., Tomasella M., Wouters K. Gusmeroli S., Haller S. *Internet of things strategic research roadmap*, IoT European Research Cluster: Brussels, Belgium, 2009.

25.	Pujolle, G. *An Autonomic-Oriented Architecture for the Internet of Things.* In Proceedings of IEEE John Vincent Atanasoff International Symposium on Modern Computing (JVA 2006), Sofia, Bulgaria, pp. 163–168, 3–6 October 2006.

26. Wu G., Talwar S., Johnsson K., Himayat N., Johnson, K.D. M2M: From mobile to embedded internet. *IEEE Commun. Mag.*, Vol. 49, pp. 36–43, 2011.

27. Cooper J., James A. Challenges for database management in the internet of things. *IETE Tech. Rev.*, Vol. 26, pp. 320–329, 2009.

28. Abu-Elkheir M., Hayajneh M., Ali N.A. Data management for the internet of things: design primitives and solution. *Sensors*, Vol. 13, pp. 15582–15612, 2013.

29. Karthick G.S., Pankajavalli P.B. Review on human healthcare internet of things: a technical perspective. *Springer*, Vol. 1, pp. 1–19, 2020.

30. Muthunagai S.U., Anitha R. TDOPS: Time series based deduplication and optimal data placement strategy for IIoT in cloud environment. *Journal of Intelligent and Fuzzy Systems*, Vol. 43, pp. 1583–1597, 2022.

4 A Study on Trending Technologies for IoT Use Cases Aspires to Build Sustainable Smart Cities

Mangayarkarasi Ramaiah, R. Mohemmed Yousuf,
R. Vishnukumar, and Adla Padma
School of Computer Science Engineering and Information Systems,
Vellore Institute of Technology, Vellore, Tamilnadu, India

4.1 INTRODUCTION

The concept of a "smart city" has recently been criticized for being too focused on technology and driven by business goals rather than those of the community and its people. This calls for a more reasonable way of doing things. Initially, sustainability has been around longer than smart towns. It is based on social, economic, and natural sustainability [1]. In [2] a more modern definition says that a "sustainable city" doesn't use more material and energy resources than it can get from its surroundings or make more trash than it can get rid of. In practice, a city's use of land, water, and energy should be the same as or less than what it gets from the natural environment. Pollution from the city shouldn't make the environment less able to provide resources for people and other animals [3]. Even though the idea of sustainability is easy and makes sense, it has been criticized for being out of date and not fitting the highly digitized, quickly changing society of today.

But later on, the idea of a "smart city" is introduced and is relatively new to urban areas and ICT, it is quickly gaining acceptance. This famous and appealing model has been described by academics, business people, governments, institutions, and ordinary people. As these concepts evolved, a new wave of academic debates proposed a smart, sustainable city as a more balanced response to current criticisms. A thorough definition of smart and sustainable cities as provided by [4][5] is as follows: "smart cities that use ICT and other innovative technologies to enhance competitiveness, efficiency of services and urban operations, quality of life, while making sure they meet the needs of both current and future generations in terms of economic, social, environmental, and cultural issues".

The smart city concept relies on IoT technologies to address global population growth-related urban infrastructure issues [6]. These technologies would aid in the detection of infrastructure issues such as traffic congestion, energy shortages, water shortages, and security incidents. It also employs sensors to notify citizens about parking spaces, malfunctions, and electrical outages, among other issues, and would aid cities in achieving smart infrastructure, healthcare, warehouses, transportation, waste management, and community development, among other goals. In the current economic climate, time management can conserve both energy and money. Smart dwellings enable the integration and balancing of renewable energy technologies in an efficient manner [7]. This innovative and appealing IoT technologies in smart residences [8] as part of the smart city concept improve residential life as well. However, this study mainly focuses on the technical sustainability of smart cities and its operational challenges.

DOI: 10.1201/9781003407959-4

To discuss and comprehend the objectives of smart, sustainable towns, a thorough discussion is required. The organization of this chapter is as follows: Section 4.2 describes the role of IoT in the context of a smart city as well as the incurred technical difficulties. Section 4.3 presents the role of edge, fog, and blockchain-based solutions for sustainable smart cities through the existing works. Section 4.4 highlights the importance of AI in mitigating and complementing the smart environment applications especially in view of reaching technical sustainability. Finally, Section 4.5 brings a conclusion to the chapter.

4.2 IOT IN SMART CITY IN THE PERSPECTIVE OF SUSTAINABILITY

Given that the idea is constantly changing, the term "smart cities" does not yet have a widely accepted and established definition. Most definitions, however, highlight the enhancement of city life through the application of ICT. Smart cities use information and communication technologies to improve quality of life, reduce environmental impact, and increase urban service efficiency, according to [9]. According to [10], smart cities use ICT to "improve the quality of life of citizens, the sustainability of the urban environment, and the efficiency of urban services". Also, in [5, 6], "smart cities" use information and communication technologies (ICT) to collect, analyze, and share data to improve urban systems like transportation, energy, water, and waste management and engage residents in planning and decision-making. Even in [10], the flexibility to change with the times, an emphasis on sustainability and resilience, and a focus on innovation and creativity are all hallmarks of "smart cities".

There are six components that make up a "smart city", as described by [9]. These components are the following: economy, governance, mobility, environment, people, and living. Sustainable smart cities use cutting-edge technology to lessen their negative effects on the natural world without sacrificing the standard of living for their residents. In order to maximize utility and minimize waste and energy consumption, these cities leverage data and technology. The representation of the various components in smart cities is presented in Figure 4.1.

4.2.1 SMART CITY CHALLENGES

Smart cities face a range of challenges, including but not limited to the exorbitant cost of implementing and maintaining information and communication technology (ICT) infrastructure, the imperative need for robust privacy and security protocols, and the potential for social exclusion. Furthermore, governments, corporations, and citizens all need to work together to bring smart city ideas to fruition, which can be difficult due to competing interests and objectives [10]. The Internet of Things (IoT) has enormous potential to aid in the creation of sustainable smart cities. However, the realization of this potential is hindered by a number of obstacles that must be addressed. For instance, IoT sensors can track and optimize a building's energy usage to cut down on waste and save money for both occupants and owners alike. Several unresolved issues have emerged during the endeavor to establish sustainable smart cities.

The considerable volume of data generated by IoT devices holds the promise of significantly augmenting the operational efficiency of intelligent household systems. However, if it is not secured properly, private information could be compromised. It's safe to assume that you're not among the tech-savvy majority. The elderly have a negative view of smart mobility solutions due to their experiences as victims of cybercrime. To ensure the sustainable endurance and secure operation of intelligent residences, the implementation of blockchain technology presents advantageous features with regard to data integrity and transparency.

IoT device mismanagement can increase energy use. Solar and wind energy are the answer for smart houses. Energy management systems in smart homes allow for off-peak task scheduling, temperature regulation, and device power off. Smart mobility solutions haven't been standardized yet,

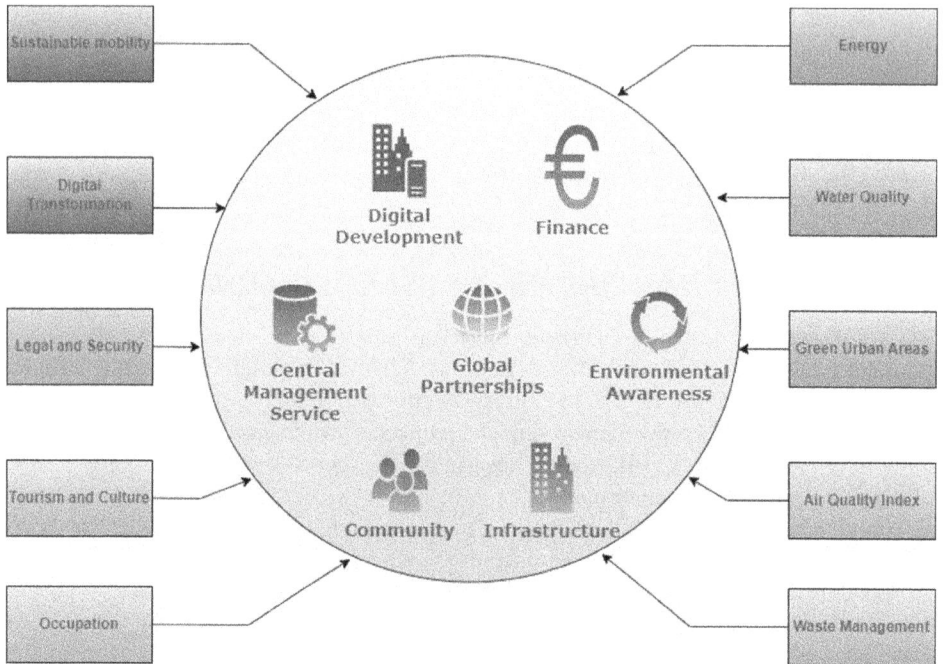

FIGURE 4.1   Illustration of smart city components.

despite the exponential development of technology. Smart mobility is difficult and many regional approaches fall short. To comprehensively attain the advantages of smart mobility, it is crucial to scrutinize inventive and sustainable procedures, uncomplicated interfaces, equitable pricing, limited data consumption, and transparent regulations pertaining to open data.

Fog and edge computing frameworks' lack of security increases privacy concerns. Such devices cannot implement a strong security system because of their limited processing power and storage. Because of this, cybercriminals and other criminals aim at them specifically. They can be hacked via distributed denial of service, side channel, or eavesdropping. There is an emphasis on automation in fog and edge computing. This guarantees quicker processing and solutions in near real-time.

However, problems can arise due to things like network failures, foreign attacks, natural disasters, etc. It is challenging to align smart transportation with older infrastructure. There now exists no product on the market that delivers universally compatible smart mobility solutions. Sustainable smart houses also struggle to communicate.

The connectivity among diverse IoT devices and technologies holds significant implications for their functionality and usability. The lack of inter-connectivity among these entities has the potential to undermine their intended operations. Intelligent buildings must integrate inter-technical communication protocols and standards to solve this problem.

It is necessary to organize the data collected by different sensors first. Smart mobility is made more difficult by sensor data and application cases. Challenges to mobility solutions in cities are vastly different from those in rural or semi-urban areas. Ideally, data storage capacity would increase at an exponential rate. Data integrity must be preserved even after massive data scaling. User adoption is a final consideration for long-term smart homes. While there are numerous advantages to having a smart home, many people are put off by factors such as price, complexity, and privacy concerns. Figure 4.2 shows several risks of creating suitable sustainable intelligent urban centers.

Yet, several metropolises throughout the globe have launched sustainable "smart city" programmes. In order to distribute renewable energy effectively, cities like Amsterdam have installed "smart grid"

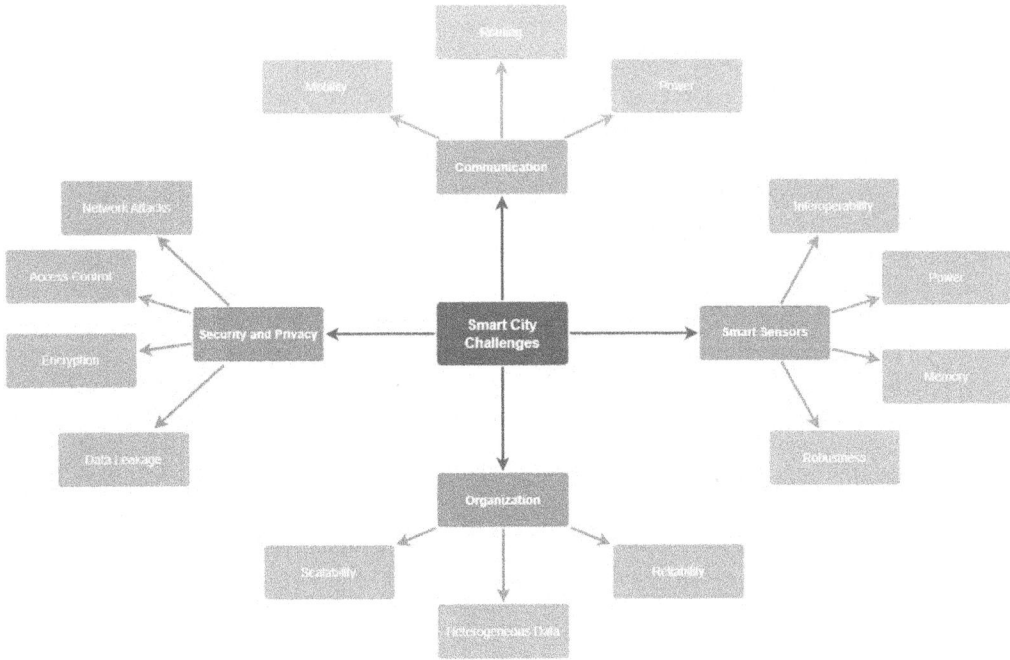

FIGURE 4.2 Security issues of a sustainable smart city.

systems. IoT sensors track usage and adjust power distribution in this system. The implementation of LED lights and IoT sensors has been observed in Barcelona's smart lighting system, aimed at mitigating expenses linked to energy consumption while concurrently enhancing the quality of illumination, as corroborated in a previous study.

4.3 TRENDING TECHNOLOGY-BASED SOLUTIONS FOR SMART CITIES

The great majority of people on Earth make their homes in urban regions. A smart city may have seemed like science fiction in the early 20th century, but advances in computer power and networking, along with responsible human use, have shown that such a place is now a reality. All of that changed as information and communication technology combined to create the intelligent urban life. What we have encountered up to this point. In addition to helping with more conventional tasks like traffic management, housing and population management, smart cities also enable analysis, monitoring, and planning for bettering the quality of life for city dwellers. The latest definition of "smart cities" is "urban agglomerations where a wide variety of Internet of Things devices and sensors are employed to gather data and use it to efficiently manage resources". Smart city technology cannot handle traffic control, real-time manufacturing, autonomous vehicles, fire alerts, or patient health monitoring. Thus, the use of fog computing strategies that are time-, location-, and security-sensitive in smart cities has resulted in positive results for the cities.

Many services that improve residents' quality of life may be enabled in a city that operates on fog computing and supports IoT applications. The study might be devoted to figuring out how to harness technologies like hybrid cloud and fog computing, the IoT, and mobile devices in light of the challenges and issues that smart cities must overcome. Specifically, mobility-related services like intelligent car commuting and increases in energy efficiency. The advantages of fog computing may be used to a variety of fields, including healthcare services, network kinds, and citizens' telecommunications.

4.3.1 EDGE-BASED SOLUTIONS FOR SMART CITIES

Computer networks and real-world objects form the "Internet of Things". The IoT lets you remote-control many devices. The Internet of Things can be used for intelligent transportation, environmental protection, government work, public safety, secured homes, intelligent fire control, industrial and environmental surveillance, elderly care and health monitoring, flower cultivation, water system monitoring, food traceability, enemy investigation, and intelligence gathering. There are 11 possible uses for these. Even though the Internet is the basis for the IoT, there are other technology requirements that must be met. It also needs more network connections and Internet growth, as well as more positioning devices and satellite remote sensing. The Internet of Things will enable remote control and precise location of each thing. It can be used in a lot of different ways, from building and keeping up power lines to designing and keeping up bridges and highways.

Even in its simplest form, the Internet of Things can improve people's lives and smarten their homes. Smart towns may initially use the Internet of Things. Most intelligent cities have four layers: application, support platform, network, and awareness. The city's future depends on "smart city" concepts. It usually affects business operations, logistics, transportation, healthcare, tourism, and other city sectors. Smart towns will advance science and technology and smarten cities. It boosts city growth. The city needs the Internet of Things to improve. Smart towns will improve science, technology, and cities. It boosts city growth. Long-term, smarter cities will boost commerce, tourism, industry, and other sectors. A smart city is organized, intelligent, and perceptive, which is essential for information transmission.

4.3.2 FOG-BASED SOLUTIONS FOR SMART CITIES

Cisco, a network manufacturer, coined "fog computing" in 2012. The Open Fog Consortium defines fog computing as "a system level horizontal architecture that brings user closer by distributing storage, computing, networking tasks, and control along a cloud-to-thing continuum". Horizontal platforms let platforms and sectors share computing services and processes. Managers and users can adapt to the fog computing platform. Fog has unique benefits [11]. We listed fog-based existing works in Table 4.1.

Security holes and risks shrink, and security monitoring and authentication methods work best. This is because data and information can be sent over a very short distance, and because all fog nodes use the same policy, procedure, and control. In a fog setting, latency is cut down by a lot

TABLE 4.1
Existing Research for Fog-Based Smart Cities

Study	Published Year	Main Focus	Methodology	Taxonomy	Challenges	Timeline
[12]	2017	Smart city based on fog	Vague	No	Discussed	Not Presented
[13]	2020	Sustainable smart cities based on fog	Vague	Yes	Discussed	Not Presented
[14]	2019	Fog-based 5G smart cities	Vague	No	Discussed	Not Presented
	2020	Smart city based on edge	Vague	Yes	Discussed	Not Presented
[15]	2017	Smart city based on cloud	Vague	No	Discussed	Not Presented
[16]	2019	Challenges and technologies on smart city	Vague	Yes	Not Discussed	Not Presented
[17]	2020	Smart city applications based on fog	Clear	Yes	Discussed	2013–2018
[18]	2020	Smart city approaches based on fog	Clear	Yes	Discussed	2013–2020

because fog nodes are close to end devices. So, it is a given that fog computing works well for apps that depend on time and place. Fog computing may be better integrated with end-device systems when end-points are closer to the end-devices. This can improve total performance and efficiency. In mission-critical and cyberphysical systems, this trait is very important. Fog architecture's widely spread fog nodes are in the best position to figure out where to store data, control functions, and do computations because they know how user needs change depending on where they are. SaaS, PaaS, and IaaS—the three most common cloud services—are also applicable in fog environments. Next, we'll talk more about what makes fog different from other ways of computing [19].

- Fog computing, unlike centralized cloud computing, is sensitive to delay and aware of position. Therefore, it chooses the method with the least possible delay and responds to or analyzes data much faster. This is because fog nodes can figure out logically where they fit in the whole network.
- Fog environments process data at network edges. Control tasks and latency-sensitive applications are closer to the user.
- Heterogeneity: Fog computing can work with many different kinds of network architectures, and each may need a different way to process or receive data.
- Federated and compatible systems are necessary for a lot of services to be offered. Because of this, all services need to be linked together.
- The design of cloud computing is centralized, but fog computing is spread out over several places.
- Fog computing can adapt to resource integration, elastic computing, network conditions, and data loads.

4.3.2.1   Fog Architecture

Fog computing is a horizontal multi-tier design with storage, networking, and control functions that are spread out across the network topology and grow as the network grows. Figure 4.3 shows how fog computing is made up of three levels. Each part is explained in more depth below. The cloud layer makes sure that the various data centers that make up this tier quickly and effectively gather, store, and process data. The fog layer has fog nodes near the human device layer. Fog layer

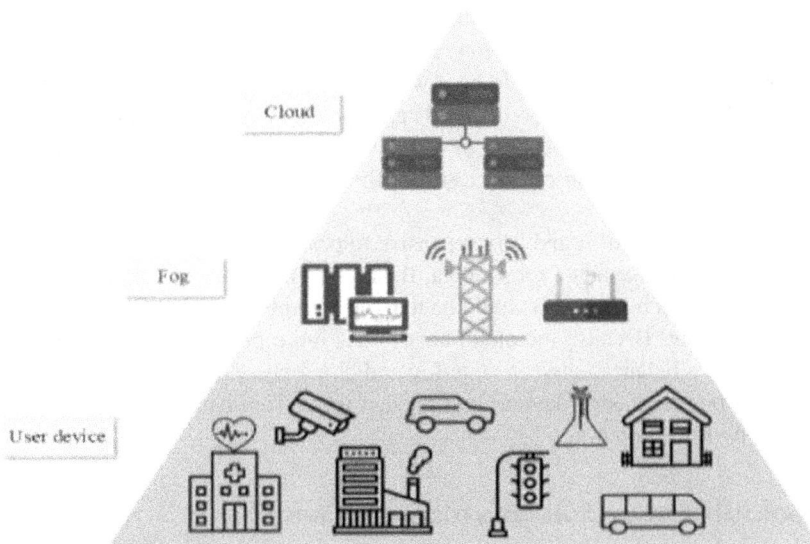

FIGURE 4.3   Three-tiered framework for fog-based smart cities.

data handling improves localization. The end-device and cloud layers use many types of hardware. Examples of these components include switches, virtual computers, routers, and gateways.

There are numerous items that improve people's lives in the user device layer. Data can be moved from the lower levels where it is found to the higher levels where it can be processed and saved by a variety of devices, including cell phones, cars, sensors, cameras, etc. This three-tiered design connects fog nodes to user devices and sensors via Bluetooth, 5G (5th generation), 4G (4th generation), wireless fidelity, ZigBee, and hardwired systems. Fog nodes can also connect to one another in a variety of ways, including wirelessly and over wires [4].

Fog networking sends data to fog nodes that are spread out all over, where processing, analysis, and storage are done. This is different from centralized network topologies, which increase network traffic in a bad way. The core layer of a fog architecture is made up of real or virtual parts like VMs, switches, servers, gateways, and routers. Between the user's device and the cloud, they must offer some kind of service or app for managing information and communicating. In a fog environment, fog nodes have the following qualities and traits: they can handle and accept a wide range of data types that are used by different networks. Fog nodes are doable common or routine jobs that are done without a person's help, thanks to computerized planning and management. In a fog system, each server or cluster can work on its own and make decisions at the local level. There is openness to customization by a wide range of customers, service providers, and network experts. After a network starts to work together, fog nodes not only provide a number of services, but they also help with hierarchical grouping.

4.3.2.2 Characteristics of Fog Computing

This section demonstrates how fog computing functions. When an application initiates a task, like checking the health of a medical device over time, the cloud-based IoT solution must decipher the command, send it to a cloud server for processing, and then display the results graphically. Businesses can save time and resources by implementing edge computing-based solutions to locally store and process IoT data streams. Some businesses that can benefit greatly from less latency are supply chain and logistics, intelligent transportation, and digital healthcare.

Some estimates predict more than 30 billion IoT devices in a few years. These devices generate 2.5 billion bytes daily. Edge analysis of IoT data reduces network traffic and maximizes bandwidth. Daily cloud storage of state data is possible. AI and machine learning could help edge data analysis.

Edge and fog computing greatly improve data security by analyzing private information at a gateway or fog node inside the company rather than at a remote data center. The security holes in the network are caused by IoT devices' low computer power, bad hardware design, and hard to update firmware. These differences could be a way for hackers to get into a device. When data is sent between IoT devices and the cloud, it could cause problems. The network level security of IoT systems needs to be strengthened because of this. The network edge may be the ideal location to implement security changes because it is closest to the devices that could be attacked and the rest of the network [13].

Because there is so much of it and it is expensive to process in the cloud, many businesses find it challenging to gather and analyze sensor data. Businesses may start utilizing smart IoT solutions once they realize how much simpler it would be to gather, combine, and analyze their IoT data with edge and fog computing. Because the cloud always has more processing and storage power than edge devices, finding a balance between edge/fog and cloud processing and storage is crucial. The lack of processing power in edge devices that prevent them from handling in-depth data could be compensated for by fog computing.

4.3.3 Blockchain Enabled Solutions for Smart Cities

Many towns have upgraded their infrastructure in the last few decades to make it more intelligent and practical. This includes smart transportation, smart parking, and smart cars, among other things.

In order for people to live in a smart city, all of these things are necessary. ICT and IoT dominate innovative city environments. To address the challenges of expanding urbanization, it develops, puts into practise, and supports sustainable development in smart applications like computing power, communication bandwidth, and others. ICT is a complex system of devices and objects. Wi-Fi, Bluetooth, GPS, and cellular transmission are needed to send data to the cloud. Applications for sustainable smart cities can help with energy distribution, air quality, and vehicle movement in urban areas.

A blockchain-based energy-efficient smart parking system would address centralization, communication bandwidth, power efficiency, integrity, security, and privacy issues with conventional smart parking systems. Blockchain and the elliptic-curve cryptography (ECC) algorithm secure smart parking spaces and vehicle data. Virtualization allows cloud-layer virtual machines to use most of the server's resources.

This makes smart parking a place to park that saves energy and is good for the environment. At the analysis layer, a deep learning LSTM network looks at cloud data to give the driver the best parking spot based on their needs, such as time, distance, and money. Using blockchain, a safe framework for energy-saving smart parking in a sustainable city setting.

- RSU-based blockchain networks secure parking zone and driver data. This improves privacy and security.
- Energy-saving virtualization stores smart parking zone and driver data in cloud-based virtual machines.
- Deep learning predicts the best parking spot based on intelligent cloud data and tells the driver where, when, and how much.
- Compare the proposed framework to time, storage, energy, security, and privacy metrics [21].

4.3.3.1 Blockchain-Based Services for Sharing: A Step Towards Smart Cities

The Triangle framework of service orientation can be used to sum up the parts of blockchain-based shared service management and computing. This made it possible to add the blockchain solution. There are a total of six different service connections between people, technology, and organizations. With the blockchain-based method, no one has to believe anyone else. *The Economist* called blockchain "the trust machine" because it makes it easier for people to trust each other. In other words, because the blockchain-based economic system works without people, each transaction can be seen as "trust-free". In the past, businesses relied on trust, which meant that a trusted third party had to be involved. Blockchain technology makes it possible to get rid of the need for middlemen, which could save a sharing business money and make it run more smoothly. Blockchain technology allows for new kinds of digital connections on trustworthy sharing platforms. This means that some of the most basic business deals in the world can be done in very different ways.

Computing is a key part of sharing tools that people can trust in terms of security for blockchain-based services. Confidentiality, integrity, and availability are the three bases of security, and all three must be present for there to be full security. (1) It can only be used for allowed purposes; (2) it is separate and unique; and (3) because it is decentralized, blockchain data is not controlled by a central authority. Encryption with private and public keys, which is built into a blockchain's system, makes it almost impossible to argue about privacy. The stability of the blockchain can be ensured in this way because it can be thought of as a distributed file system in which members store copies of data and agree on changes through consensus. Bitcoin and Ethereum, two blockchain-based apps that have been around for a long time, have shown that blockchain-based service computing is reliable and getting safer [22]. If we look at the term "smart" in the word "smart city" from the point of view of the "sharing economy", we can better understand what it means and how adopting new technologies might help build smart cities. In theory, the features of blockchain could help smart towns grow by making it easier for people to share services.

4.4 AI-ENABLED IOT APPLICATIONS

This section shows how important AI is for IoT-based applications and how it helps the smart environment. AI has been very important in fixing problems in the real world over the past 10 years. AI is the study of making machines smart enough to do jobs that have usually been done by smart people. AI-based systems are getting better and better at being useful, adaptable, quick to process, and able to do more. Machines are getting better at doing things that aren't regular. Human intelligence is "taking" the right choice at the right time, while AI is "choosing" it [23]. Smart sensors and gadgets use machine learning to find patterns and outliers in data like temperature, pressure, humidity, air quality, vibration, and sound. Business intelligence tools that use numeric thresholds to make operational predictions are 20 times slower and less accurate than machine learning methods.

4.4.1 MITIGATION OF SMART CITY CHALLENGE THROUGH AI

The most advanced smart cities use AI, IoT, and big data to promote sustainable growth, meet rising citizen expectations, and build resilience, while other cities have yet to improve their infrastructure. Figure 4.4 shows how smart settings can be used in different ways. Applications for a smart world are based on technology and data. It uses them to improve performance all around, which gives its people a better level of living in the long run. The construction of a smart city necessitates an intricate understanding of the optimal utilization of data, followed by its effective implementation towards the alteration of governing policies and planning strategies. The most progressive smart cities use AI, ML, IoT, and big data analysis to enable sustainable expansion, meet rising resident demands, and improve city resilience. In a smart city, the effects of technology and data on daily life can be seen in a number of important areas [24]. Smart cities still have problems that need to be solved.

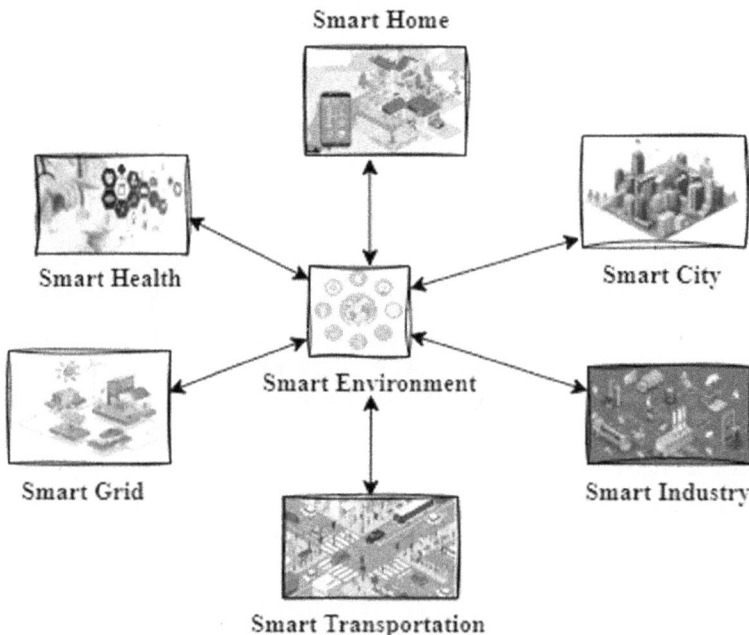

FIGURE 4.4 Smart environment applications.

Integration and interoperability: Smart cities combine different technologies and systems, which can be hard to do because there aren't any norms or ways for different systems to work together.

Data management: Data management in smart cities raises big concerns related to data collection, storage, processing, and security.

Sustainability and resilience: Smart cities must be structured with a central emphasis on sustainability and resilience while considering the enduring environmental, social, and economic effects that arise from the implementation of these endeavors. Along with the focus on ensuring the sustainability and resilience of smart city solutions.

Citizen engagement: Ensuring meaningful citizen engagement in smart city initiatives is crucial for their success to ensuring equitable access, transparency, and accountability in citizen engagement processes.

Funding and financing: Implementing smart city solutions can be costly, and finding adequate funding and financing mechanisms is a major challenge for cities.

Legal and regulatory frameworks: The inception of smart city initiatives has engendered various legal and regulatory predicaments, predominantly centered around data privacy, security, and ownership. These challenges necessitate a robust framework to enable the ethical deployment of technology and data in smart cities.

Technology adoption and diffusion: Smart city solutions often require significant changes in technology adoption, diffusion, readiness, adoption, and diffusion among citizens, businesses, and government agencies.

Digital divide: The implementation of smart city endeavors bears the possibility of amplifying preexisting social and economic disparities, thereby engendering a dichotomy between individuals who possess technological access and those who lack it. To ensure equitable access to smart city solutions and bridge the digital divide to ensure that all citizens benefit from these initiatives.

Human-centered design: The development of smart city solutions requires a human-centric approach that incorporates the needs, preferences, and values of urban citizens as primary considerations. This ensures that smart city solutions are designed with a user-centric approach, including issues related to user experience, usability, and accessibility.

ML-based solution: This study [25] demonstrates the potential application of machine learning in assisting municipalities to tackle challenging issues such as transportation, waste management, public security, and civic engagement. This article covers supervised, unsupervised, and reinforcement machine learning. The present article discusses the utilization of various forms of data, including but not limited to social media, sensor data, and traffic data, in machine learning-based intelligent urban environments. It also stresses how important it is to deal with ethical and privacy issues when using ML in smart cities. It also points out some of the problems and limits of current ML-based smart city solutions, such as the need for a lot of data and computational resources and the fact that they don't work well with each other or follow any standards. A study showed that there is also a need for the possible addition of knowledge from different fields.

DL-based solution: When DL and IoT were brought together, they brought more [26][27] benefits. And they greatly improve how the smart world works, such as by making it more efficient, more environmentally friendly, and better for people's quality of life. The integration of deep learning has been implemented to facilitate various applications, including traffic management, energy management, and public safety within smart cities. The IoT has also enabled data aggregation and analysis to aid decision-making. CNN and RNN have shown through their techniques how technical sustainability has been put into place. Still, security needs to be a priority.

On the way to making smart cities technically sustainable through the use of federated learning-based solutions [28][29], there is also a record of another effort. Possible benefits of federated learning include more privacy and less time spent on conversation. Most of these things are likely to happen in smart city apps that can benefit from federated learning, like traffic management, energy management, and public safety. Federated averaging and federated optimization are shown in smart settings, but they have some problems because their data sources are complicated and different.

In recent years, intelligent communities have integrated explainable artificial intelligence (XAI) [30][31]. AI-powered smart city technologies foster public trust and endorsement by making smart contexts clear and understandable. Clear smart city apps were made using XAI. XAI techniques expand and understand smart city applications.

AI and blockchain can improve transparency, security, and productivity in smart cities. A study in [32][33] shows how AI can be used to make predictive analytics and optimization possible in smart city apps, and how blockchain can be used to make data sharing and transactions safe and clear. Lastly, there is a need for AI and blockchain frameworks that are more standardized and can work with each other, as well as the development of new AI and blockchain methods that can handle complex and different types of data sources in smart cities. The possible benefits of AI [34][35] for smart city applications like traffic management, energy efficiency, and public safety, as well as the challenges of using AI in smart cities, such as issues with data privacy, ethics, and regulations. AI can be used in apps for smart cities, to analyze data, and to make decisions. The active participation of citizens in the development and execution of AI-based intelligent urban applications is imperative for their effective functioning. In [36][37], a smart IoT framework for sustainable towns is shown that uses blockchain and AI technologies. The framework is made to deal with problems like energy use, trash management, and air pollution that are linked to sustainability in urban areas. Through the different modules, smart sensors, data analytics, decision-making, blockchain, and AI technologies, the benefits of making sure security, openness, and efficiency have been shown. The job of an AI-enabled framework is to make sure that sharing data and making decisions are done in a transparent way. So, the combination of AI and intelligent technologies helps a lot to make sure that smart apps are technically sustainable [38].

4.5 CONCLUSION

In this chapter, we focused on the IoT's function in relation to smart environments. Due to memory constraints, IoT performance suffers when there is a large amount of data present. Numerous studies have also demonstrated that due to the resource limitations of Internet of Things technologies, data processing can be done significantly more effectively in the cloud than on-body computing. Since the risks associated with security that using the cloud entails are its major drawback, we have offered a variety of works targeted at improving cloud security. However, there is currently no standard that can be put into place quickly in a wearable, IoT-based healthcare system. Access control guidelines and encryption were found to considerably increase security. Thus, the benefits of edge, fog, blockchain, and AI have secured technical sustainability. Various application cases gained from edge, fog, blockchain, and AI technologies are illustrated in order to help readers understand how these trendy and smart technologies contribute to the sustainability challenge. Blockchain-enabled frameworks have made it possible to address problems with centralization and communication bandwidth. The energy-efficient smart parking in a sustainable city environment framework will be improved to protect privacy and include a hybrid federated learning algorithm at the cloud layer. This chapter provides a summary of the traits of emerging technologies and the challenges they face in order to assist with various services related to the technological sustainability of smart cities.

REFERENCES

1. Imperatives, S. Report of the World Commission on Environment and Development: Our Common Future, Vol. 10., 1987, Available online: www.askforce.org/web/Sustainability/BrundtlandOur-Common-Future-1987-2008.pdf

2. Kennedy, C., Cuddihy, J., Engel-Yan, J. The changing metabolism of cities. *J. Ind. Ecol.*, 11, 43–59, 10.1162/jie.2007.1107, 2007.

3. Zhang, K., Ni, J., Yang, K., Liang, X., Ren, J., Shen, X.S. Security and privacy in smart city applications: Challenges and solutions. *IEEE Commun. Mag.*, 55, 122–129, 2017.

4. Dameri, R.P. Searching for smart city definition: A comprehensive proposal. *Int. J. Comput. Technol.*, 11, 2544–2551, 10.24297/ijct.v11i5.1142, 2013.

5. Bibri, S.E., Krogstie, J. Smart sustainable cities of the future: An extensive interdisciplinary literature review. *Sustain. Cities Soc.*, 31, 183–212, 2017.

6. Janik, A., Ryszko, A., Szafraniec, M., Scientific landscape of smart and sustainable cities literature: A bibliometric analysis. *Sustainability*, 12(3), 779, 2020.

7. Stavrakas, V., Flamos, A., A modular high-resolution demand-side management model to quantify benefits of demand-flexibility in the residential sector. *Energy Convers. Manag.*, 205, 112339, 2020.

8. Moniruzzaman, M., Khezr, S., Yassine, A., Benlamri, R., Blockchain for smart homes: Review of current trends and research challenges. *Comput. Electr. Eng.*, 83, 106585, 2020.

9. Giffinger, R., Fertner, C., Kramar, H., Meijers, E. City-ranking of European medium-sized cities. *Cent. Reg. Sci. Vienna UT*, 9(1), 1–12, 2007.

10. Caragliu, A., Del Bo, C., Nijkamp, P. Smart cities in Europe. *Journal of Urban Technology*, 18(2), 65–82, 2011.

11. Amsterdam Smart City. Amsterdam Smart City ~ About ASC, 2015, Retrieved from https://web.arch ive.org/web/20150427024215/http:/amsterdamsmartcity.com/about-asc

12. Ni, J., Zhang, K., Lin, X., Shen, X. Securing fog computing for internet of things applications: Challenges and solutions. *IEEE Communications Surveys & Tutorials*, 20 (1), 601–628, 2017.

13. Yousefpour, A., Fung, C., Nguyen, T., Kadiyala, K., Jalali, F., Niakanlahiji, A., ... Jue, J. P. All one needs to know about fog computing and related edge computing paradigms: A complete survey. *Journal of Systems Architecture*, 98, 289–330, 2019.

14. Chiang, M., Ha, S., Risso, F., Zhang, T., Chih-Lin, I. Clarifying fog computing and networking: 10 questions and answers. *IEEE Communications Magazine*, 55(4), 18–20, 2017.

15. Iorga, M. Fog Computing Conceptual Model-Recommendations of the National Institute of Standards and Technology. US Department of Commerce, 2018.

16. Songhorabadi, M., Rahimi, M., Farid, A.M.M., Kashani, M.H. Fog computing approaches in smart cities: A state-of-the-art review. *Journal of Network and Computer Applications*, 153, 102531, 2020. https://doi.org/10.1016/j.jnca.2020.102531

17. Badidi, E., Mahrez, Z., Sabir, E. Fog computing for smart cities' big data management and analytics: A review. *Future Internet*, 12(11), 190, 2020.

18. Nowicka, K. Smart city logistics on cloud computing model, *Procedia - Social and Behavioral Sciences*, 151, 266–281, 2014. ISSN 1877-0428, https://doi.org/10.1016/j.sbspro.2014.10.025. (www. sciencedirect.com/science/article/pii/S1877042814054676)

19. Zigurat Global Institute of Technology. Barcelona Smart City: Most Remarkable Example of Implementation, 2021, www.e-zigurat.com/blog/en/smart-city-barcelona-experience.

21. Perera, C., Qin, Y., Estrella, J.C., Reiff-Marganiec, S., Vasilakos, A. V. Fog computing for sustainable smart cities: A survey. *ACM Computing Surveys (CSUR)*, 50(3), 1–43, 2017.

22. Zahmatkesh, H., Al-Turjman, F. Fog computing for sustainable smart cities in the IoT era: Caching techniques and enabling technologies-an overview. *Sustainable Cities and Society*, 59, 102139, 2020.

23. Khan, M.A. Fog computing in 5G enabled smart cities: Conceptual framework, overview and challenges. In *2019 IEEE International Smart Cities Conference (ISC2)*, 438–443, 2019.

24. Khan, L.U., Yaqoob, I., Tran, N.H., Kazmi, S.A., Dang, T.N., Hong, C. S. Edge-computing-enabled smart cities: A comprehensive survey. *IEEE Internet of Things Journal*, 7(10), 10200–10232, 2020.

25. Javed, A.R., Shahzad, F., ur Rehman, S., Zikria, Y.B., Razzak, I., Jalil, Z., Xu, G. Future smart cities requirements, emerging technologies, applications, challenges, and future aspects. *Cities*, 129, 103794, 2022.

26. Petrolo, R., Loscri, V., Mitton, N. Towards a smart city based on cloud of things, a survey on the smart city vision and paradigms. *Transactions on Emerging Telecommunications Technologies*, 28(1), e2931, 2017.

27. Band, S.S., Ardabili, S., Sookhak, M., Theodore, A., Elnaffar, S., Moslehpour, M., ... Mosavi, A. When smart cities get smarter via machine learning: An in-depth literature review. *IEEE Access*, 2022. doi:10.1109/ACCESS.2022.3181718

28. Sánchez-Corcuera, R., Nuñez-Marcos, A., Sesma-Solance, J., Bilbao-Jayo, A., Mulero, R., Zulaika, U., Almeida, A. Smart cities survey: Technologies, application domains and challenges for the cities of the future. *International Journal of Distributed Sensor Networks*, 15(6), 1550147719853984, 2019.

29. Rajyalakshmi, V., Lakshmanna, K. A review on smart city-IoT and deep learning algorithms, challenges. *International Journal of Engineering Systems Modelling and Simulation*, 13(1), 3–26, 2022.

30. Javadzadeh, G., Rahmani, A.M. Fog computing applications in smart cities: A systematic survey. *Wireless Networks*, 26(2), 1433–1457, 2020.

31. Pandya, S., Srivastava, G., Jhaveri, R., Babu, M.R., Bhattacharya, S., Maddikunta, P.K.R., ... Gadekallu, T.R. Federated learning for smart cities: A comprehensive survey. *Sustainable Energy Technologies and Assessments*, 55, 102987, 2023.

32. Singh, S.K., Pan, Y., Park, J.H. Blockchain-enabled secure framework for energy-efficient smart parking in sustainable city environment. *Sustainable Cities and Society*, 76, 103364, 2022.

33. Humayn Kabir, M., Fida Hasan, K., Kamrul Hasan, M., Ansari, K. Explainable artificial intelligence for smart city application: A secure and trusted platform, 241–263, 2022. doi: 10.1007/978-3-030-96630-0_11

34. Ramaiah, M., Chithanuru, V., Padma, A., Ravi, V. A review of security vulnerabilities in industry 4.0 application and the possible solutions using blockchain. *Cyber Security Applications for Industry*, 4, 63–95, 2023.

35. Singh, J., Sajid, M., Gupta, S.K., Haidri, R.A. Artificial intelligence and blockchain technologies for smart city. *Intelligent Green Technologies for Sustainable Smart Cities*, 317–330, 2022.

36. Sun, J., Yan, J., Zhang, K.Z. Blockchain-based sharing services: What blockchain technology can contribute to smart cities. *Financial Innovation*, 2(1), 1–9, 2016.

37. Herath, H.M.K.K.M.B., Mittal, M. Adoption of artificial intelligence in smart cities: A comprehensive review. *International Journal of Information Management Data Insights*, 2(1), 100076, 2022.

38. Ahmed, I., Zhang, Y., Jeon, G., Lin, W., Khosravi, M.R., Qi, L. A blockchain-and artificial intelligence-enabled smart IoT framework for sustainable city. *International Journal of Intelligent Systems*, 37(9), 6493–6507, 2022.

5 Hydro-Meteorological Disaster Prediction Using Deep Learning Techniques

P. Kaviya[1] and P. Chitra[2]
[1] Department of Information Technology, Kamaraj College of Engineering and Technology, Virudhunagar, Tamilnadu, India
[2] Department of Computer Applications, Thiagarajar Engineering College, Tamilnadu, India

5.1 INTRODUCTION

Any natural occurrence, including floods, earthquakes, droughts, hurricanes, landfalls, heat waves, or cold waves, that results in significant property damage or human casualties is considered as a natural disaster. Some of the climatic factors like temperature, humidity, air density, wind speed and direction, pressure and precipitation have influences on the rate of occurrence of HMD. These disasters result in excessive or insufficient rainfall, rapid snowmelt, cyclones, tsunamis, and collisions of subterranean earth plates [1].

Predicting abnormal climatic conditions has been carried out for the past decades, which remains a challenging issue even today. Various social bodies like hydrology, remote sensing, and meteorology departments are investigating various climatic parameters to increase the accuracy of future natural disaster predictions. In the period of big data, numerous scientific and technological fields that predict climatic events produce enormous amounts of data. It is extremely difficult to analyze and interpret such a large amount of data and come up with useful results. The emerging technologies like artificial intelligence and big-data analytics offer a flexible means of processing ample data and provide scalable solutions for the data-driven applications. They serve as suitable forecasting technology for future events.

A subset of artificial intelligence called deep learning (DL) addresses the shortcomings of various real-time predictive systems currently in existence and provides accurate solutions to them. The working principle of the DL algorithm is analogous to the operations of the human brain. It uses many neural network layers for huge data processing and computing as per the researcher's will. Besides, it handles structured as well as unstructured data and acquires the features of them automatically.

The traditional forecasting systems lack in inferring the effective attributes so that many researchers are turning to deep learning algorithms to develop an efficient and accurate forecasting system. The systems must excel at handling spatiotemporal data and provide the most precise outcomes when using deep learning techniques in HMD forecasting system [2].

HMD forecasting models are essential to provide precise information about upcoming disasters. The development of such deep learning models can greatly reduce the frequency of unforeseen fatalities and the devastation of infrastructure under the worst climatic conditions [3–5]. If the deep learning models outperform the conventional weather predicting technologies, to save livelihood using better disaster emergency procedures, which can be planned and executed.

DOI: 10.1201/9781003407959-5

The following describes how the paper is organized: The various natural disasters are described in Section 5.2. Section 5.3 presents the natural disaster management cycle. The extreme climatic events forecasting using numerical weather prediction and deep learning techniques are discussed in Section 5.4. Section 5.5 elaborates a DL-based study design that predicts an abnormal weather condition with a suitable example.

5.2   NATURAL DISASTERS

Natural disasters are catastrophic events, which include droughts, earthquakes, floods, hurricanes, and landslides. These catastrophes may result in fatalities, property loss, and environmental and societal upheaval. These natural disasters are classified into three categories namely hydrometeorological, geological, and biological [1]. Figure 5.1 illustrates how natural disasters are categorized.

5.2.1   HYDRO-METEOROLOGICAL DISASTERS

The most frequent natural disasters are hydrometeorological events, which include floods, droughts, heat waves, and cold waves [6]. These events occurred due to climate changes like long-term shifts in temperatures and weather patterns.

5.2.1.1   Flood

When water spills over onto normally dry land, it creates a flood. Flooding occurs when rushing water from a lake, river, or the ocean obliterates homes, commercial buildings, and road infrastructure. Loss of human life, damage to property and destruction of crops are immediate impacts of flooding.

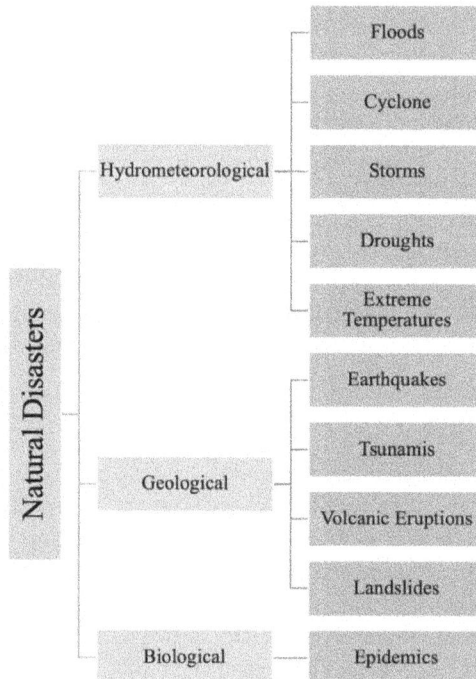

FIGURE 5.1   Classification of natural disasters.

5.2.1.2 Cyclone

A low-pressure area around atmospheric disturbances causes cyclones. It is symbolized by the swift destructive air circulation. Cyclones cause storms and unfavorable weather conditions. When there is a cyclone, the air moves in the southern hemisphere in a clockwise direction and anticlockwise in the northern hemisphere. The different types of cyclones are given as follows:

(i) *Extra tropical cyclones:* Although it occurs in polar regions, it can also be found in high latitude areas and temperate zones.

(ii) *Tropical cyclones:* A tropical cyclone originates over tropical oceans as a rapid rotating storm from where it draws the energy to develop. Strong winds, a low-pressure center, a spiraling cluster of thunderstorms that produce torrential rain, squalls, and restricted low-level air circulation are all signs of it. Depending on where it is and how intense it is, tropical cyclones are called a variety of names, including tropical storms, hurricanes, tropical depressions, typhoons, cyclonic storms, and simply cyclones.

Low pressure systems in the Arabian Sea and the Bay of Bengal are categorized by the Indian Meteorological Department (IMD) according to their potential for harm. Cyclone benchmarks of cyclones are shown in Table 5.1.

Table 5.2 shows the five distinct orders of cyclones based on wind speed and their capacity to cause damage.

5.2.1.3 Droughts

Drought is a prolonged dry period due to the lack of precipitation, resulting in a water shortage. Droughts are classified as follows:

(i) *Meteorological drought:* Based on the degree of dryness and how long the dry period has occurred.

TABLE 5.1
Benchmarks of Cyclone

Wind Speed in km/hr	Type of Disturbances
Less than 31	Low Pressure
31–49	Depression
49–61	Deep Depression
61–88	Cyclonic Storm
88–117	Severe Cyclonic Storm
More than 221	Super Cyclone

TABLE 5.2
Distinct Orders of Cyclone

Cyclone Category	Damage Capacity	Wind Speed (km/hr)
01	Minimal	120–150
02	Moderate	150–180
03	Extensive	180–210
04	Extreme	210–250
05	Catastrophic	250 and above

(ii) **Agricultural drought:** Refers to a condition that results in reduced crop and forage yields to forage failure.

(iii) **Hydrological drought:** Occurs when the water level in sources like reservoirs, aquifers, and lakes drops below the regional significant threshold level.

5.2.1.4　Heat Waves

An extended period of extremely hot weather that is hotter than the typical high summer temperature is known as a heat wave. When the temperature is ≥ 30°C for hilly areas and ≥ 40°C for plains, it is a heat wave. A heat wave is in effect if the temperature is 4.5°C to 6.4°C above average or if the actual maximum temperature at a station is 45°C or higher. The presence of a severe heat wave is indicated by a temperature that is >6.4°C above normal or ≥ 47°C.

Heat waves normally occur from March to June, though they can occasionally last into July in unusual circumstances. During the month of May, the intensity of heat waves reaches its peak value. The inhabitants of these areas either experience physiological stress or die due to the intense heat.

5.2.1.5　Cold Waves

A cold wave is a period of extreme cold weather that persists longer than the winter's moderate low temperature. When the temperature is ≤ 0°C for hilly areas and ≤ 10°C for plains, it is considered to be a cold wave. A cold wave is justified as if the temperature difference from the norm is between –4.5°C and –6.4°C or if a station's actual maximum temperature is recorded as being ≤ 4°. A severe cold wave is considered when the temperature difference from normal is < –6.4°C or the temperature is ≤ 2°C.

Cold waves normally happen from November to March. During the month of February, the intensity of cold waves reaches its peak value. The people living in these regions undergo hypothermia and abnormally low body temperature under extreme cold temperatures.

5.2.2　Geological Disasters

Geological disasters are caused due to shifting and movement of tectonic plates and solid mass [7]. Some of the geographical disasters are discussed as follows.

5.2.2.1　Earthquakes

An earthquake is an event that happens without any alert and results in violent shaking of the ground. Due to the movement of lithospheric plates, earthquakes occur. The structure of the earth's surface created, altered, and moved by internal forces are factors in tectonics, which cause earthquakes. The movement of earth plates causes violent earth tremors. Numerous people die because of an earthquake's effects in a populated area, and there is also significant property damage.

5.2.2.2　Tsunamis

Large earthquakes that occur near or under the ocean trigger the ocean waves that may reach up to 100 feet height onto land is known as a tsunami. These dreadful waves are caused by an underwater or coastal landslide or a volcanic eruption. The effects of tsunami cause complete property damage and loss of life.

5.2.2.3　Volcanic Eruptions

A volcano is a portion of the Earth's crust where hot magma, rocks, ash, and gases erupt. *Glowing avalanche*, a dangerous type of eruption caused when freshly erupted magma cascades down a volcano's sides. They can reach 1,200 degrees Fahrenheit. Volcanic eruptions can persist for days, months, or even years. Many people have died out of volcanic blasts. Additional threats of volcanic eruptions are health issues, mudslides, power outages, drinking water contamination, and forest fires.

5.2.2.4 Landslides

The movement of a mass of rock, debris under the direct influence of gravity that makes the earth fall down a slope, is known as a landslide. Landslides are caused due to precipitation, earthquakes, and volcanoes that make the slope unstable. Landslides can result in fatalities, property damage, the depletion of natural resources, and the destruction of wildlife habitat.

5.2.3 BIOLOGICAL DISASTERS

Different biological agents like pathogenic microorganisms, toxins, and other bioactive substances are examples of biological vectors that spread biological disasters. It harms the environment, results in fatalities, illnesses, or other adverse health effects, damages property, eliminates livelihoods, and disrupts the economy. Infestation, plagues of insects or other animals, plant or animal contagion, and epidemic diseases are examples of biological disasters.

Disasters caused by biological agents: In a population, community, or area, an epidemic is a disease that simultaneously affects many people. A few examples of epidemic diseases are cholera, acute encephalitis syndrome, and plague.

An epidemic known as a pandemic that spreads existing, newly emerging, or previously eradicated diseases and pestilences across several nations or continents. A few examples are influenza H1N1 (swine flu) and corona virus sisease (COVID-19).

5.3 NATURAL DISASTER MANAGEMENT

It is the process by which people respond to the unanticipated natural disasters that are discussed above.

5.3.1 NATURAL DISASTER MANAGEMENT CYCLE

Natural disaster management cycle [1] is depicted in Figure 5.2.

5.3.1.1 Prevention

Being proactive is the optimum way of addressing natural disasters. Recognizing potential risks and creating safety measures to lessen their effects. These stable measures help to minimize disaster risk and at the same time everyone must accept that disasters cannot always be prevented.

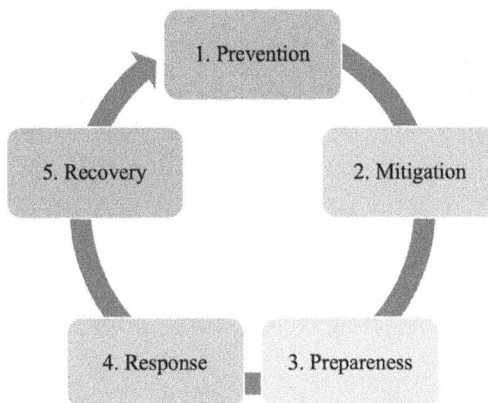

FIGURE 5.2   Natural disaster management cycle.

5.3.1.2 Mitigation

The mitigation phase aids in reducing the number of lives lost because of a natural disaster. Both structural and non-structural actions must be taken during the mitigation phase.

- *Structural measures:* The physical traits of a building or an area are modified to lessen the effects of a disaster.
- *Non-structural measures:* Applying or changing building codes to improve the safety of all upcoming building projects.

5.3.1.3 Preparedness

The ongoing process of preparing people, communities, businesses, and organizations for disasters through planning and training. It ensures that the highest level of readiness by providing extant training and taking corrective and preventive measures. This state includes the following key points to prepare for a potential hazard.

1. Identify the potential risks.
2. Determine the necessary needs.
3. Schedule the needs.
4. Depute the responsibilities.
5. Supply of resources.
6. Finalize the plan.

5.3.1.4 Response

Reaction after a disaster occurs is known as response. It includes both long-term and short-term responses. In this phase, to restore personal and environmental safety the disaster management leader will plan the allocation of resources to minimize the possibility of any further property damage. Hazards that are still present are eliminated during the response phase.

5.3.1.5 Recovery

The disaster management cycle's fifth stage is recovery. It might take months or even years to complete this. It entails restoring all crucial community functions to improve the affected area. Recovery is based on the order of priority. Food, potable water, utilities, healthcare, and transportation are the basic services that must be initially restored. In the end, depending on the effects of the disasters, this stage ultimately assists individuals, communities, businesses, and organizations in returning to normal life.

5.3.2 Developing Skills for Disaster Management

For effective coordination in the natural disaster management cycle, disaster management leaders must possess some demanding skills [1]. The following skills are necessary for each phase of the cycle.

5.3.2.1 Prevention

During this stage, leaders can find potential threats, hazards, and high-risk areas with the aid of strong analytical skills. Finding the optimal ways to stop or reduce the likelihood of catastrophic events requires the use of problem-solving skills.

5.3.2.2 Mitigation

The person in charge of disaster management needs to develop approaches and structural adjustments that can assist in resolving potential threats. Increasing public awareness is crucial at this stage. It is necessary to inform the community's residents of the precautions they can take to be ready for any eventuality.

5.3.2.3 Preparedness

During the preparation stage, it is crucial to have skilled trainers on hand to respond to emergencies. It is important to stay organized and to ensure readiness for any upcoming disasters. In worst-case scenarios, oral and written communication skills are required to laypeople for taking any necessary actions against disasters.

5.3.2.4 Response

Making prompt decisions is crucial during the response stage because it is time sensitive. Another crucial skill to use during this phase is delegating important tasks to additional volunteers or rescue crews.

5.3.2.5 Recovery

To aid their communities in recovering from disasters, disaster management leaders need to possess the three most crucial skills of empathy, understanding, and relationship building.

5.4 HYDRO-METEOROLOGICAL DISASTER PREDICTION

In this section, deep learning applications in HMD prediction are focused for five representative factors: flood, cyclone, drought, heat wave, and cold wave. The deep learning models aim to solve stated problems in the general progress of hydro-meteorological hazard analysis.

HMD forecasting plays a primary role in rescuing the human livelihood. The goal of meteorological scientists has always been to accurately predict the weather conditions. There are two paradigms for predicting HMD. They are the traditional theory driven approach, i.e., numerical weather prediction (NWP) and data-driven approach, i.e., deep learning-based weather prediction (DLWP). Figure 5.3 shows the procedures involved for HMD prediction using NWP and DLWP. Initially, the meteorological data are collected from various data sources like satellites, spot observation, and model simulations. Then these data are used to predict the weather events using either NWP or DLWP.

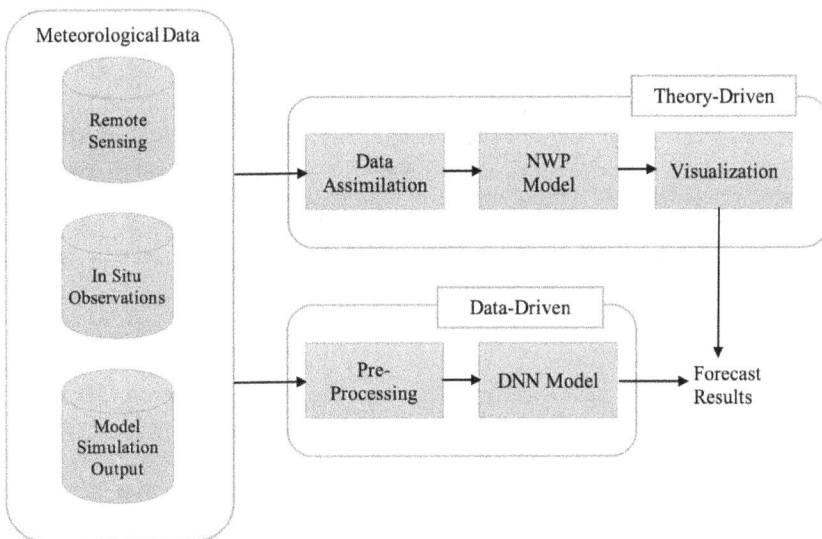

FIGURE 5.3 Paradigms for hydro-meteorological disaster prediction.

For many years, meteorological scientists have used a theory-driven approach to increase forecast accuracy by comprehending physical mechanisms. Due to the meteorological data's explosive growth in multiple sources, different scale, and high dimensionality, it has evolved into a standard big spatiotemporal data. Data analysts have recently attempted to use a data-driven approach to explore intricate spatiotemporal relationships between various climatological data items. DLWP has emerged as a popular research field, and it is anticipated that it will be able to address the data problems that traditional theory driven approaches face.

5.4.1 NUMERICAL WEATHER PREDICTION

The conventional method combines the partial differential equations (PDEs), which read present weather conditions as an input to predict the forthcoming weather conditions. The chemical, dynamic, thermodynamic, and radiative processes of atmosphere are represented by nonlinear PDEs [8, 9]. The PDEs of NWP must be numerically inferred using spatiotemporal quantization, due to the continuous nature of the physical process. The various climatic parameters like temperature, rainfall, wind speed, wind direction, and average sea level pressure are predicted by the NWP models.

The following are the primary steps in NWP: (i) acquiring the datasets from reliable data sources include remote sensing, instrument observation measures and historical NWP model results; (ii) pre-processing the collected raw datasets to ensure data quality through the data integration process; (iii) feeding the pre-processed data to the weather prediction equations in the dynamic model; and (iv) post-processing and visualizing the predicted outputs.

However, the traditional theory-driven NWP methods face several difficulties, including the need for powerful computing resources, an absence of knowledge of physical mechanisms and the difficulties of knowledge extraction from the massive observation data.

5.4.2 DEEP LEARNING-BASED WEATHER PREDICTION

The data-driven DL techniques have been used successfully in several real-time applications, including time serious prediction, computer vision, natural language processing (NLP), and e-commerce. Additionally, it has been demonstrated that using deep learning, it is possible to efficiently extract spatial and temporal features from spatiotemporal data. A classical huge geospatial data set is meteorological data. Meteorological data is a classical huge geospatial dataset. DLWP is anticipated to be a strong companion to the traditional approach. Many researchers are currently attempting to integrate data-driven DL techniques into weather forecasting, and they have seen some promising outcomes [3, 10–12].

DLWP uses a data-driven decision-making strategy. The collected datasets are given to the DNN models as inputs, with the goal of identifying the characteristics of changing weather through a massive volume of data and extracting relationships from the input data. Based on the features of the climatical data, it is necessary to examine the best DNN models for the pertinent data types. The following discussion includes some of the well-known deep learning models that can be applied to HMD prediction.

5.4.2.1 Multi-Layer Perceptron

A substitute for feed-forward neural networks is the multi-layer perceptron (MLP). As depicted in Figure 5.4, it has three different types of layers, including an input layer, a hidden layer, and an output layer. The data that needs to be processed is given to the input layer. The output layer is performing the required tasks, such as classification and prediction. The hidden layers are the computational part of the MLP are located between the input and output layer. An MLP's data flow from the input layer to the output layer resembles a feed forward network. To train the neurons in the MLP, back propagation is used. MLPs are developed to be capable of approximating any continuous

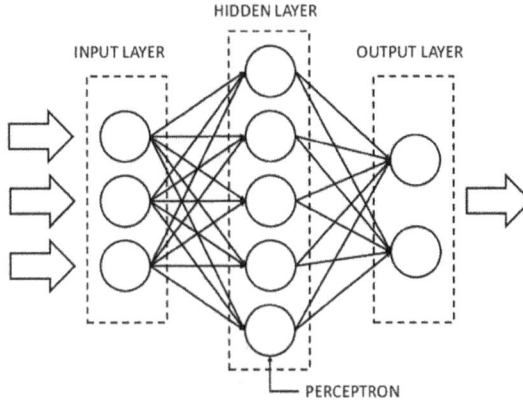

FIGURE 5.4 Multi-layer perceptron with a single hidden layer.

function and to address issues that cannot be resolved linearly. Some of the applications of MLP include pattern approximation, recognition, classification, and prediction.

Each neuron in the hidden and output layers performs the following calculations:

$$h(y) = \phi(y) = s\big(W(1)y + b(1)\big) \tag{5.1}$$

$$O(y) = G\big(W(2)h(y) + b(2)\big) \tag{5.2}$$

where
$b(1)$, $b(2) \rightarrow$ bias vectors
$W(1)$, $W(2) \rightarrow$ weight matrices
G and $s \rightarrow$ activation functions
$\theta = \{W(1), b(1), W(2), b(2)\} \rightarrow$ set of parameters
The *tanh* or *logistic sigmoid* functions can be as s, which are represented as follows:

$$\tanh(x) = (e^x - e^x) / (e^x + e^{-x}) \text{ or } \text{sigmoid}(x) = 1/(1 + e^x).$$

5.4.2.2 Convolution Neural Networks

A development of MLP with biological inspiration is convolutional neural networks (CNNs). CNNs are frequently used for object detection in images, image clustering, and image classification. Aside from that, they are used for NLP and optical character recognition. In addition to image-based applications, CNNs can also be used to analyze text, sound, and graphical data. CNN is successful in many fields because of its superior efficiency to its base algorithms.

As shown in Figure 5.5, CNN finds features by employing kernels, which are also known as filters. A weighted matrix of values used as a filter that has been trained to recognize specific features. The convolution operation is the product and sum of features between two matrices, is carried out by the filter. CNN training is accelerated by reducing input feature redundancy.

Consequently, the network's memory usage is also decreased. A popular method for doing this is called "max pooling", in which input data is passed through a window, and the highest value is collected into an output matrix. To increase the effectiveness of the algorithm for feature extraction, a greater number of convolutional layers and max pooling operations are concatenated. These deep layers process the data to produce feature maps, which are then converted into feature vectors by

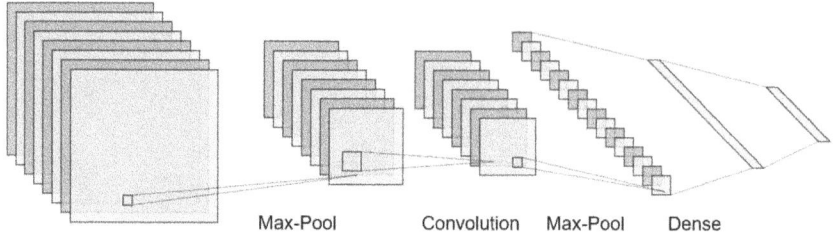

Max-Pool   Convolution   Max-Pool   Dense

FIGURE 5.5   Convolutional neural network.

running through an MLP. In the developed model, the high-level reasoning is carried out by a fully connected (FC) layer.

The feature map h is calculated using the *tanh* activation function as given as follows:

$$h_{ij}^k = \tanh\left(\left(W^k * x\right)_{ij} + b^k\right) \tag{5.3}$$

where
    $h^k \rightarrow k^{th}$ feature map
    $W^k \rightarrow$ weights
    $b^k \rightarrow$ bias
The FC layer output contains likelihood of each class. The output is categorized based on the class with the highest likelihood. Gradient back propagation is used to update the algorithm's weights and improve it.

5.4.2.3   Recurrent Neural Networks

Modeling sequence data with MLP is not possible. A powerful tool for modeling sequence data, such as sound, text, and time-series data, is the deep learning algorithm known as recurrent neural networks (RNN). In RNN, the output of $(n-1)^{th}$ step is connected to a feedback loop, which impacts the results of nth step. The RNN differs from conventional feed-forward networks due to this feedback loop. The mathematical representation for carrying memory forward is given as follows:

$$h_t = \phi\left(Wx_t + U * h_{t-1}\right) \tag{5.4}$$

where
    $h_t \rightarrow$ hidden state at time step t
    $x_t \rightarrow$ input at time step t
    $U \rightarrow$ transition matrix
    $W \rightarrow$ weight matrix
    $h_{t-1} \rightarrow$ hidden state at previous time step $t-1$
The input in the current state is modified by adding W with h_{t-1} and multiplied by its own h_t to U. The weight matrices define the relative weights to be given to the current input and the prior hidden state. The weights are updated using back propagation to reduce the error function. The activation function ϕ is used to reduce the total weight inputs and hidden state. The entire process of an unfolded RNN is shown in Figure 5.6.

The transition matrix's weights have a big impact on how the RNN learns during the gradient back propagation (GBP) phase. The gradient signal might become too small because of the matrix's small weights. As a result, the model's learning rate drastically slows down or even stops. It is referred to as the vanishing gradient problem; finding long-term relationships

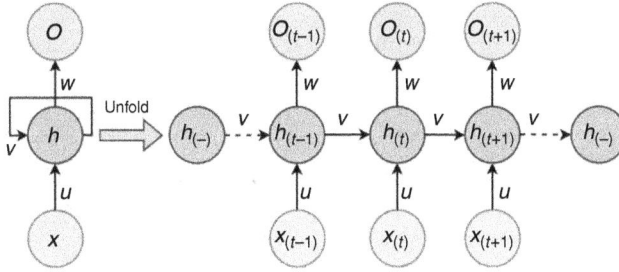

FIGURE 5.6 Recurrent neural networks.

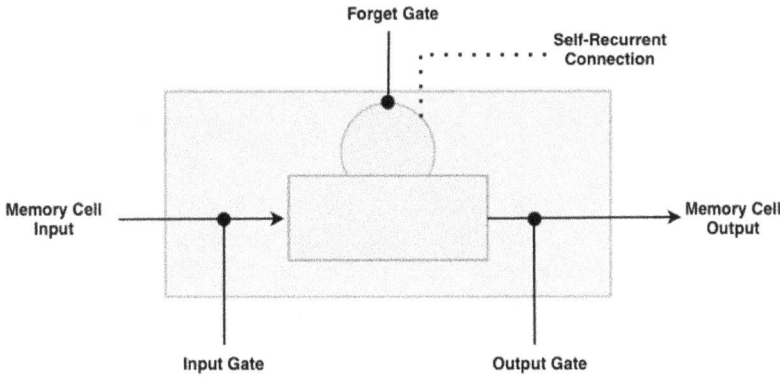

FIGURE 5.7 LSTM – memory cell.

in the inputs becomes extremely difficult as a result. Alternatively, exploding gradient problem; the learning may diverge, if the transition matrix has large weights because this can result in a strong gradient signal. The LSTM model is used to solve the vanishing and exploding gradient problems by adding a new component called a memory cell. An input gate, a forget gate, an output gate, and a self-recurrent connected neuron are the four main parts of a memory cell, as shown in Figure 5.7.

The following equations describe the updating of memory cells at each time step t:

$$i_t = \sigma\left(W_i x_t + b_i + U_i h_{t-1}\right) \tag{5.5}$$

$$\widetilde{C}_t = \tanh\left(U_c h_{t-1} + W_c x_t + b_c\right) \tag{5.6}$$

$$f_t = \sigma\left(W_f x_t + U_f h_{t-1} + b_f\right) \tag{5.7}$$

$$C_t = f_t * C_{t-1} + i_t * \widetilde{C}_t \tag{5.8}$$

$$O_t = \sigma\left(W_o x_t + U_0 h_{t-1} + b_0\right) \tag{5.9}$$

$$h_t = \tanh(C_t) * O_t \tag{5.10}$$

where

$x_t \rightarrow$ input to the memory cell layer at time t

$W_i, W_f, W_c, W_o, U_i, U_f, U_c$, and $U_o \rightarrow$ weight matrices

b_i, b_f, b_c and $b_o \rightarrow$ bias vectors

5.4.2.4 Generative Adversarial Networks

It is very easy for generative adversarial networks (GANs) to imitate any type of data dispersion. GAN consists of two networks that are positioned in opposition to one another. Each network plays a distinct role. The generator and discriminator are the two neural networks which generate and assess new data instances, respectively. GANs typically require a lot of time to train the data. The generator values do not change when the discriminator is trained, whereas the discriminator values do not change when the generator is trained. By engaging in combat, the generator and discriminator train using a static opponent.

The mathematical representation of GANs is given as follows:

$$V(D,G) = E_{a \sim P_{data}(a)} * \left[\log D(a) \right] + E_{c \sim P_c(c)} * \left[\log\left(1 - D\left(G(c)\right)\right)\right] \tag{5.11}$$

During training, it is typical for one side of the GAN to dominate the other. In other words, a discriminator that is too good will make it difficult for the generator to read the gradient, and a generator that is too good will keep exploiting the discriminator's shortcomings, leading to false negatives. To solve this problem, their individual learning rates can be adjusted. Figure 5.8 depicts the general structure of GANs.

5.4.2.5 Deep Autoencoders

An unusual deep learning algorithm called a "deep autoencoder" is shown in Figure 5.9. With four or five shallow layers, it consists of two symmetrical deep belief networks. The network is divided into two halves, with the first half handling the encoding and the second half the decoding. Autoencoders are a subset of neural networks. They resemble principal components analysis (PCA) very closely, but they are more responsive. In contrast to PCA, autoencoders can be more adaptable because they can encode data in both linear and non-linear transformations whereas PCA can perform only linear transformation. The autoencoders recognize important key features that are present in the data by decreasing the reconstruction error in the input and output data. The various kinds of autoencoders are given as follows:

- Denoising autoencoders
- Sparse autoencoders
- Deep autoencoder

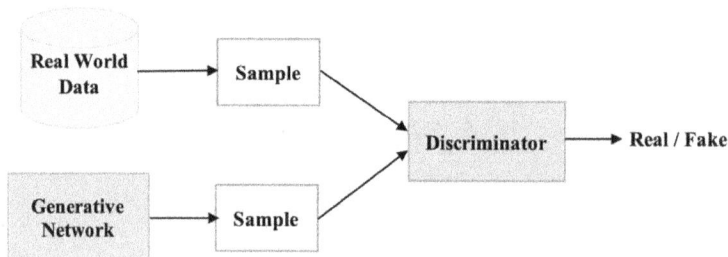

FIGURE 5.8 Generative adversarial networks.

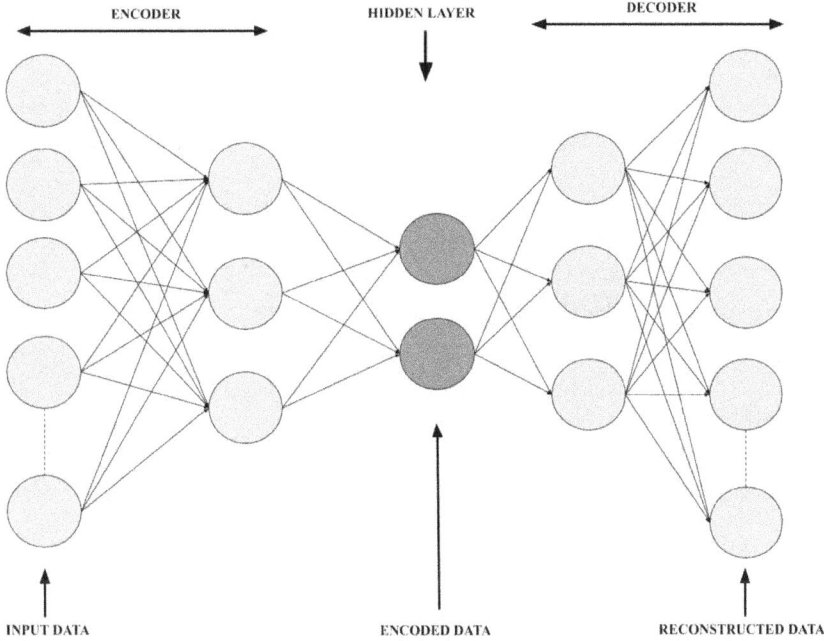

FIGURE 5.9 Deep autoencoders.

- Contractive autoencoders
- Undercomplete autoencoders
- Convolutional autoencoder
- Variational autoencoder

5.4.2.6 Long-Short Term Memory Networks

A subtype of RNN called long-short term memory (LSTM) is useful for solving problems involving sequence prediction because it can learn order dependence. It addressed the problem of long-term dependence of RNNs, which can forecast words using recent data but not those stored in long-term memory. RNN struggles to operate effectively as the gap length increases. By default, the LSTM retains data for a very long time. It is employed to analyze, forecast, and classify time-series data.

There are feedback connections in LSTMs as opposed to conventional feed-forward neural networks. It is capable of handling images, audios, and videos. LSTM can be applied for speech or handwriting recognition.

In a chain structure, the LSTM is composed of multiple memory cells and four neural networks. An ordinary LSTM unit consists of four components, namely, an input gate, a forget gate, an output gate, and a cell. The information flow in and out of the cell is controlled by three gates, and the cell stores values for arbitrary lengths of time.

The LSTM algorithm performs well for categorizing, investigating, and forecasting time-series data with uncertain duration. Figure 5.10 depicts a schematic illustration of the LSTM.

The gates control memory, whereas the cells store information. There are three entrances:

- **Input gate:** It determines the input values that change the memory. The *sigmoid* function selects a value of 0 or 1 to pass. By ranking the significance of the input data on a scale from −1 to 1, the tanh function assigns weight to the data.

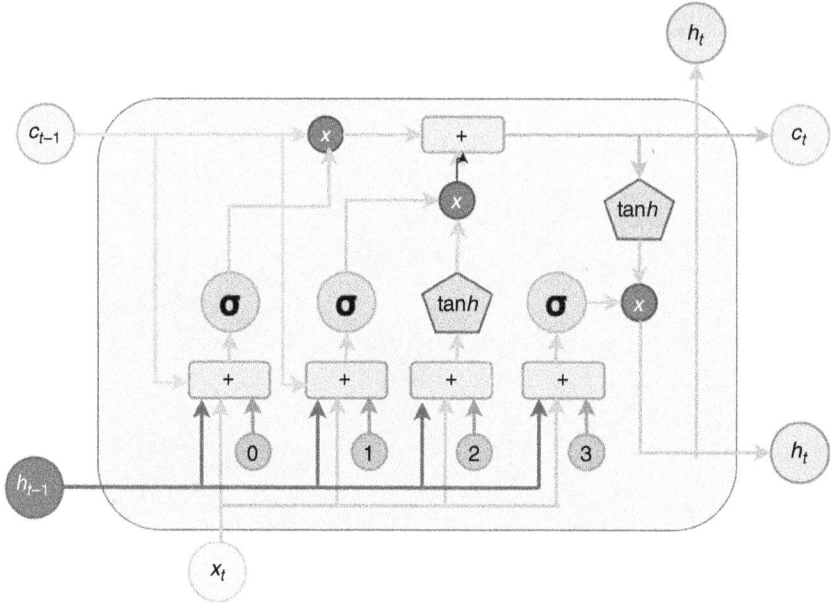

FIGURE 5.10   Long-short term memory.

$$i_t = \sigma\left(\left[h_{t-1}, x_t\right] * W_i + b_i\right) \tag{5.12}$$

$$C_t = \tanh\left(\left[h_{t-1}, x_t\right] * W_c + b_c\right) \tag{5.13}$$

- **Forget gate:** A *sigmoid* function identifies the data in the block that needs to be removed. It examines the input (x_t) and the previous state (h_{t-1}), for each cell state C_{t-1} and outputs a value between 0 and 1. If the outcome is 0, don't include the value.

$$f_t = \sigma\left(\left[h_{t-1}, x_t\right] * W_f + b_f\right) \tag{5.14}$$

- **Output gate:** The input and memory of the block are used to calculate the output. The *sigmoid* and *tanh* functions decide whether to pass values 0 or 1. By ranking the significance of the input data on a scale from −1 to 1, the tanh function assigns weight to it. Then it multiplies that weight with the sigmoid function's output.

$$O_t = \sigma\left(\left[h_{t-1}, x_t\right] * W_o + b_0\right) \tag{5.15}$$

$$h_t = \tanh(C_t) * O_t \tag{5.16}$$

The RNN makes use of LSTM blocks to give context to the software's input and output processes. The four steps in the LSTM cycle are depicted in Figure 5.11.

- The forget gate identifies the information to be forgotten from a previous time step.
- The input gate and tanh function are looking for new information to update the cell state.

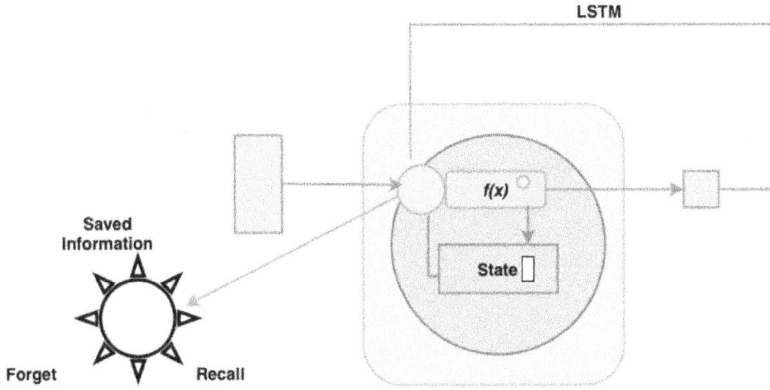

FIGURE 5.11 LSTM cycle.

- The data outputs from the input gate and forget gate are used to update the cell state.
- Both the squashing operation and the output gate offer insightful information.

A dense layer receives the LSTM cell's output. The output of the dense layer is supplied to the "*softmax*" activation function at the output stage for further classification.

5.4.3 CHALLENGES IN DEEP LEARNING-BASED WEATHER PREDICTION

In this section, the challenges in applying DL models to the analysis of HMD are discussed since huge voluminous data are collected from multiple heterogeneous sources.

The availability of training data is crucial to the effectiveness of DL techniques because they are induced by large amounts of data. Additionally, a decline in the volume or reliability of training data may result in a decline in the model's performance. However, in many situations, there is a limited supply of training data for building the model. The challenges in gathering enough training data are typically related to the cost of data collection, which restricts the range of monitoring devices that can gather authentic location data. More challenges remain in collecting large-scale, high-dimensional, and high-quality data sets.

In DL, it has always been difficult to represent data in a way that a model can understand. Data inputs from an image, an audio or a video file or a time-series recording, etc., can be represented as a vector or tensor of an entity. Inputs with high-level features are necessary for DL models to perform better.

In disaster warning systems, to analyze the disasters most deep learning models are deployed. Reliability is an important metric in determining whether these models can be used in practical applications. A significant challenge is approving trustworthy results from deep learning models. Deep learning models for forecasting natural disasters must be developed to be more generic and repeatable.

Table 5.3 shows the research works related to HMD prediction using deep learning techniques.

5.5 PROPOSED METHOD

The general framework for HMD prediction using deep learning is shown in Figure 5.12. Initially the dataset must be collected from official and authentic sources like fifth generation ECMWF atmospheric reanalysis dataset (ERA5) and IMD, which has various spatiotemporal climatic conditions include latitude, longitude, time-index, surface temperature, pressure, u-wind, v-wind,

TABLE 5.3
Summary of Research Work on Hydro-Meteorological Disaster Prediction

Reference (Year)	Extreme Climatic Event Prediction	Techniques Used
Fister et al. (2023) [13]	Heatwave	CNN
Li et al. (2023) [14]	Heatwave	Graph Neural Network (GNN)
Boussioux et al. (2022) [15]	Hurricane	CNN-encoder GRU-decoder
Chen et al. (2022) [16]	Flood	ConvLSTM
Danandeh Mehr et al. (2022) [17]	Drought	CNN + LSTM
Jacques-Dumas et al. (2022) [18]	Heatwave	CNN
Khan et al. (2022) [19]	Daily Maximum Temperature Heatwave	Conv1D + LSTM
Xu et al. (2022) [20]	Typhoon Intensity	Spatial Attention Fusing Network (SAF-Net)
Wu et al. (2022) [21]	Tropical Cyclone	GAN
Yang et al. (2022) [22]	Urban Temperature	LSTM
Dabhade et al. (2021) [23]	Cyclone	ConvLSTM
Dar et al. (2021) [24]	Drought	Empirical Orthogonal Function (EOF)
Lee et al. (2021) [25]	Heatwave	Bayesian Inference
Lowe et al. (2021) [26]	Flood	U-NET
Maity et al. (2021) [27]	Drought	Conv1D
Panahi et al. (2021) [28]	Flood	CNN & RNN
Varalakshmi et al. (2021) [29]	Tropical Cyclone	CNN + Genetic Algorithm
Jung et al. (2020) [30]	Heatwave (Sea Surface Temperature)	LSTM, ConvLSTM
Khan et al. (2020) [31]	Rainfall	Conv1D + MLP
Puttinaovarat et al. (2020) [32]	Flood	MLP, Radial Basis Function (RBF)
Dawood et al. (2019) [33]	Hurricane	Deep CNN
Hu et al. (2019) [34]	Flood	LSTM + Reduced Order Model (ROM)
Khan et al. (2019) [35]	Heatwave	Quantile Regression Forests (QRF)
Matsuoka et al. (2018) [36]	Tropical Cyclone	CNN
Liu et al. (2016) [37]	Tropical Cyclones, Atmospheric Rivers, Weather Fronts	Deep CNN

relative humidity, and rainfall [38, 39]. During pre-processing, a massive amount of data will be processed for filtering outliers and formatted appropriately for modeling. This step involves conducting a correlation analysis to find the relationship among the attributes, as well as its strength and course of action. The input is organized into a tensor that contains the location data in a way that is like how the locations are organized in space. As a result, the data from two related locations are close together. The tensor mimics a time distributed multi-variate 3D structure once it has been organized [40].

The pre-processed data is given as an input to the spatiotemporal model to analyze the abnormal weather conditions for predicting HMD. The spatiotemporal modeling can be of (i) discriminative modeling; (ii) generative modeling; and (iii) probabilistic modeling. Finally, the model's performance is evaluated using a range of metrics, such as mean absolute error (MAE), mean absolute scaled error (MASE), mean absolute percent error (MAPE), mean squared error (MSE), root mean square error (RMSE), and R-squared (R^2).

5.5.1 SPATIOTEMPORAL MODELING

Spatiotemporal models begin to emerge when data are gathered across time and space, as well as having at least one spatial and one temporal property. An event in a spatiotemporal dataset refers to

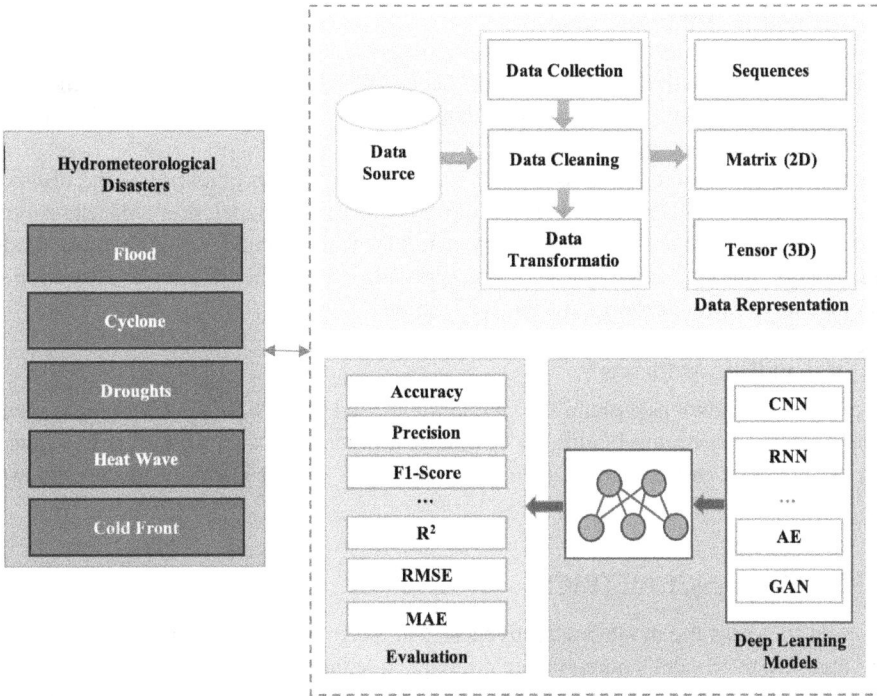

FIGURE 5.12 General framework for hydro-meteorological disaster prediction.

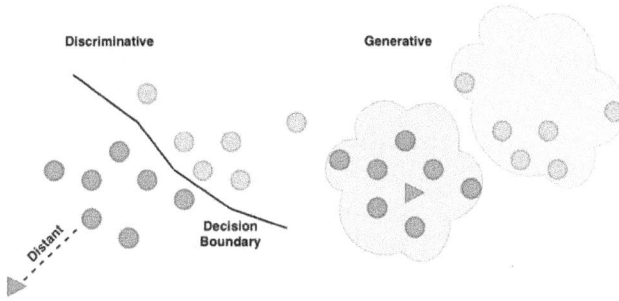

FIGURE 5.13 Discriminative versus generative models.

a temporal and spatial phenomenon that exists at a specific time t and place x, respectively. An illustration of discriminative and generative models is shown in Figure 5.13.

5.5.1.1 Discriminative Modeling

Supervised machine learning makes use of the discriminative model. It determines the boundaries between labels or classes in a dataset and is also known as a conditional model. It generates new instances using maximum likelihood and probability estimates. However, they are unable to produce new data points. The main objective of this model is to distinguish one class from another.

A discriminative model, denoted as $P(B|A = a)$, the conditional probability of the target B, given an observation A. Typical examples of the discriminative model are logistic regression, decision tree, support vector machines (SVMs), random forest, CNN, and conditional random fields (CRFs) [31, 40].

5.5.1.2 Generative Modeling

A subset of statistical models called generative models creates fresh data instances. Unsupervised machine learning employs these models to carry out tasks like estimating probability and likelihood, modeling data points, and classifying entities using these probabilities. These models depend on the Bayes theorem to determine the joint probability.

A generative model, represented as $P(A|B=b)$, the conditional probability of the observable A, given a target b. This model explains, in terms of the probabilistic model, how a dataset is produced. This model can be sampled to produce new data. Typical examples of generative models are Bayesian network, hidden Markov model, autoregressive model, latent Dirichlet allocation (LDA), and generative adversarial networks (GANs) [25, 35, 40].

5.5.1.3 Probabilistic Modeling

A statistical method known as probabilistic modeling is used to consider the effect of random events or actions when predicting the likelihood of potential outcomes in the future. It is a quantitative modeling method that projects several possible outcomes that might even go beyond what has happened recently.

5.5.2 PERFORMANCE METRICS

It is important to assess the model's effectiveness and the precision of its predictions. The metrics chosen to measure the model's outcome are crucial, since they are influencing the conclusion. The values between observed and predicted values can be calculated using the performance indicators listed below.

5.5.2.1 Mean Absolute Error (MAE)

The term "mean absolute error" refers to the average of the absolute differences between the predicted and actual results. The MAE is calculated using Eqn. (5.17).

$$\text{MAE} = \frac{1}{n}\sum |z - z'| \tag{5.17}$$

where
    z is the actual output
    z' is the predicted output
    n is the number of data points

5.5.2.2 Mean Squared Error (MSE)

The term "mean squared error" describes the squared deviation between the true value and the predicted value. Eqn. (5.18) is used to calculate the MSE.

$$\text{MSE} = \frac{1}{n}\sum (z - z')^2 \tag{5.18}$$

5.5.2.3 Root Mean Squared Error (RMSE)

The RMSE calculates the square root of the average squared deviation between the actual and the predicted output, as given in Eqn. (5.19). It is used when the error is noticeably nonlinear. The RMSE is good for prediction accuracy because it shows the average number of errors in the forecasted data.

$$\text{RMSE} = \sqrt{\frac{1}{n}\sum_{j=1}^{n}\left(z_j - z_j'\right)^2} \tag{5.19}$$

5.5.2.4 *R*-squared (*R²*)

The percentage of variation in a dependent variable is explained by one or more independent variables in a model is expressed statistically as *R*-squared. It is calculated using Eqn. (5.20).

$$R^2 = 1 - \frac{X}{Y} \tag{5.20}$$

where
R^2 is a coefficient of determination
X is the sum of squares of residuals
Y is the total sum of squares

5.6 CONCLUSION

The hydro-meteorological disaster (HMD) prediction can be accomplished using the data-driven deep learning approach, i.e., DLWP, which is expected to be a substitute for the traditional theory driven method, i.e., NWP. Deep learning techniques can be used to predict extreme climate events and also, they can recognize complex features from various types of data automatically. Besides, it is less dependent on domain knowledge. The effectiveness of the deep learning-based HMD prediction model can be assessed using various metrics like mean absolute error (MAE), mean absolute scaled error (MASE), mean absolute percent error (MAPE), mean squared error (MSE), root mean square error (RMSE), and *R*-squared (*R²*). DLWP, unlike NWP technique, utilizes minimal computational resources and robust data processing thereby allowing for adopting better disaster management strategies that can mitigate the ill effects of natural events on lives, environment and assets. Furthermore, it is expected that studies on DLWP-based HMD predication can be a boon to the scientific community that works to create a novel disaster management paradigm from which human beings and other life forms can be benefitted indirectly or directly.

REFERENCES

[1] Prasad, A.S., Francescutti, L.H., Natural disasters. *International Encyclopedia of Public Health*, 2017. https://doi.org/10.1016/b978-0-12-803678-5.00519-1

[2] Ren, X., Li, X., Ren, K., Song, J., Xu, Z., Deng, K., Wang, X., Deep learning-based weather prediction: A survey. *Big Data Research*, 2021. https://doi.org/10.1016/j.bdr.2020.100178

[3] Chattopadhyay, A., Nabizadeh, E., Hassanzadeh, P., Analog forecasting of extreme-causing weather patterns using deep learning. *Journal of Advances in Modeling Earth System*, 2020. https://doi.org/10.1029/2019ms001958

[4] Fang, W., Xue, Q., Shen, L., Sheng, V.S., Survey on the application of deep learning in extreme weather prediction. Atmosphere, 2021. https://doi.org/10.3390/atmos12060661

[5] Pham, B.T., Luu, C., Dao, D.V., Phong, T.V., Nguyen, H.D., Le, H.V., von Meding, J., Prakash, I., Flood risk assessment using deep learning integrated with multi-criteria decision analysis. *Knowledge-Based Systems*, 2021. https://doi.org/10.1016/j.knosys.2021.106899

[6] National Disaster Management Authority, India – https://ndma.gov.in

[7] Ma, Z., Mei, G., Deep learning for geological hazards analysis: Data, models, applications, and opportunities. *Earth-Science Reviews*, 2021. https://doi.org/10.1016/j.earscirev.2021.103858

[8] Hu, S., Xiang, Y., Zhang, H., Xie, S., Li, J., Gu, C., Sun, W., Liu, J., Hybrid forecasting method for wind power integrating spatial correlation and corrected numerical weather prediction. *Applied Energy*, 2021. https://doi.org/10.1016/j.apenergy.2021.116951

[9] Randriamampianina, R., Bormann, N., Køltzow, M.A.Ø., Lawrence, H., Sandu, I., Wang, Z.Q., Relative impact of observations on a regional Arctic numerical weather prediction system. *Quarterly Journal of the Royal Meteorological Society*, 2021. https://doi.org/10.1002/qj.4018

[10] Dalmau, R., Perez-Batlle, M., Prats, X., Estimation and prediction of weather variables from surveillance data using spatio-temporal Kriging. IEEE/AIAA 36th Digital Avionics Systems Conference (DASC), 2017. https://doi.org/10.1109/dasc.2017.8102132

[11] Danandeh Mehr, A., Rikhtehgar Ghiasi, A., Yaseen, Z.M., Sorman, A.U., Abualigah, L., A novel intelligent deep learning predictive model for meteorological drought forecasting. *Journal of Ambient Intelligence and Humanized Computing*, 2022. https://doi.org/10.1007/s12652-022-03701-7

[12] Dikshit, A., Pradhan, B., Santosh, M., Artificial neural networks in drought prediction in the 21st century – A scientometric analysis. *Applied Soft Computing*, 2022. https://doi.org/10.1016/j.asoc.2021.108 080

[13] Fister, D., Pérez-Aracil, J., Peláez-Rodríguez, C., Del Ser, J., Salcedo-Sanz, S., Accurate long-term air temperature prediction with Machine Learning models and data reduction techniques. *Applied Soft Computing*, 2023. https://doi.org/10.1016/j.asoc.2023.110118

[14] Li, P., Yu, Y., Huang, D., Wang, Z., Sharma, A., Regional heatwave prediction using graph neural network and weather station data. *Geophysical Research Letters*, 2023. https://doi.org/10.1029/2023g l103405

[15] Boussioux, L., Zeng, C., Guénais, T., Bertsimas, D., Hurricane forecasting: A novel multimodal machine learning framework. *Weather and Forecasting*, 2022. https://doi.org/10.1175/waf-d-21-0091.1

[16] Chen, C., Jiang, J., Liao, Z., Zhou, Y., Wang, H., Pei, Q., A short-term flood prediction based on spatial deep learning network: A case study for Xi County, China. *Journal of Hydrology*, 2022. https://doi.org/10.1016/j.jhydrol.2022.127535

[17] Danandeh Mehr, A., Rikhtehgar Ghiasi, A., Yaseen, Z.M., Sorman, A.U., Abualigah, L., A novel intelligent deep learning predictive model for meteorological drought forecasting. *Journal of Ambient Intelligence and Humanized Computing*, 2022. https://doi.org/10.1007/s12652-022-03701-7

[18] Jacques-Dumas, V., Ragone, F., Borgnat, P., Abry, P., Bouchet, F., Deep learning-based extreme heatwave forecast: Frontiers in climate. *Frontiers Media SA*, 2022. https://doi.org/10.3389/fclim.2022.789 641

[19] Khan, M.I., Maity, R., Hybrid deep learning approach for multi-step-ahead prediction for daily maximum temperature and heatwaves. *Theoretical and Applied Climatology*, 2022. https://doi.org/10.1007/s00704-022-04103-7

[20] Xu, G., Lin, K., Li, X., Ye, Y., SAF-Net: A spatio-temporal deep learning method for typhoon intensity prediction. *Pattern Recognition Letters*, 2022. https://doi.org/10.1016/j.patrec.2021.11.012

[21] Wu, Y., Geng, X., Liu, Z., Shi, Z., Tropical cyclone forecast using multitask deep learning framework. *IEEE Geoscience and Remote Sensing Letters*, 2022. https://doi.org/10.1109/lgrs.2021.3132395

[22] Yang, J., Yu, M., Liu, Q., Li, Y., Duffy, D.Q., Yang, C., A high spatiotemporal resolution framework for urban temperature prediction using IoT data. *Computers and Geosciences*, 2022. https://doi.org/10.1016/j.cageo.2021.104991

[23] Dabhade, A., Roy, S., Moustafa, M.S., Mohamed, S.A., El Gendy, R., Barma, S., Extreme weather event (cyclone) detection in India using advanced deep learning techniques. 9th International Conference on Orange Technology (ICOT), 2021. https://doi.org/10.1109/icot54518.2021.9680663

[24] Dar, J., Dar, A. Q., Spatio-temporal variability of meteorological drought over India with footprints on agricultural production. *Environmental Science and Pollution Research*, 2021. https://doi.org/10.1007/s11356-021-14866-7

[25] Lee, O., Seo, J., Won, J., Choi, J., Kim, S., Future extreme heat wave events using Bayesian heat wave intensity-persistence day-frequency model and their uncertainty. *Atmospheric Research*, 2021. https://doi.org/10.1016/j.atmosres.2021.105541

[26] Löwe, R., Böhm, J., Jensen, D.G., Leandro, J., Rasmussen, S.H., U-FLOOD – Topographic deep learning for predicting urban pluvial flood water depth. *Journal of Hydrology*, 2021. https://doi.org/10.1016/j.jhydrol.2021.126898

[27] Maity, R., Khan, M.I., Sarkar, S., Dutta, R., Maity, S.S., Pal, M., Chanda, K., Potential of deep learning in drought assessment by extracting information from hydrometeorological precursors. *Journal of Water and Climate Change*, 2021. https://doi.org/10.2166/wcc.2021.062

[28] Panahi, M., Jaafari, A., Shirzadi, A., Shahabi, H., Rahmati, O., Omidvar, E., Lee, S., Bui, D. T., Deep learning neural networks for spatially explicit prediction of flash flood probability. *Geoscience Frontiers*, 2021. https://doi.org/10.1016/j.gsf.2020.09.007

[29] Varalakshmi, P., Vasumathi, N., Venkatesan, R., Tropical Cyclone prediction based on multi-model fusion across Indian coastal region. *Progress in Oceanography*, 2021. https://doi.org/10.1016/j.pocean.2021.102557

[30] Jung, S., Kim, Y.J., Park, S., Im, J., Prediction of sea surface temperature and detection of ocean heat wave in the south sea of Korea using time-series deep-learning approaches. *Korean Journal of Remote Sensing*, 2020. https://doi.org/10.7780/KJRS.2020.36.5.3.7

[31] Khan, M.I., Maity, R., Hybrid deep learning approach for multi-step-ahead daily rainfall prediction using GCM simulations. *IEEE Access*, 2020. https://doi.org/10.1109/access.2020.2980977

[32] Puttinaovarat, S., Horkaew, P., Flood forecasting system based on integrated big and crowdsource data by using machine learning techniques. *IEEE Access*, 2020. https://doi.org/10.1109/access.2019.2963819

[33] Dawood, M., Asif, A., Minhas, F. ul A.A., Deep-PHURIE: Deep learning based hurricane intensity estimation from infrared satellite imagery. *Neural Computing and Applications*, 2019. https://doi.org/10.1007/s00521-019-04410-7

[34] Hu, R., Fang, F., Pain, C.C., Navon, I.M., Rapid spatio-temporal flood prediction and uncertainty quantification using a deep learning method. *Journal of Hydrology*, 2019. https://doi.org/10.1016/j.jhydrol.2019.05.087

[35] Khan, N., Shahid, S., Juneng, L., Ahmed, K., Ismail, T., Nawaz, N., Prediction of heat waves in Pakistan using quantile regression forests. *Atmospheric Research*, 2019. https://doi.org/10.1016/j.atmosres.2019.01.024

[36] Matsuoka, D., Nakano, M., Sugiyama, D., Uchida, S., Deep learning approach for detecting tropical cyclones and their precursors in the simulation by a cloud-resolving global nonhydrostatic atmospheric model. *Progress in Earth and Planetary Science*, 2018. https://doi.org/10.1186/s40645-018-0245-y

[37] Liu, Y., Racah, E., Prabhat, Correa, J., Khosrowshahi, A., Lavers, D., Kunkel, K., Wehner, M., Collins, W., Application of deep convolutional neural networks for detecting extreme weather in climate datasets (Version 1). arXiv, 2016. https://doi.org/10.48550/ARXIV.1605.01156

[38] European Centre for Medium-Range Weather Forecasts: ERA5 Dataset – www.ecmwf.int/en/forecasts/datasets/reanalysis-datasets/era5

[39] India Meteorological Department, Ministry of Earth Sciences, Government of India – https://mausam.imd.gov.in

[40] Das, M., Ghosh, S.K., A probabilistic approach for weather forecast using spatio-temporal inter-relationships among climate variables. 9th International Conference on Industrial and Information Systems (ICIIS), 2014. https://doi.org/10.1109/iciinfs.2014.7036528

6 Assessment of ICT for Sustainable Developments with Reference to Fog and Cloud Computing

H.K. Shilpa[1], D.K. Girija[2], M. Rashmi[3], and N. Yogeesh[4]
[1] Faculty in Computer Science, Mandya University, Mandya, Karnataka, India
[2] Department of Computer Science, Government First Grade College, Madhugiri, Karnataka, India
[3] Faculty in Computer Science, GFGC, Vijayanagar, Bengaluru, Karnataka, India
[4] Department of Mathematics, Government First Grade College, Tumkur, Karnataka, India

6.1 INTRODUCTION

When it comes to progress, ICT (Information and Communication Technology) is seen as both a means and an aim in and of themselves. It's no surprise that many developing countries have made ICT their top priority; the services sector accounts for around two-thirds of the global economy. Meanwhile, countries like India and the Philippines have emerged as major participants on the international stage of information technology. As a result, ICT is playing an increasingly vital role in the manufacturing and industrial sectors as well [1]. Because there is so much information that needs to be kept track of in every given business or industry. The capability to collect and investigate data in actual time allows for greater oversight and management of vital business functions. Businesses in the modern day have access to a plethora of information, with the potential to consistently collect vast amounts of data through sensors. These days, cloud storage is the norm for such information. Clouds are vast, freely accessible, and utilizable collections of shared, virtualized resources, such as computers, programming environments, and service offerings. The adaptability of these resources allows for continuous reorganization to meet ever-evolving needs, allowing for optimal resource utilization [2]. Typically, this pool of resources is used in conjunction with a pay-as-you-go pricing model, with the infrastructure provider providing guarantees in the form of individualized Service Level Agreements (SLAs). As defined by the SLA among the service provider and the customers, the "cloud" stands a sort of distributed computing architecture in which a collection of servers is dynamically supplied and made available to consumers as a single or group of shared computing resources.

6.2 EDGE COMPUTATION

The term "edge computing" refers to the practise of processing, analyzing, and storing data closer to the location where it is generated. This enables rapid analysis and response in a time frame that is nearly real-time. In recent years, some businesses have streamlined their operations by storing data and performing computing tasks in a centralized location using cloud computing. Figure 6.1 shows the overview of edge computing.

DOI: 10.1201/9781003407959-6

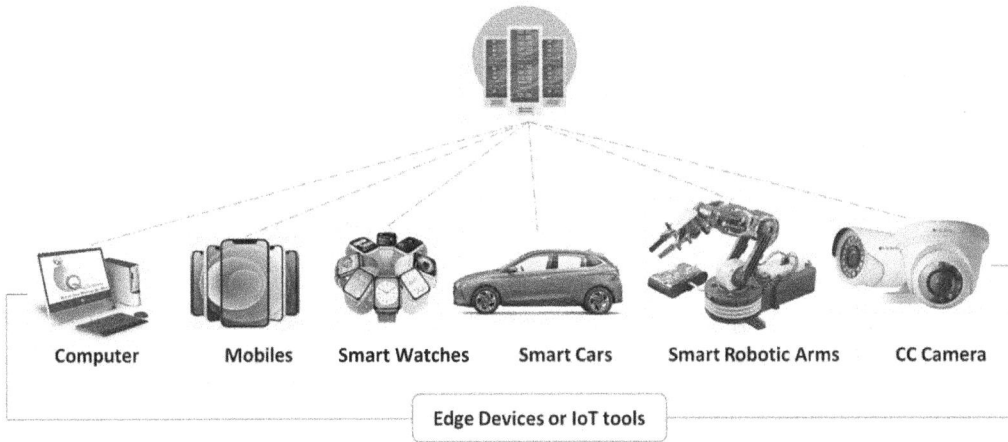

FIGURE 6.1 An overview of edge computation.

6.2.1 WHEN SHOULD I EMPLOY EDGE COMPUTING?

Any business can benefit from using edge computing. Edge computing can deliver quick and reliable service whenever a steady flow of data is required.

- **Self-driving vehicles:** Decisions based on data must be made extremely quickly by autonomous cars. If a person suddenly darts out in front of your automobile, you don't have time to send an emergency request to the cloud and wait for it to be returned to your local network. In contrast to cloud services, edge services don't have to transmit requests to the cloud before making a call, allowing for quicker processing times. Additionally, while the car is not online, edge computing IoT still offers a real-time data stream.
- **Healthcare:** The software used in healthcare must process data in real-time, independent of the speed of the user's Internet connection. The gadget needs to have instant and error-free access to a patient's medical record. Computing at network's edge ("edge") exists online-capable then, like autonomous vehicles, offers lightning-fast server response times due to its proximity to the network's endpoints.
- **Safety:** When time is of the essence, edge computing architecture is preferable to solutions hosted in the cloud for ensuring data protection. Instead of going through the data center, the requests are handled locally. As a result, security services can respond to threats more quickly and accurately predict future hazards.
- **Finance:** Edge computing enables instantaneous data updates for the Internet of Things (IoT) and artificial intelligence (AI) apps on mobile devices. Because the crucial data is saved locally on the user's device, they will be able to manage their finances, retrieve documents, and see operations even when they are not connected to the Internet.
- **Smart speakers:** The user's input should be instantly processed by the speaker in order to carry out the desired actions. Once again, they shouldn't depend on the speed of your Internet connection. With edge computing, data is stored reliably and requests from consumers are processed quickly [3].

6.2.2 EDGE COMPUTING'S OPERATIONAL PROCEDURE

In order to be effective, edge computing must gather and investigate data as near to the data's or event's original point of origin as possible. It uses sensors, computers, and machines to gather information, which is then sent to localized or remote storage facilities. Depending on the goal, this information could be used to power analytics and machine learning systems, facilitate the introduction of automation features, or provide insight into the current functioning of a product or system.

The majority of data calculations done today take place in a data center or in the cloud. The deployment of edge servers, gateway devices, and other equipment that shorten the time and distance needed for computing tasks and connect the entire infrastructure is necessary, as enterprises move to an edge model with IoT devices. Smaller edge data centers situated in small towns or even rural areas might be a component of this architecture, as could cloud containers that are easily transferable between clouds and systems [2].

However, edge data centers are not the only method of data processing. In other instances, IoT devices may perform calculations onboard or communicate the data to a smartphone, an edge server, or a storage device. In reality, a multitude of technologies can be used to construct an edge network. These include mobile edge computing that operates over wireless channels, fog computing that combines infrastructure that leverages clouds and other storage to place data in the optimal position, and so-called cloudlets that function as ultra-small data centers.

Edge frameworks provide the essential flexibility, agility, and scalability for an increasing variety of business use cases. For instance, a sensor might offer real-time information regarding the temperature at which a vaccine is stored and whether it has been transported at the proper temperature.

With the help of edge devices and sensors, connected cars and other IoT devices can monitor traffic in real-time to reveal previously unknown information about bottlenecks and other routes. In addition, artificial intelligence (AI) algorithms can be built into motion sensors to detect the onset of an earthquake, giving residents and business owner's ample time to turn off gas lines and other potentially explosive systems [4].

6.2.3   GAINS FROM USING EDGE COMPUTING

Ringing some data processing, analysis, and storage quicker to the point of fact generation and away from the cloud can have the following major advantages,

- **Enhanced speed and decreased latency:** Moving data processing and analysis to the edge improves system response time, allowing for quicker transactions and better user experiences that may be essential in near real-time applications, such as autonomous car operation.
- **Better control of network traffic:** The bandwidth and expenses associated with transmitting and storing huge volumes of data can be decreased by limiting the amount of data transported over the grid to the cloud.
- **Greater dependability:** Networks can only send a certain quantity of data at once. The capacity to store and process data locally, even in areas with poor Internet connections, increases resilience in the event that the cloud connection is lost.
- **Increased security:** By limiting the transmission of data over the Internet, an edge computing solution may improve data security when properly implemented [5].

6.2.4   LIMITATION OF EDGE COMPUTING

- **Complexity:** When dealing with a huge quantity of devices or a widespread geographic zone, it might be difficult to set up and maintain an effective edge computing system.
- **Limited resources:** Edge devices sometimes have limited processing power, storage space, and network bandwidth, limiting their utility for some tasks.
- **Dependence on connectivity:** Edge computing can't function properly without network access. The functionality of the system could be compromised if the connection is lost.
- **Security concern:** Malware, hacking, and physical interference are all potential security issues for edge devices.

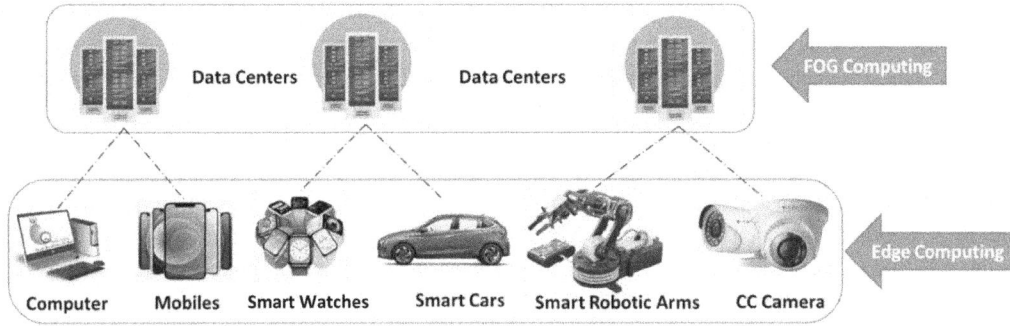

FIGURE 6.2   An overview of FOG computation.

6.3   FOG COMPUTING

The flexibility and decentralization of fog computing set it apart from its cloud computing counterpart, the cloud. Fog computing stays a dispersed computing system that exists among the cloud and data generating procedures. It is similarly recognized as fog networking or fogging. To maximize efficiency, users can organize their apps and the data they generate in this flexible framework. Below Figure 6.2 shows the overview of the fog computing.

The framework's goal is to position core analytical functions closer to the network's periphery, where they will be most useful. This improves network performance and overall efficiency by reducing the distance across which data must be transferred by users. Consumers can gain benefits from worries about cloud computing security. Using the fog computing paradigm, users can boost network security by separating sensitive data from general traffic [4]. Many of the desirable qualities of its progenitor, cloud computing, are present in fog computing as well. By adopting a fog computing strategy, users can continue to store data and apps remotely while simultaneously incurring costs associated with cloud updates and maintenance. Data, for example, can still be accessed remotely by their employees.

6.3.1   When Should I Employ Fog Computing?

Fog computing remain able to be rummage-sale in the succeeding circumstances:

- When some individual facts need to be uploaded to the cloud, it is employed in certain situations. This particular data has been selected for durable storing and is retrieved by the host on a less frequent basis than other data.
- It is utilized in situations in which the data must be evaluated in a very short amount of time, often known as having a low "latency".
- Whenever there is a requirement to supply a big number of services across a vast area in a variety of geographical locations, this method is utilized.
- Fog computing is required to be used by any devices that are required to perform intensive computations and processing.
- IoT devices (like the Car to Car Confederation in Europe), strategies with sensors then cameras (IIoT -Industrial Internet of Things) then additional uses are examples of real-world applications of fog computing [6].

6.3.2   Fog Computing's Operational Procedure

In fog computing, nodes known as "fog nodes" are used. These fog nodes have greater processing and storage capacity since they are closer to the data sources. Rather than transmitting the request to the coud for centralized processing, the data can be processed at the fog nodes.

The proliferation of Internet-connected gadgets is causing congestion in the cloud. Using fog computing for IoT devices is now a need due to the limitations of cloud computing. It is capable of processing the massive amounts of data produced by these gadgets. Once in place, fog-enabled devices provide real-time analysis on data, such as alarm status, device status, defect alert, and more. The delay is reduced and serious damage is avoided as a result. Fog computing has the potential to significantly lessen the bandwidth required, which in turn would expedite communication with the cloud and numerous sensors. As an illustration of fog computing, consider the following scenario: a user on a mobile device wishes to watch the most recent footage from a locally located IoT security camera. While this could take some time, it is unnecessary with fog computing because users can just connect to a fog node in their immediate area to stream videos much more quickly [4].

6.3.3   GAINS FROM USING FOG COMPUTING

- **Privacy:** With fog computing, you have more say over how much data stays private. If the user has any sensitive data, it can be evaluated on their local machine instead of being sent to some remote server in the cloud. IT departments will be capable to observe and manipulate each device in question in this way. In addition, data can be transmitted to the cloud if a portion of it has to be evaluated.
- **Productivity:** Customers can use fog applications to customize machine behavior to their specifications. When developers have access to the correct resources, creating these fog applications is a breeze. They can roll out the finished product whenever they like after development is complete.
- **Security:** The capability to connect several devices to the same network is one of the benefits of fog computing. As a result of this, the operations are not carried out in one centralized location but rather at a number of different end points inside a complicated dispersed environment. Because of this, it is much simpler to recognize possible dangers before they have an effect on the entire network.
- **Bandwidth:** Depending on available resources, the required bandwidth for data transmission can be costly. Due to the fact that selected data can be processed nearby instead of existence sent to the cloud, bandwidth requirements are minimal. This bandwidth decrease will be especially advantageous as the amount of IoT devices increases.
- **Latency:** A further gain of dealing with selected data nearby is the decrease in delay. The facts can be handled by those that are geographically closest to the home of the user. This can result in immediate replies, particularly for time-sensitive services [7].

6.3.4   LIMITATION OF FOG COMPUTING

- Fog computing negates the specifics of the "anytime/anywhere" assistances of cloud computing due to its reliance on a specific physical location.
- Internet Protocol (IP) address deceiving and Man-in-the-Middle (MitM) occurrences are two examples of potential security vulnerabilities with fog computing.
- Fog computing remains a system that makes use of edge and cloud properties together, therefore initial expenses involved in the hardware remain.
- Although fog computing has been around for a while, its precise definition remains unclear, with different providers using varying terminology to describe it.

6.4   CLOUD COMPUTING

Cloud computing discusses the rehearsal of keeping and regaining data over a grid of isolated servers that are kept proceeding the Internet. Above the Internet (similarly identified by way of "the

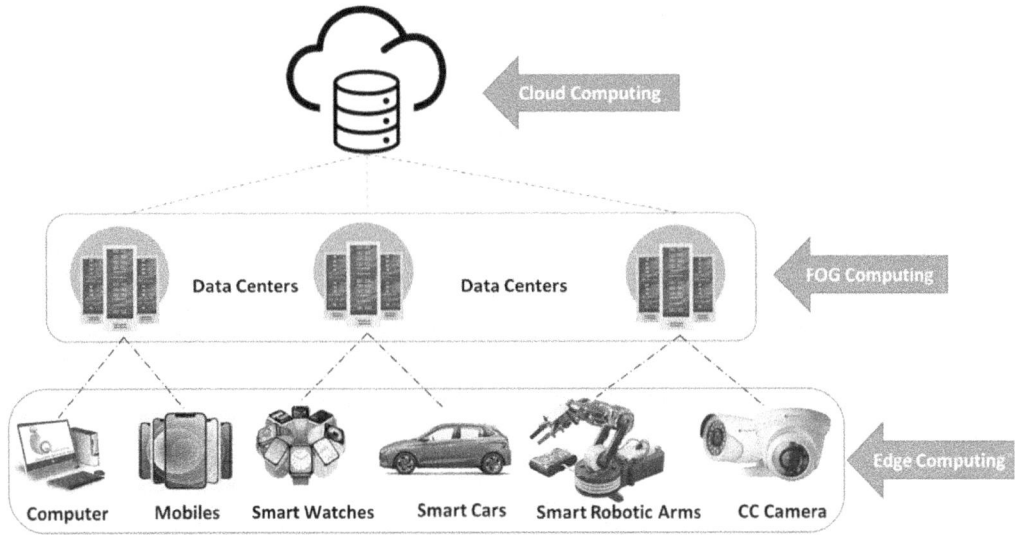

FIGURE 6.3   An overview of cloud computation.

cloud"), a number of information technology services, containing servers, databases, software, virtual storage, networking, and intelligence, are made available. These services allow for more rapid innovation, flexible resource utilization and economies of scale, among other benefits. It might be defined in layman's languages as a practical stage that ensures not enforcing any limitations on the quantity of facts you can accumulation or in what way you are simply able to access the facts over the Internet [8]. The data could be whatever, including files, photos, forms, audial, video, and a variety of other media types. Figure 6.3 below shows the overview of cloud computing.

Cloud providers are organizations that supply all of the above services to its customers. They make it possible for you to store data, retrieve it when necessary, run applications, and manage all of these things using configuration portals. Amazon Web Services and Microsoft Azure exist currently dual of the most reliable and reputable cloud service companies accessible. If you use a cloud service, you will often only be billed for that service. This can help you cut costs in several areas, including operations and IT, streamline your infrastructure management, and adapt to the changing needs of your business [9].

6.4.1   Why Should I Employ Cloud Computing?

There are a number of compelling arguments in favor of adopting cloud computing now. I won't go into great detail, but I will say that the points I've made here should assist in justifying my position.

- Many companies, both big and small, around the world have embraced cloud computing because of the dramatic financial benefits it provides. As previously mentioned, cloud computing has led to a significant decrease in the price of hardware and software, as well as other server resources. This is because all of our workload data, applications, and processes can now be run online, remotely, over the Internet, as opposed to using physical hardware and software.
- Cloud service providers handle routine tasks including server upkeep, software/hardware installation, and license renewal.
- With cloud computing, we can get to our files and programmers from any Internet-connected device, at any time. With a single click, you may update and install any one of 100 preconfigured applications.

- With cloud computing, your data is stored and managed remotely, and in the event of a crash, it is automatically backed up and can be restored.

Many cloud service providers now provide flexible access and billing plans that allow users to select the services they need and pay only for the time those services are actually in use [8]. Best suited to expanding companies with a need for more bandwidth. In the end, using the cloud can help you save both time and money.

6.4.2 Cloud Computing's Operational Procedure

When it comes to storing and retrieving information, "the cloud" (also known as "the Internet") is the ideal location because of the ubiquity of remote servers. Cloud computing can be broken down into frontend and backend components for easier comprehension [10].

The architecture design of the cloud computing is shown in Figure 6.4. Which majorly includes the frontend and backend. The frontend is what a user sees when they log into their cloud account or launches a cloud-based application to access their data, front end contains web servers (Chrome, Firefox, Opera, etc.), clients are also mobile devices. While there are many parts of cloud computing, the backend is primarily responsible for securing data and information. It consists of various computer systems, databases, and mainframes. The protocols that the central server adheres to make operations easier. It employs software known as middleware to provide constant communication between computers and other gadgets that are connected via the cloud. Many companies that offer cloud computing services back up their customers' data multiple times in case of disaster.

The file system is distributed using cloud computing, which uses several different servers and hard drives. Data is never kept in a single location, and in the event that unique unit miscarries, the other resolve take over immediately. On the distributed file system, employer disc universe is assigned, and a resource allocation algorithm is another crucial element. Strong distributed environments are essential for cloud computing, which primarily relies on solid algorithms [11].

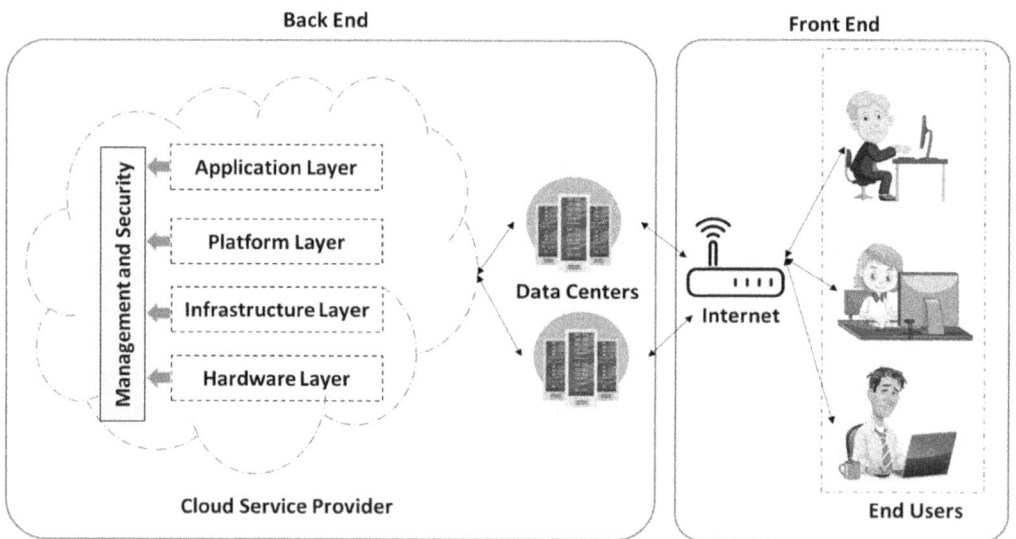

FIGURE 6.4 A cloud architecture design.

6.4.3 Gains From Using Cloud Computing

Today's organizations are increasingly adopting cloud computing due to the many advantages offered by cloud platforms. Some of the most notable advantages include the following:

- Quicker development cycles and releases are possible thanks to the easiness through which innovative occurrences are able to be produced then ancient ones closed down. By removing hardware constraints and sluggish procurement processes, cloud computing facilitates the testing and design of novel applications.
- Cloud computing provides your company with greater scalability and adaptability. You may easily add more capacity for processing and storing data without adding any new physical servers or storage.
- Businesses can avoid the costs and hassle of investing in infrastructure that would be requisite to hold even top usage. Likewise, if resources aren't being operated, they might be rapidly compact.
- Savings can be realized because of the fact that cloud service models only charge for the resources that are really used. This protects your money by preventing avertable data center building and extension, and it frees up your IT team for higher-value tactical advantages.
- Data may be accessed from any device, at any time, improving cooperation. Users are no longer limited to a distinct device or physical position, but can use any Internet-connected device to contact data from any location in the world.
- Superior protection: Conflicting to common trust, cloud computing can increase your safety posture by virtue of the extensive security structures, automated upkeep, and centralized administration.
- Cloud services from reputable companies will also have more powerful protection because they will engage top security professionals and use cutting-edge technologies.
- If you are concerned about losing data, you may rest assured that cloud services include backup and disaster recovery options. Keeping information on the cloud rather than on a local device helps protect it from being lost in the event of a disaster.

Although it's still in its infancy, cloud computing is already being adopted by a wide range of enterprises and organizations, from multinational conglomerates to local start-ups, from NGOs to the public sector, and even by individual consumers.

6.4.4 Limitations of Cloud Computing

- For instance, the fact that cloud services require constant access to the Internet is often cited as a major downside of this model of data storage and management. In the past, PCs had to be tangibly connected to a server or other loading devices in demand to regaining facts. It's possible that if you have a poor Internet assembly, you won't be able to use cloud computing to access the data or programmers you require.
- Top cloud service providers are not immune to outages caused by things like natural disasters or unexpected technological issues that can slow down service and even disrupt it entirely. It's possible that your access to cloud services will be temporarily disabled while the issue is investigated.

The worthy news is that the majority of these problems may be overwhelmed with enough preparation and also investigation into available cloud service providers and their service models. The majority of cloud migration problems may be traced back to customers' misunderstanding of the services on offer, the associated costs, and the security responsibilities that remain their obligation

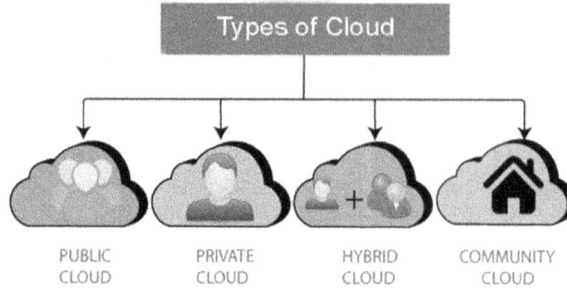

FIGURE 6.5   Classification of the cloud.

after a cloud migration has been completed. If we go through an exposed cloud policy, you'll have superior leeway wherever you set up a workshop and the services you use.

6.5   CLOUD DEPLOYMENT MODELS

Location is an important consideration in cloud computing deployment strategies. Determining your organization's needs will identify the optimal deployment strategy. Four distinct cloud deployment models exist. There are numerous types of cloud infrastructures, containing the (i) Public Cloud; (ii) Private Cloud; (iii) Hybrid Cloud; and (iv) Community Cloud as shown in Figure 6.5 and described below.

6.5.1   PUBLIC CLOUD

If you're looking for a hosting plan that offers cloud services to the whole public, you'll want to look into "public cloud" options. Customers have no say in where the necessary infrastructure is placed. All users contribute to the total price, which may be zero or take the procedure of a license strategy such as a per-user payment. Organizations with a need to manage both the host application and the numerous user applications can benefit greatly from using a public cloud. As a rule, a public cloud service will give its customers with high-bandwidth network connectivity in addition to owning and operating fact centers, hardware, and structures on which their capacities operate.

Example: IBM Cloud, Microsoft Azure and Amazon Web Services.

6.5.2   PRIVATE CLOUD

This one is a cloud structure utilized absolutely for particular enterprise. It grants enterprises greater control over security and data controlled internally and protected by a firewall. It can be introduced locally or remotely. Private clouds stay ideal for enterprises with stringent security, management, and availability requirements. An organization's own data center is usually the location of a private cloud. It is also possible to host a private cloud to build unique out of hardware located in a remote data center or on the servers of a third-party cloud supplier.

Example: Elastra-private cloud, HP Data Centers, Microsoft and Ubuntu.

6.5.3   HYBRID CLOUD

This type of cloud utilizes both private and public clouds, yet they can exist as distinct entities. Internal or external providers can manage and deliver the necessary resources. A hybrid cloud provides excellent scalability, adaptability, and security. This is illustrated by a company that may use the open cloud to engage through clients though securing their facts with an isolated cloud. This cloud is toward establishing a combination of private and public cloud properties, with an arrangement among them that agrees an organization to select the optimal cloud for each one application or capacity and to easily move workloads among the binary clouds as environments change. This helps organizations to achieve practical and commercial goals more successfully and cost-efficiently than they could through either a private or public cloud alone.

Example: Microsoft, Amazon, Google, NetApp and Cisco.

6.5.4 COMMUNITY CLOUD

That one is an infrastructure that organizations that are a part of a certain community share with one another. Concerns about security, performance, and privacy are frequently shared by community members. A communal cloud at a bank, the administration of nation or a transaction company is an illustration of this. A community cloud can be hosted and managed both internally and externally. For businesses engaged in collaborative ventures that require centralized cloud computing capabilities for planning, developing, and carrying out their projects, a community cloud is a useful option.

Example: Health Care community cloud.

6.6 CLOUD-BASED SERVICES

Software as a Service (SaaS), Platform as a Service (PaaS), Infrastructure as a Service (IaaS), Serverless Computing make up the bulk of cloud computing's service offerings. These are stacked one on top of another, hence the name "stack" for cloud computing. Being aware of what these are and how they vary will help your business succeed.

6.6.1 SOFTWARE-AS-A-SERVICE (SAAS)

This one enjoys the same notoriety as other cloud-based applications. To access and utilize this software, which is stored and managed in the cloud, you will need to make use of a net browser, a desktop customer or an application programming interface (API). It is compatible with your PC or mobile Operating System. Customers of SaaS typically remuneration a subscription fee on a periodic or annual basis; however, assured vendors may deliver "pay as you go" estimating based on authentic usage. In addition to reduced expenses, faster time to value, and more scalability, SaaS also has these benefits.

- **Automatic renovations:** With SaaS, you may yield the benefit of fresh capabilities as rapidly as the worker offers them, short of consuming to coordinate an advancement on premises.
- **Defense against facts loss:** Since your request data is stored on the cloud, you are not unable to find data if your device smashes or malfunctions.

From extremely specialized industry and division apps to corporate software catalogues and AI (artificial intelligence) software, hundreds of thousands of SaaS solutions are at your disposal.

Examples of SaaS: Salesforce, Basecamp, Google Docs, Office 365, etc.

6.6.2 PLATFORM-AS-A-SERVICE (PAAS)

It gives software designers on-demand stage hardware, a whole software load, organization, and even advance tools for functioning, generating, and handling requests shorn of the expenditure, difficulty and obstinacy of keeping an on-premises platform. With PaaS, the cloud service provider takes care of housing all of physical components and instruction required to run presentation, together with servers, networks, storage, OS, middleware, and catalogs. Developers are able to "spin up" the servers and settings desired to track, construct, arrange, keep, update, and measure requests by making a few selections from a drop-down menu.

Examples of PaaS: Cloud Foundry, Google App Engine, Engine Yard, etc.

6.6.3 INFRASTRUCTURE-AS-A-SERVICE (IAAS)

Internet-based IaaS provides pay-as-you-go admission to critical figuring incomes like servers (both physical and virtual), networks (both public and private), and data storage (clouds). Infrastructure as a facility is another name for this offering (IaaS). With IaaS, customers can expand and contract their use of resources on demand, eliminating the need for costly upfront investments in infrastructure or excess purchasing of resources to handle temporary spikes in demand. As an added bonus, with IaaS, customers no longer have to overspend on infrastructure to handle occasional peaks in demand.

6.6.4 SERVERLESS COMPUTING

Serverless computing is a model of cloud computing, which includes unloading all of the backend organization management responsibilities, such as provisioning, scaling, arrangement, and repairing, to the cloud worker. This allows designers to give their full attention and time to the code and commercial reason that is explicit to their requests. Serverless computing is also referred to as simply serverless.

In addition to this, serverless computing only executes application code when it is specifically requested, and it automatically adjusts the size of the supporting infrastructure in accordance with the capacity of these needs. Clients never remuneration for vacant capacity when spending serverless computing meanwhile they only pay for the properties that are being utilized when the presentation is vigorous.

6.6.5 FUNCTION-AS-A-SERVICE (FAAS)

It stays frequently incorrect through serverless computing, despite the information that it is a subgroup of serverless. In response toward specified events, FaaS enables designers to execute sections of presentation code (called functions). Everything but the programmer virtual machine operating system, physical hardware, and web server software organization is spontaneously provided and decommissioned through the cloud service provider simultaneously as the code performs, introducing initiates when performance begins and conclusions when performance concludes.

6.7 COMPARATIVE INVESTIGATION OF EDGE, FOG, AND CLOUD COMPUTATION

IT services and resources can be made available whenever needed thanks to cloud computing. Cloud computing is the provision of an online computer service on demand. Users can get to their data and applications whenever they need to thanks to cloud computing.

Every second, more and more information is created on the planet. To expand, businesses gather and analyze this information from their customers. There could be data traffic if many companies accessed their data at once from remote servers located in data centers. Accessing data may be delayed, bandwidth may be reduced, etc. due to network congestion. However, cloud computing

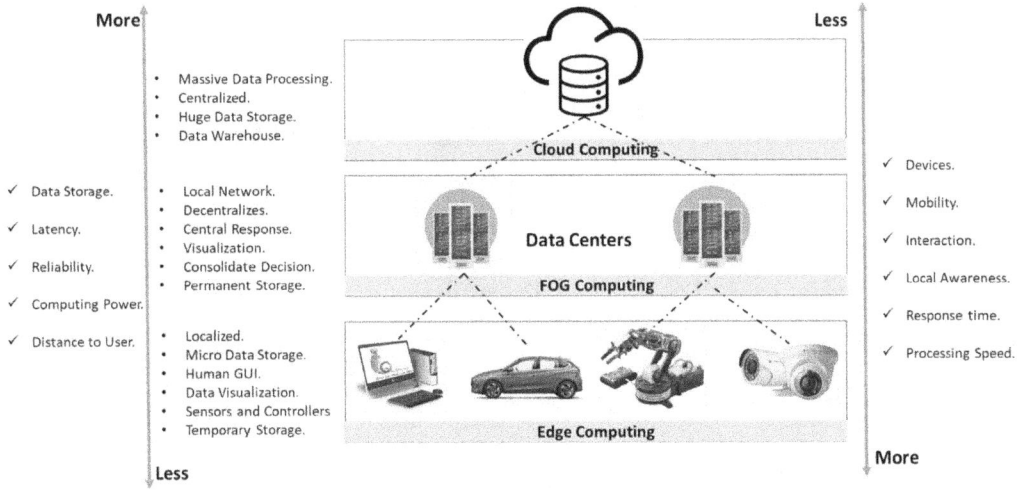

FIGURE 6.6 Comparative investigation of edge, fog, and cloud computing.

technology by itself is insufficient to efficiently store, process, and respond to enormous data volumes.

Data-driven businesses are increasingly adopting cloud, fog, and edge computing architectures. Organizations can benefit from the IIoT and other computational and data storage resources by adopting such an architecture. Although they have some similarities, cloud, fog, and edge computing are actually distinct IIoT layers. By allowing processing to take place locally at many decision points, edge computing for the IIoT can help keep networks more manageable. High-performing industrial applications can be developed with the help of WINSYSTEMS' knowledge of industrial embedded computer systems and the power of the IIoT.

Organizations can take advantage of a wide variety of computing resources and data storage assets with the help of cloud, fog, and edge computing infrastructures. These three computing resources share certain similarities, but they actually represent distinct tiers of information technology. This distinct tier of different computing can be implemented with their features in Figure 6.6.

6.8 WORLDWIDE STATISTICAL ANALYSIS OF REVENUE OF PUBLIC CLOUD COMPUTING

Cloud computing is a relatively new subfield and infrastructure component of the IT industry. Computing in the cloud is a form of virtualization made possible via the practice of the Internet and related facilities. In the realm of IT, cloud computing is expanding at a breakneck pace. An effective virtual IT infrastructure relies on this. Distributing an organization's IT resources to its satellite offices and other affiliates is a key function that may greatly benefit from cloud computing. Due to its many advantages, cloud computing is quickly gaining traction in the public and government sectors as well. Table 6.1 and Figure 6.7 represent the global public cloud service increases in billions every year. Virtualization technology is another name for cloud computing. Software, demonstrations, platforms and operating systems, hardware and organization, content and material, etc. are just some of the various service models and kinds that are evolving in support of cloud computing. Big data, analytics, the Internet of Things (IoT), etc. are all examples of additional technologies that are on the rise and often intertwined with cloud computing [9]. Many factors influence the cloud industry, but one crucial fact is that many new, specialized cloud service providers are fast developing in addition to the already established IT organizations and corporations. In this chapter

TABLE 6.1
Yearwise Revenue in a Public Cloud

Year	Revenue of a Public Cloud in Billions
2011	$91.4
2012	$109
2013	$131
2014	$140
2015	$146
2016	$150
2017	$153.5
2018	$186.4
2019	$214.3
2020	$249.8
2021	$289.1
2022	$331.2

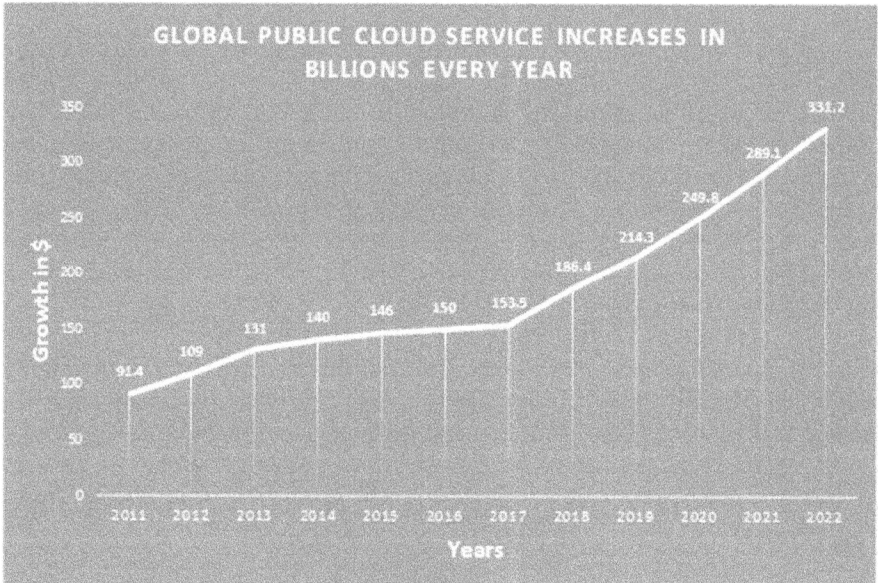

FIGURE 6.7   Global revenue of public cloud service market.

we concentrate on the Revenue Trends in the Market for Public Cloud Computing Services globally and also how it rapidly increases across the years.

6.9   CLIMATE ANALYSIS USING CLOUD COMPUTING

Temperature, precipitation, wind speed and direction, and cloud cover are the elements that make up an area's climate. A climate is overall weather conditions over a stretched period of time (yearly conditions). The climate is a dynamical structure that is not just severe due to large external variables like lunar energy or geography of exterior of the hard ground, then it can similarly be excessive due to essentially insignificant facts like butterflies flapping their wings [9]. Temperature and amount of rainfall are the two most important aspects that go into establishing a region's climate. Latitude, height, and the presence of major quantities of water, such as oceans and lakes, as well as the currents within those bodies of water, as well as the distance that land areas are from those bodies of water, are the primary determinants of temperature. Prevailing winds, the existence of mountains, and seasonal winds that change with the amount of energy that each hemisphere receives from the sun are the primary factors that influence the quantity of precipitation that falls at any given location. A microclimate is a localized region with climate conditions that are unique in comparison to those of the surrounding areas [12].

Lack of long-standing observations, especially in locations through climatic and land-cover diversity, are able to be a barrier to including climate facts at suitable geographical scales for supervisory and investigation. In order to better serve the needs of land managers and scientists, many gridded weather, climate and remote sensing produces have been established [13]. These products have contributed to expanding scientific understanding and improving early warning systems. Assuming the processing loads of huge data, though, these facts continue to be generally unreachable for a wider segment of people.

Since the 1970s, RS organizations have been gathering huge dimensions of data that are difficult to handle and analyze using standard software and desktop computers. Consequently, Google established a cloud computing stage so-called Google Earth Engine (GEE) to discourse the issues of analyzing large document sets. The ability to process enormous amounts of geo data across wide regions and maintain a continuous watch over the environment is greatly facilitated by this platform [14]. Although this platform has been available since 2010, its potential uses have just recently been explored, and it is only now being used for RS applications. Many other disciplines have also found useful applications for GEE, including Land-living Shield/Land-living Usage categorization, hydrology, city planning, analysis of usual disasters and environment, and image handling. On the whole, it was practical that the quantity of GEE periodicals has increased suggestively over the earlier few centuries besides it anticipated more people from many sectors would use GEE to address their issues with processing big data.

Publicly available satellite and climate data can be utilized to generate maps and charts instantly with the help of Google Earth Engine and the free web tool Climate Engine (ClimateEngine.org). Simply open it in your preferred web browser to get started. In addition to climatic data like temperature and precipitation, Climate Engine also provides access to geospatial data on plants, snow, and water from all over the world. Users are spared the hassle of downloading and storing massive data files on their local machines by having access to and working with datasets hosted on the cloud. Climate Engine was developed by researchers from the Desert Research Institute (DRI), the University of Idaho, and Google.

With Google Earth Engine, Google's similar cloud computing platform, users may quickly and easily analyze, visualize, transfer, and share environment and remote detecting datasets. To illustrate how cloud computing could be used to handle massive data rapidly and well, we briefly outline the development and architecture of the Climate Engine software application. Secondly, many case studies are provided to demonstrate the versatility and speed with which Climate Engine can be used to conduct a variety of climate-related studies and develop a variety of climate-related applications, including those pertaining to dearth, intensity, ecology, and cultivation [10]. Archives of weather types and remote recognizing data are now reachable through on-demand parallel cloud computing. Many previously unavailable avenues for keeping tabs on and learning from our planet's natural resources have been made possible by this.

6.10 USAGE OF CLOUD COMPUTING IN MEDICAL AND HEALTHCARE

The term "cloud-based healthcare" is the practice of incorporating cloud computing technology in healthcare services for the purpose of lowering healthcare costs, facilitating data sharing, providing personalized treatment, facilitating the use of tele-health applications, and other advantages. When

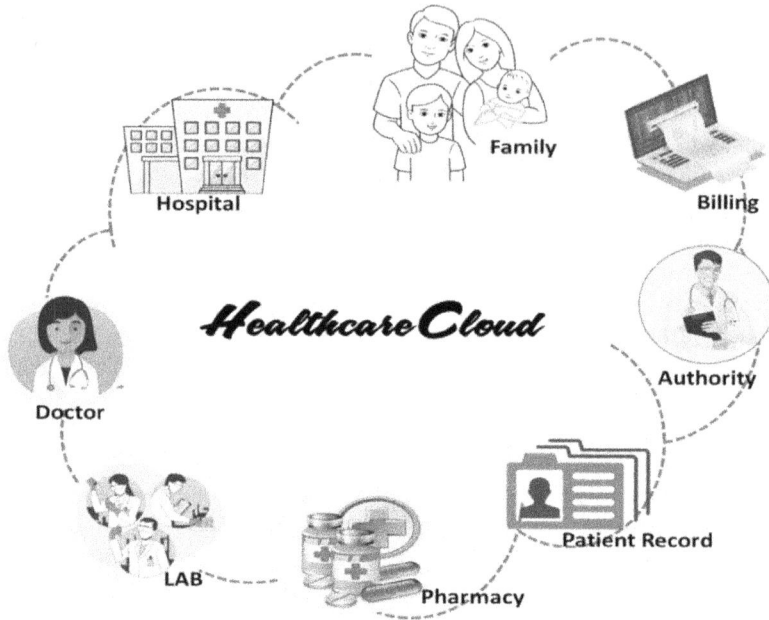

FIGURE 6.8 An illustration of healthcare using cloud.

it comes to patient records, many hospitals and clinics now utilize cloud-based systems for backups, archiving, and easy access to electronic files. Figure 6.8 represents the use of cloud computing in healthcare for different purposes to store health data. The tenure "cloud computing" discusses the use of Internet-connected remote servers for storing, managing, and processing healthcare-related data [5]. As opposed to setting up a local data center with servers or keeping the information on a single computer, this option allows for greater flexibility and scalability.

In order to collection massive volumes of data in a harmless and reliable atmosphere accomplished by IT experts, cloud storage provides a versatile answer that can be leveraged by healthcare professionals and hospitals.

The Electronic Medical Records (EMR) Mandate has resulted in healthcare providers all over the United States adopting cloud-based healthcare solutions for the storage and security of patient records. BCC research projects that the worldwide healthcare cloud computing market will expand from its current $11.6 billion in 2017 to $35 billion in 2022. Despite this, 69% of 2018 survey respondents reported that their hospital did not have a plan for migrating current data centers to the cloud [5–7, 15–27]. In healthcare, cloud computing mitigates the aforementioned concerns. Additionally, they do the following:

- **Make data operations simpler:** Healthcare businesses rely on data, including records of patients' appointments, diagnoses, treatments, insurance coverage, and financial transactions, which are essential to the practice's smooth functioning. It is challenging to manually structure and process. Making mistakes has real, material consequences in every industry. Cloud-based analytics solutions have made work more readily available, creative, and insightful. AI and machine learning abilities have a substantial impact on enduring care and operative efficiency. As additional cloud platforms include AI and ML keen on their offerings, healthcare organizations may take advantage of cloud computing to better manage their massive data sets.
- **Help items reach the market quickly:** The experience of endemic clearly determines this benefit. Throughout 2020, Spoedtest corona has to meet a very tight deadline for the delivery

of a cross-platform web app while aiming to provide safe and inexpensive testing chains in the Netherlands and Belgium. The software would reduce face-to-face contact and make testing secure for patients and medical staff [15]. By deciding to use Amazon Web Application services, they were able to launch the app inside the allotted binary week time surround and observe rigorous obedience standards aimed at handling delicate personal besides health information.

- **Increasing access to healthcare services:** Telemedicine is a current healthtech trend that enables obtaining medical care remotely. It has become a vital resource in the fight against the COVID pandemic. Remote consultations allow those who reside in rural areas or who are disabled to avoid making long trips to urban areas to get medical care. The ability to maintain constant connection with a health specialist is unique and one of the most important benefits of cloud computing in healthcare, especially in the wake of surgery or an operation. One may extend this logic to many other situations, such as getting a timely reminder to take your medication. With the help of wearable technology connected to the cloud, doctors may even be able to monitor their patients' health without ever leaving their own homes. This one particular function benefits medical professionals in making snap judgments and so saving patients' lives.
- **Improve how clinics and patients talk to each other:** Healthcare cloud computing assistance surgeons and other medical supervisers give patients the best facilities and messages possible. They are able to get the results of tests and analyze them on online, keep track of how the treatment is going, and be notified of any changes. Also, it adds a new level of safety and keeps people from getting too many prescriptions because doctors can look at their medical records.
- **Improve the way doctors work together**: Loading data in the cloud means that authorities from dissimilar corporations can admission it at any time without having to talk to each other or the patients. Because of this, the diagnosis is more likely to be right, and the chance of getting it wrong goes down.

6.11   EDGE-DRIVEN INTELLIGENCE FOR WEARABLE DEVICE

Wearable gadgets are a rapidly evolving technology that will have an influence on individual healthcare for both society and economy. Because of their unique ability to "wear while using", smart wearable devices are becoming increasingly popular in our daily lives [7]. However, the demands of bright load and compressed size result in a scarcity of device incomes in the best wearable goods, impeding the advancement of wearable technology. Edge computing allows wearable devices to access more resources while remaining within weight and size limits. Due to the well-known deployment of instruments in ubiquitous and dispersed networks, future smart wearable devices will need to excel in areas such as low power consumption, fast computation, and adaptable architecture. Researchers have begun to envisage and forecast how computing can be carried to the edge through smart sensors with the end objective of providing adaptive extreme edge computing. We describe in depth the hardware and theoretical solutions for smart wearable devices in order to guide future research in this age of ubiquitous computing. In the arena of neuromorphic computing, which is used in implantable sensors, we offer a number of models that are medically plausible and may be used for continuous learning. In order to demonstrate this point, we give a detailed analysis of potential low influence and low latency circumstances of wearable instruments in neuromorphic devices. Using CMOS and other memory technologies, we investigate promising future directions for neuromorphic computers (e.g., memristive devices). We also assess the necessary size, power, latency, and data capacity for edge computing in wearable procedures. Outside of neuromorphic computing hardware, algorithms and policies, we also explore possible obstacles to adaptive edge computing in clever wearable devices.

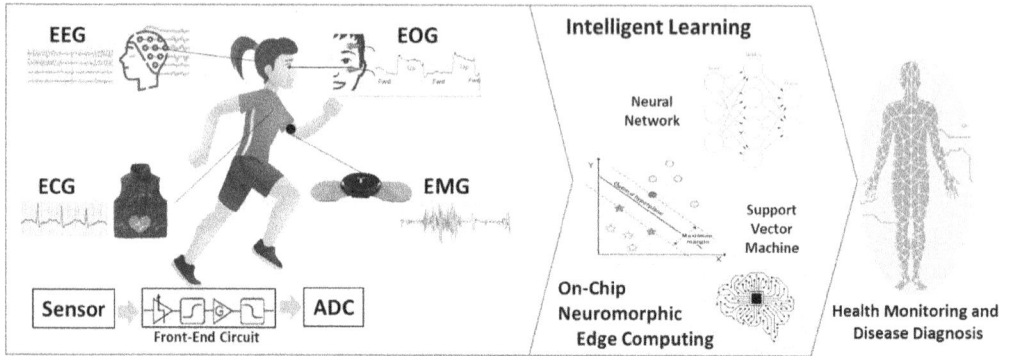

FIGURE 6.9 Idea of adaptive edge computing in the context of a wearable health device.

Figure 6.9 shows the most usable wearable device and its functionalities. Wearable sensors are able to track vitals like heart rate, breathing, movement, and brain activity in humans. Miniaturized devices equipped with a variety of sensors may monitor, predict, and analyze the human body's corporeal performance, biological status, biochemical conformation, and emotional attentiveness [16]. Modern wearable devices face a variety of challenges, such as limited processing power, high power consumption, a vast quantity of facts that need to be sent and a slow data transmission speed, regardless of advances in new resources that can recover sensor determination and understanding. Data collected by wearable sensing expedients is typically transmitted to remote servers aimed at off-chip computation and dealing out. The information bottleneck that comes from this approach is a major hurdle in the quest to reduce power consumption and increase the speed with which sensing devices function. In addition, processing this temporal real-time sensing data using conservative inaccessible servers and conservative signal handling procedures is computationally demanding, leading to high power ingesting and physical component livelihood [6]. In this case, the advantages of the edge computing prototype, which normally contains complete grids in which processing knot is not in the cloud, have been apparent. When the processing unit is physically close to the sensing unit, the system uses less energy to do both. In order to be considered "extreme edge computing" capable, a system must perform data processing in close proximity to the sensor, on the same device.

Recent years have seen the publication of multiple surveys concerning edge computing and wearable gadgets. There was an in-depth analysis of wearables with regards to processing power, energy savings, and communication safety. However, the authors neglected to include the usage of MEC in wearable applications and instead focused on the few works that integrate MCC with wearables. In healthcare, wireless body area networks were the subject of a recent poll. For this wearable-based healthcare system, which is essentially an edge computing system, the authors envisioned a four-layer architecture. However, the bulk of their evaluation is devoted to exploring and contrasting the remaining low-power message equipment, aimed approriatelyt at cutting-edge wearables in a domestic context. Offloading techniques and edge computing architecture have been discussed. Studying edge computing was done (Figure 6.10). This article provides a synopsis of the studies that have been conducted on edge computing for wearables thus far [19]. Regarding frameworks, advantages, uses, state-of-the-art (SOTA) study, applications, safety, and so on. However, the aforementioned publications don't discuss the overlap between wearables and edge computing; rather, they each focus on one or the other. We feel that a review of works combining wearables with edge computing would be of substantial interest to many academics, given the enormous potential for merging these two technologies.

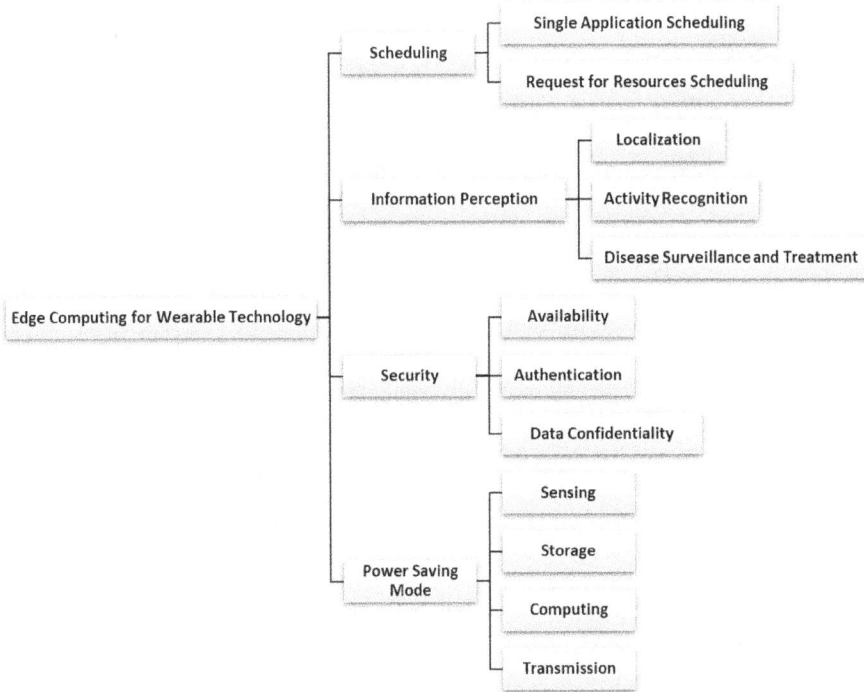

FIGURE 6.10 Overview of existing properties of wearable technology.

This is the first thorough introduction to edge computing for wearables. Arrangement, fact perception, energy savings, and safekeeping are the four major areas of study in which we have divided the existing literature on edge computing for wearables (see Figure 6.10).

6.11.1 SOME MAJOR WEARABLE DEVICE

Higher levels of complexity and accuracy in sensory units are required to characterize physical things, making them essential for use in intelligent applications and high-performance systems. This allows for the algorithm to return more precise findings. In Figure 6.9, we see six different sensors, four biopotential sensors and two common wearable sensors arranged in order of where their signals are acquired. Sensors with direct electrodes may read biopotential signals produced by the body [20]. Ionic currents are produced by electrochemical action of cells in nervy, muscle, and glandular flesh. Ionic current is transformed into electric current via an electrode-electrolyte transducer in the input stage of the circuit. In order to release metal ions and electrons, electrolytes can oxidize metal electrodes. In the same way that they can oxidize anions to neutral atoms and free electrons, electrolytes can do the same with anions. Electrode current is generated by the movement of free electrons. The electrochemical surface potential of the cell can be measured by the electrode. Biosignals picked up by an electrode are weak and distorted by background noise. In order to convert analog signals into digital ones through ADC, an analog frontend is required. Biopotential electrode frontend design needs to meet the following criteria. (i) In elevation communal mode refusal ratio; (ii) strong signal-to-noise ratio; (iii) low authority ingesting; (iv) sign purifying; and (v) programmable improvement.

- **Electrocardiography (ECG)**

  An electrocardiogram (ECG) sensor records the electrical impulses generated by the electrochemistry of the heart's tissues. Due to its morphological or numerical qualities, ECG is responsible for comprehensive data aimed at investigating and establishing cardiovascular syndromes. Ordinary, ventricular, supraventricular, a union of normal and ventricular and unknown beats are the five types of ECGs of interest, as defined by the connotation aimed at the advancement of medical instrumentation. Methods producing > 90% correctness and kindliness for the five classes when tested against the existing ECG database are necessary for the future of cardiovascular health observing [21–26]. In the context of wearable technology, Hossain and Muhammad (2016) and Yang et al. (2016) offer methods that quantify ECG and communicate it to the cloud for grouping and fitness observing. Self-diagnosis of common cardiovascular issues like atrial fibrillation is also possible with the help of ECG sensors integrated into some commercially obtainable products, like the Apple watch (Apple Inc.) [26].

- **Electrooculography (EOG)**

  Emotional and environmental factors combine to influence eye movement, which can result in peripheral eye modifications known as EOG [23]. It generates a feeble electrical energy (0.01–0.1 mV) and a little regularity (0.1–10 Hz). While traditional eye tracking uses a cinematographic camera and electromagnetic technology, EOG offers a wearable, low-cost alternative [25]. The wearable HMI is the most widely used, especially for assisting those with quadriplegia. EEG and EOG provide analogous understood evidence, such as fatigue and intellectual strength, therefore subsequent studies have merged them to grow the grade of signal liberty and improve organization consistency [26]. Furthermore, EOG can be used as a complement to an EEG system to provide features or instructions.

6.12 CONCLUSION

In the summary of this chapter, we conclude that nowadays Information and Communications Technology (ICT) plays a vital role in all fields especially in the arena of computer science. In first part of this chapter, we contrast several cloud computing services, including those based in the cloud, on the edge, and in the fog, and give a comparative analysis of the fog, edge, and cloud computing. It's quite clear that cloud computing is a paradigm shift that can help businesses make the most of their IT budgets. Many established and thoroughly investigated ideas form its foundation. This chapter also concentrates on many trending fields of cloud computing, such as climate analysis by using GEE and Climate Engine and healthcare. The ultimate goal of entrenching sensors, a frontend track edge, a neuromorphic workstation, and memristive procedures into a wearable device for smart sensing and edge computing is dependent on a custom platform. A detailed explanation of some wearable devices which use cloud computing like ECG and EOG is given.

REFERENCES

[1]   Bhisikar, A. Singh J., "Innovative ICT through Cloud Computing". ResearchGate, The IUP Journal of Computer Sciences, 7, 37–52, 2015.

[2]   Shi, W., Cao, J., Zhang, Q., Li, Y. , Xu, L., "Edge Computing: Vision and Challenges" Internet Things IEEE, 3(5), 637–646, 2018. doi: 10.1109/JIOT.2016.2579198

[3]   Mach, P., Becvar, Z., "Mobile Edge Computing: A Survey on Architecture and Computation Offloading". IEEE Communications Surveys & Tutorials, 19(3), 1628–1656, 2017. doi: 10.1109/COMST.2017.2682318

[4]   Paul, P. K., Aremu, B., "A Study on Cloud Computing and Service Market: International Context with Reference to India". Asian Journal of Managerial Science, 9(1), 52–56, 2020.

[5]   Ben Ali, A. J., Hashemifar, Z. S., Dantu, K., "Edge-SLAM: Edge-Assisted Visual Simultaneous Localization and Mapping", ACM MobiSys. 22(1), 1–32, 2020.

[6]   Shilpa., H. K., "Analyzing the Bank Scam's Financial Fraud and Its Technological Repercussions Using Data Mining", Second International Conference on Electronics and Renewable Systems (ICEARS), Tuticorin, India, 1553–1559, 2023, doi: 10.1109/ICEARS56392.2023.10085354

[7]   de Quadros, T., Lazzaretti, A. E., Schneider F. K., "A Movement Decomposition and Machine Learning-Based Fall Detection System Using Wrist Wearable Device". IEEE Sensors Journal, 18(12), 5082–5089, 2020.

[8]   Cisco, "Cisco Global Cloud Index: Forecast and Methodology 2016–2021" 2018.

[9]   Montes, D., Añel, J. A., Wallom, D. C. H., Uhe, P., Caderno, P. V., Pena, T. F., "Cloud Computing for Climate Modelling: Evaluation, Challenges and Benefits". Springer, 9(2), 52, 2020. https://doi.org/10.3390/computers9020052

[10] Yogeesh, N. "Mathematical Approach to Representation of Locations Using K-Means Clustering Algorithm". International Journal of Mathematics and Its Applications, 9(1), 127–136. 2021. Retrieved from http://ijmaa.in/index.php/ijmaa/article/view/110

[11] Hassan, S., El-Shirbeny, M. A., "Cloud Computing for Water and Climate Big Data Analysis". ResearchGate, 2(3), 1–5, 2022.

[12] De La Prieta, F., Corchado, J. M., "Cloud Computing and Multiagent Systems, a Promising Relationship". Springer, 14, 143–161, Cham, 2016.

[13] Ghamari, M., Janko, B., Sherratt, R. S., Harwin, W., Piechockic, R., Soltanpur, C. "A Survey on Wireless Body Area Networks for eHealthcare Systems in Residential Environments, Sensors", IEEE, 16(6), 831, 2019. https://doi.org/10.3390/s16060831

[14] Sultan, N. "Making Use of Cloud Computing for Healthcare Provision: Opportunities and Challenges", Science Direct, 34(2), 177–184, 2014.

[15] Covi, E., Donati, E., Liang, X., "Adaptive Extreme Edge Computing for Wearable Devices". Frontiers in Neuroscience, 15, 611300, 2021. doi: 10.3389/fnins.2021.611300

[16] Mauldin, T. R., Canby, M. E., Metsis, V., Ngu, A. H. H., Rivera, C. C., "SmartFall: A Smartwatch-Based Fall Detection System Using Deep Learning". Sensors, 18(10), 3363, 2018.

[17] Abbas, N., Zhang, Y., Taherkordi, A., Skeie, T., "Mobile Edge Computing: A Survey", IEEE, 5(1), 450–456, 2018.

[18] Dinh, H. T., Lee, C., Niyato, D., Wang, P., "A Survey of Mobile Cloud Computing: Architecture, Applications, and Approaches", Wireless Communications and Mobile Computing, 13, 1587–1611, 2013.

[19] Yazicioglu, R. F., Van Hoof, C., Puers, R., "Biopotential Readout Circuits for Portable Acquisition ystems". Springer Science & Business Media, 2008.

[20] Isakadze, N., Martin, S. S., "How Useful Is the Smartwatch ECG?" Trends in Cardiovascular Medicine, 30(7), 442–448, 2021.

[21] Yogeesh, N., "Graphical Representation of Mathematical Equations Using Open Source Software". Journal of Advances and Scholarly Researches in Allied Education, 16(5), 2204–2209, 2019, www.ignited.in/p/304820.

[22] Thakor, N. V., Biopotentials and Electrophysiology Measurements, in Telehealth and Mobile Health. 1st edition, 1–19. CRC Press, 2015.

[23] Duchowski, A., Eye Tracking Methodology Theory and Practice. Springer, Cham, 2007.

[24] William, P., Yogeesh, N., Vimala, S., Gite, P., "Blockchain Technology for Data Privacy Using Contract Mechanism for 5G Networks". 3rd International Conference on Intelligent Engineering and Management (ICIEM), 461–465, 2022. doi: 10.1109/ICIEM54221.2022.9853118

[25] Girija, D. K., Shashidhara, M. S., Giri, M. "Data Mining Approach for Prediction of Fibroid Disease Using Neural Networks". International Conference on Emerging Trends in Communication, Control, Signal Processing and Computing Applications (C2SPCA), Bangalore, India, 1–5, 2013, doi: 10.1109/C2SPCA.2013.6749370

[26] Mishra, S., Choubey, S., Choubey, A., Yogeesh, N., Durga Prasad Rao, J., William, P., "Data Extraction Approach Using Natural Language Processing for Sentiment Analysis", International Conference

on Automation, Computing and Renewable Systems (ICACRS), Pudukkottai, India, 970–972, 2022, doi: 10.1109s/ICACRS55517.2022.10029216

[27]   Girija, D. K., Varshney, M., "A Comparative Analysis of the Performance of Multiple Data Mining Classification Approaches Using the Kn Fold Validation", Journal of Pharmaceutical Research International, 34(11B), 25–32, 2022.

7 Explainable Artificial Intelligence (XAI) for Computational Sustainability

Concepts, Opportunities, Challenges, and Future Directions

B. Prabadevi[1,2], M. Pradeepa[1], and S. Kumaraperumal[3]
[1] School of Computer Science Engineering and Information Systems, Vellore Institute of Technology, Vellore, India
[2] Research Fellow, INTI International University, Malaysia
[3] Master of Management Studies (MMS), St. John College of Engineering & Management, Palghar, Maharashtra, India

7.1 INTRODUCTION TO XAI

Artificial intelligence models have been widely adopted in all sectors. Still, their adoption has been tapering due to the transparency concerns required in mission-critical applications or applications relying on sensitive information. In most cases, artificially intelligent models remain a "black box" where even the programmers do not know how the final decision is attained. Various stakeholders of these systems, like programmers, users, and decision-makers, are puzzled at the prediction results and are concerned about the sensitive information fed to them. The systems must enforce transparency in decision-making to enhance stakeholders' trust and guarantee faster adoption of artificially intelligent systems. The evolution of XAI has changed the prevailing trustworthiness concerns in adopting the AI models.

XAI is decisive for an organization to build confidence and self-reliance when implementing AI models. More explainable AI helps organizations to adopt a liable approach to AI expansion. XAI is a new technology encompassing various processes and approaches that allow human users and research practitioners to comprehend better and trust AI algorithms' outcomes and experimental results. Though explainability and interpretability are often used interchangeably while conversing about AI, they differ technically. The later interpretable AI models interpret human decisions without supplementary information in considerable time. Also, human users can replicate the results with considerable data and time. Whereas the former, an explainable model, is intricate, requiring humans to lay extra effort into understanding and comprehending the decisions arrived. Sometimes it may require an analogy to help humans understand the model's interpretations and the results. But, explainable models cannot be replicated even with infinite time and enormous data.

7.1.1 XAI, How It Differs from AI

Machine learning (ML) algorithms are often alleged as black boxes that are difficult to deduce by what means the results have arrived. Some ML models, like logistic regression and decision trees, are transparent and easier to interpret as they derive conclusions based on the relationship between input features and target value [1]. However, this is not the case with all applications; ML algorithms

DOI: 10.1201/9781003407959-7

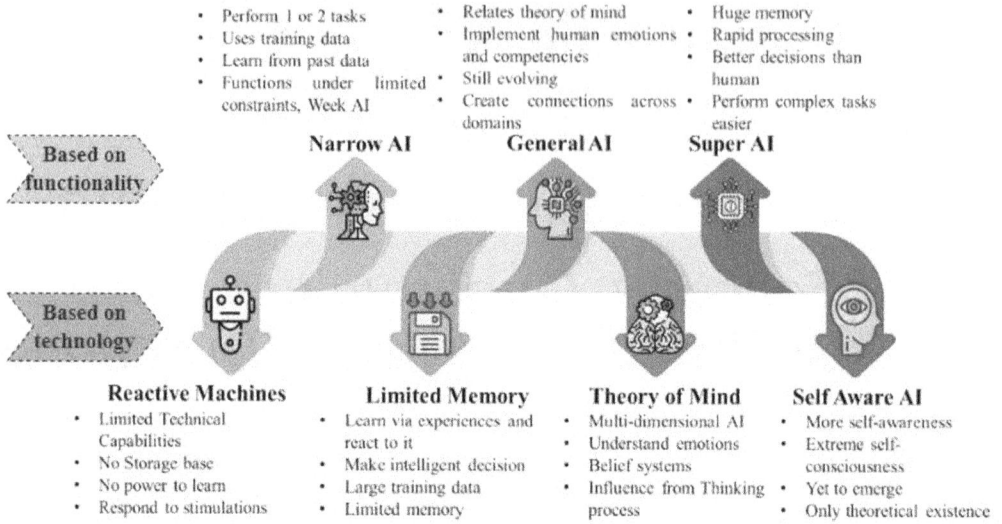

FIGURE 7.1 Different categories of artificial intelligence.

are too complex to interpret. The replica of the human brain and the neural networks applied in deep learning mimic the human brain's functionality, which is very hard for a human to comprehend. The artificially intelligent systems can predict and anticipate future happenings based on the input data analysis for modeling some real-world scenarios. AI has been classified differently based on technology and its functionality, as shown in Figure 7.1. The different categories of AI describe the way it has evolved from reactive machines, which cannot learn by themselves but only responds as it has no memory, to super AI, which can perform complex tasks in a shorter period possessing huge memory capacity and can earn and react based on past experiences. The other forms are AI that is in evolution are one with limited memory capacity to get itself trained, theory of mind (which can understand different emotions as humans, is yet to evolve), self-aware AI (more self-consciousness and self-awareness, theoretically exists), narrow AI (most of today's AI models), and general AI (makes more accurate decisions, still evolving) [2].

Biased data, such as data related to or focussed on the human race, age, gender, or location, has been a long-lasting risk associated with training AI models. Certainly, AI model performance can degrade or show deviated results because production data contrasts with training data. Also, the General Data Protection Regulation (GDPR) act of the European Union provides guidelines on the usage, processing, and storing of personally identifiable information. It assures data security and privacy of the individual's data [3].

AI is the superset of ML, and deep learning algorithms, if harnessed appropriately, will produce more accurate results. Advancement in AI algorithms has led to its extensive adoption in almost all domains and business organizations, as AI can improve efficiency in decision-making and nurtures predictive maintenance. Furthermore, leveraging AI can make error-free decisions progressively through results-based comprehension. AI has been changing today's world since its inception. AI's most advanced and powerful variant generative pre-trained transformer (GPT), has harnessed most of the applications through its multimodal learning capability, trained through numerous data from the internet. GPT can perceive data from internet sources, learn from human feedback, and generate automated text results for human queries [4]. As AI progressively advances daily, it has been a greater challenge for human users, practitioners, and data scientists to interpret and comprehend the internal logic of AI algorithms in deriving the results. The internal computation involved in deriving the specific results remains a "black box" for the developers or the creators of the models. The AI model's "black-box" nature remains an obstacle to its far-reaching and trustworthy adoption in data-sensitive decision-making applications like healthcare analysis, financial transactions, and other applications utilizing personally identifiable information. With its advent, XAI has evolved as a crucial feature reliable deployment of AI. With its explainability, XAI helps humans and other practitioners fully understand the AI models' decisions and manage artificially intelligent systems more effectively [5]. Thus, it promotes end-user trust and productive utilization of AI for a better impact on business adoption. Furthermore, it alleviates AI adoption's legal, compliance, reputational, and ethical issues.

The key implementation requirement of responsible AI is XAI, which aids in the robust implementation of AI in large-scale organizations with a high level of transparency, fairness, accountability, and model explainability. To implement these responsible AI systems in organizations, organizations must embed ethical principles related to data trust in the construction process of AI systems. XAI attracts the AI researcher community and practitioners with characteristics such as resilience, accountability, transparency, and lack of machine bias. It has witnessed greater attention in finance, healthcare, criminal justice, and human resources management [6]. XAI helps the users understand and comprehend the specific output attained and assures the developers on expectance from the system's performance, by meeting all legal and regulatory requirements. This promotes the widespread adoption of AI systems.

7.1.1.1 Model Interpretability

As all data science, machine learning, and AI applications are considered black boxes due to the bias maintained in the data and are not transparent in the decisions arrived. In addition to the developer, an intuitive understanding of these models' pipelines must be known to humans to get better insight into the decisions, but this is practically difficult. So, the model's interpretability should be provided understandably for any type of user concerned. XAI's model interpretation helps to provide detailed steps and final decisions an AI model takes while making predictions. The black boxes of AI models can be understood differently, helping humans know the answers to "Which attributes are important for the model prediction?" and "Why has the model arrived at this decision?". The interpretation results may be like feature summary statistics or visualization, the model's intrinsic behavior, data point, or a combination of these, based on the type of interpretation method. XAI's model interpretability taxonomy is given in Figure 7.2. It is classified based on the level of interpretability and content held by the model to be interpreted as follows [7]:

Based on the complexity of models: This category of machine learning models is based on the model type. Some models may exhibit a simple structure, so interpretations can be directly derived from their intrinsic behavior, like decision trees or linear regression, called intrinsic model

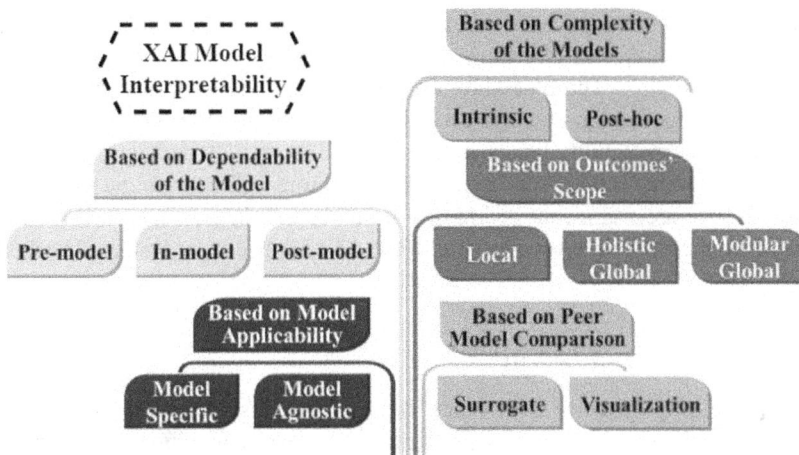

FIGURE 7.2 XAI model interpretability.

interpretability. On the other hand, some models may be complex; their interpretability and possible explanations could be arrived at by employing some interpretation methods after the model is trained. Such an approach is termed a post hoc model interpretability

Based on dependability of the model: This category is based on the nature of the model, classified as pre-model, in-model, or post-model. Pre-model interpretability methods are independent or generic and can be employed on any model, such as principle component analysis. Some model interpretability methods with in-built interpretability integrated into the model's architecture are called in-model. Some methods are employed after the model is developed, focussing on the interpretability of what happened during model training, i.e., the knowledge acquired by the model during training.

Based on the outcome's scope: These approaches are about individual prediction or the full model behavior is concerned. These are named local model interpretability and global model interpretability. The local interpretability is for the whereabouts of every single prediction or group of similar predictions, i.e., for a single input data given to the model. This approach explains why the particular decision has arrived and how these may affect the overall outcome. Thus providing a detailed and more accurate local explanation. In contrast, global interpretability is employed in the trained model and provides explanations based on a holistic view of the features taken, internal machinery (learned components), and the model's structure. Global model interpretability details the feature selection, relationship among the features, and predicted outcome with feature contribution. Global interpretability comes in two flavors: one focuses on the whole model level, whereas the other focuses on the modular level of a model. The former, holistic global model interpretability is practically difficult to implement as it concerns all features, their combinations, and outcomes incurring more storage constraints. The latter one, modular global model interpretability, helps to understand models at some modular level. But only some models can be inspected at modular levels.

Based on model applicability: These approaches are classified as model-specific and model-agnostic. As the names indicate, the model-specific approach is specifically employed or applied to a single model or particular group of models, concerned about the inner workings and machinery of the specific model used to make the final decision. Some of the examples are regression weight interpretation in linear regression and bias, or weights in neural networks. It has a different evaluation framework or criteria to determine a specific model's interpretability. Whereas model-agnostic approaches use the same evaluation framework to analyze any models and are functional after the model is trained. This approach functions by analyzing the attribute's input-output pairs but doesn't access the internal structural machinery like the weights of a model. So, it guarantees consistency when comparing different models.

Based on peer model comparison: These methods are surrogate and visualization methods. Surrogate methods are composed of different ensemble models, which can be used for comparing any other black-box models. The black-box models can be well interpreted through these comparison results. The visualization methods do not approach different visualization techniques used to view the model's interpretability like heat maps.

These model interpretability approaches are derived through different logical intuitions and are not exclusive, as there are significant overlaps. For an instant, the post hoc method overlaps with the model-agnostic method. So the explainability of AI models can be further classified based on attributions [8].

7.1.2 Generic Applications of XAI

XAI has widespread employability in domains like computer vision, natural language processing, and time-series predictions encompassing applications: image recognition, video analysis, visual query response, video analysis, speech processing, sentiment analysis, and text classification.

Furthermore, XAI has been adopted in various applications and analyses of medical, biomedical, and scientific reports like human gait analysis, histopathology, automobiles, judicial systems, insurance, medical image analysis, radiology, toxicology, therapy prediction, meteorology, and so [9]. In the recent era of AI, it has shown dramatic improvements in medicine, but due to its black-box nature, AI applications in healthcare trials were restricted. XAI, with its capability of generating understandable human terms for explaining AI's decision-making, has alleviated this issue and promoted the adoption of AI in medical applications more effectively [10,11]. Some of the common usage requirements of user groups on AI explanations are for improving the model through debugging, assessing models' capability in other domains, devising regulatory compliance on transparency and trust, inculcating appropriate trust in models for making informed decisions, and seeking AI in life-changing factors.

7.1.3 Need for XAI in Sustainable Computing

Computational sustainability is an interdisciplinary approach amalgamating computer science, information science, operations research, statistics, and applied mathematics to promote sustainable development by harmonizing ecological, financial, and communal needs. Computational sustainability brings together experts like computer scientists, data scientists, biologists, operations researchers, economists, statisticians, policymakers, environmental scientists, and more to provide sustainability. Computationally intelligent models are used in the complex real-world decision-making process. AI has practical applicability for promoting computational sustainability. Humanitarian society is still facing challenges in the economy and natural calamities, and the livelihood of future generations is also endangered due to the depletion of Earthern resources. Though AI and other computing technologies have bought many groundbreaking transformations [12], the impact of these technologies is uneven. Most profitable sectors are only beneficial, whereas minimal benefits are foreseen in the environmental and societal sectors, exacerbating the inequalities. So, the major focus of computational sustainability is utilizing computer technology advancements to mitigate environmental and societal challenges, thereby assuring a sustainable livelihood for future generations and protecting the green planet [13]. In pursuing a sustainable future, computational sustainability combines the various efforts of assorted computer scientists to balance the needs of the environment, society, and economy. Computational sustainability is an interdisciplinary setting, enabling collaboration among a large group of computer technologists with an even larger group of domain experts from natural, social, and environmental sciences. As the 1987 United Nations World Commission's seminal report records the interconnectedness of societal, environmental, and economic issues toward sustainability, it was suggested that sustainable development must aim at delivering the needs of today with the ability of forthcoming generations to meet them. Most of the computational research today focuses on sustainability and computational themes. The former includes balancing socioeconomic needs with the natural environment, sustainable – renewable energy and materials, transportation, conservation, and biodiversity. Later one includes data and prediction learning, policy and action learning, optimized dynamic models with simulation, and multi-agent crowdsourcing [14]. Fisher reviewed various research on advances in AI for computational sustainability, focusing on the aforementioned themes [15]. In addition, he has also discussed the third dimension concentrating on the decision-making process. It was observed from the study that most of the AI models were built from the data without interventions to the applications, some focused on existing transportation designs with interventions, whereas others focused on open designs for new road transportation networks. The general application areas of computational sustainability include mitigating poverty, wildlife conservation, biodiversity, transportation, renewable energy resources management, and material science. Few AI models towards computational sustainability were focused on the decision-making of the natural environment, conservation planning, or protection for wildlife species, classification of bird species, green security games used for protecting aircraft, fluctuating electricity pricing, traffic control in urban settings, reactive interventions of autonomous vehicles and socioeconomic maps creation from varied sources [15].

These models should be computationally sustainable, and the results are interpretable and provide confidence about the decision made. Therefore, decision-making in this computational sustainability should include varied data from integrating sources and sensitive information like asset maintenance in an automated environment [16]. Furthermore, some AI models are non-linear in nature and very complex. They may take high-dimensional input, which makes it more tedious for the data scientists, engineers, and the human to understand the reason for the arrived solution. Transparency of data acquired from the integrating sources will entrust the stakeholders with faster model adoption. Therefore, XAI will be a mandate option for models focussing on computational sustainability as it amalgamates data from three amigos and should provide a balanced solution. This will provide greater insight for the policymakers to trust the sustainable decisions made by the AI models. XAI models take heterogeneous data as input and make detailed interpretations or explanations about the decision summary. The results of the interpretation method may be summary statistics of each feature, visualization of the summary feature, or learned weights or data points. Based on the type of model and the inputs, interpretability is chosen. Hence, XAI with computational sustainability will provide effective decision-making models with more data processing details. XAI can be used as a service to render more trustworthy artificially intelligent systems for computational sustainability.

7.1.4 ORGANIZATION OF THE CHAPTER

This chapter analyzes the utilization of XAI for computational sustainability. Section 7.1 discusses the introduction to XAI, how XAI differs from AI, model interpretability, generic applications of XAI, and the need for XAI in sustainable computing. Section 7.2 discusses the various XAI models for handling heterogeneous data, black-box models, and existing XAI methods. Section 7.3 addresses the utilization of XAI as a service, XAI for decision support in computational sustainability, challenges in adoption, and Section 7.4 discussed beyond XAI, challenges and the future research directions in adopting XAI for computational sustainability. Section 7.5 concludes the XAI for sustainable computation.

7.2 XAI MODELS FOR HANDLING HETEROGENEOUS DATA

In the current digital era, a huge amount of data (text, image, audio, video, etc.) is in existence and continuously generated by various sources. These data have to be analyzed and processed for various purposes. Handling huge amounts of data needs efficient techniques to meet the computational requirement and provide explanations for the decision made on those data.

Pre-modeling explainability refers to a sequence of data processing methods, which is applied to gain an understanding of the datasets collected and further used to train the machine learning models [17]. Some of the pre-modeling tools are data analysis, transformations, data summarization, and data squashing.

Data analysis: The data analysis is performed on various statistical information of a dataset, such as dimensionality, mean, standard deviation, range, and missing samples, etc. [18].

Transformations: In the current scenario, the number of datasets is available and published without adequate information, so it is challenging for users to analyze these data. Interactions do the transformation of this data with the creators. This process is one of the solutions to the problem such as data bias, misuse of the data, etc. [19]. Various transformation methods are proposed by researchers, such as data statements, datasheets, nutrition labels, etc. [20]. Each approach provides different solutions to define data formation, data handling, and data collection, legal and ethical consideration.

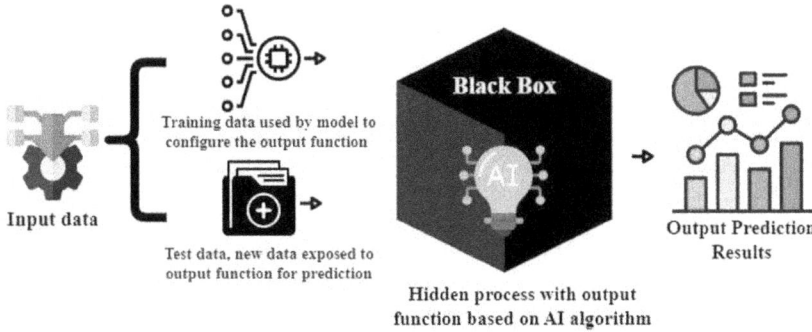

FIGURE 7.3 Black-box model.

Data summarization: Data summarization is a process to find a minimal subset from the huge original dataset, which are a representative sample of the entire dataset. Various techniques are used for data summarization, such as K-medoid clustering, K-means clustering [21], and Bayesian-based teaching model, etc. [22].

Data squashing: Data squashing is similar to data summarization, but the samples in the subset are given weights. An example for data squashing is Bayesian corsets specified in the Bayesian learning atmosphere [23].

7.2.1 BLACK-BOX MODELS

In computational intelligence, a black box is a device, system, or object, which produces a useful conclusion without revealing any internal process. The explanations for its conclusions remain opaque. A black-box model, shown in Figure 7.3, accepts inputs and produces outputs but its internal functions are not interpretable.

Advances in computing power, artificial intelligence, machine learning, and deep learning capabilities are triggering the spread of black-box models in many professions. The application of black-box models is continuously increasing in the area of financial markets, business analysis, and other decision-making requirements. Advancement in computational intelligence is providing precise conclusions through black-box models, particularly when the complex analysis is limited to the human brain. For example, deep learning algorithms generate hundreds of thousands of non-linear relationships between inputs and outputs. The complexity of the relationships makes it challenging for a human to explain which specific features led to the output.

The black-box model is undesirable in critical applications for many reasons, because of the unexposed closed internal process of the model. So, the dependability of the output for decision-making is questionable, it is highly challenging to understand whether the AI model conclusion is biased or with errors. And also, it is not easy to pinpoint the attribute which is accountable for the flawed conclusion. The usage of the black-box model may be dangerous in critical applications.

To overcome this significant challenge in the black-box model, the process of the AI system needs to be transparent and interpretable. Converting the black-box AI into explainable AI (XAI) model will enrich the reliability and integrity of the system.

7.2.2 EXISTING XAI MODELS

Understanding the terms interpretability and explainability are important to discuss various XAI methods. In a machine learning model, interpretability means how accurately a cause is associated

TABLE 7.1
XAI Methods

XAI Methods	Explainability	Machine Learning Model
Intrinsic	Interpretable	Linear Regression, Logistic Regression, Generalized Linear Model (GLM) and Generalized Additive Model (GAM), Decision Tree, Decision Rules, Naive Bayes Classifier, K Nearest Neighbor (KNN)
Post hoc	Test Explanation	Neural Networks
	Visual Explanation	Ensemble Methods, Classifier system, SVM, and Neural Network
	Global and Local Explanation	Decision Tree, Rule-Based learners, Neural Networks
	Example-Based Explanation	Decision Tree, KNN, Neural networks
	Explanation by Simplification	Linear Models, Decision Tree, Rule-Based Learners, SVM, Probabilistic Methods
	Feature Interaction	Tree Ensembles, SVM, Neural Networks
	Feature importance/Relevance	Tree Ensembles, Classifier Systems, SVM, Neural Networks

with an effect where, and explainability is used to explain the behavior of the ML model in human-understandable terms.

There are various approaches to explain the closed process of the black box. For different end users, the explanation requirement differs. Based on the end-user requirement and the model applied, the methods of explanation (XAI methods) are classified into two types, intrinsic and post hoc [24].

The intrinsic method of explanation applies to all interpretable machine learning models such as linear regression, logistic regression, generalized linear model (GLM), generalized additive model (GAM), decision tree, decision rules, naive Bayes classifier, K nearest neighbor (KNN), and other interpretable models. Interpretable models' behavior in decision-making can be understood without additional support or techniques. By observing the model parameters, humans can easily understand how the model makes predictions because the interpretable models provide their own explanations.

Some models are not interpretable, so humans cannot easily understand how the model makes predictions. So, additional techniques are required to understand the behavior of the model and how they make predictions. The post hoc methods specify various methods of explainability such as test explanation for neural network, visual explanation for ensemble methods, classifier system, SVM, and neural networks, global and local explanation for the decision tree, rule-based learners, neural networks, explanation by example for decision tree, KNN, and neural networks, explanation by simplification for linear models, decision tree, rule-based learners, SVM, probabilistic methods, Feature interaction for tree ensembles, SVM and, neural networks, feature importance/relevance for tree ensembles, classifier systems, SVM and, neural networks. The XAI methods are given in Table 7.1 [25].

The interpretability and the various models are represented in Figure 7.4 [26].

7.3   XAI AS A SERVICE

XAI as a service implies the development of a deployable XAI tool, namely AutoML, which will make automated decisions with an explanation [27]. This deployable XAI tool is a pipeline connecting the supply chain process from input or raw material to the outcome. These are model-agnostic approaches. Similarly, this AutoML tool can help in generating automated explanation steps like calculating the feature importance, surrogate model construction, plotting partial dependence, etc. The more advanced Auto XAI is further designed to extract collective variables, the semantics of each variable, and statistical formulas involved in generating the collective variable, generating

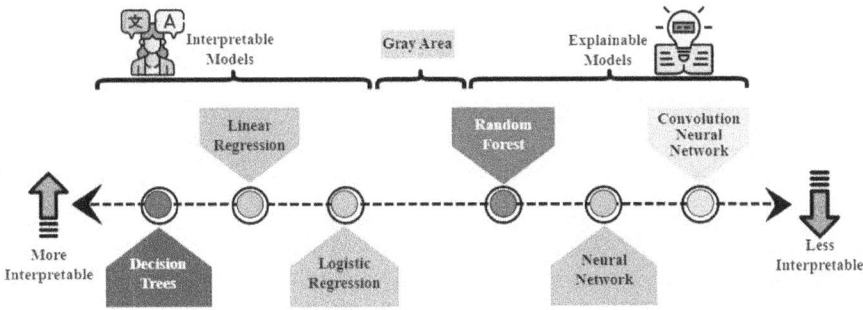

FIGURE 7.4   Interpretability spectrum.

explanations from these variables and formulas. Some other AutoML tools are AutoML H2O [28] and MLJAR AutoML [29], generating automated explanations. In the future, Auto XAI or XAI as a service can be offered by integrating with these AutoML tools. This section provides various decision support services offered by XAI.

7.3.1   XAI for Decision Support System

Decision-making is one of the crucial activities of concern, irrespective of the domain. In most cases, decision-making characterizes the organization's goal and operations and helps make strategic choices. Decision support system (DSS) can afford informed decisions, enhance efficiency in major operations, and allow timely problem-solving decisions. DSS plays a major role in all organizations requiring speedy and informed decisions. An effective DSS can improve control, predict futuristic decisions, and utilize communication technologies, data, models, and documents to complete decision-making tasks. Furthermore, DSS provides a graphical representation of information using artificially intelligent agents replicating the task of knowledge workers. The type of information DSS would present includes information assets (legacy and relational data sources), data figures comparison, projected figures based on assumption, and different decision alternatives based on history. Based on the type of service DSS renders, it is classified as follows:

7.3.1.1   Communication-Driven DSS
These DSS are concerned with smoother collaboration and communication among different internal stakeholders and assist them in making decisions during internal meetings.

7.3.1.2   Data-Driven DSS
These DSS assist mid-level managers in making high-level decisions based on the data retrieved from the databases or other repositories like a warehouse for a specific query.

7.3.1.3   Document-Driven DSS
Typical searches to the web pages using key terms or words specified by the users from heterogeneous documents residing at different sources are assisted by document-driven DSS.

7.3.1.4   Knowledge-Driven DSS
These DSSs serve a crucial purpose in assisting the end users and internal users having a stake in the decisions, covering wide-ranging systems, for making decisions in choosing the product or service based on the knowledge acquired. Most of the top-level managers and consumers will be benefitted from these DSS.

7.3.1.5 Model-Driven DSS

These DSSs are pretty complex, providing decisions by analyzing the different data models related to statistics, scheduling, or financing. Third parties involved in the business, managers, and internal staff of the organization would be benefitted from DSS.

7.3.1.6 Intelligent DSS

Any DSS with an embedded AI component utilizing ML algorithms for processing the results from larger datasets are intelligent DSS. As well, this DSS is preferred in applications requiring human-consultant-like services. Some common uses of intelligent DSS include managers, and medical diagnosticians for identifying trends and patterns in making decisions. Leveraging the power of AI such as case-based reasoning, fuzzy logic, artificial neural networks, evolutionary computing, and intelligent agents with DSS, helps to make quicker and the most effective decisions for any complex problem categorized by a huge amount of data, complex reasoning, and knowledge.

7.3.1.7 Manual DSS

This DSS is purely done by a human without any computer or information system assistance. Though these DSSs are slower than computerized DSS, certain applications like economists still require human intervention or assisted decision in all steps.

7.3.1.8 Hybrid DSS

This DSS is designed by combining one or more aforementioned DSS or parts of them. For instance, the most critical applications like healthcare and finance may require both data-driven and knowledge-driven DSS for making informed decisions. Hybrid DSS may require assisted computation to bring together the functionality of different DSS to work. Most of the clinical DSSs are hybrid.

Apart from this, DSS used for personal decision-making are called personal DSS, DSS used for collaborative decisions involving multiple entities are called group DSS, and those that are specific for organizational decision-making are called organizational DSS [30].

7.4 BEYOND XAI

7.4.1 CHALLENGES

Various challenges faced by the XAI system are as follows:

a. XAI often focuses on explaining solutions that are understandable to the users [31]. XAI assumes that users have a certain level of knowledge of the model and the explanation. However, for many reasons, users will simply lack that expertise and cannot assess the solution provided by the AI model.

b. The main challenge related to the method of design is explainability. Post hoc explanation methods are recommended to approximate the behavior of models. Sometimes, this approximation is unsatisfactory, and the explanation may be unsuccessful. Even when explanations are highly reliable, the method fails to characterize the model behavior under normal conditions. Without cautious design, global and local explanations may activate the artefacts of ML models, rather than provide significant explanations [32].

c. Scalability is another major challenge in explainable models [33]. For example, each instance is explained locally using local explanation by the LIME explainable model [34] and becomes hectic if the model uses many instances. Similarly, when calculating Shapley values [35], all combinations of variables will be used for computing variable contributions. For such cases, the computations are expensive for problems that involve more variables.

d. Explaining AI-based decisions is not a neutral process. Decisions are made based on complex operations performed on the data and applied multiple algorithms by the ML model. While providing the explanation, the methods may provide biased explanations based on certain features. Moreover, no single explanation explains all decisions made by an algorithm.

e. The model is not always static. The model learns from its own decisions by incorporating new data [36]. So, the model is dynamic. If the model is more dynamic, the more challenging XAI will be. Hence the explanation provided by the model will be obsolete tomorrow.

f. If the models are used to solve some sort of wicked problems in which traditional rule-based systems are not suitable and the approach has to search for non-visible patterns [36]. Problem definitions are incomplete for these cases and varying requirements are difficult to recognize. Hence explanation using the general cause-effect relationship is not appropriate for wicked problems [37] and, as such, cannot be used for explaining them.

g. The XAI approach can explain the causality behind an AI-based decision. Reliability of the explanation is obtained by verifying in practice by regular audits and examining the working of the algorithm. Various explanations may be needed for the same algorithms over time. For humans, altering the explanations over time is neither understandable nor acceptable.

7.5   FUTURE RESEARCH DIRECTIONS

This section provides various research directions for enhancing the performance of the XAI models.

7.5.1   ONTOLOGIES FOR EXPLANATIONS

Ontology is a philosophy that can be used for interpreting explanations of entities based on their classification. As the future XAI models must be generic and employable for heterogeneous complex data types, ontologies can be utilized for generating human-understandable explanations for complex data types. Ontologies are used in several model-agnostic XAI and post hoc XAI models, which has improved the quality of the explanations by the domain knowledge specified by ontology. As ontology provides the user's domain conceptualization, new design patterns or methodologies can be developed for ontology-based XAI systems [38]. This will provide qualitative explanations, leading to computationally sustainable systems.

7.5.2   HUMAN OR MACHINE USER-CENTRIC AND DYNAMIC EXPLANATIONS

As discussed earlier, the current XAI provides a single explanation that generically fits all the users at stake. Certainly, the explanations may be to another machine, a target end user, or an expert in XAI. To effectively serve the purpose of explanations, these explanations must be meaningful from the user's or machine's context. Poorly framed explanations with too little or more information may cause distrust among the users. Therefore, researchers must understand the entity and its context to successfully communicate explanations to the entities at stake [39, 40]. To ensure user-friendly XAI solutions, some approaches can be adopted for generating explanations like contextual querying, cognitive task understanding, mental model elicitation, and stakeholder participatory design. Furthermore, stakeholders and the developers' team can conduct a codesign workshop to understand the user requirement for explanations effectively.

Dynamic Explanation

Since the user's requirements might not be static, it keeps changing with varied dimensions. Therefore, in addition to user-centric explanations, the XAI model must generate dynamic explanations according to the user's changing needs. Hence dynamic explanations with interaction from the user or based on historical data predictions with explanations can be provided.

7.5.3 Explainable Security

Security issues for XAI models come from different aspects like hiding biases in data for fooling post hoc explanations and manipulation of explanations of images through perturbations in the image domain [41]. As the explanations might be shared between machines or among different users at stake, these explanations can be encrypted or digitally signed for secured explanations.

7.5.4 Standardization and Measurements for XAI

Like the GDPR act for private data protections and the right to explanations, standards for XAI development and adoption must be enforced for generic model development and deployment. Furthermore, various metrics must be devised for measuring the quality of the XAI approaches from different dimension and their visualization. Future research must focus on different methods for conceptualizing the XAI from stakeholders with low to high literacy rates [40]. On the other hand, various measurements for evaluating the XAI behavioral effects like transparency, trust, understandability, usability, and impact of XAI on different stakeholders of the XAI systems and its explanations still remain unexplored completely.

7.6 CONCLUSION

The advent of artificial intelligence has embarked almost all fields from finance to healthcare and more critical defense systems. But its widespread applicability led to various questions unanswered by the experts involved in the development too. This has slowed down the adoption of artificial intelligence. The most prevalent concern with AI adoption is trust and transparency of data used and clarity in the decisions arrived. Explainable artificial intelligence has paved the way for these adoption issues by providing meaningful explanations for how the data has been interpreted to derive the decisions. Explainable artificial intelligence is an active field of research, with abundant methodologies continuously introduced every year. The chapter discussed the need of XAI for sustainable computing, an overview of various XAI methods, utilization of XAI as a service, and XAI for a decision support system to promote computational sustainability. It also addressed the challenges in adoption, and future research directions. This will help the research community to understand the existing methods and issues in XAI, in order to explore new directions to make reliable AI.

REFERENCES

1. Zhou, J., Khawaja, M. A., Li, Z., Sun, J., Wang, Y., & Chen, F. (2016). Making machine learning useable by revealing internal states update-a transparent approach. *International Journal of Computational Science and Engineering, 13*(4), 378–389.
2. Mitrou, L. (2018). Data protection, artificial intelligence and cognitive services: Is the general data protection regulation (GDPR) "artificial intelligence-proof"? http://dx.doi.org/10.2139/ssrn.3386914
3. Bessen, J. E., Impink, S. M., Reichensperger, L., & Seamans, R. (2020). *GDPR and the Importance of Data to AI Startups.* NYU Stern School of Business. http://dx.doi.org/10.2139/ssrn.3576714
4. Floridi, L., & Chiriatti, M. (2020). GPT-3: Its nature, scope, limits, and consequences. *Minds and Machines, 30*, 681–694.
5. Arrieta, A. B., Díaz-Rodríguez, N., Del Ser, J., Bennetot, A., Tabik, S., Barbado, A., ... & Herrera, F. (2020). Explainable artificial intelligence (XAI): Concepts, taxonomies, opportunities and challenges toward responsible AI. *Information Fusion, 58*, 82–115.
6. Gunning, D., Stefik, M., Choi, J., Miller, T., Stumpf, S., & Yang, G. Z. (2019). XAI – Explainable artificial intelligence. *Science Robotics, 4*(37), eaay7120. doi: 10.1126/scirobotics.aay7120
7. Christoph, M. (2019). *Interpretable machine learning: A guide for making black box models explainable.* URL: https://christophm.github.io/interpretable-ml-book

8. Singh, A., Sengupta, S., & Lakshminarayanan, V. (2020). Explainable deep learning models in medical image analysis. *Journal of Imaging, 6*(6), 52.

9. Mathews, S. M. (2019). Explainable artificial intelligence applications in NLP, biomedical, and malware classification: A literature review. In *Intelligent Computing: Proceedings of the 2019 Computing Conference, 2* (pp. 1269–1292). Springer International Publishing.

10. Doshi-Velez, F., & Kim, B. (2017). Towards a rigorous science of interpretable machine learning. *arXiv preprint arXiv:1702.08608.*

11. Singh, A., Sengupta, S., & Lakshminarayanan, V. (2020). Explainable deep learning models in medical image analysis. *Journal of Imaging, 6*(6), 52.

12. Zhang, Y., Weng, Y., & Lund, J. (2022). Applications of explainable artificial intelligence in diagnosis and surgery. *Diagnostics, 12*(2), 237.

13. Lässig, J., Kersting, K., & Morik, K. (Eds.). (2016). *Computational Sustainability* (Vol. 645). Springer International Publishning AG Switzerland.

14. Gomes, C., Dietterich, T., Barrett, C., Conrad, J., Dilkina, B., Ermon, S., ... & Zeeman, M. L. (2019). Computational sustainability: Computing for a better world and a sustainable future. *Communications of the ACM, 62*(9), 56–65.

15. Fisher, D. H. (2016). Recent advances in AI for computational sustainability. *IEEE Intelligent Systems, 31*(04), 56–61.

16. Turner, C., Okorie, O., & Oyekan, J. (2022). XAI sustainable human in the loop maintenance. *IFAC-PapersOnLine, 55*(19), 67–72.

17. Minh, D., Wang, H. X., Li, Y. F., & Nguyen, T. N. (2022). Explainable artificial intelligence: A comprehensive review. *Artificial Intelligence Review, 55*(5), 1–66.

18. Zhuang, Y. T., Wu, F., Chen, C., & Pan, Y. H. (2017). Challenges and opportunities: From big data to knowledge in AI 2.0. *Frontiers of Information Technology & Electronic Engineering, 18*, 3–14.

19. Anysz, H., Zbiciak, A., & Ibadov, N. (2016). The influence of input data standardization method on prediction accuracy of artificial neural networks. *Procedia Engineering, 153*, 66–70.

20. Bender, E. M., & Friedman, B. (2018). Data statements for natural language processing: Toward mitigating system bias and enabling better science. *Transactions of the Association for Computational Linguistics, 6*, 587–604.

21. Mohit, Kumari, A. C., & Sharma, M. (2019). A novel approach to text clustering using shift k-medoid. *International Journal of Social Computing and Cyber-Physical Systems, 2*(2), 106–118.

22. Yang, S. C. H., & Shafto, P. (2017, December). *Explainable artificial intelligence via bayesian teaching.* In NIPS 2017 workshop on teaching machines, robots, and humans (Vol. 2), 1–11.

23. Campbell, T., & Broderick, T. (2019). Automated scalable Bayesian inference via Hilbert coresets. *The Journal of Machine Learning Research, 20*(1), 551–588.

24. Zhang, Y., Weng, Y., & Lund, J. (2022). Applications of explainable artificial intelligence in diagnosis and surgery. *Diagnostics, 12*(2), 237.

25. Belle, V., & Papantonis, I. (2021). Principles and practice of explainable machine learning. *Frontiers in Big Data*, 39. https://doi.org/10.3389/fdata.2021.688969

26. Rawal, A., McCoy, J., Rawat, D. B., Sadler, B. M., & Amant, R. S. (2021). Recent advances in trustworthy explainable artificial intelligence: Status, challenges, and perspectives. *IEEE Transactions on Artificial Intelligence, 3*(6), 852–866.

27. Molnar, C. (2022). *Interpretable Machine Learning: A Guide for Making Black Box Models Explainable (2nd ed.).* christophm.github.io/interpretable-ml-book/

28. Samek, W., Montavon, G., Lapuschkin, S., Anders, C. J., & Müller, K. R. (2021). Explaining deep neural networks and beyond: A review of methods and applications. *Proceedings of the IEEE, 109*(3), 247–278.

29. Saeed, W., & Omlin, C. (2023). Explainable ai (xai): A systematic meta-survey of current challenges and future opportunities. *Knowledge-Based Systems, 263*, 110273.

30. Keen, P. G. (1980). *Decision support systems: A research perspective.* In Decision Support Systems: Issues and Challenges: Proceedings of an International Task Force Meeting (pp. 23–44).

31. Swartout, W. R., & Moore, J. D. (1993). Explanation in second generation expert systems. *In Second Generation Expert Systems* (pp. 543–585). Springer Berlin Heidelberg.

32. Du, M., Liu, N., & Hu, X. (2019). Techniques for interpretable machine learning. *Communications of the ACM, 63*(1), 68–77.

33. Ahmad, M. A., Eckert, C., & Teredesai, A. (2018, August). *Interpretable machine learning in healthcare.* In Proceedings of the 2018 ACM International Conference on Bioinformatics, Computational Biology, and Health Informatics (pp. 559–560).
34. Shapley, L. S. (1997). A value for n-person games. *Classics in Game Theory*, 69, 307–317.
35. Štrumbelj, E., & Kononenko, I. (2014). Explaining prediction models and individual predictions with feature contributions. *Knowledge and Information Systems*, 41, 647–665.
36. Jordan, M. I., & Mitchell, T. M. (2015). Machine learning: Trends, perspectives, and prospects. *Science*, 349(6245), 255–260.
37. Rittel, H. W., & Webber, M. M. (1973). Dilemmas in a general theory of planning. *Policy Sciences*, 4(2), 155–169.
38. Tudorache, T. (2020). Ontology engineering: Current state, challenges, and future directions. *Semantic Web*, 11(1), 125–138.
39. Kotriwala, A., Klöpper, B., Dix, M., Gopalakrishnan, G., Ziobro, D., & Potschka, A. (2021). *XAI for operations in the process industry-applications, theses, and research directions.* In AAAI Spring Symposium: Combining Machine Learning with Knowledge Engineering, 1–12.
40. Haque, A. B., Islam, A. N., & Mikalef, P. (2023). Explainable Artificial Intelligence (XAI) from a user perspective: A synthesis of prior literature and problematizing avenues for future research. *Technological Forecasting and Social Change*, 186, 122120.
41. Kuppa, A., & Le-Khac, N. A. (2020, July). *Black box attacks on explainable artificial intelligence (XAI) methods in cyber security.* In 2020 International Joint Conference on Neural Networks (IJCNN) (pp. 1–8). IEEE.

8 Edge Computing-Based Intrusion Detection Systems

A Review of Applications, Challenges, and Opportunities

Posham Uppamma and Sweta Bhattacharya*
School of Information Technology and Engineering, VIT Vellore,
Tamil Nadu, India

8.1 INTRUSION DETECTION SYSTEM

An intrusion detection system (IDS) is a system that observes data traffic and anomalous activities related to data theft and generates alarms to ensure privacy, security, and data integrity is sustained. It is a distinctive software program that scans for security breaches and other unwanted antics on the network. This system mainly focuses on any harmful activities raised through the network; these get notified to the network administrator or stored in the SIEM (security information and event management) system. This SIEM system combines data from many sources and employs an alert filtering technique to differentiate between fake and malicious alerts [1]. Figure 8.1 shows how the intrusion detection system worked in the network.

Maintaining secure transactions in business environments has become increasingly challenging in recent years. The intrusion detection system monitors and notifies of several security attacks and threats to the communication channel for confidentiality, integrity, and availability [2, 3]. IDS characteristics are as follows:

- It is possible to analyze malware signatures in system files.
- During the scanning process, potentially harmful pattern indications are found and detected.
- Monitoring user activity helps identify malicious intent.
- The configurations and settings of the system are being monitored.

8.1.1 CLASSIFICATION OF INTRUSION DETECTION SYSTEM

The IDS system is mainly classified into two major categories, one is based on location, and the other is based on data [4]. The following Figure 8.2 represented various classification levels of IDS.

Location-based intrusion detection system
Location-based IDS defends the entire site where it has been installed, whether it has a network or a host, against attackers and intrusions. Based on this situation, again, it would be further categorized in two ways as follows:

Host intrusion detection system (HIDS)
A host-based intrusion detection system monitors the data traffic, and network interfaces, and monitors the internal activities of the system. The HIDS is installed in a computer system that detects malicious data while working with operating system files. Once the HIDS is installed in the host, it controls the components of authorized access and restricts known system files [5].

DOI: 10.1201/9781003407959-8

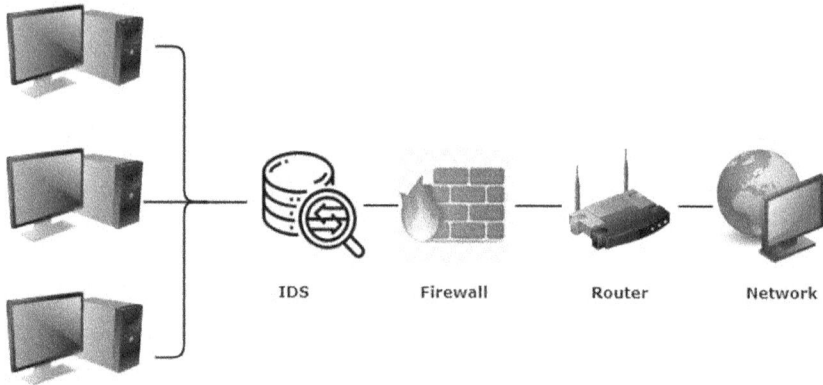

FIGURE 8.1   Components of the intrusion detection system.

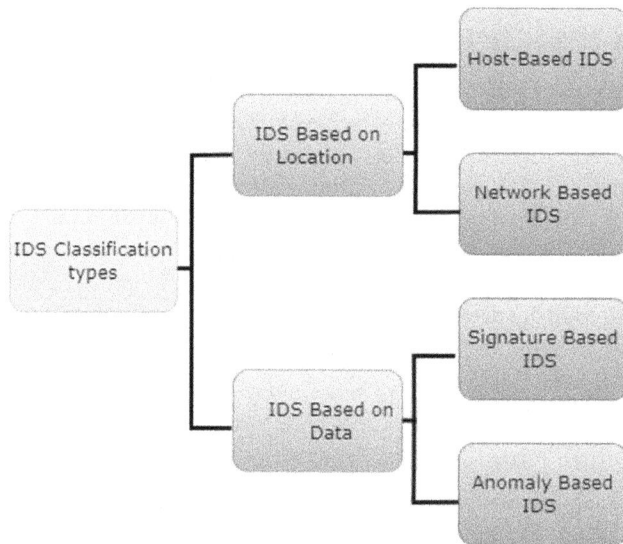

FIGURE 8.2   Classification of the intrusion detection system.

The advantages of HIDS are as follows:

- HIDS can read encrypted data packets and detect breaches that become more effective.
- Both the application and system programs must observe audit log information.
- In contrast to NIDS system, it is a low-cost setup.
- We have fewer packets to examine the encryption traffic.

Disadvantages

- It has single device maintenance to visibility issues.
- Maintaining too many organizations as a group for handling and configurations of each one is a difficulty.
- This system cannot detect proper network-related attacks.

8.1.1.1 Network Intrusion Detection System (NIDS)

The network intrusion detection system examines the data traffic that is connected to all the systems in a network, as well as how it enters and exits the network. It operates by monitoring the various traffic channels on a subnet, which correspond to the associated traffic and can be examined to determine if there is additional traffic for which an alarm should be issued [6]. The malicious attack can be reported to the administrator once it should be recognized.

Advantages of NIDS

- Ensures the whole network for IDS security.
- Parts of the network system that are most susceptible to traffic are blocked.
- A device that is not interconnected and does not impact its availability or throughput.
- It is simple to secure and conceal from intruders.
- The huge network system is monitored with just a few duly positioned NIDS systems.

Disadvantages

- The major problem is the high cost of introducing it into the network.
- Within the encrypted traffic channel to identify the threats is a big task.
- In switch-based networks it cannot be adjusted properly.

Based on DATA

8.1.1.2 Signature-Based Intrusion Detection System

A SIDS keeps track of packets transmitted over a network and compares them to a database of known attack signatures or properties. The specific patterns relevant to instruction or byte are identified associated security concerns are handled by the system. The major advantage of this SIDS is that it uses undefined, secure signatures to oppose attackers. SIDS is an efficient and effective way of tracing network traffic through many sources [7].

8.1.1.3 Anomaly-Based Intrusion Detection System

This system compares the data patterns to the base systems so that network traffic is observed at any time. It looks for harmful patterns on the network more than it looks for specific patterns. This system can work with machine learning approaches for trusted system design elements such as bandwidth, various ports, and devices. In case of an anomaly-based system, the whole network organization is considered and any suspicious activities are detected and easily prevented. By using artificial intelligence and machine learning approaches to design a trusted environment in those networks for the secure transmission of data [8].

8.1.2 Challenges of Intrusion Detection System

Intrusion detection systems work like a tool to maintain every organization's protection from malicious attacks [9]. But IDS still has some challenges that organizations face when using its methods. These IDS challenges are represented in Figure 8.3.

False alarm rates: These false-positive reports are notified regularly while using an intrusion detection system. These alerts may place a burden on internal groups of organizations. Most organizations do not have consistent space for maintenance and do not have the time and resources to analyze every alert. In such cases, some malicious attack signals are combined in the notifications.

Low detection rate: It can depend on the false-positive report. It needs an efficient analyzer to understand more suspicious attacks. The signature-based and network-based intrusion systems

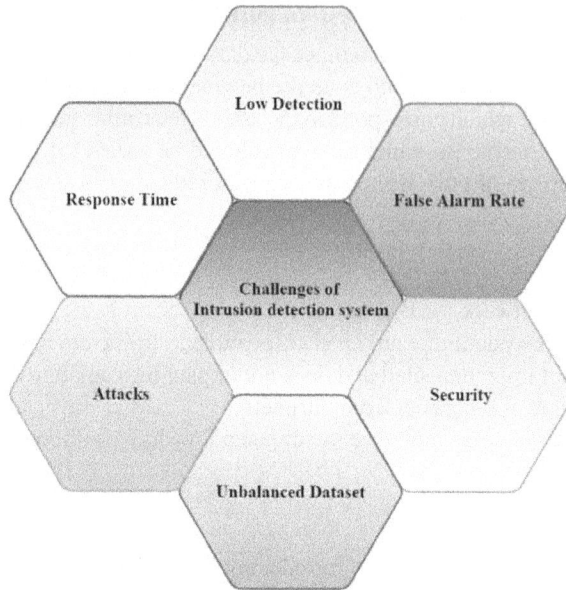

FIGURE 8.3 Challenges of the intrusion detection system.

struggle to find the malicious data present in their networks. Finally, the low detection rate leads to suspicion about the attacks.

Unbalanced dataset: Working with large datasets to classify the data patterns in IDS is not an effective method. If unrelated data is to be presented, an intruder takes a lot of time to verify the attacks in the network.

Response time: In an IDS system, response time is considered as a major issue. An incident of a data breach could be analyzed by organizational experts. But, in most cases, understanding the problem requires both more resources and time constraints to resolve.

- An IDS process takes a lot of ensuring security for organization safety. Nowadays, IDS provides automation processes like router access control. If malicious attacks are found, it alerts the system. It is complex to maintain complete data protection.
- An IDS implementation takes place at an organization for data to be presented in either a host-based or network-based system. It completely depends on the organizational behavior to determine how the process can be worked. In the technology aspect, NIDS is a better way to organize the various devices at a time, as well as there is no need to maintain a tool or program when compared to the HIDS process. For this reason, many organizations have chosen the NIDS.
- To maintain the various types of sensors that can be used by an organization to detect different types of errors, if sensors are not placed properly, it fails to generate accurate results.
- The IDS technology works with some organizations that use the signature-based IDS process for protecting against information attacks. This process completely depends on the attack's signatures for detecting signature threats. If any new signature attack is identified, it is updated in the signature database.
- When we want to present the IDS process in the network environment, it is important to make sure that the organization uses switches for completion. In this channel, traffic monitoring between inbound and outbound as detected by ports cannot be efficient.

- Switch-based channel spanning and mirroring of ports are very much needed for malicious activity detection. But in real-time, it's hard to maintain due to some harmful attacks that present error reports.

8.2 EDGE COMPUTING

Edge computing is a distributed environment that concentrates on network communication and brings data processing to place to reduce bandwidth and network latency. In simple words, edge computing technology is referred to mainly as cloud data, and these related data can be processed into local momentum concerning an edge server, IoT devices, and user computers [10]. Network edge communication minimizes computational time and focuses on very long-distance interactions between a client and server while sharing data. This edge computing is related to cloud computing, fog computing, and IoT environments, which relate to the distributed communication channel to increase the computational speed and response time. These are all focused on storage resources from a digital perspective and easy sharing of resources. The standardized architecture of edge computing is specified in Figure 8.4. These can be differentiated as follows:

Cloud computing

Cloud computing provides a lot of services and various applications within the internet for rapid business development. The data can be stored in the cloud or transmitted remotely. In this system, there is no need to place the user in a specific location to access files or applications [11]. The major cloud services are databases, servers, storage media, software, and networking-related applications. These services could be maintained both publicly and privately; public services are accessed by paying a fee online, but private services are organized and maintained by clients.

Fog computing

The information can be stored closer to the databases in the fog computing environment for real-time data analytics and optimal usage. This process mainly works between cloud and edge computing environments. The local data can be stored in fog nodes that are near the databases. This data can be

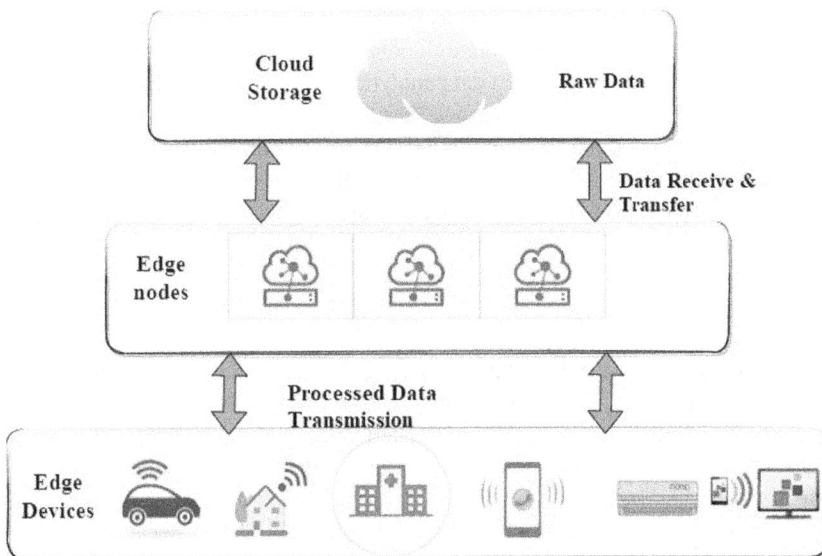

FIGURE 8.4 Layered architecture of edge computing.

maintained in local momentum if necessary to utilize the data. The same situation raised in IoT is the system to solve the issues by using edge computing [12]. All three approaches are concentrated on data storage, repository relations, and the deployment of physical data in a distributed environment. That computing environment followed a layered approach to transmitting data between clients and cloud data sources subsequently [13].

Benefits of edge computing

- **Autonomy**
  In edge computing, unknown sources or untrusted bandwidth can be presented and reverted based on their condition. Once it is found during the transmission of data to the clients locally, the sharing of resources can be stopped, and less latency is required to be maintained.
- **Data sovereignty**
  Maintaining a huge amount of data transmission is a big challenge and has raised many security issues in global communication. In edge computing, the data can be arranged closer to the data sources within the region, following data sovereignty rules. The main advantage of the edge is that even raw data is processed within the boundaries of secure data before being sent everywhere in the cloud.

Limitations and challenges of edge computing

Various challenges and issues are faced by edge computing to share the data resources between the cloud and various client computing environments.

Limitations

- **Bandwidth**
  In general, bandwidth can be defined as the minimum amount of network data transmitted in a given period. It is based on the network, which is termed bits per second. It states that there is a limited amount of data transmission for limited devices over the network. If we present more data and devices over the network for huge data transmission, it is very difficult to maintain the network because of the huge cost and related issues raised.
- **Latency**
  It refers to the time it takes to transfer data between two nodes on a network. In communication, speed plays a vital role in rapid transmission. When long distances exist over the network, it takes more time to complete the data transmission process. Ultimately, it delays everything like decision-making and data analytics over the network.
- **Congestion**
  It refers to the network point of view; in a cloud environment, data change is presented globally. Every day increases the communication medium because of the number of devices that are presented. Finally, it yields a low transmission rate because of huge data congestion.

Challenges of edge computing

- **Security**
  Security is the major issue while the edge uses IoT devices for the transmission of data over the network. To maintain the required sensor-based system for secure computing results as well as system maintenance like software-related issues and data source communication. Still, there are some issues when IoT device services are needed by the cloud.
- **Connectivity**
  Edge computing reduces the network restrictions to communicate at the minimum level. When communication can be lost, then data sharing cannot be possible. This is one of the major concerns about edge device connections.

- **Limited capacity**
  In a cloud environment, more resources are available. Presenting the specific application resources at the edge computing level cannot be more effective since there are limited communication services and devices presented.
- **Data lifecycles**
  In recent years, data maintenance has played a vital role in cloud storage. There are a lot of unnecessary data to be recorded for future usage, but it's a big task to analyze which is to be presented or removed from the source.

8.3 ROLE OF INTRUSION DETECTION SYSTEM (IDS) IN EDGE COMPUTING

The edge computing process involves the communication of real-time data between the cloud and storage devices through edge devices. These edge nodes use limited bandwidth and low latency principles to transfer the data efficiently. In recent years, the major challenging issue has been privacy and security in the edge network for different layers of transmission [14]. In this situation, the intrusion detection system takes over to detect unauthorized access from the intruders. In addition to that, this approach can easily detect malicious nodes in a distributed edge network architecture. Along with this, various machine learning and deep learning techniques are applied to find out the results of the advanced network traffic detection strategies accurately [15]. The role of edge computing in terms of intrusion detection systems is represented as in Figure 8.5.

In [16], the authors implemented the network architecture with changes to the most effective design, which is the deployment of IDS in fog and cloud computing. The security and privacy issues are solvable by using the intrusion detection system in fog computing. We also discuss the existing solutions for the challenges faced by intrusion detection systems. Using IDS, the computational power, storage resources, and bandwidth become manageable in the architecture. While using the edge of things computing [17], authors came across privacy and security issues in the

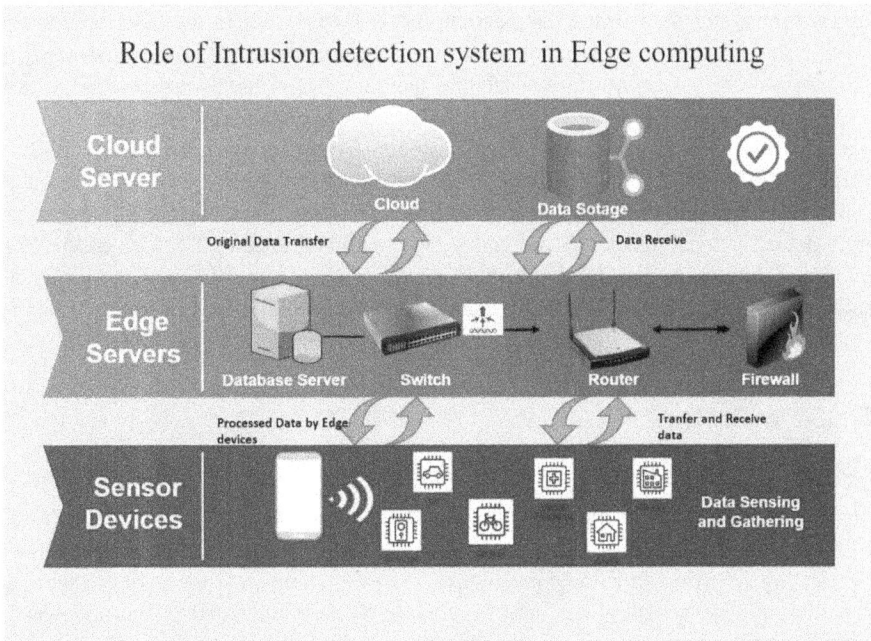

FIGURE 8.5 Role of the intrusion detection system (IDS) in edge computing.

architecture. The intrusion detection system is used to eliminate them. Nowadays, the edge of things is an emerging technology in data processing, storage, and service providers between devices based on the Internet of Things. Here, we propose a deep belief network based on an intrusion detection system. The different studies based on existing detection models and the performance of the proposed methodology are reviewed. There are instances of privacy leakage at the time of data transmission in smart devices. Though the edge nodes are used, there is still the possibility of attacks. The proposed method works on the IDS to be arrayed on the edge nodes to convert the network traffic to images and apply it to convolutional neural networks (CNN), which categorize network traffic. The proposed scheme is effective for smaller categories, but its accuracy can be improved. In addition, the accuracy of binary classification is beyond that of normal and abnormal situations [18]. To predict edge computing network attacks, the authors presented a new intrusion detection model [19]. It uses convolutional-LSTM-based edge computing and uses the optimal intrusion detection deployment strategy in terms of intrusion detection revenue. It also analyzes the interactions between an attacker and an intrusion detection system to obtain an optimal decision tree for a set of defensive actions. The proposed method uses Nash equilibrium for attacks and gets profited in our game model to optimize cost and intrusion detection service strategy. Moreover, to improve the intelligent false alarm in a distributed environment created on edge computing devices in terms of response time and energy efficiency, a framework is proposed. Using this framework, the central server workload was reduced and the delay was shortened when compared to other frameworks [20].

An intrusion detection and prevention system architecture for VANETs was used because of its disastrous results for applications [21]. One of the highlights of the architecture is that it targets high detection accuracy by applying reinforcement learning throughout the architecture to handle the dynamics of the VANET and makes the right decisions based on the current state of the VANET. As IDPS VANETs are latency-sensitive, especially for security applications, an architecture has been deployed in edge computing to achieve low latency detection with high processing efficiency. Here [22], the authors discuss the intrusion detection system used in various applications with machine learning and deep learning techniques. And using different datasets and detailed descriptions of the recently used datasets, in the present world, IoT devices in edge computing have increased a lot in various applications, and their usage has become more widespread. In the case of security issues in edge computing, an IDS plays a major role. Similarly, we discussed how the SDMMF (single-layer dominant max-min fair) allocation scheme handles the problem of equitable allocation of multiple sibling resources and how the MDMMF (multi-layer dominant max-min fair) allocation scheme handles the tiered resource allocation problem. Going forward, we plan to prototype the proposed evaluation approach either in a closed laboratory environment or in collaboration with real organizations [23]. The authors stated in [24] that there is scope for identifying known attacks but that existing detection systems cannot recognize new unknown attacks. Because of this drawback, we proposed a technique as an edge-based hybrid intrusion detection framework to detect known and unknown attacks with a low false alarm rate. Moreover, this framework used various machine learning techniques for noticing the traffic in mobile edge computing for better accuracy.

8.4   EDGE-ENABLED APPROACHES

8.4.1   In Smart Healthcare

A smart healthcare system works through sensors that capture information. The information is transmitted to smart devices like IoT, sensors, smartwatches, and detection mechanisms that are used in a cloud computing environment. These are internet-assisted to monitor the patient's data, communicate with the people, various goods, and organizations interconnected to the hospital management system, and then immediately respond to the multidisciplinary healthcare system [25, 26]. Smart healthcare can facilitate interaction among all concerned parties in the healthcare sector, including facilitating the allocation of resources and the services provided on a needs basis. From a

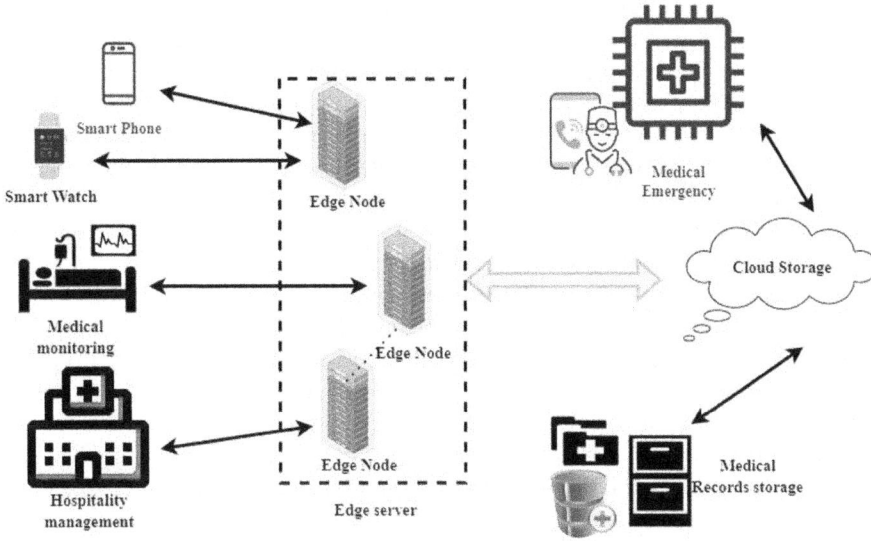

FIGURE 8.6 Edge-enabled smart healthcare system architecture.

smart industry standpoint, the healthcare sector plays a critical role in that patients require advanced healthcare services for their present lifestyle. It helps to provide efficient and effective real-time data processing with low latency using an edge computing platform. It provides effective communication between the end-users for maintaining data storage nearer to the data repository [27]. In the recent healthcare environment, dependable and effective patient care systems, such as routine services and patient condition monitoring, are required. For this purpose, maintaining up-to-date patient data should be managed via cloud and edge computing platforms. The study explained a novel approach called a resource preservation network (RPN), which connects emergency departments using cloud and edge computing. This architecture focuses on providing real-time environmental support for healthcare to meet the needs of patients [28]. The architecture of an edge-enabled smart healthcare system is represented in Figure 8.6.

Examining how a patient's length of stay (LOS) affects resource use and how to lower the average waiting time for better results is paramount. Similarly, edge devices for medical data analysis could facilitate the implementation of classification techniques and the estimation of risks. As part of the artificial intelligence technique of LogNNET, for easy access to medical information with low resources [29], the study of [30] suggests that edge devices can be used for monitoring the data processing of fall detection in older people. This study can analyze the person's motion and tendency by using sensor nodes to identify the fall. The proposed framework stated an image forgery detection system for the identification of patients' original images that are not altered. In this scenario, the primary role of edge computing is to transmit real-time data with low resource consumption in a limited environment that works efficiently. The combination of edge and cloud computing results in effective communication and low latency by taking real-time inputs [31].

8.4.2 In Smart Home

In recent days, a part of human life gets connected with computers and sensing devices. These sensors and computers can be used to do efficient work humanely and comfortably. A smart environment is defined as a limited environment in which various smart devices and sensors get connected to lead a more convenient lifestyle. Smart environments are specifically designed for reducing the physical manpower and continuous work done in daily activities and decreasing unrelated risks.

Moreover, it provides safety for a person using various supportive services. These services afford direct assistance with the user's requirements and the actions needed for house maintenance. To provide the security of the home for services like home appliance control, avoiding energy consumption, and solar plant energy maintenance using an intrusion detection system to detect unauthorized access [32]. According to the authors in [33], IoT is a part of smart home-based activities to provide safety measures and increase security comfortably. Similarly, a new approach is emerging as Vigilia works on IoT-enabled security systems. This system could be used to restrict the denial of network attacks on the smart home. Moreover, the system supports standalone data for communication, which tends to escape runtime firewall attacks in an effective environment [34].

The requirement of smart home intelligence technology takes place for various activities like reducing computational power and cost. Along with that, safety and security measures are presented for a smart, integrated system. The proposed system works for enriching the security measures in various smart home-based devices. Moreover, an edge-enabled approach can be implemented to accomplish real-time data processing. These edge devices present effective results in low latency and bandwidth [35]. Similarly, the state-of-the-art contribution of edge computing with IoT-enabled services is to provide data security using authentication mechanisms. In this scenario, all the edge-enabled smart home devices maintain the authentication mechanism to avoid intrusion attacks [36]. In [37], specific development of edge computing applications in a smart construction environment is discussed. The construction management has various supporting activities, like a timely response to supply chain activities and observing each work performed in various locations. To provide the quality of services necessary to monitor the real-time, up-to-date information and communication channel through the edge devices efficiently. The architecture of the edge-enabled smart home network system is represented in Figure 8.7.

According to [38], the development of the smart home environment frequently raises security-related issues. Those issues are intruders hacking the information or collapsing the user's private information. In general, data can be taken from cloud servers, which increases the chances of stealing the user's information during transmission. Here, edge computing takes place to maintain the edge nodes nearer to the cloud center for easy access with low bandwidth. To implement the intrusion detection system mechanism to identify network traffic-related attacks, train the data using CNN (convolutional neural network) for classification of anomaly detection. Similarly, a smart surface inspection system could monitor the surface texture geometrically. The proposed method uses R-CNN (regional conventional neural networks), especially for object detection in a wide area.

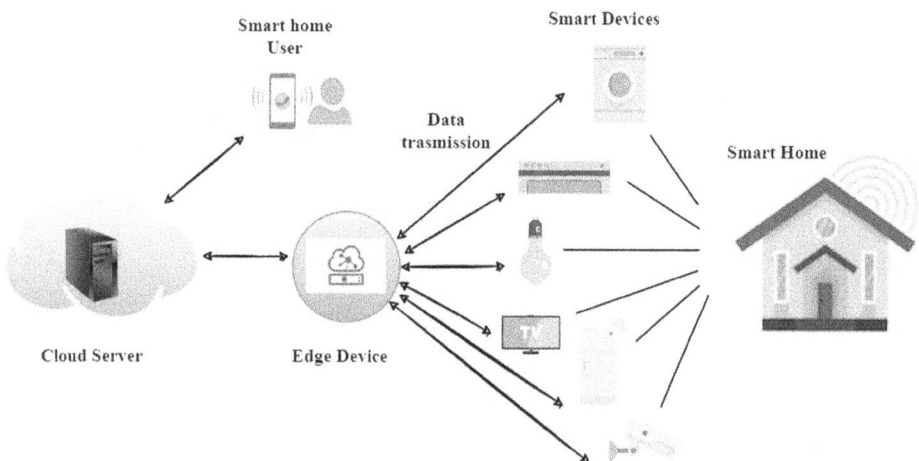

FIGURE 8.7 Edge-enabled smart home network system architecture.

These methods require more computing power, and at the movement edge, devices can be presented to monitor real-time communication channels. It results in inefficient and effective communication through the cloud and edge computing frameworks [39].

8.4.3 SMART GRID

A smart grid is one of the Internet of Things (IoT)-assisted technologies that work on electricity networks. This technology was implemented through two-way communication for handling the pieces of machinery. To overcome the drawbacks of unidirectional networks, smart grids are developed based on supply-chain management systems [40]. The smart grid system is mainly used for decreasing the consumption of electricity, reducing costs, and increasing efficiency. This system is mainly focused on identifying the path of electricity consumption for specific organizations like institutions, hospitals, and industries. The process of connecting data to the cloud centers is inefficient for information transmission over long distances. In this scenario, the edge computing framework came into the picture for reliable and effective communication media through edge devices [41]. Deep learning methods have been studied in recent years by researchers to solve transmission delay problems in edge devices. To increase the speed and computational power while transferring data between centralized servers to target resources, edge-enabled types of equipment are providing easier access [42]. Similarly, in the medium of smart grid tracking, various network traffic-related issues are raised. These issues are identified and solved by implementing an intrusion detection system based on artificial intelligence [43]. The architecture of the edge-enabled smart grid system is in Figure 8.8.

Another approach [44] is for smart grid monitoring systems using wireless sensor networks. In this area, smaller sensor items form a single wireless sensor communication over the wide-area networks. These sensors provide highly realistic and secure energy sources, but it is a cost-effective procedure to maintain wireless sensor networks. However, in a smart grid communication environment, various security vulnerabilities were observed. The proposed system completely works on the identity-based signature procedure, which specifies impersonally secured key protocol generation for uniquely known authorities in smart grids. While using this protocol, check if the communication channel is valid or not. Moreover, this framework increases the efficiency and performance of smart grid communication, but it is cost-oriented [45]. Similarly, a data protection framework is introduced through edge-enabled devices for effective communication at a minimal distance. The real-world data can be produced via an IoT-based system on the smart grid connection. This

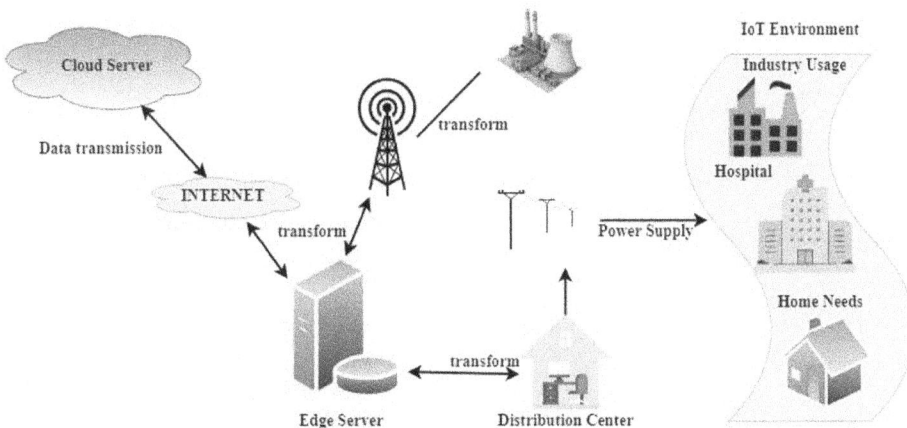

FIGURE 8.8 Edge-enabled smart grid architecture.

framework provides an advanced metering concept that specifies compelling electricity charges, and responses are produced based on the user's need [46]. Smart grid systems still have some data transmission issues and challenges. The problems are raised due to the highly standardized data presented for high energy requirements. IoT and edge-enabled services on smart grids need to improve the quality of sharing energy resources [47].

8.4.4 SMART CITY

The idea of a "smart city" includes smart devices that are connected to the Internet of Things (IoT). Information and communication technologies (ICT) are used in modern city operations to improve the services and operations communicated to residents. Various electronic techniques, sensors, and voice-assisted activation techniques connect these communication channels. Smart city technologies assist individuals and city officials to check how the city develops and what happens in the community. The major benefits of implementing the smart city concept are easier communication, decreased costs, and better utilization of resources to increase communication between government bodies and citizens. These smart devices sometimes provide specific information transmission to other individuals during movement; edge computing works when highly secured information is communicated over the network [48–50]. The smart city environment is involved in various levels as shown in Figure 8.9.

Edge computing is a decentralized communication framework for various individual applications like health, social, cognitive aid, and intelligent transportation system. The proposed model for a secure and trustworthy framework between the distributed edge servers. These servers communicated through the edge nodes along with help of an intrusion detection system for reliable and secure application development. In this context, highly secured information is transmitted

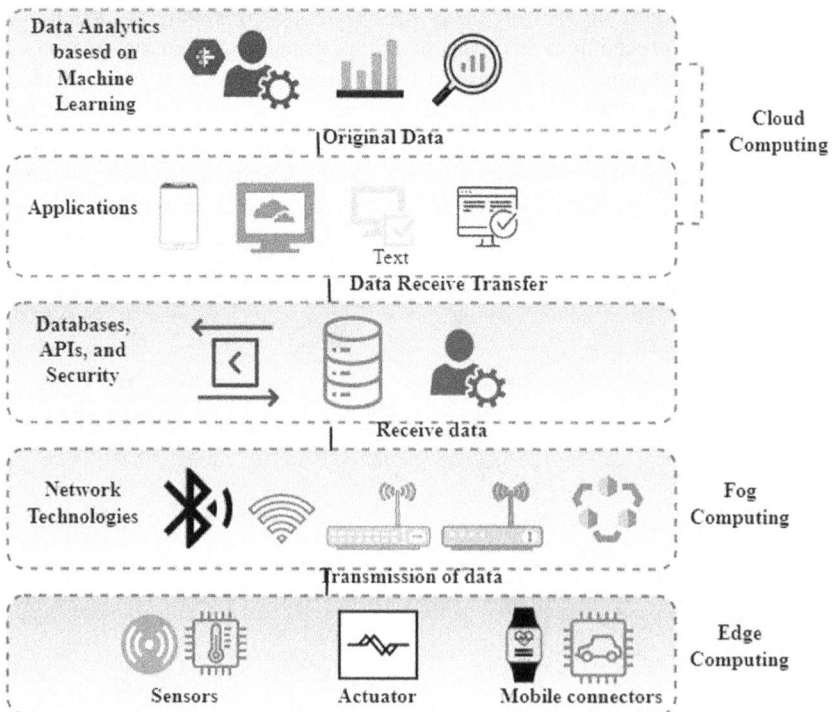

FIGURE 8.9 Smart city layered approach.

through the edge cluster servers. Smart vehicles are major communication and transportation systems with advanced refinements. Cloud facilitates incessant accessibility of vehicular communication in smart cities place a vital role and needs to increase the vehicular communication method [51]. Studies discuss another framework which provides a secure cloud incessant accessibility of smart vehicles against security attacks solved by intrusion detection system mechanisms. This mechanism works on specific cluster communication for security by using trusted third-party modules. Intrusion detection mechanism for identification of third-party service providers and requests over false requests. The study also implemented deep belief and decision tree techniques for classification and data reduction issues [52]. Similarly, it highlighted an intelligent transportation system that handles traffic-related issues and captures transport information on heavy roads using the multiple-object vehicle tracking system. A vehicle detection algorithm is trained with a large extent of traffic-related data, which guarantees the efficiency of the edge devices. The new era of electronic vehicles communicated through the hybrid cloud and edge computing in a blockchain platform. The security-related information in-vehicle interactions are maintained using consensus energy and data coins [53].

8.5 USE CASES OF IDS IN EDGE COMPUTING

Use case 1: In the present era, it is difficult to face security issues with the principles of biological immune systems. Here [54], we implemented a gene-immune detection algorithm. An edge intrusion detection system has been embedded into the GIDA algorithm because of the dynamic data problem efficiently. The proposed system has the highest TP rate and the lowest FP rate. The system here presents better dynamic improvements in contradiction of attacks.

Use case 2: GLIDE is a mechanism used to detect the behavior of an attack and also a defense mechanism for attacks based on game theory. The authors used the Nash equilibrium of attack and defense income in the game model and also the Nash equilibrium condition under different conditions to analyze the strategy of an intrusion detection service based on multi-redundancy edge calculation. Implementation of the GLIDE model requires up to 500 edge computing terminals for simulation. As a result, this is an optimal deployment strategy for an intrusion detection system based on the probability of attacks but also has a high cost [55].

Use case 3: E-GraphSAGE [56] is used to capture the features of edge computing and the topological pattern of a network flow graph and can implement the detection of malicious attacks. The E-GraphSAGE application is used for IoT network intrusion detection. The experimental results were mainly created on four IoT NIDS datasets, which show that E-GraphSAGE-based NIDS performs extremely well and overall outperforms the state-of-the-art ML-based classifiers.

Use case 4: Due to the high usage of IoT devices in the present environment, there are a large number of security issues in edge home networks. For this purpose, we proposed an edge smart gateway installed on a Raspberry Pi that runs an SDN controller and Open-VSwitch (OVS) to perform traffic monitoring, anomaly detection, and traffic filtering. To divide the traffic between the devices and to identify interruptions in the networks, the authors used the machine learning algorithm, i.e., Decision Tree J48. Simulation results show that the model has a high accuracy of intrusion and can effectively ensure the security of home IoT interactions [57].

Use case 5: The application layer has attacks like Advanced Persistent with Ransom DoS, botnets, and application-based DDoS flood attacks. For this type of attack, a distributed lightweight real-time DDoS threat analytics and response framework (DTARS) had been introduced. The performance of the DTARS framework is enhanced when compared to other use cases. The detection speed is 10 times faster and more accurate compared to older ones in terms of attacks and end-to-end connections [58].

Use case 6: AI@EDGE [59], the new era of artificial intelligence, provides multi-service next-generation internet (NGI). The proposed method worked to secure, reuse, and safeguard the data in machine learning models. This technique is mainly used to protect and provide a secure real-time path for AI-enabled networks and vehicular communication systems without resistance.

Use case 7: The proposed model named a generative adversarial network (GAN) for detecting the problem of the intrusion detection system in collaborative edge computing is based on the social Internet of Things. Mainly, this method is used to detect attacks like DoS attacks, man-in-the-middle attacks, unauthorized access, and packet sniffing. The proposed method comprises three phases. First, the flow should be pre-processed and feature extraction performed. Second and third, an intrusion detection algorithm directed at a single attack and multiple attacks based on GAN results are more efficient when compared to other methods [60].

Use case 8: The main aim of the proposed method is to calculate the performance of pass-ban IDS in terms of threat detection accuracy and the computational resources needed for executing the IoT edge devices. In this method, authors detected the attacks by using two one-class classification techniques, i.e., iForest and LOF. Moreover, the passband can be implemented even on cheap IoT gateway boards [61].

8.6 OPEN CHALLENGES AND FUTURE SCOPE

8.6.1 OPEN CHALLENGES

In a distributed environment, data processing makes it possible for more devices to connect to a large network. These data servicing devices are managed over very long distances from sources like organizations, and companies. This situation raised network visibility and control problems for the main sources. The traditional approach to edge computing is supported by all device's security protocols and routinely up-to-date information using an intrusion detection system. In this scenario, the device-level and network gateway-level protection systems detect malicious attacks on edge-enabled IoT devices. The major concern of malignant attacks on edge nodes with remote access is when the edge nodes turn out to be damaged. In the supply chain management system, these edge devices are deployed in public areas to communicate an intelligent smart environment, which leads to damage to the edge network system [62]. Finally, the intruders altered the way to transfer the correct information along with the edge devices, which leads to the smart IoT communication giving false information. These various challenges associated with IDS are mentioned in Figure 8.10.

To improve the data transmission speed and low latency in edge devices, an intrusion detection system that provides specific security concerns for decreasing device failures was applied. This system still observed various challenges that were specified as follows:

- **Encrypted traffic channel**: In the intrusion detection system, if any malignant attacks may arise during transmission of the day through the edge device channels, you get alerted or alarmed. In case the data transmission is mined into an encryption format, the IDS cannot find the false data because it won't work with standardized encryption. Due to that reason, edge-enabled IoT devices need to refine secure data transmission operations.
- **Variability of high resource usage:** The edge nodes specify high computational resources that are maintained for specific hardware resources like the Raspberry Pi. The intrusion detection system needs various computing methods that require high computational power to identify fault detections in IoT devices. In this situation, several requested resources are applied to edge nodes, and the IDS system computes continuous cycles. Finally, the total framework is going to be prevented from being executed because of the consistent usage of resources. In other situations, plenty of resources are available for cost-effective edge devices.

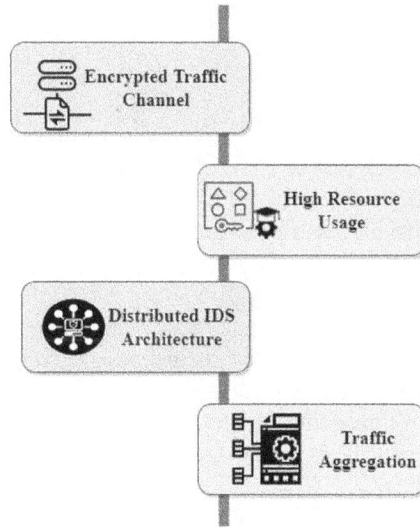

FIGURE 8.10 Challenges of IDS in edge computing.

- **Distributed IDS architecture:** Based on the availability of resources when compared to the edge and IoT devices, implementation was distributed to some extent on the network medium. In reality, the IDS worked efficiently on a single system, but when it was applied to multiple subsystems, its performance decreased. The major challenge is maintaining the distributed environment using IDS to minimize the complexities.
- **Traffic aggregation:** One of the challenging issues based on the internet protocol (IP) information is the edge, and the IDS performs the data transmission. While data packet transmission, the edge, and the IDS system are suspected, the destination cannot be identified by IoT devices.

8.6.2 Opportunities

Intelligent communication has been a key part of the smart computing environment in recent years. Smart environments are specifically designed to reduce physical labor (constant work is done in daily activities) and also decrease unrelated risks. Moreover, it provides safety for a person using various supportive services. Information security and data privacy in the modern ecosystem are hot topics in the current research trend. Similarly, edge devices provide security from intrusion attacks using various artificial intelligence techniques in IDS systems. In [63], the authors surveyed especially for data privacy and information security in edge-enabled IoT devices. The major contribution of the edge-enabled approaches against security attacks and proposed possible solutions to them.

In edge computing with IDS, it is most important to work on the privacy preservation of data in real-time. From a remote location, this mechanism should focus on dynamic data communication and the unique identity of the edge device for each user. Also, this system requires creating the belief to compute the efficient and evaluation mechanism for creating the new edge devices in edge-enabled IoT devices to communicate without middlemen [64]. In consideration of the low resources to be presented on the limited edge devices. This mechanism only concentrates on queuing techniques in IDS devices for specific users' privacy. This framework can be implemented in hospital management in the future. Similarly, the intrusion detection system framework, while implemented into the real-world edge platform, detects anomalies. To overcome the signature- and anomaly-based attacks in IDS using various machine learning techniques for efficient network communication on edge devices.

8.7   CONCLUSION

Edge computing makes the data or information available to nearby data centers and endpoint devices. The advantage of providing data sharing between stakeholders is to decrease the time it takes to process it. Moreover, it offers intensified security and more privacy because of the edge devices placed nearer to the central cloud servers. Still, these devices need reforms to prevent information theft and protect the data. An intrusion detection system provides a great extent of security protection from various intruders using deep learning techniques. These techniques offer the ability to classify and detect various unknown attacks and data traffic channels easily. Smart homes, smart cities, smart healthcare, and smart grids are among the edge-enabled approaches presented here. Moreover, it provides various challenges and future scope to develop an efficient and effective framework using intrusion detection systems on edge devices in a distributed manner.

REFERENCES

1. Sarmah, A. (2001). Intrusion detection systems: Definition, need and challenges. *Rapport Technique, SANS Institute,* 2001, 343.
2. Hatamian, A., Tavakoli, M. B., & Moradkhani, M. (2021). Improving the security and confidentiality in the Internet of medical things based on edge computing using clustering. *Computational Intelligence and Neuroscience*, 2021, 1687–5265.
3. Aljanabi, M., Ismail, M. A., & Ali, A. H. (2021). Intrusion detection systems, issues, challenges, and needs. *International Journal of Computational Intelligence Systems*, 14(1), 560–571.
4. Liao, H. J., Lin, C. H. R., Lin, Y. C., & Tung, K. Y. (2013). Intrusion detection system: A comprehensive review. *Journal of Network and Computer Applications,* 36(1), 16–24.
5. Liu, M., Xue, Z., Xu, X., Zhong, C., & Chen, J. (2018). Host-based intrusion detection system with system calls: Review and future trends. *ACM Computing Surveys (CSUR)*, 51(5), 1–36.
6. Ring, M., Wunderlich, S., Scheuring, D., Landes, D., & Hotho, A. (2019). A survey of network-based intrusion detection data sets. *Computers & Security,* 86, 147–167.
7. Ioulianou, P., Vasilakis, V., Moscholios, I., & Logothetis, M. (2018). A signature-based intrusion detection system for the Internet of Things. *Information and Communication Technology Form.* Graz, Austria: The Institute of Electronics, Information and Communication Engineers.
8. Alsoufi, M. A., Razak, S., Siraj, M. M., Nafea, I., Ghaleb, F. A., Saeed, F., & Nasser, M. (2021). Anomaly-based intrusion detection systems in iot using deep learning: A systematic literature review. *Applied Sciences,* 1(18), 8383.
9. Mell, P., Hu, V., Lippmann, R., Haines, J., & Zissman, M. (2003). An overview of issues in testing intrusion detection systems. NIST Interagency/Internal Report. National Institute of Standards and Technology.
10. Khan, W. Z., Ahmed, E., Hakak, S., Yaqoob, I., & Ahmed, A. (2019). Edge computing: A survey. *Future Generation Computer Systems,* 97, 219–235.
11. Wang, T., Yang, Q., Shen, X., Gadekallu, T. R., Wang, W., & Dev, K. (2021). A privacy-enhanced retrieval technology for the cloud-assisted internet of things. *IEEE Transactions on Industrial Informatics*, 18(7), 4981–4989.
12. Yousefpour, A., Fung, C., Nguyen, T., Kadiyala, K., Jalali, F., Niakanlahiji, A., ... & Jue, J. P. (2019). All one needs to know about fog computing and related edge computing paradigms: A complete survey. *Journal of Systems Architecture,* 98, 289–330.
13. Kumar, P., Gupta, G. P., & Tripathi, R. (2021). A distributed ensemble design based intrusion detection system using fog computing to protect the internet of things networks. *Journal of Ambient Intelligence and Humanized Computing,* 12, 9555–9572.
14. Arshad, J., Azad, M. A., Amad, R., Salah, K., Alazab, M., & Iqbal, R. (2020). A review of performance, energy and privacy of intrusion detection systems for IoT. *Electronics*, 9(4), 629.

15. Chen, J., Li, K., Deng, Q., Li, K., & Philip, S. Y. (2019). Distributed deep learning model for intelligent video surveillance systems with edge computing. *IEEE Transactions on Industrial Informatics,* 5,4(2019), 1060–1072.

16. Raponi, S., Caprolu, M., & Di Pietro, R. (2019). Intrusion detection at the network edge: Solutions, limitations, and future directions. In Edge Computing–EDGE 2019: Third International Conference, Held as Part of the Services Conference Federation, SCF 2019, San Diego, CA, USA, June 25–30, 2019, Proceedings 3 (pp. 59–75*). Springer International Publishing.*

17. Almogren, A. S. (2020). Intrusion detection in edge-of-things computing. *Journal of Parallel and Distributed Computing,* 137, 259–265.

18. Ramaiah, M., Chithanuru, V., Padma, A., & Ravi, V. (2023). A review of security vulnerabilities in industry 4.0 application and the possible solutions using blockchain. *Cyber Security Applications for Industry,* 1st Edition, Chapman and Hall/CRC, Volume 4, 63–95.

19. Khan, M. A., Karim, M. R., & Kim, Y. (2019). A scalable and hybrid intrusion detection system based on the convolutional-LSTM network. *Symmetry,* 11(4), 583.

20. Meng, W., Wang, Y., Li, W., Liu, Z., Li, J., & Probst, C. W. (2018). Enhancing intelligent alarm reduction for distributed intrusion detection systems via edge computing. In Information Security and Privacy: 23rd Australasian Conference, ACISP 2018, Wollongong, NSW, Australia, July 11–13, 2018, Proceedings 23 (pp. 759–767). *Springer International Publishing.*

21. Xiong, M., Li, Y., Gu, L., Pan, S., Zeng, D., & Li, P. (2020). Reinforcement learning empowered IDPS for vehicular networks in edge computing. *IEEE Network*, 34(3), 57–63.

22. Resende, P. A. A., & Drummond, A. C. (2018). A survey of random forest based methods for intrusion detection systems. *ACM Computing Surveys (CSUR),* 51(3), 1–36.

23. Singh, A., Chatterjee, K., & Satapathy, S. C. (2022). An edge based hybrid intrusion detection framework for mobile edge computing. *Complex & Intelligent Systems,* 8(5), 3719–3746.

24. Hayyolalam, V., Aloqaily, M., Özkasap, Ö., & Guizani, M. (2021). Edge intelligence for empowering IoT-based healthcare systems. *IEEE Wireless Communications*, 28(3), 6–14.

25. Oueida, S., Kotb, Y., Aloqaily, M., Jararweh, Y., & Baker, T. (2018). An edge computing based smart healthcare framework for resource management. *Sensors*, 18(12), 4307.

26. Tian, S., Yang, W., Le Grange, J. M., Wang, P., Huang, W., & Ye, Z. (2019). Smart healthcare: Making medical care more intelligent. *Global Health Journal*, 3(3), 62–65.

27. Hartmann, M., Hashmi, U. S., & Imran, A. (2022). Edge computing in smart health care systems: Review, challenges, and research directions. *Transactions on Emerging Telecommunications Technologies,* 33(3), e3710.

28. Oueida, S., Aloqaily, M., & Ionescu, S. (2019). A smart healthcare reward model for resource allocation in smart city. *Multimedia Tools and Applications,* 78, 24573–24594.

29. Velichko, A. (2021). A method for medical data analysis using the LogNNet for clinical decision support systems and edge computing in healthcare. *Sensors*, 21(18), 6209.

30. Vimal, S., Robinson, Y. H., Kadry, S., Long, H. V., & Nam, Y. (2021). IoT based smart health monitoring with CNN using edge computing. *Journal of Internet Technology*, 22(1), 173–185.

31. Ghoneim, A., Muhammad, G., Amin, S. U., & Gupta, B. (2018). Medical image forgery detection for smart healthcare. *IEEE Communications Magazine*, 56(4), 33–37.

32. Yu, B., Zhang, X., You, I., & Khan, U. S. (2021). Efficient computation offloading in edge computing enabled smart home. *IEEE Access,* 9, 48631–48639.

33. Gomez, C., Chessa, S., Fleury, A., Roussos, G., & Preuveneers, D. (2019). Internet of things for enabling smart environments: A technology-centric perspective. *Journal of Ambient Intelligence and Smart Environments*, 11(1), 23–43.

34. Trimananda, R., Younis, A., Wang, B., Xu, B., Demsky, B., & Xu, G. (2018, October). Vigilia: Securing smart home edge computing. In 2018 IEEE/ACM Symposium on Edge Computing (SEC) (pp. 74–89). IEEE.

35. Yar, H., Imran, A. S., Khan, Z. A., Sajjad, M., & Kastrati, Z. (2021). Towards smart home automation using IoT-enabled edge-computing paradigm. *Sensors*, 21(14), 4932.

36. Li, X., Chen, T., Cheng, Q., Ma, S., & Ma, J. (2020). Smart applications in edge computing: Overview on authentication and data security. *IEEE Internet of Things Journal*, 8(6), 4063–4080.

37. Kochovski, P., & Stankovski, V. (2018). Supporting smart construction with dependable edge computing infrastructures and applications. *Automation in Construction,* 85, 182–192.

38. Yuan, D., Ota, K., Dong, M., Zhu, X., Wu, T., Zhang, L., & Ma, J. (2020, June). Intrusion detection for smart home security based on data augmentation with edge computing. In ICC 2020–2020 IEEE International Conference on Communications (ICC) (pp. 1–6). IEEE.

39. Wang, Y., Liu, M., Zheng, P., Yang, H., & Zou, J. (2020). A smart surface inspection system using faster R-CNN in cloud-edge computing environment. *Advanced Engineering Informatics, 43*, 101037.

40. Feng, C., Wang, Y., Chen, Q., Ding, Y., Strbac, G., & Kang, C. (2021). Smart grid encounters edge computing: Opportunities and applications. *Advances in Applied Energy, 1*, 100006.

41. Yang, C., Chen, X., Liu, Y., Zhong, W., & Xie, S. (2019, May). Efficient task offloading and resource allocation for edge computing-based smart grid networks. In ICC 2019–2019 IEEE International Conference on Communications (ICC) (pp. 1–6). IEEE.

42. Qin, N., Li, B., Li, D., Jing, X., Du, C., & Wan, C. (2021, February). Resource allocation method based on mobile edge computing in smart grid. In *IOP Conference Series: Earth and Environmental Science* (Vol. 634, No. 1, p. 012054). IOP Publishing.

43. Zhang, Y., Wang, L., Sun, W., Green II, R. C., & Alam, M. (2011). Distributed intrusion detection system in a multi-layer network architecture of smart grids. *IEEE Transactions on Smart Grid, 2*(4), 796–808.

44. Chhaya, L., Sharma, P., Bhagwatikar, G., & Kumar, A. (2017). Wireless sensor network based smart grid communications: Cyber attacks, intrusion detection system and topology control. *Electronics, 6*(1), 5.

45. Mahmood, K., Li, X., Chaudhry, S. A., Naqvi, H., Kumari, S., Sangaiah, A. K., & Rodrigues, J. J. (2018). Pairing based anonymous and secure key agreement protocol for smart grid edge computing infrastructure. *Future Generation Computer Systems, 88*, 491–500.

46. Chen, S., Wen, H., Wu, J., Lei, W., Hou, W., Liu, W., ... & Jiang, Y. (2019). Internet of things based smart grids supported by intelligent edge computing. *IEEE Access, 7*, 74089–74102.

47. Mehmood, M. Y., Oad, A., Abrar, M., Munir, H. M., Hasan, S. F., Muqeet, H. A. U., & Golilarz, N. A. (2021). Edge computing for IoT-enabled smart grid. *Security and Communication Networks, 2021*, 1–16.

48. Khan, L. U., Yaqoob, I., Tran, N. H., Kazmi, S. A., Dang, T. N., & Hong, C. S. (2020). Edge-computing-enabled smart cities: A comprehensive survey. *IEEE Internet of Things Journal, 7*(10), 10200–10232.

49. Gheisari, M., Wang, G., & Chen, S. (2020). An edge computing-enhanced internet of things framework for privacy-preserving in smart city. *Computers & Electrical Engineering, 81*, 106504.

50. Bhattacharya, S., Somayaji, S. R. K., Gadekallu, T. R., Alazab, M., & Maddikunta, P. K. R. (2022). A review on deep learning for future smart cities. *Internet Technology Letters, 5*(1), e187.

51. Jararweh, Y., Otoum, S., & Al Ridhawi, I. (2020). Trustworthy and sustainable smart city services at the edge. *Sustainable Cities and Society, 62*, 102394.

52. Zhou, Z., Liao, H., Gu, B., Huq, K. M. S., Mumtaz, S., & Rodriguez, J. (2018). Robust mobile crowd sensing: When deep learning meets edge computing. *IEEE Network, 32*(4), 54–60.

53. Aloqaily, M., Otoum, S., Al Ridhawi, I., & Jararweh, Y. (2019). An intrusion detection system for connected vehicles in smart cities. *Ad Hoc Networks, 90*, 101842.

54. Zhang, Y., Wei, J., & Wang, K. (2020). An edge IDS based on biological immune principles for dynamic threat detection. *Wireless Communications and Mobile Computing, 2020*, 1–15.

55. Li, Q., Hou, J., Meng, S., & Long, H. (2020). GLIDE: A game theory and data-driven mimicking linkage intrusion detection for edge computing networks. *Complexity, 2020*, 1–18.

56. Lo, W. W., Layeghy, S., Sarhan, M., Gallagher, M., & Portmann, M. (2022, April). E-graphsage: A graph neural network based intrusion detection system for iot. In NOMS 2022–2022 IEEE/IFIP Network Operations and Management Symposium (pp. 1–9). IEEE.

57. Jiang, C., Kuang, J., & Wang, S. (2019, December). Home IOT intrusion prevention strategy based on edge computing. In 2019 IEEE 2nd International Conference on Electronics and Communication Engineering (ICECE) (pp. 94–98). IEEE.

58. Krishnan, P., Duttagupta, S., & Achuthan, K. (2019). SDNFV based threat monitoring and security framework for multi-access edge computing infrastructure. *Mobile Networks and Applications, 24*, 1896–1923.

59. Riggio, R., Coronado, E., Linder, N., Jovanka, A., Mastinu, G., Goratti, L., ... & Pistore, M. (2021, June). AI@ EDGE: A secure and reusable artificial intelligence platform for edge computing. In 2021

Joint European Conference on Networks and Communications & 6G Summit (EuCNC/6G Summit) (pp. 610–615). IEEE.

60. Nie, L., Wu, Y., Wang, X., Guo, L., Wang, G., Gao, X., & Li, S. (2021). Intrusion detection for secure social internet of things based on collaborative edge computing: A generative adversarial network-based approach. *IEEE Transactions on Computational Social Systems*, 9(1), 134–145.

61. Eskandari, M., Janjua, Z. H., Vecchio, M., & Antonelli, F. (2020). Passban IDS: An intelligent anomaly-based intrusion detection system for IoT edge devices. *IEEE Internet of Things Journal*, 7(8), 6882–6897.

62. Umer, M. A., Junejo, K. N., Jilani, M. T., & Mathur, A. P. (2022). Machine learning for intrusion detection in industrial control systems: Applications, challenges, and recommendations. *International Journal of Critical Infrastructure Protection*, 38, 100516.

63. Alwarafy, A., Al-Thelaya, K. A., Abdallah, M., Schneider, J., & Hamdi, M. (2020). A survey on security and privacy issues in edge-computing-assisted internet of things. *IEEE Internet of Things Journal*, 8(6), 4004–4022.

64. Yadav, N., Pande, S., Khamparia, A., & Gupta, D. (2022). Intrusion detection system on IoT with 5G network using deep learning. *Wireless Communications and Mobile Computing,* 2022, 1–13.

9 Recent Advancements in IoT Security-Based Challenges

A Brief Review

Suranjeet Chowdhury Avik[1], Abdullahi Chowdhury[2],
Ranesh Naha[3], Shahriar Kaisar[4], Arunkumar Arulappan,[5]
and Aniket Mahanti[6]

[1] School of Computer Science, Bangladesh University of Textiles,
Dhaka, Bangladesh, India
[2] School of Computer Science, The University of Adelaide,
South Australia, Australia
[3] Centre for Smart Analytics, Federation University Australia, Melbourne,
Australia
[4] Department of IS and Business Analytics, RMIT University, Melbourne,
Victoria, Australia
[5] School of Information Technology, VIT University, Vellore,
Tamil Nadu, India
[6] School of Computer Science, The University of Auckland, New Zealand

9.1 INTRODUCTION

IoT has acquired widespread recognition and appeal as the standard solution with limited resources. "Things," or devices embedded with sensors, can be interconnected via private or public networks, enabling remote control and management. The network is then used to exchange information amongst the devices using the established communication protocols. Connected "Things" range from modest wearables to massive machinery that contains sensor chips [1]. Smart shoes like the Lenovo ones have chips in them that let you track and analyze your exercise data. Remote monitoring of security cameras deployed at a spot is possible from any location in the world [2].

Furthermore, the IoT can contribute to the betterment of society. A wide range of smart gadgets, including those that monitor surgeries, measure weather conditions in cars, and identify animals using biochips, are already being used to meet the specific demands of the community [3]. The various aspects of IoT networks are illustrated in Figure 9.1. Enhancing the overall system efficiency can be achieved through real-time processing of the data collected by these devices.

Due to its extensive integration into daily life, the IoT holds a promising future. However, to address the issue of underutilization of the frequency spectrum, measures need to be taken [4, 5], and hardware solutions like cognitive radio-based networks are being used to increase bandwidth. Wireless sensor networks (WSN) and machine-to-machine (M2M) or cyber-physical systems (CPS), as well as the wider term "IoT," have grown as an integral part of the literature. The security concerns associated with WSN, M2M, or CPS persist within the realm of IoT as the IP protocol remains the dominant standard for connectivity. As a result, all aspects of the IoT deployment architecture must be protected from attacks that could disrupt or threaten data security, privacy, integrity, and the integrity of IoT services. Concerns about security are built into the IoT model because it comprises different networks and devices that are all linked together. The security of IoT is made

DOI: 10.1201/9781003407959-9

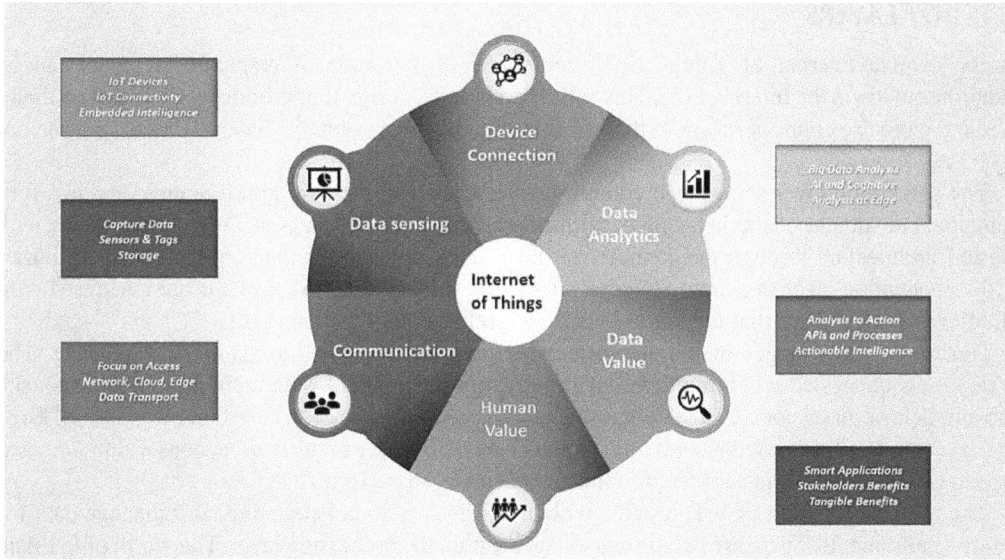

FIGURE 9.1 Functions and benefits of IoT.

more intricate by the constrained resources of small devices or sensor-equipped "Things" that possess limited power and memory.

Consequently, any proposed security measures must be customized to suit these restricted architectures. In recent years, the challenges posed by IoT security have been extensively studied. Various solutions have been proposed, with some focusing on specific security layers and others striving to secure IoT comprehensively across all layers. This chapter will concentrate on the security concerns related to application, architecture, communication, and data. For IoT security, this new taxonomy differs greatly from the traditional layered architecture presented in [6]. IoT protocols are also discussed and analyzed in terms of their security. Comparative access control systems are examined in that work. Comparative evaluation of intrusion detection systems is also a goal of the authors. The security concerns associated with fog computing are scrutinized in [7]. The authors emphasize trust management, authentication, privacy, data, and network security, as well as intrusion detection systems. On the other hand, [8] addresses identification and authentication, access control systems, network security, and trust management issues, while [9] [10] focus on fault tolerance for edge computing-based paradigms. For cloud-based IoT, [11] addresses several security issues and their possible solutions.

Unlike existing literature reviews, our primary contributions are as follows:

- A parametric examination of security risks and their correlation with potential IoT solutions.
- Explanation of the fundamental traits of the machine learning and deep learning-based intrusion detection systems and evaluation of their efficacy for IoT security.
- A summary of the suggested solutions for preventing attacks and guaranteeing the security of IoT devices.
- Future directions, outlining potential answers to unresolved IoT security issues.

The subsequent sections of this chapter are structured as follows. Section 9.2 delineates the IoT device layers. Section 9.3 categorizes the main components of the IoT ecosystem. In Section 9.4, we discuss the methods for IoT threat detection. Proposed mitigation approaches for ensuring IoT security are presented in Section 9.5. In Section 9.6, future research scopes related to IoT securities are discussed.

9.2   IOT LAYERS

In the diverse Internet of Things (IoT) ecosystem, a multitude of disparate elements can be interconnected via the Internet [12]. This calls for a layered design that is both versatile and resilient. While the precise count of layers in the IoT is still under discussion, we have abstracted seven core layers as follows:

The perception layer serves as a conduit between the tangible and digital realms, and is a fundamental constituent of the IoT's physical infrastructure [13]. Physical devices communicate with the IoT architecture through the connectivity layer. Data collected is then collated and scrutinized at the application layer to extract valuable business insights. Here, IoT systems are interfaced with middleware or software that facilitates better comprehension of the data [14].

During the early stages of IoT network expansion, latency poses a significant challenge. The system gets congested when multiple devices attempt to connect to the main hub simultaneously, causing delays. In response, edge computing accelerated the pace of IoT system progression. Edge IoT layers now allow systems to process and analyze data closer to the source, conserving time and resources. This leads to faster response times and enhanced performance.

The primary function of IoT systems within this layer is to collect, store, and manage data for future requirements. There are two primary stages within the processing layer. The worth of IoT data is realized only when utilized strategically within business planning and operations. Every organization, through data analysis, aspires to meet a specific set of goals and objectives. Business owners and stakeholders can now forecast the future with remarkable precision by analyzing past and present data. In the current business environment, data analysis has emerged as a precious resource. Companies are in fierce competition to access, analyze, and make decisions based on a vast amount of data. There is heightened interest in software, CRM, and business intelligence systems due to their superior performance.

Security stands as one of the most critical issues in the present IT environment. Concerns related to security breaches, malware tracking, and hacking rise significantly when integrating IoT technologies. The devices themselves form the first line of defense in the IoT security layers. For IoT integration, the majority of manufacturers comply with rigorous security guidelines that need to be incorporated both in the firmware and the hardware.

9.3   COMPONENTS OF THE IOT ECOSYSTEM

IoT ecosystem components are highlighted in this portion of the report. There are five sections to the entire ecosystem [15]:

9.3.1   IoT DEVICES

The IoT system is built on the foundation of IoT devices. The IoT architecture includes these devices at every level. Open-source or proprietary IoT devices exist. Accessible IoT devices are ones that are available to the public [16]. It's possible to study, modify, and reproduce open-source devices to suit your needs, whereas proprietary IoT devices are licensed, and their specifications are kept secret. Due to a lack of internal storage, memory, processing power, and battery life, IoT devices have a limited range of applications [17].

9.3.2   IoT GATEWAYS

Gateways connect networks of sensors, high-end IoT devices, and cloud platforms across the Internet. In IoT architecture, gateways serve as a crucial link between different networks. To increase performance and create a solid foundation for the Internet of Things, gateways keep track of both them and the terminal nodes they're connected to [18]. When the IoT ecosystem is vast, gateways can operate as high-end devices, but more commonly, they act as middle-end devices when there are only a few low-end devices in the ecosystem. The hardware of gateways is designed to withstand a wide range of environmental conditions. To overcome failure, they can address the communication barrier between lower-end devices and controllers. Operating systems for gateways are smaller and more limited in terms of memory, processing power consumption, and power consumption [19].

9.3.3 IoT OS

Operating systems (OS) are a collection of programs that serve as a link between software and hardware. The operating system is used to design and operate programs. The OS is installed on IoT devices so that programs may run on the IoT device and the device can be managed. As a result, the OS can be characterized as a component of the hardware that makes an IoT device complete [20]. In other words, OS oversees managing and monitoring power consumption, enabling the execution of instructions, programming IoT devices for their attributes, and enabling the connection between devices. An IoT ecosystem that is both scalable and trustworthy relies heavily on its OS. An OS must be able to adapt to the unique needs of each sub-system in the IoT ecosystem. Capacity, compute, energy capacity, and memory capacity are some of the key properties of IoT devices. In the future, IoT devices should have the ability to customize these three attributes [21]. OSs are most suited if they can operate, connect, and communicate with IoT devices as their properties evolve.

9.3.4 IoT Communication

IoT devices have limited storage. Protocols and interfaces are needed to connect many devices to the Internet. If IoT devices connect directly to the Internet, an IP stack is required. In terms of resources, the IP stack consumes a lot of both power and memory. IoT devices also use a variety of IoT interfaces, such as GPIO (general purpose input output), ADC, DAC, Inter IC (I2C), serial peripherals (SPI), universal asynchronous receiver transmitter (UART), and so on. GPIO is a low-level I/O interface that directly connects processors to peripheral devices. Its flexibility, ease of implementation, and portability make it an excellent choice. The ADC/DAC functions as a voltage converter, converting analog voltage to digital with the ADC and digital voltage to analog with the DAC. Whether an ADC or DAC is necessary now depends on the present signal format and the input and output signal formats required [22]. For accurate signal transmission, ADC is necessary, while DAC is necessary for high-speed signal transmission. Interface for SPI. Serial data interchange is made possible by the use of interfaces of this type. Full-duplex operation of SPI allows data to be sent in both directions at the same time. Processors with peripherals make use of this. Short-range communication is made possible through SPI. Most of the time, these kinds of interfaces are employed in the context of a master–slave relationship. Integrated circuits can communicate with each other via the I2C serial bus, which is a bidirectional two-wire serial connection. It is utilized in devices where each chip can assume the initiator role. Serial communication between a master and a slave is supported.

9.3.5 IoT Middleware

In the IoT ecosystem, most of the IoT middleware sends raw data to users via web applications. A variety of IoT-related applications require data obtained from IoT devices to be transformed into a usable format. It is possible to overcome the difficulties by using IoT middleware [23]. However, its data interpretation and integration capabilities are severely constrained. Inter-device connection, data collection and access, and combining data from numerous devices to generate a flexible composition are major challenges. Scalability, adaptability, flexibility, security, and open source are all requirements for IoT middleware [24]. Using this type of middleware, researchers and developers could create new apps and add new IoT devices without having to rewrite the entire IoT ecosystem at a low level.

9.4 METHODS FOR IOT THREAT DETECTION

9.4.1 THREAT DETECTION USING MACHINE LEARNING (ML)

Machine learning (ML) algorithms, including decision trees, support vector machines, Bayesian and K-nearest neighbor (KNN) algorithms, random forests, association rules, and ensemble learning, are evaluated in this study in relation to their strengths, limitations, and potential applications in IoT security [25].

Decision tree (DT) classifiers categorize data samples into various groups using data attributes. The classifier learns decision rules from training datasets, thus establishing class labels from a target variable. The classifier utilizes two methods—information gain and gini index—to identify the optimal feature set. The primary functions of the DT classifier are to facilitate classification and construct decision trees. A class label's position as the tree's root is determined by its Gini index or information gain [26].

Support vector machines (SVM) utilize hyper-planes to segregate data into two or more classes. The hyperplane scans all data points to maximize the margin between the hyperplane and its closest points [27, 28]. Naive Bayes (NB) leverages the Bayes theorem to calculate the posterior probability of an occurrence, given its prior probability of belonging to a particular category [29]. For instance, the NB technique can identify probes in network traffic. The NB method calculates the probability of a specific class using Bayes' theorem and examines a given input sample's feature set to determine if it falls within a predefined class [30]. The NB classifier's simplicity and straightforward implementation make it a favorite choice, especially in intrusion detection systems (IDS), where there is a strong correlation between features and input variables.

The k-nearest neighbor algorithm is a non-parametric method that addresses classification and regression problems with a user-defined k [31]. All the sample class labels and vectors are stored in the training data. Random forest (RF) can classify data in binary and multi-class formats [32]. An accurate classification model for a specific subject is built using multiple decision trees in an RF tree. The random forest algorithm creates random trees trained to predict the class of individual input samples, with the most common class label assigned to the final product [33].

The RF classifier uses mutual information to determine feature importance and select the most relevant attributes for data categorization. However, this method struggles with extensive datasets due to its time-consuming nature. These algorithms can help detect distributed denial of service (DDoS) attacks in IoT networks, and the RF algorithm can also identify unauthorized IoT devices [34].

Association rule (AR) mining techniques uncover the unknown variable in a dataset by exploring the relationships between other variables in the set. The AR method seeks patterns amongst variables in large datasets while simultaneously constructing a predictive model [35, 36, 37]. However, AR algorithms are less commonly used in IoT applications compared to other ML methods, as they might overemphasize correlations between variables and occurrences, reducing their effectiveness in security applications [38].

Ensemble learning (EL) approaches excel at stream categorization [39]. They can easily incorporate drift detection methods, facilitating dynamic adjustments such as feature removal or the addition of new classifiers [40]. EL leverages multiple methodologies to reduce variance and hypothesis bias, demonstrating adaptability [41]. However, they are more time-consuming than single-classifier systems because they utilize multiple classifiers [42]. EL has been deployed in intrusion and malware detection [43] and can detect network attacks in IoT networks using a tree classifier ensemble [44].

9.4.2 THREAT DETECTION USING DEEP LEARNING (DL)

Given the significant volume of data generated by IoT devices in the network, deep learning (DL) techniques are fitting for such networks. DL methods can extract features from the data, providing a more efficient data representation. Owing to their layered structure, layered design methods enhance connectivity between IoT network devices and applications, removing the need for human intervention [45]. The intelligence of an IoT-based house stems from the automation integrated into each device. DL techniques utilize multiple layers of data abstraction within their computational structure, which encompasses numerous processing stages. Unlike traditional ML methods, DL techniques can handle more complex problems due to their non-linear processing layers, which discover and represent learning patterns in data through generative or discriminative feature extraction and representation. The deep architecture of the DL approach enables data to be represented in a hierarchical structure [45], which is why DL methods are sometimes referred to as hierarchical learning. Ensemble DL refers to the integration of multiple DL techniques.

Convolutional neural networks (CNNs) were developed to reduce data attributes that could be utilized in standard neural networks [46]. A typical CNN design consists of a convolutional layer and a pooling layer. Convolutional layers use kernel functions of the same size to process data attributes. These layers also conduct sampling to facilitate subsequent levels. Through max-pooling, clusters don't overlap, and the maximum value is selected for each one. The average pooling layer takes an average value from each cluster, while the activation layer of a CNN contains non-linear activation functions for each feature [47].

DL techniques like the recurrent neural network (RNN) can construct directed graph structures. RNNs are particularly adept at processing sequential data that demonstrates dynamic behavior, being able to ingest inputs and retain their prior state. Applications of RNN algorithms include text classification and wireless sensor networks [48].

Restricted Boltzmann machines (RBMs) were designed as deep generative models for shallow learning of the input data collection's probability distribution. In RBMs, nodes in the same layer are not interconnected. There are considerably more hidden variables than input variables in the visible layer. Each layer in the RBM hierarchy builds upon the features learned in the preceding layer [49]. Constructing an RBM-based intrusion detection model presents challenges due to the dataset's multiple components and irregular nature. A network anomaly detection model is created using discriminative RBM to address these issues. However, the model proved ineffective when a different network dataset was used, indicating a need for a more generalized classifier [50]. Multiple RBMs can be stacked to construct a DBN, which addresses the limitation of a single RBM's feature learning capability.

DBN models, built on RBMs, employ unsupervised hierarchical training to enhance performance [51]. The DBN uses layers of pre-trained RBM for pre-training, followed by the formation of a general feedforward network. A softmax layer is then applied to fine-tune features based on labeled samples [52]. DBN-based anomaly detection has been implemented in secure mobile edge computing and performed well when compared to other ML techniques [53]. A self-adaptive model has been developed to handle different types of network anomalies using a genetic algorithm and DBN [54]. A hybrid technique based on DBN and SVM has been used for detecting abnormalities by analyzing network traffic and payload properties [55].

Generative adversarial networks (GANs), a DL technique, can simultaneously train two models: generative and discriminative [56]. The generative model creates data samples and analyzes the data distribution, while the discriminative model evaluates these samples. During training, GANs are presented with dataset data until satisfactory accuracy is achieved. A GAN-based distributed IDS has been proposed, which operates without a centralized entity and keeps watch for internal and external attacks [57]. GANs can also generate samples faster than other DL techniques, but training high-dimensional data using a GAN model is challenging due to the instability and complexity of the training model [58].

9.5 THREAT MITIGATION APPROACHES FOR ENSURING IOT SECURITY

9.5.1 SDN-BASED SECURITY FEATURES FOR IoT

Increasingly, security researchers are turning to SDNs for their studies. In this part, we'll look at the significant SDN capabilities that can be used to secure IoT systems [59]. High-level concepts are explained in detail, along with the relevant answers and real-world applications.

The findings of the current study also shed light on the deployment of SDN-based security solutions in a variety of IoT networking environments (Figure 9.2).

9.5.2 NFV SECURITY FEATURES IN THE IoT ECOSYSTEM

New security solutions that consider the heterogeneity of IoT devices and their predicted wide-spread deployment could be offered by the NFV paradigm to cope with IoT vulnerabilities. There are several other services available to consumers of Telco providers, including the security-as-a-service model, which provides on-demand preventative and defense measures. The NFV paradigm has various advantages for IoT contexts, as detailed in the following analysis of the primary parts of NFV-based security techniques [60]. Organizations and end-users can benefit from CSA standards for cloud-delivered defense solutions. These standards have been set by CSA. Because virtualized security services can be deployed directly on the forwarding path, the NFV method offers significant benefits when hosting in faraway cloud data centers. We'll focus on the most critical components of the NFV paradigm for IoT systems in the following sections. NFV security aspects are being presented to conclude how to safeguard IoT devices using NFV-based security methods. SDN features were reviewed in the preceding section.

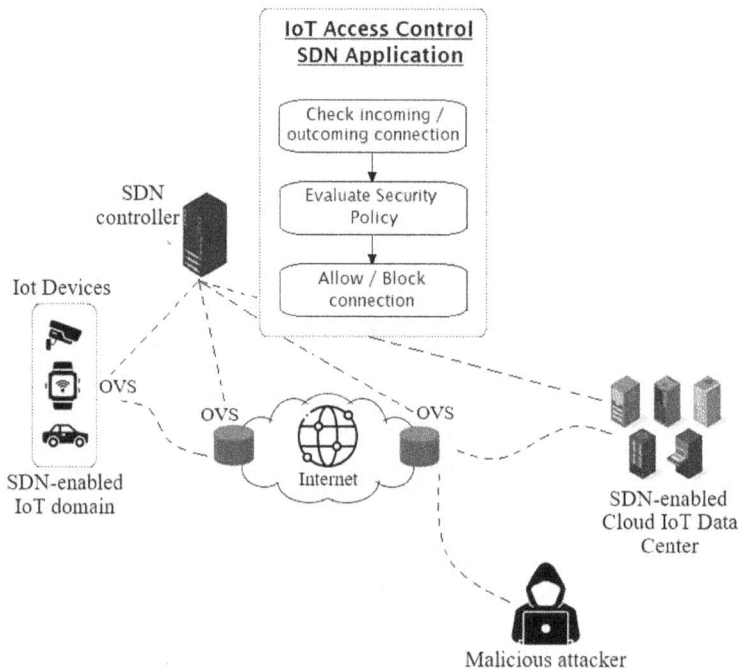

FIGURE 9.2 SDN-based access control features in IoT domains.

9.5.3 Learning Automata-Based DDoS Defense

Learning automaton-based solutions to prevent DDoS attacks on IoT networks that use service oriented architecture (SOA) as a system model [61]. It also proposes an alternative approach for preventing DDoS attacks at all network architecture levels, including detecting attacks, identifying the attacker, and countering them. A threshold value is set to detect the extent of the attack, and a learning automata idea is used to determine the optimal sample rate. The server rejects requests received during a random sampling process, effectively nullifying the DDoS attack.

9.5.4 Honeypots-Based DDoS Defense

Honeypots are used as a trap for attackers and to collect information about their attack methods (Figure 9.3). In the case of a suspected DDoS attack, intrusion detection systems are used to detect anomalies in incoming requests, and the honeypot is used instead of the main server [62]. The collected information about the suspect is stored in the database. Upon identifying a suspicious request, the primary server sends a request to verify the client's legitimacy. If the request is considered spam, the client is blocked, and legitimate requests are served by the main server.

9.5.5 Intrusion Detection Based on a Multivariate Model

Models based on correlations between two or more measurements are an expansion of the average and standard deviation models rather than the mean and standard deviation models [63]. Rather than relying on a single statistic to detect unusual occurrences, combining correlation measurements provides better precision. In a Markov process model-based intrusion detection technique, the state transition matrix represents the frequency of migration between states rather than individual states or audit records (Figure 9.4). If the reported new event is lower than anticipated for a given prior state and matrix, it's called abnormal. Odd instructions or event sequences can be spotted using the model rather than single events, which is its advantage. Once the flow of events has been represented using vectors, clustering algorithms can be used to classify them into different behavior categories. Normal or abnormal conduct is defined by allocating an activity to one of two categories: normal or abnormal.

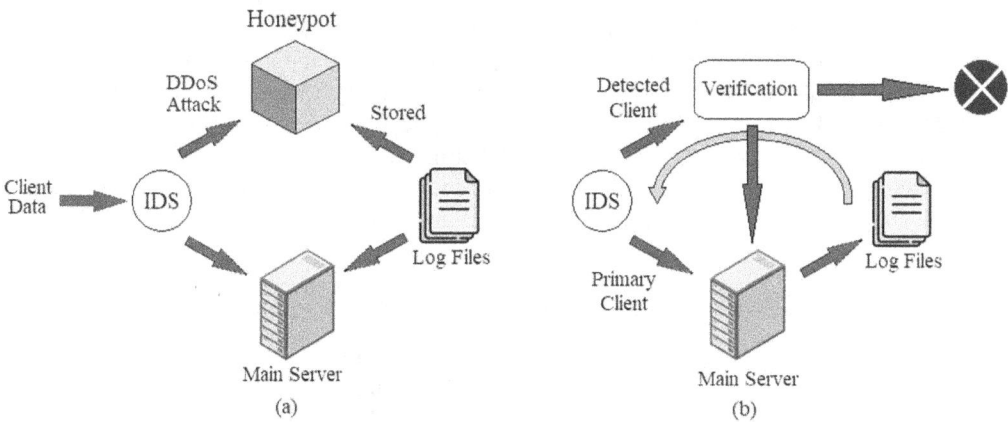

FIGURE 9.3 Honeypots collect DDoS attack features into log files and match them against suspected invasions.

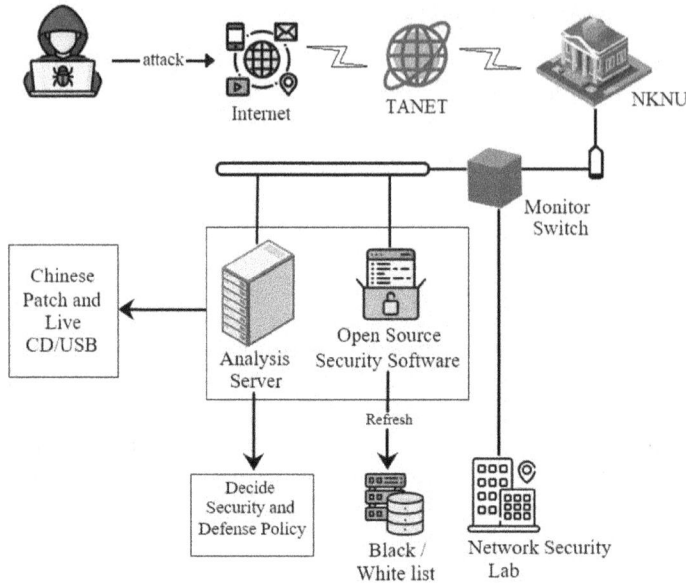

FIGURE 9.4 Intrusion detection process.

9.5.6 SODA: A SOFTWARE-DEFINED SECURITY FRAMEWORK FOR IoT ENVIRONMENTS

An IoT network where each device has its own set of security policies is the target of SODA's design [64]. The software-based security scheme has been implemented on a flexible and programmable device. This makes it possible for the IoT network to have a security scheme that is set up by the user and fits their needs. In an IoT environment, the security policy and service management are done by SODA's different parts, including Datapath, which connects IoT devices to the control plane, function plane, and data plane.

9.5.7 LIGHTWEIGHT NFC PROTOCOL FOR PRIVACY PROTECTION IN MOBILE IoT

Mobile IoT has a different structure than traditional IoT [65]. NFC tags, NFC phones, NFC watches, special NFC readers, and many more NFC intelligent devices are all part of the system for the NFC mobile IoT network (Figure 9.5). In this case, we must emphasize the importance of NFC mobile phones. The NFC mobile phone can operate in three different modes: as a tag, as a reader, and as a file-sharing device between two phones. It's possible that devices have authentication requirements, making the connection perplexing [66]. A reticular structure can be found in such a system. The mobile Internet of Things (IoT) must examine how to ensure mutual authentication between devices while offering security services to facilitate the interconnection of devices [67].

9.5.8 SECURE DATA PROVENANCE IN IoT NETWORK USING BLOOM FILTERS

Ciphertext-policy attribute-based encryption (CP-ABE) is used to protect the provenance process (Figure 9.6). In an IoT-based cloud context, the CP-ABE implementation model is utlized [68]. By outsourcing encryption and decryption to the edge nodes (CP-ABE algorithm), load sharing has been established in [69, 70]. To construct the partial ciphertext, the IoT devices use the distributed CP-ABE algorithm [71]. This algorithm has a low computing overhead. To produce the rest of the ciphertext, which has a substantial computational burden, fog/edge nodes would have to be used.

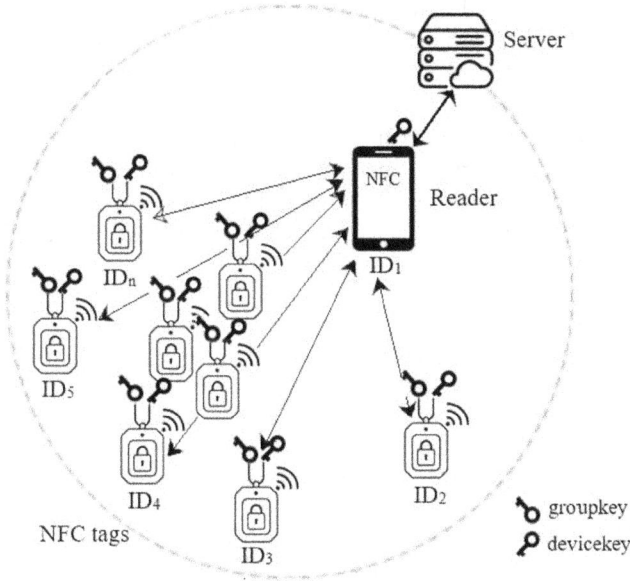

FIGURE 9.5 Key-management model for mobile IoT authentication.

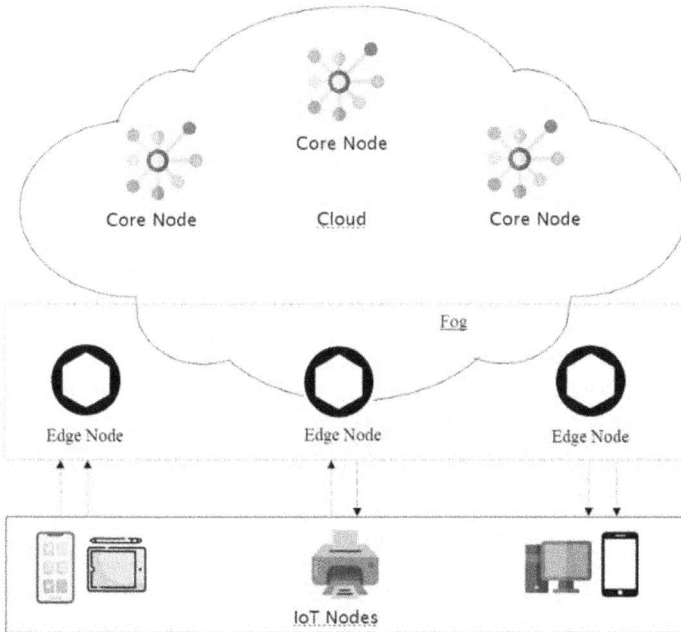

FIGURE 9.6 Conceptual model-IoT-based cloud/fog computing scenario.

9.6 FUTURE RESEARCH SCOPE

IoT is widely used due to its capacity to connect physical equipment from multiple application domains to consumers over the Internet. Security flaws in IoT networks have emerged, however, due to the interconnectedness of IoT and the devices' ability to communicate with one another.

In this research, we have reviewed several approaches to protecting IoT networks. We start with a high-level introduction to IoT technology, then move on to a taxonomy of the various tiers and building blocks that make up the IoT. After that, the chapter goes through numerous security risks and obstacles that IoT networks face.

Furthermore, machine learning and deep learning strategies for IDS monitoring are given. The use of machine learning algorithms to analyze real-time data is explored here. Many research processes have been implemented to ensure IoT security. With the advancement of engagement of more complex ML algorithms in IoT, security issues will mitigate more in the future.

REFERENCES

[1] D. Giusto, A. Iera, G. Morabito, and L. Atzori, *The internet of things: 20th Tyrrhenian workshop on digital communications.* Springer Science & Business Media, 2010.

[2] B. Heater, *"Lenovo shows off a pair of intel-powered smart shoes, 2016,"* URL https://techcrunch.com/2016/06/09/lenovo-smart-shoes

[3] M. Rouse, *"Internet of things (IoT),"* [online] available at: http://Internetofthingsagenda.techtarget.com/definition, 2016.

[4] A. A. Khan, M. H. Rehmani, and A. Rachedi, "Cognitive-radio-based Internet of things: Applications, architectures, spectrum related functionalities, and future research directions," *IEEE Wireless Communications*, vol. 24, no. 3, pp. 17–25, 2017.

[5] F. Akhtar, M. H. Rehmani, and M. Reisslein, "White space: Definitional perspectives and their role in exploiting spectrum opportunities," *Telecommunications Policy*, vol. 40, no. 4, pp. 319–331, 2016.

[6] M. A. Sadeeq, S. R. Zeebaree, R. Qashi, S. H. Ahmed, and K. Jacksi, *"Internet of things security: A survey,"* in 2018 International Conference on Advanced Science and Engineering (ICOASE). IEEE, 2018, pp. 162–166.

[7] J. Granjal, E. Monteiro, and J. S. Silva, "Security for the internet of things: A survey of existing protocols and open research issues," *IEEE Communications Surveys & Tutorials*, vol. 17, no. 3, pp. 1294–1312, 2015.

[8] R. Roman, C. Alcaraz, J. Lopez, and N. Sklavos, "Key management systems for sensor networks in the context of the internet of things," *Computers Electrical Engineering*, vol. 37, no. 2, pp. 147–159, 2011.

[9] S. Yi, Z. Qin, and Q. Li, *"Security and privacy issues of fog computing: A survey,"* in International Conference on Wireless Algorithms, Systems, and Applications. Springer, 2015, pp. 685–695.

[10] Y. Wang, T. Uehara, and R. Sasaki, "Fog computing: Issues and challenges in security and forensics," in 2015 IEEE 39th annual computer software and applications conference, vol. 3. IEEE, 2015, pp. 53–59.

[11] I. Al Barazanchi, A. Murthy, A. A. Al Rababah, G. Khader, H. R. Abdulshaheed, H. T. Rauf, E. Daghighi, and Y. Niu, "Blockchain technology-based solutions for IoT security," *Iraqi Journal for Computer Science and Mathematics*, vol. 3, no. 1, pp. 53–63, 2022.

[12] S. Panchiwala and M. Shah, "A comprehensive study on critical security issues and challenges of the IoT world," *Journal of Data, Information and Management*, vol. 2, no. 4, pp. 257–278, 2020.

[13] A. V. Vijayalakshmi and L. Arockiam, "A study on security issues and challenges in IoT," *International Journal of Engineering Sciences & Management Research*, vol. 3, no. 11, pp. 34–43, 2016.

[14] M. Frustaci, P. Pace, G. Aloi, and G. Fortino, "Evaluating critical security issues of the IoT world: Present and future challenges," *IEEE Internet of Things Journal*, vol. 5, no. 4, pp. 2483–2495, 2017.

[15] S. Bansal and D. Kumar, "IoT ecosystem: A survey on devices, gateways, operating systems, middleware and communication," *International Journal of Wireless Information Networks,* pp. 1–25, 2020.

[16] N. Singh and M. Vardhan, "Distributed ledger technology based property transaction system with support for IoT devices," *International Journal of Cloud Applications and Computing (IJCAC)*, vol. 9, no. 2, pp. 60–78, 2019.

[17] E. Baccelli, C. Gundo̎gan, O. Hahm, P. Kietzmann, M. S. Lenders,̎ H. Petersen, K. Schleiser, T. C. Schmidt, and M. Wahlisch, "Riot: An̎ open source operating system for low-end embedded devices in the IoT," *IEEE Internet of Things Journal*, vol. 5, no. 6, pp. 4428–4440, 2018.

[18] P. Desai, A. Sheth, and P. Anantharam, *"Semantic gateway as a service architecture for IoT interoperability,"* in 2015 IEEE International Conference on Mobile Services. IEEE, 2015, pp. 313–319.

[19] G. Aloi, G. Caliciuri, G. Fortino, R. Gravina, P. Pace, W. Russo, and C. Savaglio, "Enabling IoT interoperability through opportunistic smartphone-based mobile gateways," *Journal of Network and Computer Applications*, vol. 81, pp. 74–84, 2017.

[20] A. Musaddiq, Y. B. Zikria, O. Hahm, H. Yu, A. K. Bashir, and S. W. Kim, "A survey on resource management in IoT operating systems," *IEEE Access*, vol. 6, pp. 8459–8482, 2018.

[21] D. Balsamo, A. Elboreini, B. M. Al-Hashimi, and G. V. Merrett, *"Exploring arm mbed support for transient computing in energy harvesting IoT systems,"* in 2017 7th IEEE International Workshop on Advances in Sensors and Interfaces (IWASI). IEEE, 2017, pp. 115–120.

[22] S. Iraji, P. Mogensen, and R. Ratasuk, "Recent advances in m2m communications and internet of things (IoT)," *International Journal of Wireless Information Networks*, vol. 24, no. 3, pp. 240–242, 2017.

[23] A. H. Ngu, M. Gutierrez, V. Metsis, S. Nepal, and Q. Z. Sheng, "IoT middleware: A survey on issues and enabling technologies," *IEEE Internet of Things Journal*, vol. 4, no. 1, pp. 1–20, 2016.

[24] M. A. da Cruz, J. J. Rodrigues, A. K. Sangaiah, J. Al-Muhtadi, and V. Korotaev, "Performance evaluation of IoT middleware," *Journal of Network and Computer Applications,* vol. 109, pp. 53–65, 2018.

[25] A. Thakkar and R. Lohiya, "A review on machine learning and deep learning perspectives of ids for IoT: Recent updates, security issues, and challenges," *Archives of Computational Methods in Engineering*, vol. 28, no. 4, pp. 3211–3243, 2021.

[26] B. Gupta, A. Rawat, A. Jain, A. Arora, and N. Dhami, "Analysis of various decision tree algorithms for classification in data mining," *International Journal of Computer Applications*, vol. 163, no. 8, pp. 15–19, 2017.

[27] H. Xiao, B. Biggio, B. Nelson, H. Xiao, C. Eckert, and F. Roli, "Support vector machines under adversarial label contamination," *Neurocomputing*, vol. 160, pp. 53–62, 2015.

[28] E. A. Shams and A. Rizaner, "A novel support vector machine based intrusion detection system for mobile ad hoc networks," *Wireless Networks*, vol. 24, no. 5, pp. 1821–1829, 2018.

[29] A. S. A. Aziz, E. Sanaa, and A. E. Hassanien, "Comparison of classification techniques applied for network intrusion detection and classification," *Journal of Applied Logic*, vol. 24, pp. 109–118, 2017.

[30] N. Ashraf, W. Ahmad, and R. Ashraf, "A comparative study of data mining algorithms for high detection rate in intrusion detection system," *Annals of Emerging Technologies in Computing (AETiC),* Print ISSN, vol. 2, no.1, pp. 2516–0281, 2018.

[31] G. Serpen and E. Aghaei, "Host-based misuse intrusion detection using pca feature extraction and knn classification algorithms," *Intelligent Data Analysis*, vol. 22, no. 5, pp. 1101–1114, 2018.

[32] G. Biau and E. Scornet, "A random forest guided tour," *Test*, vol. 25, no. 2, pp. 197–227, 2016.

[33] A. L. Buczak and E. Guven, "A survey of data mining and machine learning methods for cyber security intrusion detection," *IEEE Communications Surveys & Tutorials*, vol. 18, no. 2, pp. 1153–1176, 2015.

[34] Y. Meidan, M. Bohadana, A. Shabtai, M. Ochoa, N. O. Tippenhauer, J. D. Guarnizo, and Y. Elovici, "Detection of unauthorized IoT devices using machine learning techniques," *arXiv preprint arXiv:1709.04647*, 2017.

[35] J. Hussain and P. Kalita, "Understanding network intrusion detection system using olap on nsl-kdd dataset," *IUP Journal of Computer Sciences*, vol. 9, no. 3, p. 59, 2015.

[36] S. Elhag, A. Fernandez, A. Bawakid, S. Alshomrani, and F. Herrera, "On the combination of genetic fuzzy systems and pairwise learning for improving detection rates on intrusion detection systems," *Expert Systems with Applications*, vol. 42, no. 1, pp. 193–202, 2015.

[37] A. Tajbakhsh, M. Rahmati, and A. Mirzaei, "Intrusion detection using fuzzy association rules," *Applied Soft Computing*, vol. 9, no. 2, pp. 462–469, 2009.

[38] D. Kim, H. Yu, H. Lee, E. Beighley, M. Durand, D. E. Alsdorf, and E. Hwang, "Ensemble learning regression for estimating river discharges using satellite altimetry data: Central congo river as a testbed," *Remote Sensing of Environment*, vol. 221, pp. 741–755, 2019.

[39] H. M. Gomes, J. P. Barddal, F. Enembreck, and A. Bifet, "A survey on ensemble learning for data stream classification," *ACM Computing Surveys (CSUR)*, vol. 50, no. 2, pp. 1–36, 2017.

[40] Y. Ren, L. Zhang, and P. N. Suganthan, "Ensemble classification and regression-recent developments, applications and future directions," *IEEE Computational Intelligence Magazine*, vol. 11, no. 1, pp. 41–53, 2016.

[41] I. H. Witten and E. Frank, "Data mining: Practical machine learning tools and techniques with java implementations," *ACM Sigmod Record*, vol. 31, no. 1, pp. 76–77, 2002.

[42]   D. Gaikwad and R. C. Thool, *"Intrusion detection system using bagging ensemble method of machine learning,"* in 2015 International Conference on Computing Communication Control and Automation. IEEE, 2015, pp. 291–295.

[43]   A. A. Aburomman and M. B. I. Reaz, "A novel svm-knn-pso ensemble method for intrusion detection system," *Applied Soft Computing*, vol. 38, pp. 360–372, 2016.

[44]   R. R. Reddy, Y. Ramadevi, and K. Sunitha, *"Enhanced anomaly detection using ensemble support vector machine,"* in 2017 International Conference on Big Data Analytics and Computational Intelligence (ICBDAC). IEEE, 2017, pp. 107–111.

[45]   V. Sze, Y.-H. Chen, T.-J. Yang, and J. S. Emer, "Efficient processing of deep neural networks: A tutorial and survey," *Proceedings of the IEEE*, vol. 105, no. 12, pp. 2295–2329, 2017.

[46]   D. Scherer, A. Muller, and S. Behnke, "Evaluation of pooling operations¨ in convolutional architectures for object recognition," *in International Conference on Artificial Neural Networks*. Springer, 2010, pp. 92–101.

[47]   E. De Coninck, T. Verbelen, B. Vankeirsbilck, S. Bohez, P. Simoens, P. Demeester, and B. Dhoedt, "Distributed neural networks for internet of things: The big-little approach," *in International Internet of Things Summit*. Springer, 2015, pp. 484–492.

[48]   P. Voigtlaender, P. Doetsch, and H. Ney, *"Handwriting recognition with large multidimensional long short-term memory recurrent neural networks,"* in 2016 15th International Conference on Frontiers in Handwriting Recognition (ICFHR). IEEE, 2016, pp. 228–233.

[49]   U. Fiore, F. Palmieri, A. Castiglione, and A. De Santis, "Network anomaly detection with the restricted boltzmann machine," *Neurocomputing*, vol. 122, pp. 13–23, 2013.

[50]   G. E. Hinton, "Deep belief networks," *Scholarpedia*, vol. 4, no. 5, p. 5947, 2009.

[51]   Q. Zhang, L. T. Yang, Z. Chen, and P. Li, "A survey on deep learning for big data," *Information Fusion*, vol. 42, pp. 146–157, 2018.

[52]   Y. Chen, Y. Zhang, and S. Maharjan, "Deep learning for secure mobile edge computing," *arXiv preprint arXiv:1709.08025,* 2017.

[53]   Y. Zhang, P. Li, and X. Wang, "Intrusion detection for IoT based on improved genetic algorithm and deep belief network," *IEEE Access*, vol. 7, pp. 31711–31722, 2019.

[54]   S. Huda, J. Yearwood, M. M. Hassan, and A. Almogren, "Securing the operations in SCADA-IoT platform based industrial control system using ensemble of deep belief networks," *Applied Soft Computing*, vol. 71, pp. 66–77, 2018.

[55]   B. A. Tama and K.-H. Rhee, "Attack classification analysis of IoT network via deep learning approach," *Research Briefs Information & Communication Technology Evolution (ReBICTE),* vol. 3, pp. 1–9, 2017.

[56]   R. E. Hiromoto, M. Haney, and A. Vakanski, *"A secure architecture for IoT with supply chain risk management,"* in 2017 9th IEEE International Conference on Intelligent Data Acquisition and Advanced Computing Systems: Technology and Applications (IDAACS), vol. 1. IEEE, 2017, pp. 431–435.

[57]   Q. Cui, Z. Zhou, Z. Fu, R. Meng, X. Sun, and Q. J. Wu, "Image steganography based on foreground object generation by generative adversarial networks in mobile edge computing with Internet of things," *IEEE Access*, vol. 7, pp. 90815–90824, 2019.

[58]   T. Salimans, I. Goodfellow, W. Zaremba, V. Cheung, A. Radford, and X. Chen, "Improved techniques for training gans," *Advances in Neural Information Processing Systems,* vol. 29, pp. 2234–2242, 2016.

[59]   I. Farris, T. Taleb, Y. Khettab, and J. Song, "A survey on emerging SDN and NFV security mechanisms for IoT systems," *IEEE Communications Surveys & Tutorials*, vol. 21, no. 1, pp. 812–837, 2018.

[60]   V. Getov, *"Security as a service in smart clouds–opportunities and concerns,"* in 2012 IEEE 36th Annual Computer Software and Applications Conference. IEEE, 2012, pp. 373–379.

[61]   R. Vishwakarma and A. K. Jain, "A survey of DDoS attacking techniques and defence mechanisms in the IoT network," *Telecommunication Systems*, vol. 73, no. 1, pp. 3–25, 2020.

[62]   M. Anirudh, S. A. Thileeban, and D. J. Nallathambi, *"Use of honeypots for mitigating DoS attacks targeted on IoT networks,"* in 2017 International Conference on Computer, Communication and Signal Processing (ICCCSP). IEEE, 2017, pp. 1–4.

[63]   D. Li, Z. Cai, L. Deng, X. Yao, and H. H. Wang, "Information security model of block chain based on intrusion sensing in the IoT environment," *Cluster Computing*, vol. 22, no. 1, pp. 451–468, 2019.

[64] Y. Kim, J. Nam, T. Park, S. Scott-Hayward, and S. Shin, "Soda: A software-defined security framework for IoT environments," *Computer Networks*, vol. 163, p. 106889, 2019.

[65] K. Fan, C. Zhang, K. Yang, H. Li, and Y. Yang, "Lightweight NFC protocol for privacy protection in mobile IoT," *Applied Sciences*, vol. 8, no. 12, p. 2506, 2018.

[66] Y. Suzuki, A. Niigata, and M. Hamada, "In-house practice of cloudbased authentication platform service focusing on palm vein authentication," *Fujitsu Scientific & Technical Journal*, vol. 52, no. 3, pp. 8–14, 2016.

[67] L. Barreto, A. Celesti, M. Villari, M. Fazio, and A. Puliafito, "An authentication model for IoT clouds," in 2015 IEEE/ACM International Conference on Advances in Social Networks Analysis and Mining (ASONAM). IEEE, 2015, pp. 1032–1035.

[68] M. S. Siddiqui, A. Rahman, and A. Nadeem, "Secure data provenance in IoT network using bloom filters," *Procedia Computer Science*, vol. 163, pp. 190–197, 2019.

[69] J. Bethencourt, A. Sahai, and B. Waters, "Ciphertext-policy attribute-based encryption." In 2007 IEEE symposium on security and privacy (SP' 2007), 20–23 May 2007, Oakland, California, USA, pp. 321–334, 2007.

[70] M. Ambrosin, A. Anzanpour, M. Conti, T. Dargahi, S. R. Moosavi, A. M. Rahmani, and P. Liljeberg, "On the feasibility of attribute-based encryption on Internet of things devices," *IEEE Micro*, vol. 36, no. 6, pp. 25–35, 2016.

[71] M. Asim, M. Petkovic, and T. Ignatenko, *"Attribute-based encryption with encryption and decryption outsourcing,"* in 12th Australian Information Security Management Conference, 1-\-3 December 2014, Perth, Western Australia, pp. 21–28, 2014.

10 An Approach to Smart Targeted Advertising Using Deep Convolutional Neural Networks

A. Gayathri¹, D. Ruby¹, N. Manikandan¹,
and T. Gopalakrishnan²
¹ School of Computer Science and Engineering, Vellore Institute of
Technology, Vellore, Tamilnadu, India
² Department of Information Technology, Manipal Institute of
Technology Bengaluru, Manipal Academy of Higher Education,
Manipal, India

10.1 INTRODUCTION

Advertisements are the means of communicating the services or products of a company to the public with the aim of influencing them to respond in a certain way towards what is advertised. The advertising industry comprises of firms that advertise their products or services, advertising agencies, media that carries the advertisements and people like brand managers, visualizers, creative heads, and designers. Advertising can be broadly categorized into two categories – online and offline.

10.1.1 OFFLINE ADVERTISING

Offline advertising tactics employ offline media channels like television, radio, print (hoardings, newspapers, bus shelters), brochures and flyers, and billboards to make the consumers aware of a firm's products or services. The strategies can be utilized by small businesses in small communities as well as by some of the most renowned firms in the world [1]. The companies that use offline advertising techniques fit a number of profiles, like – "Mom and Pop" businesses, regional, and mid-sized businesses, major national/international companies. The most suitable advertising technique for a product or service depends on its target audience and on how they can be reached in big numbers, as many times as possible, most effectively [2].

Newspaper and magazines are examples of offline advertising. Newspaper promotes the product to a wide range of consumers. However, advertisements in magazines can target the public based on interest groups (like women, children, businessmen). Magazines can serve a large area as their scope is not limited to a specific town or city. Offline advertising can also be done through radio and television. Regular advertising on the radio is a good method to approach the target audience. However, there are some drawbacks. Since there are no visuals, the impact of the ads may be lost and the listeners can find it difficult to recollect what they heard. Advertising through television is apropos if a large market in an area has to be catered to as it has a boundless reach. Television ads are very attractive and persuasive as they have the advantage of sight, sound, movement, and color. The most common mode of offline advertising is outdoor advertising (or out-of-home advertising). As the name suggests, it is done outdoors with the purpose of reaching consumers when they are outside

DOI: 10.1201/9781003407959-10

their homes. Advertisements on benches, news racks, interiors and exteriors of public transports and, the most common, billboards are a few examples.

10.1.2 Online Advertising

Online advertising is a marketing tactic that utilizes the internet as its medium to promote goods and services. The major plus point of this strategy is the rapid promotion of product and quick delivery of ads without any geographical barriers. Since the early 1990s, online advertising strategies have seen an exponential growth and now are a standard for numerous small and large organizations.

A type of online advertising is display advertising. It is a type of online paid advertising that employs attractive images and text to redirect the user's attention to the product being advertised. For example, banners, landing pages, and popups. Search engine marketing & optimization (SEM) & (SEO) are employed to advertise through search engines. In SEM, rather than paying for the actual advertisement, each time the ad is clicked on the website, the advertising have to pay and in SEO, strategies like linking, targeting keywords and meta description, and creating high-level content that other sites might link to are used to achieve a higher rank in search results. Another very popular technique is to advertise through social media. Social media advertising is increasing year by year [3]. There are 1.65 billion active mobile social accounts globally with 1 million new active mobile social users added every day. The budgets of social media advertising have grown drastically worldwide in a short period of time. Pay per click is another kind of online advertising technique. As the name is self-explanatory, the advertisers pay for such ads only when a user clicks on them [4].

10.1.3 Influence of IT on Advertisements

We have come a long way since the development of the first digital computing device. Advertisements have evolved from dull factual description to more vibrant and innovative form, synthesizing effective marketing mechanisms and a high level of creativity [5]. With rapid advances in technology, we have observed a tectonic shift in the modes and medium used for effective outreach to the customers. Information technology has brought a revolution in the way consumer attention is garnered. Media has had to adopt and develop new ways to keep up with current trends and the ever-changing consumer habits [6]. While the advent of digital marketing has made the outreach to a consumer base exponentially possible, the problems faced in creating a campaign will translate into a high viewer response, are manifold. The area where companies have struggled historically is the identification of the target audience for a highly focused and comprehensive business plan. Profiling the prospective customers ensure optimal use of resources and provide maximum impact [7]. The novel approaches using computer technology allow us to run extremely personalized campaigns to target potential customers.

10.1.4 Target-Based Advertisements

Marketing strategies aimed at learning and profiling each customer to show the most relevant advertisements and provide a better user experience have been possible due to the advanced and power data analytics and machine learning algorithms. It leads to a highly specific campaign tailored according to the needs of the customer demographics. These new methods have led to the traditional forms of advertisements such as radio and newspaper to be replaced by the modern ones powered by the internet. This has led to emergence of new advertising companies focusing predominantly on the online aspect of advertisements.

Consumers benefit from this approach by being informed and aware about products suited for their interests. This leads to the advertisers having a highly effective campaign as the targeted nature leads to a greater return on investment [8]. The trend of personalized advertisements is here to stay with our data being analyzed continuously to advertise brands and products of our preference and having a high probability of being purchased by us.

With the use of new and more powerful artificial intelligence algorithms, "hyper-personalization" is the future of marketing [9]. This is being achieved using state-of-the-art deep learning techniques.

10.2 BACKGROUND AND RELATED WORKS

In this study, we have employed deep convolutional neural networks for processing and classifying consumer classes based on the vehicles driven. We have also delved on how advancements in information technology, especially data analytics and machine learning have impacted and transformed the world of advertisements.

10.2.1 IMPACT OF ADVANCEMENTS IN TECHNOLOGY IN ADVERTISEMENTS

Rust and Oliver (1994) [10] concluded that traditional methods of mass media advertisements were on their deathbed and points out that the industry was unprepared for the interactive technologies of the future. It called for this sector to embrace and adopt the new methodologies to survive the relentless upgrades that accompany the emergence of information technology. Graepel [11] proposed a novel Bayesian click-through rate (CTR) model to design a prediction algorithm for sponsored search in Microsoft's Bing search engine. He described the algorithm used by search engines for their most revenue generating activity – advertising. The ads are displayed based on the user's search results to promote relevant advertisements. Perlich et al. [12] proposed a machine learning approach using transfer learning to automatically make predictions for each advertisement campaign. They discussed a two-stage system, which predicts whether the customer purchases the advertised product and if it could be used to make future inferences.

Deep convolutional neural networks to predict the success of the advertising strategy was implemented by Fire and Schler [13]. They used CNN to identify objects appearing in each image of the posters and extract features to be plotted on a graph. Applying a graph theory approach to the graph of vectors created yielded relevant predictions of advertisements to be promoted. A click-through-rate (CTR) model was created by Michel et al. [14]. They tweaked the ResNet-101 deep neural network to classify images of banners. The main objective was to categorize banners on the basis of whether they were effective or not and to identify the most clicked areas using heatmaps.

10.2.2 IMAGE CLASSIFICATION USING CONVOLUTIONAL NEURAL NETWORKS

Alex Krizhevsky et al. [15] published a revolutionary and path-breaking paper, which demonstrated the accuracy of deep neural networks to correctly classify images from over 1000 categories. The paper made every researcher take note of the potential of deep learning. While neural networks had been around since the 1990s, their impact was limited due to hardware limitations. However, modern CPUs and GPUs are able to train deep neural networks with relative ease and with high accuracy. The researchers commented that decreasing layers reduced accuracy and increasing depth resulted in training time increasing manifold. Simonyan and Zisserman [16] designed convolutional neural networks called *VGG16* with greater depth and smaller filter size to obtain higher accuracy in the ImageNet challenge. This method showed significant improvement on the prior state-of-the-art configurations and pushing the depth to 16–19 weight layers. It established that depth was an important parameter, which contributed to the accuracy of the model. Szegedy et al. [17] took deep neural networks to a whole new level by featuring more than 100 layers, showcasing that creatively constructed architectures are sometimes better than the traditional approach. Featuring multiple inception modules this methodology used average pooling hence reducing the number of features needed for training.

10.3 OUR METHODOLOGY

10.3.1 REVIVING BILLBOARDS

Billboards come under the category of traditional advertising. Marketers have been talking about the death of traditional advertising since the late 1990s. There is no definite indicator of death, but it can be seen that more than half of all the companies is now pursuing a digital marketing strategy. According to a study in 2015, 84% of millennials, as well as corporate brands, disfavored and distrusted traditional advertising [18]. One of the reasons for this is that putting up posters of advertisements on billboards turns out to be expensive as compared to digital billboards. Substitutes such as social media and video streaming services have become more popular and affordable. As a fallout of this, the profits of ad agencies are dropping and digital marketers are gaining in terms of market share.

Our aim is to revive one of the traditional advertising techniques, which once used to be very popular and, restoring its status. To achieve this, we combined the traditional strategy along with the new digital marketing strategy. We employed state-of-the-art techniques using deep learning to target buyers and display ads specifically just for them.

10.3.2 PROPOSED MODEL

Figure 10.1 depicts the flow of our system. It begins with a camera monitoring and capturing cars on the road at a particular instance and ends with the appropriate advertisement being displayed on the billboard. The image from the camera is processed and its size is reduced before it sent to the convolutional neural network. The trained CNN model then classifies the cars in the image into three categories – luxury, mid-level, and low level, and a count in each category is recorded. Depending on the count and weights of the categories, the decision support system computes the targeted category and then suitable advertisement is chosen and displayed on the billboard.

For the deployment of the convolutional network trained, we used a camera placed on the road, 350 meters before the digital billboard. This distance was sufficient enough to give the passing

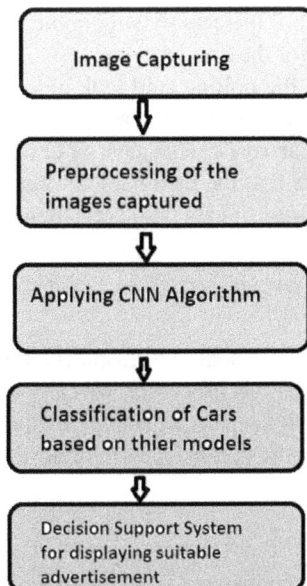

FIGURE 10.1 Flow of the proposed model.

TABLE 10.1
Categories of Cars with Corresponding Examples of Advertisements

Category	Examples
Luxury	Spa, Resorts, Farmhouses
Mid-Level	Schools, Amusements Parks, Fairs
Low-Level	Discount Offers, Sales, EMI, and Loan Options

TABLE 10.2
Categories of Cars with Corresponding Weights

Category	Weight
Luxury (w_{lux})	3
Mid-Level (w_{mid})	1.5
Low-Level (w_{low})	2

Note: The table shows the various Categories of Cars with corresponding weights.

vehicles time to cover the distance and view the billboard. We have divided our advertisements into three different categories and each category targets a specific demographic from the oncoming traffic. The advertisements based on the categorical classes are as follows. The advertisements based on the categorical classes are as follows as in Table 10.1.

Each category maintained a queue in order to ensure that all the advertisements have a chance to be aired and no particular brand or category has an unfair advantage. The task of the camera was to click high-resolution pictures of the oncoming traffic and transmit it to the processing server. The image is first processed to remove noise and blur. The image is then converted to dimensions accepted by the model we trained and is classified. Here we used the classification model we trained to classify the cars. We kept a count of the cars passing in the variables – cnt_{lux} for luxury vehicles, cnt_{mid} for mid-level cars, and cnt_{low} for the low-level cars.

After an interval of 30 seconds, the polling step took place. It involved each category polling/voting based on its weight assigned. The weights assigned were done keeping in mind the frequency of the car classes on the Indian roads. This gave a fair chance for all the classes to have their advertisements displayed and ensure that the methodology was fair to all categories of companies. Table 10.2 shows the various categories of cars with corresponding weights.

$$max\left(cnt_{lux} w_{lux},\ cnt_{mid} w_{mid},\ cnt_{low} w_{low}\right)$$

This polling step decides the advertisement to be displayed for the next 30 seconds. With the help of high-speed network connection and fast processor, we can perform this step with high speed and without much latency. This cycle will keep on repeating, leading to highly targeted advertisements and reviving the use of billboards as the dominant form of advertisements.

10.3.3 DEEP LEARNING

Deep learning is a new area of research inspired from the biological structure of the nervous system especially the structure and functioning of the brain. It involves creation of artificial neural networks as a framework for many different applications of machine learning [19]. Although neural networks have been proposed and were being researched since the 1940s, they have gained prominence only in recent times. The reason for the emergence of these algorithms and methods is the high-performance GPUs and CPUs available as commodity hardware. This has enabled us to create a deeper architecture with a greater number of layers and support for a larger number of parameters than that was possible earlier [20].

Deep learning reduces time required for feature engineering and generally outperforms the previously used algorithms and methods. Image generation, image recognition, natural language processing, or recommender systems deep learning algorithms offer the best performance and accuracy [21]. With algorithms such as convolutional neural networks, long short-term memory networks, and recurrent neural networks – the AI industry is currently undergoing a deep learning revolution [22].

Deep learning has a huge potential to be used in advertisements. Advertisements are harnessing the power of the advancements in the IT sector to create a better and more effective campaign. One of the algorithm classes that have gained popularity in recent times is deep learning. Deep learning is useful in tasks involving huge volumes of data making them highly suitable for advertisement sector tasks such as – data-driven predictive advertisement and targeted advertisement.

10.3.4 CONVOLUTIONAL NEURAL NETWORK

Convolutional neural networks are a type of deep, feedforward artificial neural networks that have huge applications in the analysis of images. Uses of CNNs include object detection and classification, image classification, and feature extraction.

The network proposed uses a smaller and denser version of the VGGNet network, developed by Simonyan and Zisserman [16]. There are certain key features of the model. The model uses a smaller filter of size 3×3 but with more layers. The increased depth of the model results in better accuracy. The training of the model is time-consuming and compute-intensive but ensures better feature detection. The max-pooling layer used leads to a reduction in the volume. There are fully connected hidden layers before the final softmax layer. The dimension of images accepted is of 96×96 with three channels – RGB.

In the convolution layer, each function f_i takes x_i and parameter vector w_i as inputs and generates x_{i+1} as output. The data $x_1,...,x_n$ are the images on which our model is trained and $w_1,...,w_n$ are the parameters that are learned from data so as to classify images. Each x_i is an array of $M{\times}N$ pixels K channels per pixel. Only the first input, i.e., x_1 is an actual image, the rest are intermediate feature maps.

$$f(x) = (f_l(...f_2(f_1(x; w_1); w_2)...), w_l)$$

The single function relation is as follows: $f : R^{MxNxK} \rightarrow R^{M'xN'xK'}$, $x \mapsto y$
y is the output of the convolution layer, a non-linear activation function (here ReLU) is used

$$y_{ijk} = max\{0, x_{ijk}\}$$

The pooling layer [23] works on singular feature channels, combining nearby feature values into one to reduce the volume. For our model we used max pooling, it can be defined as

$$y_{ijk} = max\{y_{i'j'k} : i \leq i' < i + p, j \leq j' < j + p\}$$

Normalization [24] is another crucial part for building a model. It normalizes the feature value vector at each spatial location in the input map x. The normalization function can be defined as

$$y_{ijk'} = \frac{x_{ijk}}{\left(k + \alpha \sum_{k \in G(k')} x_{ijk}^2\right)^{\beta}}$$

where $G(k) = \left[k - \left\lfloor \frac{\rho}{2} \right\rfloor, k + \left\lceil \frac{\rho}{2} \right\rceil\right] \bigcap 1, 2, \ldots K$ is a group of ρ consecutive feature channels in the input map.

10.4 EXPERIMENTAL SETUP

10.4.1 TRAINING THE MODEL

10.4.1.1 Collecting and Cleaning Our Dataset

We created our own data by scraping the web using Bing Image Search API. Using the API, we ran a python code to search the relevant term and download the images from Bing Image Search. After downloading, we had to pre-process and clean the dataset. Images were optimized to contain less noise and only the desired features, i.e., the maximum area should be covered by the car. We decided to split our dataset into three categories – luxury, family, and budget cars. The dataset we used has the following structure. Table 10.3 shows the structure of the dataset.

10.4.1.2 CNN Architecture

The architecture begins with a convolutional layer followed by a max-pooling layer. Then there are two blocks each having two convolutional layers followed by a max-pooling layer. Between each block, there is a dropout layer to prevent overfitting. All the convolutional layers use the ReLU (rectified linear unit) activation function [25]. In the end, there are two fully connected layers, the first one has 1024 channels while the second one is used for performing the classification using the SoftMax classifier.

10.4.2 OBSERVATIONS

We utilize the SoftMax classifier to get the probabilities for each class label and ensure multi-class classification occurs smoothly. Choosing the initial learning rate is very crucial for the model and we choose the initial learning rate as 1e-3. Choosing a smaller rate result in long training times while selecting a larger value led to a drop in accuracy.

TABLE 10.3
Structure of the Dataset

Luxury	Mid-Range	Budget
BMW Series 5	Honda City	Maruti Alto
Mercedes S Class	Maruti Baleno	Maruti Swift
Audi Q7	Maruti Ciaz	Maruti WagonR
Toyota Fortuner	Maruti S-Cross	Tata Bolt
--	Tata Safari	Tata Indica
--	Tata Hexa	Tata Nano

Note: The table shows the structure of the dataset.

TABLE 10.4
Observation Table

Number of Epochs	Accuracy	Training Loss	Validation Loss	Training Time (Minutes)
25	73%	0.7052	1.1978	122
50	77%	0.5104	0.8459	176
75	79%	0.3953	0.7099	252
100	85%	0.2615	0.5003	297

Note: The table shows the noted values of observed table.

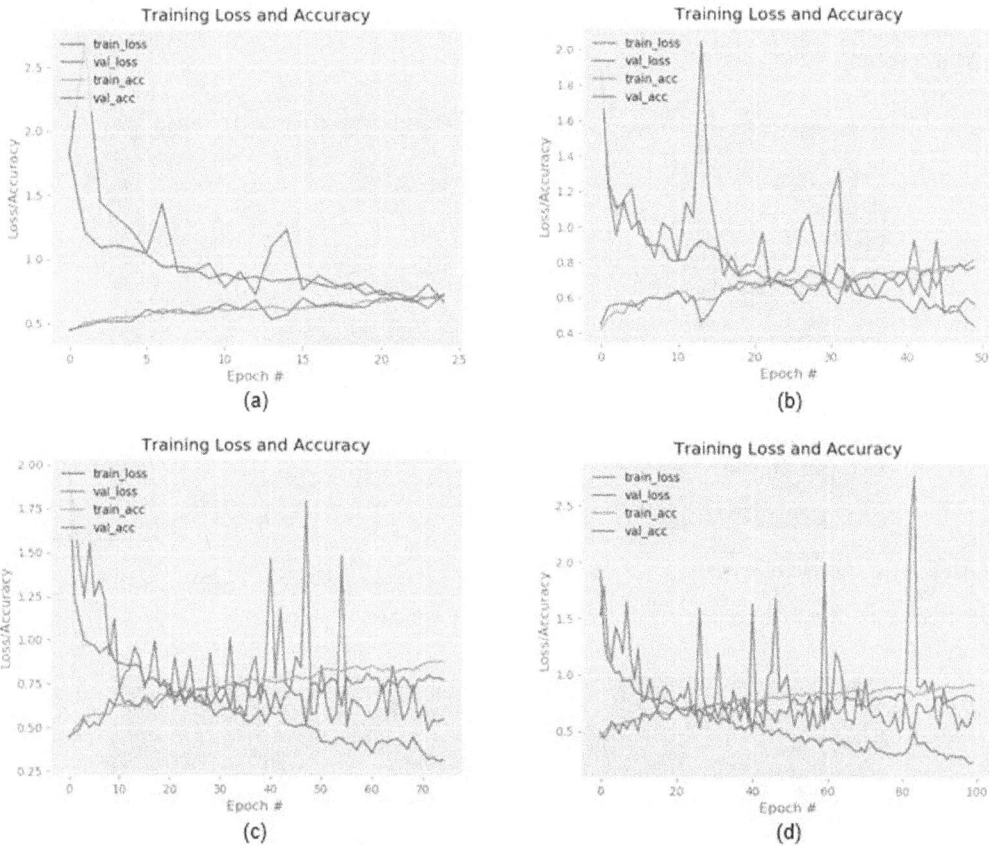

FIGURE 10.2 Graphs depicting training loss, training accuracy, value loss and value accuracy for (a) 25 epochs; (b) 50 epochs; (c) 75 epochs; (d) 100 epochs.

Note: The figure shows various graphs depicting training loss, training accuracy, value loss and value accuracy.

10.4.2.1 Epochs

One epoch refers to the passing of the entire dataset through the neural network once. We started at 25 epochs and kept increasing the number of epochs until suitable accuracy was attained. We kept the batch size uniform with a value of 32.

10.4.2.2 Loss

The loss signifies the penalty awarded by the model for a bad prediction. The aim of the training process is to find weights and biases that minimize the loss. The value of the training loss is updated after each complete epoch. Generally, loss function reduces after every epoch/iteration, however it sometimes may increase, indicating the model is being overfitted. Table 10.4 below shows the noted values of observed table.

Figure 10.2 shows various graphs depicting training loss, training accuracy, value loss, and value accuracy. The best accuracy of 85% is attained by using 100 epochs. It took 5 hours to train on our system. We can further increase the accuracy but that would require a huge amount of time and or better hardware, namely, graphics processors.

10.5 EVALUATION

Some scenarios have been computed.

Scenario 1 – the number of cars passing the camera in a 30-second time span is as follows:
Luxury cars (cnt_{lux}) = 3
Mid-level cars (cnt_{mid}) = 4

FIGURE 10.3   Output of scenario 1.

Note: The figure shows the accuracy level for luxury type of vehicles.

FIGURE 10.4   Output of scenario 2.

Note: The figure shows the accuracy level for budget type of vehicles.

Low-level cars (cnt_{low}) = 3

Therefore, using our polling equation $max\left(cnt_{lux}w_{lux}, cnt_{mid}w_{mid}, cnt_{low}w_{low}\right)$, we get our result class as luxury and therefore an advertisement of a spa resort will be displayed on the digital billboard.

The above Figure 10.3 shows the accuracy level for luxury type of vehicles. Figure 10.4 shows the accuracy level for budget type of vehicles.

Scenario 2 – the number of cars passing the camera in a 30-second time span is as follows:

Luxury cars (cnt_{lux}) = 2

Mid-level cars (cnt_{mid}) = 3

Low-level cars (cnt_{low}) = 7

Therefore, using our polling equation $max\left(cnt_{lux}w_{lux}, cnt_{mid}w_{mid}, cnt_{low}w_{low}\right)$, we get our result class as low-level and therefore an advertisement of an easy EMI scheme will be displayed on the digital billboard.

The percentages of the ads displayed are as follows (the below Figure 10.5 shows the percentage of advertisement displayed for each category; Figure 10.6 shows the frequency for each category of cars):

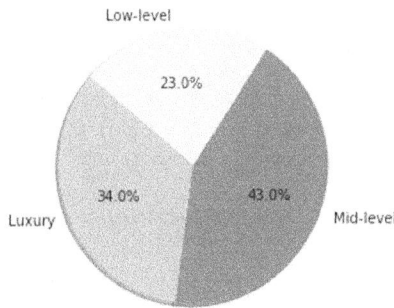

FIGURE 10.5 Percentage of advertisement displayed for each category.

Note: The figure shows the Percentage of advertisement displayed for each category.

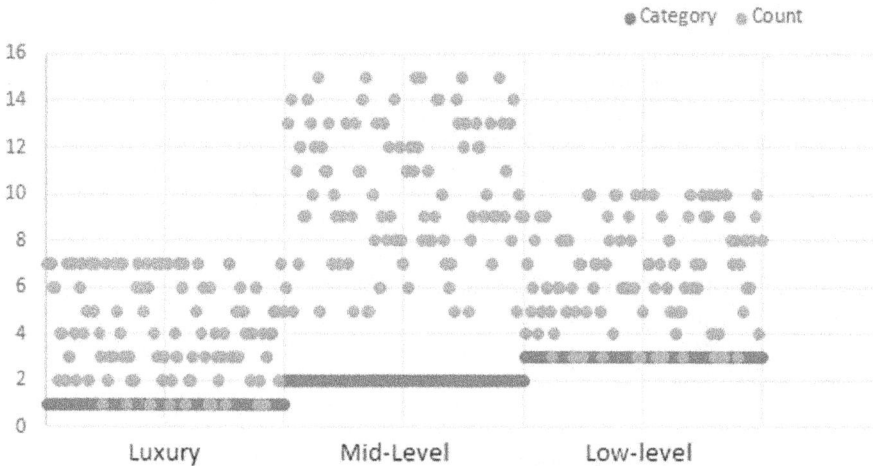

FIGURE 10.6 Frequency of each category.

Note: The figure shows the frequency for each category of cars.

10.6   CONCLUSION

In this paper, we have proposed a system for effective targeted advertisement making the use of convolutional neural networks as a decision support system. We have demonstrated a successful way of using digital billboards powered by deep learning that may have a huge commercial application and impact. This method combines outdoor marketing with the IT applications and algorithms to create an advertising strategy having a huge outreach and high return on investment for the stakeholders. Our method has the capability and capacity to increase the impact and efficacy of the advertisements.

In the future we plan on utilizing big data techniques to analysis and draw inferences from the data we will collect. This will enable us to make better decisions and help clients to analyze other aspects of traffic and advertisements. Our dataset will keep on expanding and our convolutional model will be trained periodically for the accuracy to be consistently high.

REFERENCES

[1] Casadesus-Masanell, R., and Ricart, J. E., "From strategy to business models and onto Tactics," *Long Range Planning,* 43, 2010.

[2] Bergemann, D., and Bonatti, A., "Targeting in advertising markets: Implications for offline versus online media," *RAND Journal of Economics, RAND Corporation*, 42(3), 417–443, 2011.

[3] Kaplan, A. M., "If you love something, let it go mobile: Mobile marketing and mobile social media 4x4," *Business Horizons*, 55(2), 129–139, 2012.

[4] Ryan, D. and Jones, C., *"Understanding digital marketing: Marketing strategies for engaging the digital generation,"* London and Philadelphia: Kogan Page, 2009.

[5] O'Barr, W. M., "Advertising in India," *Advertising & Society Review,* 9(3), 1–33, 2008.

[6] Rust, R. T., and Oliver, R. W., "The death of advertising," *Journal of Advertising*, 23(4), 71–77, DOI: 10.1080/00913367.1943.10673460, 1994.

[7] Lee, A. M., and Durkart, A. J., "Some optimization problems in advertising media planning," *Operational Research Quarterly*, 11, 113–122, 1960.

[8] Chen, J., and Stallaert, J., "An economic analysis of online advertising using behavioural targeting," *Mis Quarterly*, 38(2), 429–449, 2014.

[9] Rao S., Srivatsala V., and Suneetha V., "Optimizing technical ecosystem of digital marketing," In: Dash S., Bhaskar M., Panigrahi B., Das S. (eds) *Artificial Intelligence and Evolutionary Computations in Engineering Systems. Advances in Intelligent Systems and Computing*, vol 394, 691–704. Springer, 2016.

[10] Rust, R. T., and Oliver, R. W., "The death of advertising," Journal of Advertising, 23(4), 71–77, 1994. doi: 10.1080/00913367.1943.10673460

[11] Graepel, T., Candela, J. Q., Borchert, T., and Herbrich, R., "Web-scale Bayesian click-through rate prediction for sponsored search advertising in Microsoft's Bing search engine," *International Conference on Machine Learning ICML*, 2010.

[12] Perlich C., Dalessandro B., Raeder T., Stitelman O., and Provost F., "Machine learning for targeted display advertising: Transfer learning in action," *Machine Learning*, 95, 103–127, 2014.

[13] Fire, M., and Shler, J., "Exploring online ad images using a deep convolutional neural network approach," ArXiv:1509.00568v1 [cs.CV] 2 September, 2015.

[14] Michel, N., Sakata, H., Kurita, K., and Yamasaki, T., "Banner click through rate classification using deep convolutional neural network." The 32nd Annual Conference of the Japanese Society for Artificial Intelligence, 1–4, 2018.

[15] Krizhevsky, A., Sutskever, I., and Hinton, G., *"ImageNet classification with deep convolutional neural networks,"* In NIPS, pp. 1106–1114, 2012.

[16] Simonyan, K., and Zisserman, A., *"Very deep convolutional networks for large-scale image recognition,"* ICLR, 1–14, 2015.

[17] Szegedy, C., Liu, W., Jia, Y., Sermanet, P., Reed, S., Anguelov, D., Erhan, D., Vanhoucke, V., and Rabinovich, A., *"Going deeper with convolutions,"* 2015 IEEE Conference on Computer Vision and Pattern Recognition (CVPR), Boston, MA, USA, 2015, pp. 1–9, 2015. doi: 10.1109/CVPR.2015.7298594

[18]   84 Percent of Millennials Don't Trust Traditional Advertising. www.clickz.com/84-percent-of-millenni als-dont-trust-traditional-advertising, March, 2015.

[19]   Schmidhuber, J., "Deep learning in neural networks: An overview," *Neural Networks*, 61(85), 117, 2015.

[20]   LeCun, Y., Bengio, Y., and Hinton, G. "Deep learning," *Nature,* 521, 436–444, 2015.

[21]   Deng, L., and Yu, D., *"Deep learning: Methods and applications,"* NOW Publishers, 2014.

[22]   Marcus, G., *"Deep learning: A revolution in artificial intelligence,"* The New Yorker, 2012.

[23]   Scherer, D., M¨uller, A., and Behnke, S., "Evaluation of pooling operations in convolutional architectures for object recognition," In Proceedings International Conference on Artificial Neural Networks (ICANN), pp. 92–101, 2015.

[24]   Ioffe, S., and Szegedy, C., "Batch normalization: Accelerating deep network training by reducing internal covariate shift," In Proceedings of ICML, pp. 448–456, 2015.

[25]   Boutilier, C., Patrascu, R., Poupart, P., and Schuurmans, D., "Constraint-based optimization and utility elicitation using the minimax decision criterion," *Artificial Intelligence,* 170(8–9), 686–713, 2006.

11 Text Classification of Customer and Salesperson Conversations to Predict Sales Using Ensemble Models

T. Chellatamilan and Neel Rakesh Choksi
School of Computer Science & Engineering, Vellore Institute of
Technology, Vellore, Tamil Nadu, India

11.1 INTRODUCTION

Customer relationship management (CRM) is the process through which leads are converted to permanent customers. The conversations between the salesperson and the client affect the probability of conversion. However, small businesses rely on traditional data analytics methods for sales prediction, not taking into account the sentiments of conversations between the salesperson and the client. This research aims to predict the sentiments of these conversations and use them as feature columns for sales prediction using ensemble machine learning models.

A CRM application is used in cohesion with an operations management system (OMS) to build a seamless sales pipeline. The interactions between the client and the salesperson are tracked from the first call the salesperson makes to the potential client till the client is converted to a permanent account for the business. The tracked activities include conversations between the salesperson and the client, which could be regarding the product or service offerings of the business, order details, shipment and delivery status, acknowledgments, transactional details for the order payment, and after-sales services. It is essential for the manager to understand the sentiments of the salesperson and clients to make sure customer satisfaction is achieved and whether the business is fulfilling the customer requirements adequately.

To account for salesperson and client conversations in the sales prediction, it will require the integration of multiple applications including CRM, OMS, and analytics tools. A small business owner having limited technical guidance and support would prefer traditional analytics instead of complex integrations.

The operations management application is developed for small business owners having limited staff to carry out operations efficiently by reducing the lead time from sales order generation to the payment received after delivery of the product with a potential to scale. The OMS includes CRM functionality adequate to collect client and salesperson conversations. This study is aimed to find and evaluate the ensemble models for sales data prediction using client and salesperson interactions as features for the machine learning pipeline of the operations management application. Objectives of the study include preprocessing the incoming client and salesperson conversations, finding an effective text representation for the conversations, evaluating ensemble and deep learning models for text classification, modeling conversation sentiments as features in the sales data, evaluating sales prediction models, and finding the feature importance of the conversations. Models are trained using the sci-kit learn [1] library in Python.

DOI: 10.1201/9781003407959-11

The data aggregation techniques to bind the client and salesperson conversations to the orders, text data representation workflow, and integration workflow of client and salesperson conversations as features for sales prediction along with the deep learning, ensemble regression, and classification models can be used by sales executives, data engineers, data analysts, machine learning engineers, and data scientists to evaluate, structure their sales data, and make better business decisions in terms of customer relations. Models may vary in their predictions for different types of conversations in different fields of businesses, but data corpus for different fields could be built for training the models. The corpus used for training the text classification models in the study involves sentiments of movie reviews [2].

The rest of the paper is organized as follows: Section 11.2 portrays the preprocessing, text representation, ensemble text classification models, and deep learning models. Exploratory data analysis on the sales data features followed by ensemble sales prediction models. Section 11.3 shows the exploratory data analysis on the IMDB dataset and the dataset modeling and generation for sales prediction. Section 11.4 elaborates on the metrics, performance analysis, and discussions followed by Section 11.5 concluding the chapter with future work opportunities.

11.2 METHODOLOGY

11.2.1 Hybrid Text Classification Model

The text classification model pipeline shown in Figure 11.1 is built to generate sentiment predictions from sales conversation data. The models are trained using the text dataset containing movie reviews. The input text is preprocessed by eliminating unwanted and frequently repeated text, replacing abbreviations and short forms, and reducing synonymous words to a common form using stemming. The words in the sentences are represented as vectors using the word2vec model and aggregated for representing a sentence. These sentences are used to train ensemble classification models, including stacking, classification models, and deep learning models, including feed-forward and gated recurrent unit networks. Two types of predictions are expected including probabilistic and binary.

11.2.1.1 Preprocessing

The language and vocabulary used in chat can vary from the original rules. To normalize the vocabulary used and to remove the most frequently occurring words and articles, preprocessing is applied to the text before training the model. The text is converted to lowercase, HTML tags are removed, punctuations are removed, and abbreviations are replaced with grammatically correct words.

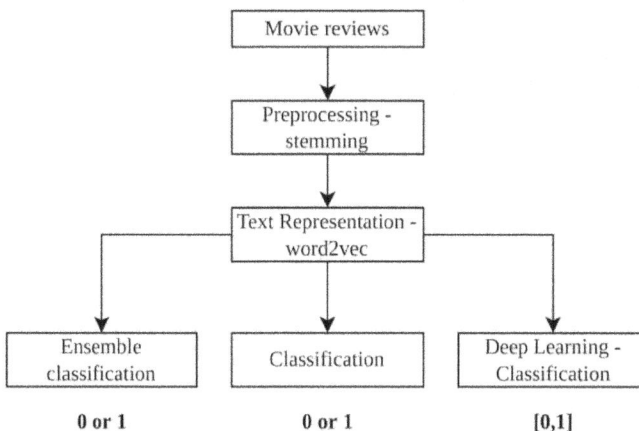

FIGURE 11.1 Text classification model pipeline.

Stopwords are removed since they do not contribute to the meaning of a sentence. Stemming is done on the words to reduce the words to their root form, which is computed algorithmically.

The Porter-Stemmer algorithm used for stemming is rule-based [3]. The rules include removing or replacing suffixes from the word. It helps to represent the words having similar meanings into one term, which reduces the size of the vocabulary and the complexity of the system. The suffix of a word is replaced based on the condition that the stem word follows. A stem word is a word without the suffix part.

11.2.1.2 Text Representation

The machine learning and deep learning models require the input data to be in numerical form. Text representation techniques used include Bag Of Words and Word2Vec. In the Bag Of Words representation, the chat text is represented as a vector. The vector elements refer to the number of times a word is occurring in the chat text. The length of the vector is equal to the vocabulary size of the corpus. Bag Of Words representation was unable to capture the ordering of words, it was overfitting and slow due to the huge size of the vocabulary, therefore, the feature columns were limited. The semantic relationship of the words is not captured by the Bag Of Words method, therefore the Word2Vec method is used. It also solves the problem of sparsity by using dense vectors [4]. The CBOW [5] method with a window of ten words is used to generate the vectors for the words in the reviews, all the vector representations of the words in a sentence are added to form the vector for the review.

11.2.1.3 Ensemble Models for Text Classification

Word2Vec text representation is used to train the models. Ensemble learning techniques, including voting, bagging, and stacking are applied to these models to enhance the overall accuracy. Accuracy scores of voting and bagging classifiers are shown in Table 11.1.

Voting is performed considering the support vector classifier (SVC), K neighbors classifier (KNC), decision tree classifier (DTC), logistic regression classifier (LRC), Gaussian naive Bayes classifier (GNB) models as base models. Hard voting is used wherein the majority vote of the base models is considered as the final prediction [6]. Models are independent and the accuracy of all the models is greater than 50%, which are the basic requirements for a voting model. Soft voting is a technique where the probabilities are averaged to get the final prediction.

Bagging is an ensemble technique where different instances of the same model are trained using bootstrapping and aggregation [7]. The LRC model is used as the base estimator. The input dataset is sampled into different sets using pasting, which is row sampling without replacement, random subspaces, which is column sampling, and random patches, which is a combination of both. Bagging reduces variance by not passing the new type of input data to one model instance at once. Logistic regression model instances are used as the base model. OOB score is also calculated, which is the testing of the bagging model based on the unused rows during the bootstrapping phase.

TABLE 11.1
Voting and Bagging Text Classification Model

Model	Accuracy
Voting	0.81
Bagging	0.865
Pasting	0.864
Random Subspaces	0.861
Random Patches	0.858
OOB Score	0.866

TABLE 11.2
Stacking and Classification Models

Model	Accuracy	MCC	F1_Score
Stacking	0.866	0.732	0.866
SVC	0.643	0.288	0.644
KNC	0.817	0.633	0.817
DTC	0.780	0.560	0.780
LRC	0.865	0.731	0.865
GNB	0.787	0.574	0.787

The models, including SVC, KNC, DTC, LRC, and GNB, are used as base estimators in the stacking model. Accuracy, Matthew's correlation coefficient (MCC), and F1 score of the stacking model and its base models are shown in Table 11.2.

Support vector classifier constructs hyperplanes to classify the data and is widely used in binary classifications [8]. Here the data is in 100 dimensions. The hyperplane is constructed with the help of the sigmoid kernel function. The loss function used involves finding the maximum distance between the two marginal planes using cross-validation, including the misclassification of data points as well as the distance of the misclassified points from the hyperplane. K neighbors classifier is a lazy learning-based algorithm that memorizes the training data to make predictions. It calculates the distances of the data point to be predicted from the existing data points and has a hyperparameter k, which determines the number of neighboring points to be considered [9]. It takes a majority vote of the k neighboring classifications and predicts the output. The decision tree classifier divides the input data into hyper cuboids using hyperplanes. The splits are decided based on the feature giving the best information gain [10]. Information gain is the measure of the reduction of entropy after the split. The maximum depth is 5 in this model, it limits the tree from splitting after the depth of the tree has reached 5, this helps in reducing the overfitting of the training data. Logistic regression classifier is used on data that is almost linearly separable. It fits a sigmoid curve on the data points using maximum likelihood [11, 12]. The curve gives the probability of occurrence of the class. If the probability is greater than 50%, the class is 1 else it is 0. It enables multidimensional feature columns, where it generates a sigmoid hyperplane. It can be used for feature selection. Gaussian naive Bayes classifier uses the Gaussian distribution function to plot the features into normal curves [13]. The probability of a data point being in a class is determined based on its prior probability and conditional probabilities for every feature. The results are transformed using the log function to avoid underflow errors. Stacking is performed considering the SVC, KNC, DTC, LRC, and GNB models as base models. Blending is used, which is a type of stacking technique wherein cross-validation is not done. The input data is divided into training and testing sets, the training set is further divided into train and test sets, and base models are trained. A meta-learner is trained using the predictions of the base models. The meta-learner uses the logistic regression model. The meta-learner is tested by passing the testing set data [14–15].

Ensemble learning models including random forest classifier (RFC), ADA boosting classifier, bagging classifier, extra trees classifier (ETC), gradient boosting decision trees (GBDT) classifier, extreme gradient boosting (XGB) classifier are evaluated using accuracy and precision scores as shown in Table 11.3.

RFC is a type of bagging technique that uses only decision trees as the base model. It uses bootstrapping and aggregation to make predictions. It reduces the variance because it distributes the new data entered over the models. A decision tree has inherently low bias and high variance, its variance is reduced by the random forest algorithm. It is not identical to bagging since random forest uses node-level column sampling whereas bagging uses tree-level column sampling [16].

TABLE 11.3
Ensemble Text Classification Models

Model	Accuracy	Precision
RFC	0.838	0.824
ADA Boost	0.829	0.817
Bagging	0.832	0.815
ETC	0.833	0.821
GBDT	0.830	0.814
XGB	0.854	0.838

ADA boosting classifier is an ensemble algorithm that uses decision trees as base estimators. The decision trees used are weak learners, which are called decision stumps, which can divide the dataset into two parts only. It is a stagewise additive method. The decision stumps are chosen based on the least entropy after the split. The misclassified points in the dataset are upsampled to increase focus on them in the next stage [17].

ETC is an algorithm that helps reduce overfitting due to end-cut preference in decision trees. The number of splits is regulated rather than all splitting on all variables, which also reduces the computational complexity. Each split is done based on a randomly selected cut point. It is used since it shows lesser variance compared to decision trees and random forest classifiers [18].

GBDT classifier uses decision trees to focus on errors made by the previous tree and upscales the errors to reach the final prediction. The initial node is found by finding probability from the log of the odds prediction of the initial dataset. Residuals are calculated and a new tree is built based on the errors, output values of the leaves are calculated, log of the odd predictions is made, which are converted to probabilities that are further used to find residuals [19]. This process is carried out till the number of estimators is reached or till the change is not significant.

XGB classifier uses decision trees built from residual values. It has an initial tree with one node. Residuals are calculated, using similarity score, cover, gain, and pruning the new tree is built on residuals [20]. Output values are found for the leaves. Log of the odds predictions is made using the log of the odds of the initial value and output values of the leaves. The probability of the log of the odds predictions is calculated. The process is repeated either till the number of estimators is reached or the residuals are insignificant.

11.2.1.4   Deep Learning Models for Text Classification

Text classification can also be done using deep learning models including ANN and GRU. The neural networks output a value ranging from 0 to 1 at the output layer. This value is a result of the sigmoid activation function. This value is used as a feature to improve the accuracy of further sales prediction models. The ANN and GRU models are evaluated based on accuracy, precision, MCC, and F1 score in Table 11.4.

ANN [21] is a feed-forward neural network. There is one hidden layer used in this neural network. The Bag Of Words representation of the review data is used to input into the network. 2517 neurons depicting the size of the vocabulary. The hidden layer consists of 128 neurons on the input side and 64 on the output side. The linear decoder decodes the 64 inputs to a range from 0 to 1 using the sigmoid activation function [22]. The loss function is calculated and based on the gradient descent, the weights are adjusted in the backward propagation, this is done for 10 epochs. GRU model is a neural network of gated recurrent units. It is a type of recurrent neural network designed to solve the vanishing gradient problem where the initial words read by the network lose importance as the sentence is processed further [23]. Word2vec embeddings are used for the input layer in this model.

TABLE 11.4
Deep Learning Text Classification Model Evaluation

Metric	ANN	GRU
Accuracy	0.908	0.884
Precision	0.881	0.882
MCC	0.819	0.768
F1_Score	0.91	0.882

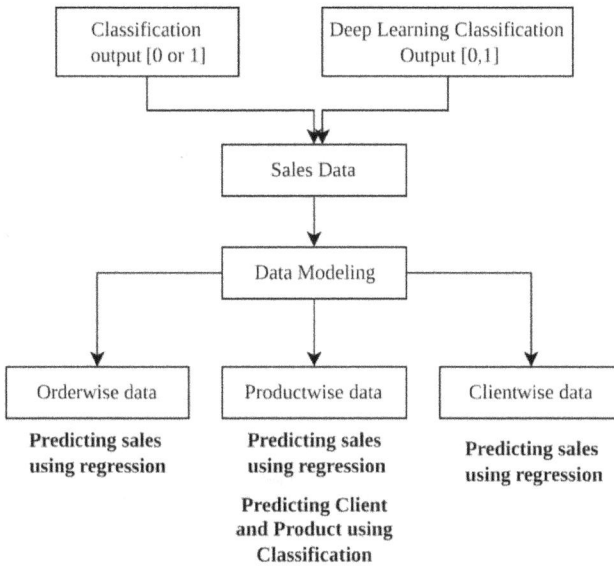

FIGURE 11.2 Sales prediction model pipeline.

The gated recurrent unit predicts the output by processing the input from the previous neuron, the new input word of the sentence using forget gates and update gates. It uses tanh and sigmoid activation functions to process the inputs and generate the prediction to be passed to the next neuron [22].

11.2.2 HYBRID SALES PREDICTION MODEL

The outputs from the classification models and deep learning classification models, which include the sentiments of client and salesperson conversations are integrated with the sales data and used as features for prediction as shown in Figure 11.2. The stacking model gives a binary output for the sentiment whereas the ANN and GRU models give a probabilistic output.

11.2.2.1 Exploratory Data Analysis of Sales Data

Order-wise data is modeled from the sales data. This data includes the features of every order entry. Figure 11.3 shows the bar plots for the categorical features in the order-wise data. Categorical features include client name, month, and day. The frequency of total orders done with clients of the small business owner is shown as a bar plot. The frequency of total orders done on different days of the month is shown starting from month number 1 indicating January. The frequency of total orders done on the specific day of the week is shown starting from day 1 indicating Monday. Most orders

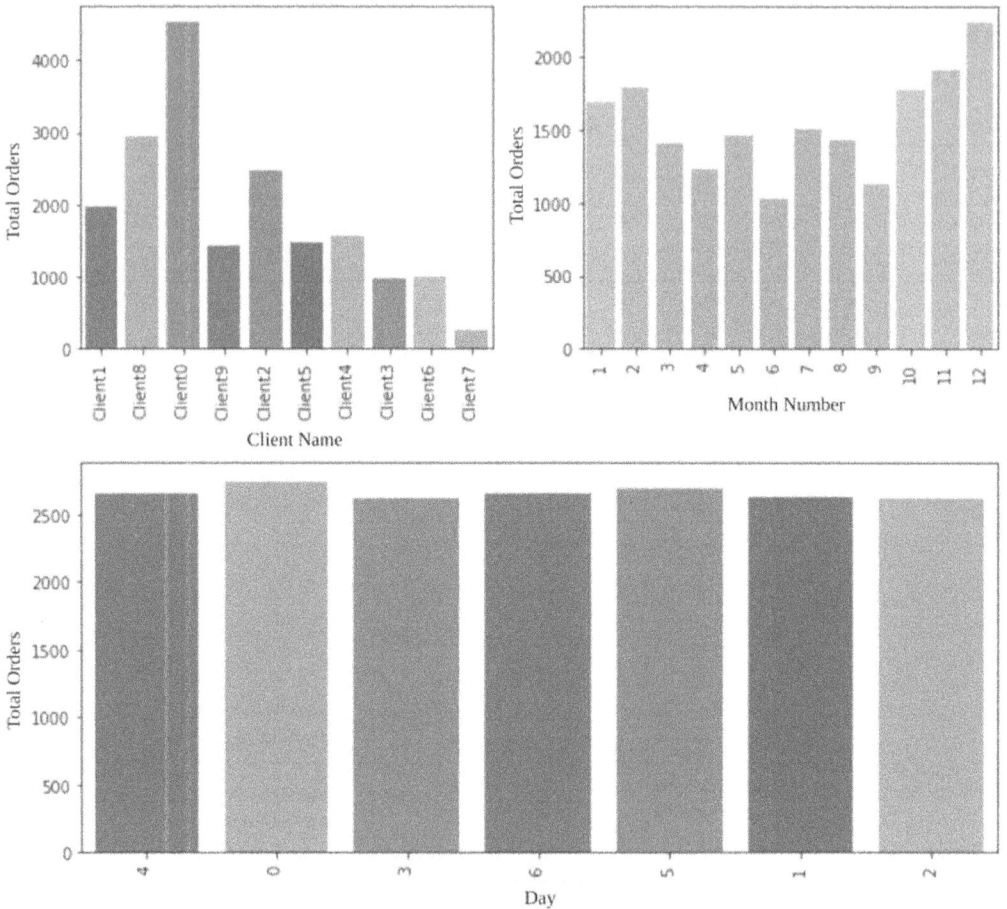

FIGURE 11.3   Order frequency versus client name, month, day.

are placed by the client, in the month and the most busy weeks could be determined for business analytics.

Figure 11.4 shows the distribution of the continuous features in the order-wise data. Continuous features include the hour of purchase, quantity ordered, and order sales. Most of the purchases were made during 11 am in the morning and 8 pm in the evening. The basket size of the order was 10 to 15 items on average. The average order value was between 5000 and 10000.

The box plot for sales is shown in Figure 11.5. The minimum value of the order can be 0. The maximum value in the dataset is 20k. There are several outliers in the range of 20k to 30k. The median order value is 7.5k. The lower quartile is 5k and the upper quartile is 12k for the sales of the order. Sales is the feature predicted by the order-wise regression model.

Sales and quantity ordered with client name and day are shown in Figure 11.6. The order value increases as the quantity increases, which shows a strong correlation between the order quantity and the sales. There are seven different clients observed, each client can be observed individually for business analytics.

Sales related to month for clients are shown in Figure 11.7. Clients portray different buying behavior according to months, which could be due to festive seasonal changes, this shows the significance of the month feature in sales prediction models. Sales according to month can be observed individually for all clients for business analytics.

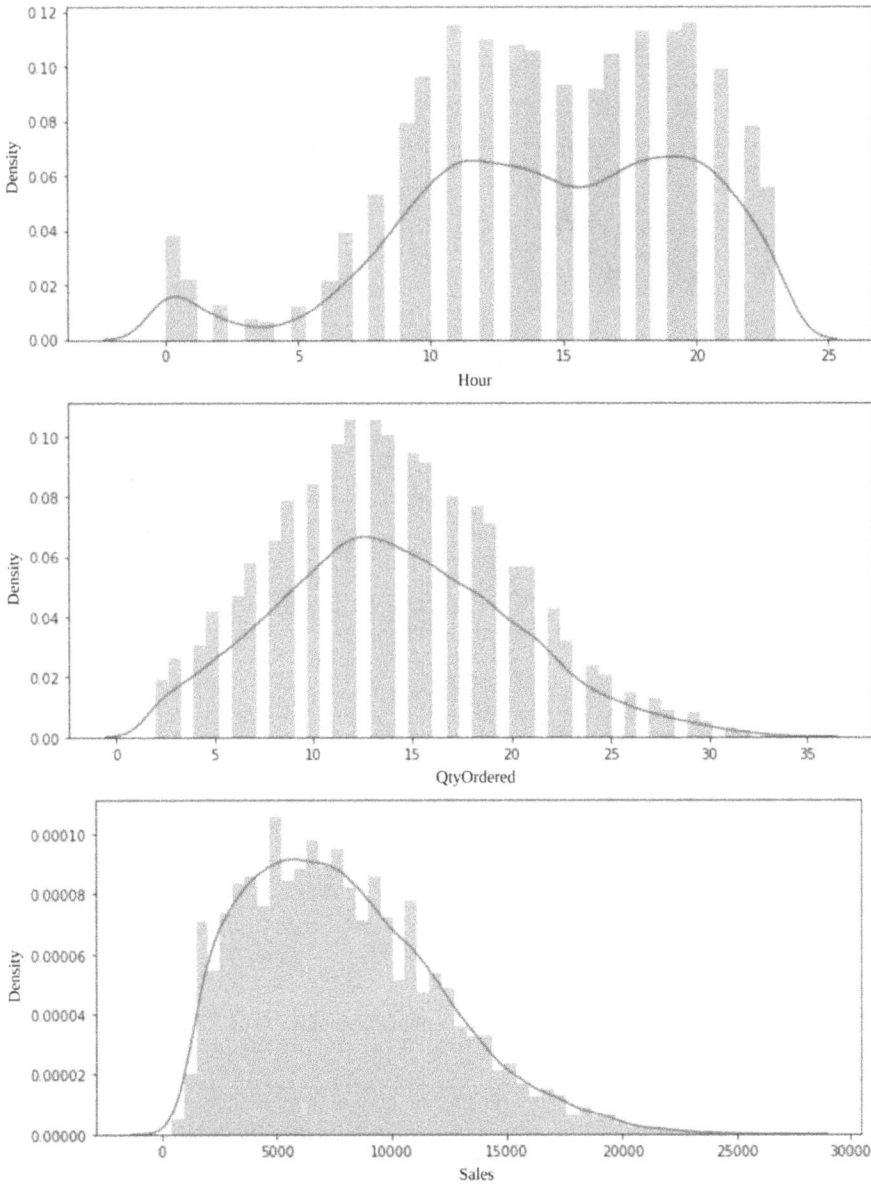

FIGURE 11.4   Distribution of hours of purchase, quantity ordered, and sales of an order.

Sales related to day for clients are shown in Figure 11.8. Clients portray different buying behavior according to the day of the week, which could be due to working and non-working days, this shows the significance of the day feature in sales prediction models. Sales according to day can be observed individually for all clients for business analytics.

Product-wise data is modeled from the sales data. This data includes the product entries from every order entry. Figure 11.9 shows the frequency of the products bought. Most frequently sold products can be found for business analytics.

Product price distribution is shown in Figure 11.10. The product price is the amount of the product entry in the respective order entry. The distribution is skewed towards the left wherein most of the product prices lie in the range of 0 to 2k. Product price becomes an important feature in determining

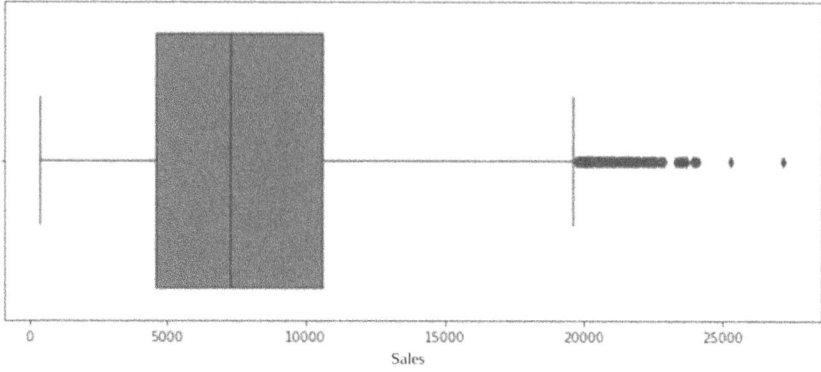

FIGURE 11.5 Box plot for sales of order.

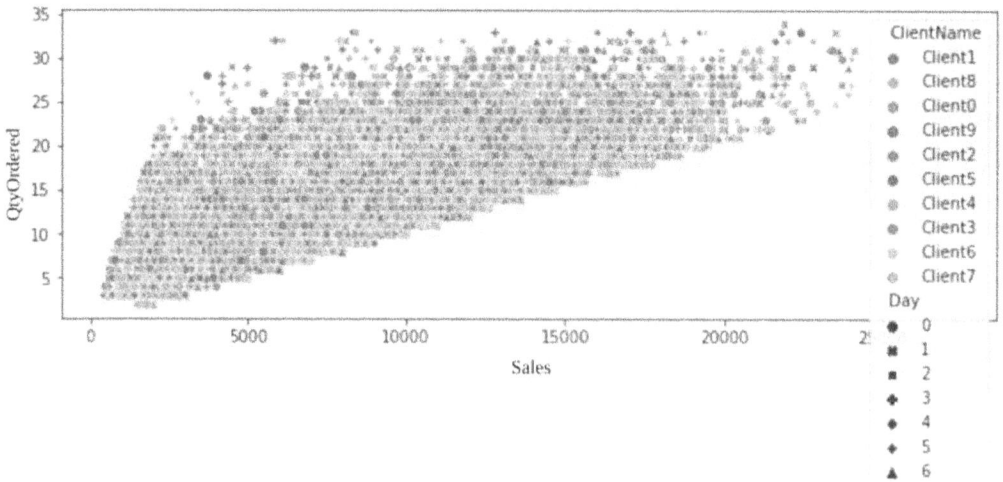

FIGURE 11.6 Quantity ordered versus sales against client name, day.

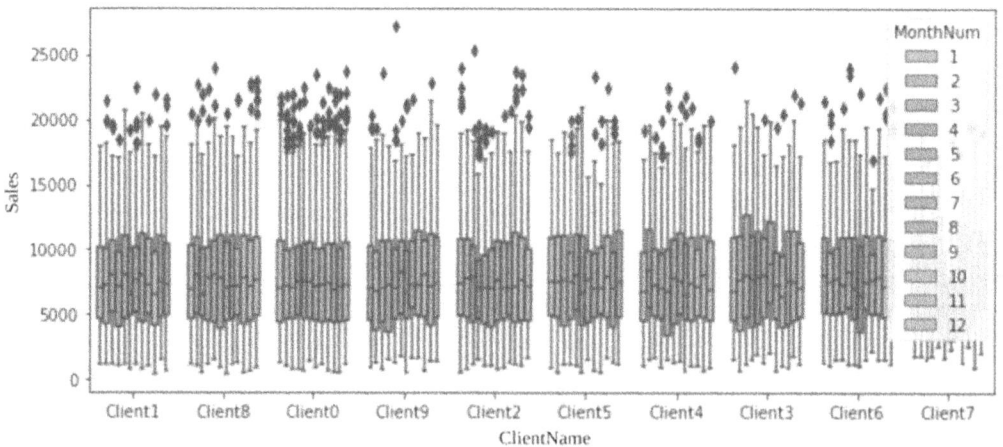

FIGURE 11.7 Box plots for sales versus client name against month.

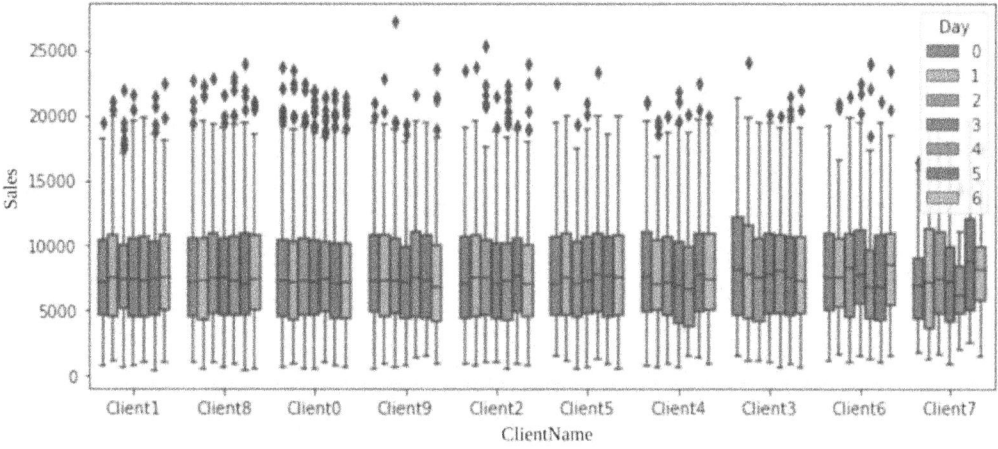

FIGURE 11.8 Box plots for sales vs client name against day.

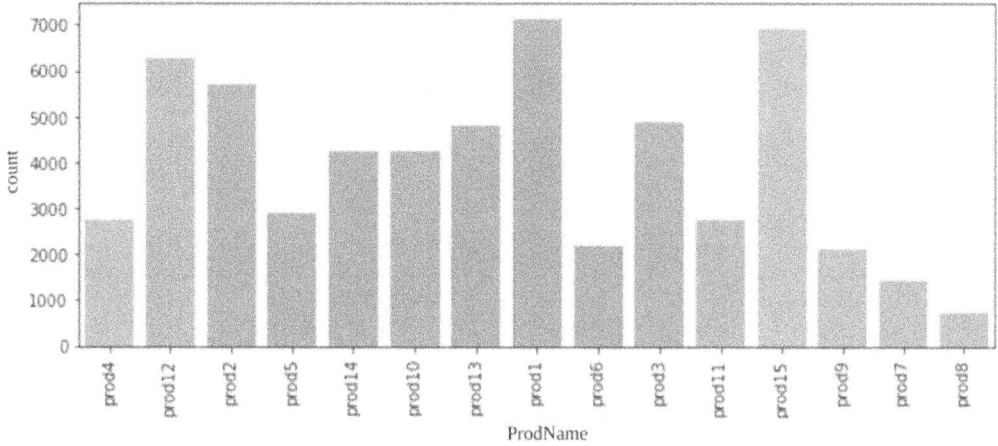

FIGURE 11.9 Product frequencies.

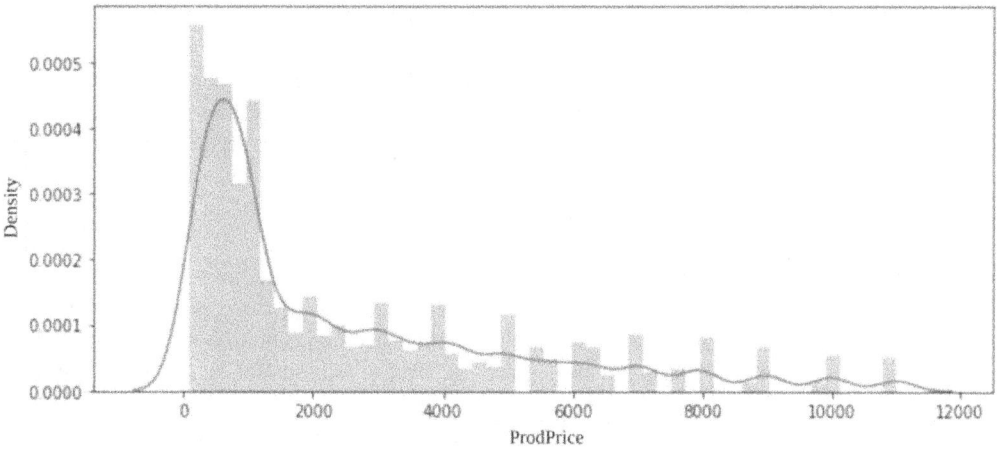

FIGURE 11.10 Distribution of product price.

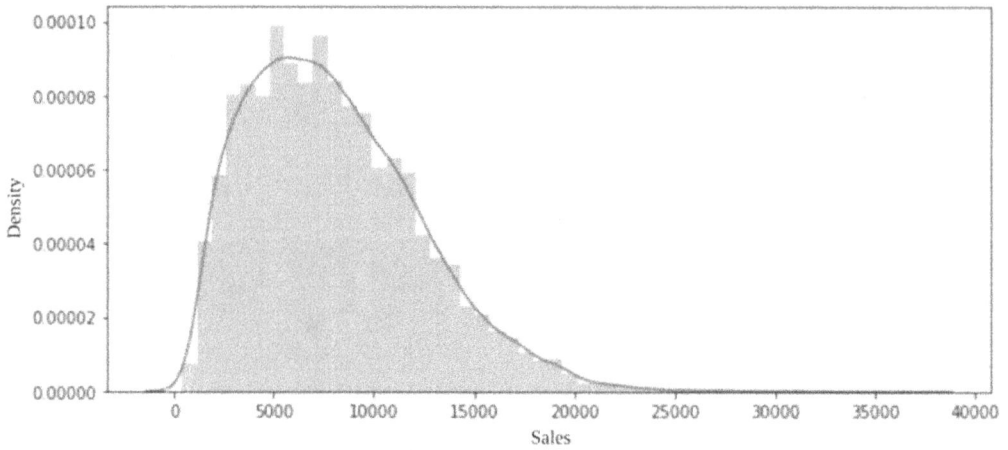

FIGURE 11.11   Distribution of client sales.

TABLE 11.5
Order-Wise, Product-Wise, and Client-Wise Sales Prediction Regression Models

Model	Order-Wise (R2_Score)	Product-Wise (R2_Score)	Client-Wise (R2_Score)
Simple Linear Regression	–	–	0.48
Multiple Linear Regression	0.488	0.215	0.482
Stochastic Gradient Descent	-1.294	-1.358	-4.53
Mini Batch Gradient Descent	-2.642	-1.446	-2.32
Polynomial Regression	0.482	0.219	0.482
Ridge Regression	0.488	0.215	0.482
Lasso Regression	0.488	0.215	0.482
ElasticNet Regression	0.488	0.215	0.482
Decision Tree Regression	0.491	0.365	0.485

the client or the product bought predicted by the product-wise classification model. It is also the feature predicted in the product-wise regression model.

Client-wise data is modeled from the sales data. This data includes the client-wise sales varying according to the month, day, and the products bought in order. Figure 11.11 shows the client sales distribution. Most of the sales were between 5k and 12.5k. It is the feature predicted using the client-wise regression model.

11.2.2.2   Regression Models for Sales Prediction

The evaluation of the regression models based on R2_Score is shown in Table 11.5. The sales data is modeled according to orders, products, and clients. The modeled data is fed into the regression models, including simple linear regression, multiple linear regression, stochastic gradient descent (SGD) regressor, mini batch gradient descent regression, polynomial regression, ridge, lasso, elasticnet regression, and decision tree regression.

Multiple linear regression is used to predict a numerical target variable when the dataset has more than one feature column [24]. The regression line in simple linear regression is converted into a plane if there are two feature columns and converted into a hyperplane if there are more than two feature columns. It is a supervised machine learning technique so the prediction made by the model is evaluated against the expected results.

SGD regressor updates the coefficients after every row of the data is processed. It selects the row randomly therefore it does not provide a steady solution and gives a different answer every time the algorithm is run. It is faster overall since fewer epochs are required and updates to the coefficients happen faster with less data processed by the hardware [25].

Mini batch gradient descent regression uses a batch of rows at a time to update the coefficients. SGD regressor is random and can only be used for nonconvex functions. At every epoch, each batch updates the coefficient. It is used to reduce the randomness of SGD regressors [25].

Polynomial regression is used when the input data is not linear. Polynomial features are extracted from the input features [26]. The degree of the polynomial is set to 2, it is a hyperparameter. The polynomial features help extract the nonlinear nature of the data.

Ridge regression is a regularization technique that helps reduce overfitting in the machine learning model by introducing bias in the model. In this case, the bias is the L2 norm, which is the square of the weights of the data points times the alpha [27]. Alpha is a hyperparameter. As the value of alpha increases, coefficients shrink but don't reach 0. Lasso regression is another type of regularization technique where the L1 norm is used in the loss function to increase the bias and reduce overfitting [27]. As alpha increases, the coefficients shrink, and they can eventually reach 0 which helps in determining the feature importance. Elasticnet regression is a combination of ridge and lasso techniques. It uses both the L1 norm and the L2 norm in the loss function [27]. It has alpha as a hyperparameter dependent on the coefficients in L1 and L2 norm. The L1 ratio is also a hyperparameter and is formed using these coefficients. It is used when the input columns portray multicollinearity.

Decision tree regression is used when the data is not linear and is clustered. It uses the decision tree algorithm, which divides the data into hyper cuboids using hyperplanes based on the information gain. The splitting criteria include the range of the values of the data points. The splitting criteria range is found by splitting one data point with another and finding the residual error of the rest of the points [28].

11.2.2.3 Ensemble Models for Sales Prediction

The evaluation of the ensemble regression models based on R2_Score is shown in Table 11.6. The sales data is modeled according to orders, products, and clients. The modeled data is fed into models including voting, bagging, random forest, and XGB.

Voting regression includes multiple base models. The prediction is calculated based on the mean of the outputs of the base models [29]. The base models include multiple linear, polynomial, ridge, lasso, elasticnet, and decision tree regression models trained earlier.

Bagging regression includes using multiple instances of the same model and feeding them different sets of sampled data to reduce the variance [30]. Hyperparameter tuning is done, which resulted in decision tree regressor as the best base estimator, row sampling without replacement,

TABLE 11.6
Order-Wise, Product-Wise, and Client-Wise Sales Prediction Ensemble Regression Models

Model	Order-Wise (R2_Score)	Product-Wise (R2_Score)	Client-Wise (R2_Score)
Voting	0.49	0.26	0.49
Bagging	0.534	0.392	0.585
Random Forest	0.54	0.412	0.562
XGB	0.633	0.454	0.711

TABLE 11.7
Product-Wise Product and Client Prediction Ensemble Model

Model	Product Name		Client Name	
	Accuracy	MCC	Accuracy	MCC
Decision Tree	0.34	0.33	0.24	0.025
Logistic Regression	0.16	–	0.24	–
KNN	0.5	–	0.18	–
Random Forest	0.5	–	0.969	0.965
Hard Voting	0.46	–	0.24	–
Soft Voting	0.54	–	0.2	–
Bagging	0.61	0.575	0.913	0.902
ADA Boosting	0.19	–	0.222	1.0
XGB	0.61	0.57	0.459	0.389

column sampling without replacement, usage of 100% of the features, usage of 50% of samples, and 100 estimators are preferred to give the best results.

Random forest regression is a bagging regressor with decision trees as the base model [31]. 500 estimators are used here. The sampling performed here is randomized at every node level of the decision tree to reduce variance. The out of box score is calculated by using the nonsampled input data points to test the model.

XGBoost regression uses unique regression trees. It has an initial value of 0.5. The residuals are calculated. The unique regression tree is built using gain and pruning. Gain is calculated using a similarity score. Regularization parameter lambda is used to prevent overfitting of the training data. Output values of the leaves are calculated, and predictions are made based on them and the prior values, and the learning rate [32]. The process is continued till the number of estimators is exhausted or the change in the predictions is insignificant.

The evaluation of the ensemble classification models based on accuracy and MCC is shown in Table 11.7. The sales data is modeled according to products. Each row consists of features related to the sales of the product in relation to the order. The client's name and product name are predicted, considering other features for training the model.

Classification models used to predict product name and client name are inspired from Section 11.2.1.3 of this chaper.

11.3 EXPERIMENTAL DESIGN

11.3.1 PREBUILT DATASET

IMDB movie reviews [2] dataset is used to train the machine learning and deep learning models. It has 50K movie reviews out of which 25K are positive and 25K are negative, which is shown in Figure 11.12.

The distribution of the number of words in positive and negative reviews is shown in Figure 11.13. The number of words in positive and negative reviews is distributed evenly across the reviews and is skewed toward the left. A positive or a negative review contains between 50 to 150 words.

The distribution of the number of characters in positive and negative reviews is shown in Figure 11.14. The number of characters in positive and negative reviews is distributed evenly across the reviews and is skewed toward the left. A positive or a negative review contains between 500 to 1k characters.

A pairplot analyzing the features including the number of characters, number of words, and number of sentences is shown in Figure 11.15. The most positive sentiment reviews have more

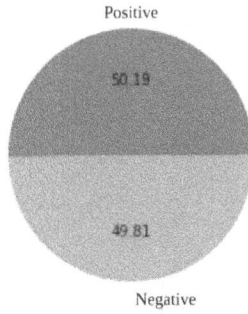

FIGURE 11.12 IMDB Movie reviews sentiment.

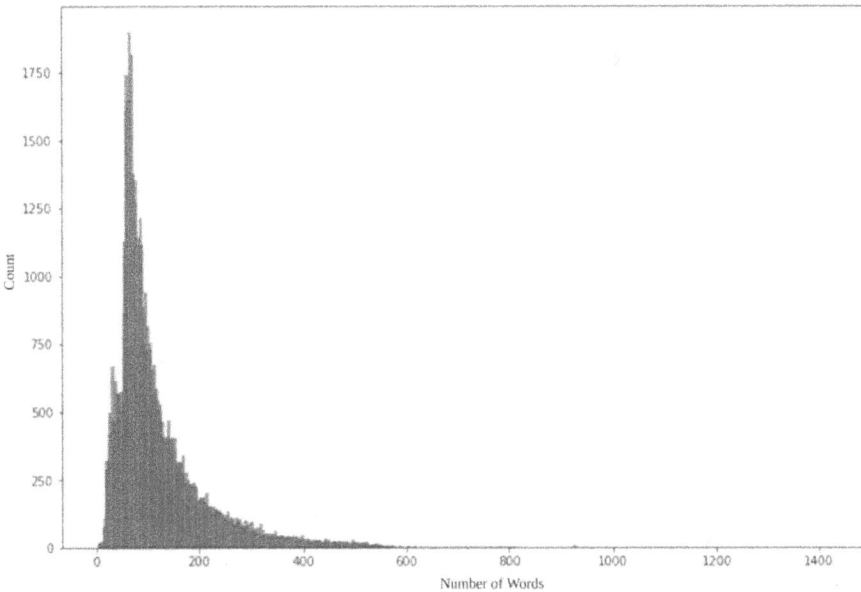

FIGURE 11.13 Distribution of the number of words in positive and negative reviews.

words and characters. There are more negative reviews having fewer words and characters than positive reviews. There are more positive reviews having more words and characters.

A correlation matrix is formed among the features including the number of characters, words, sentences, and the target, sentiment is shown in Figure 11.16. The number of words and characters are weakly correlated to the sentiment of the review and hence can be ignored as feature columns for the sales prediction data corpus.

A word cloud for positive reviews is shown in Figure 11.17. It can be used for business analytics purposes. The word cloud portrays that the words movie, film, good, time, story, love, and character are most frequently occurring in the positive reviews.

A word cloud for negative reviews is shown in Figure 11.18. It can be used for business analytics purposes. The word cloud portrays that the words movie, film, bad, time, really, much, watch, plot, and actor are most frequently occurring in the negative reviews.

Most occurring words in positive and negative reviews are shown in Figures 11.19 and 11.20, respectively. The words movie, film, one, and like are most commonly occurring in both, positive

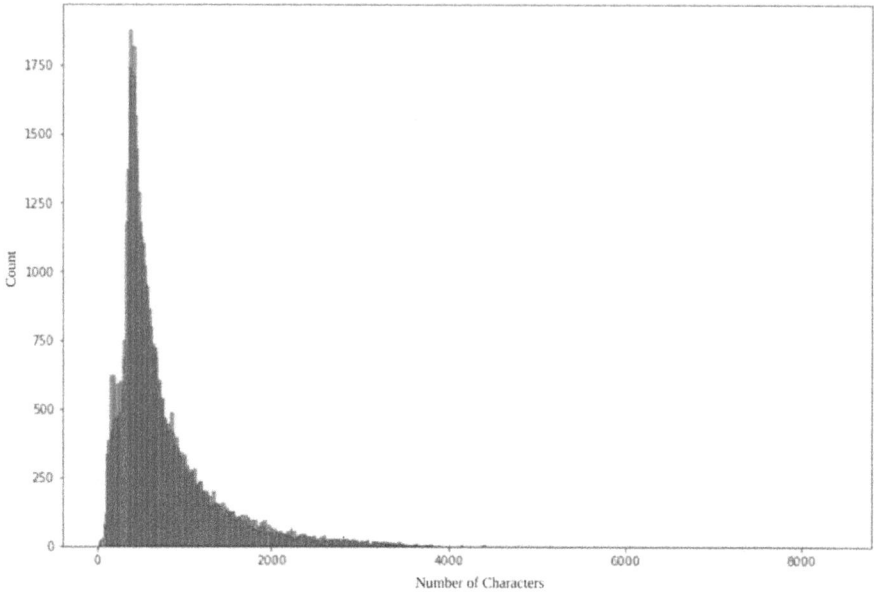

FIGURE 11.14   Distribution of the number of characters in positive and negative reviews.

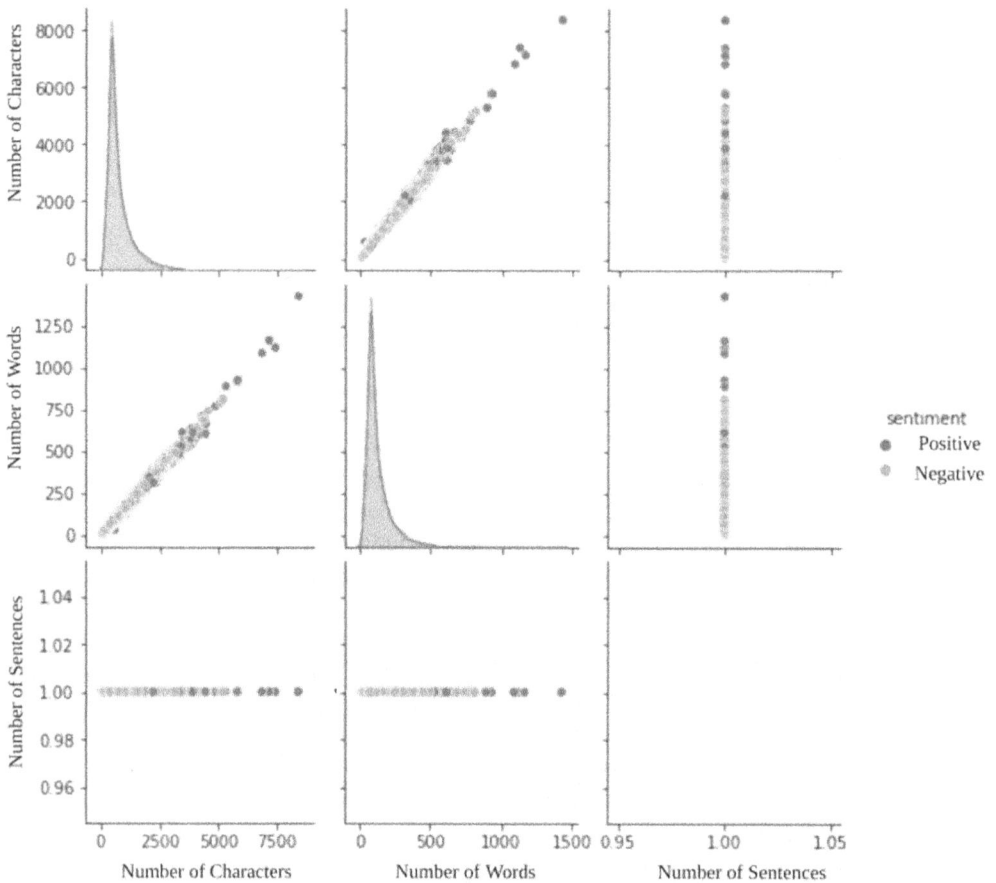

FIGURE 11.15   Pairplot for the number of characters, words, and sentences.

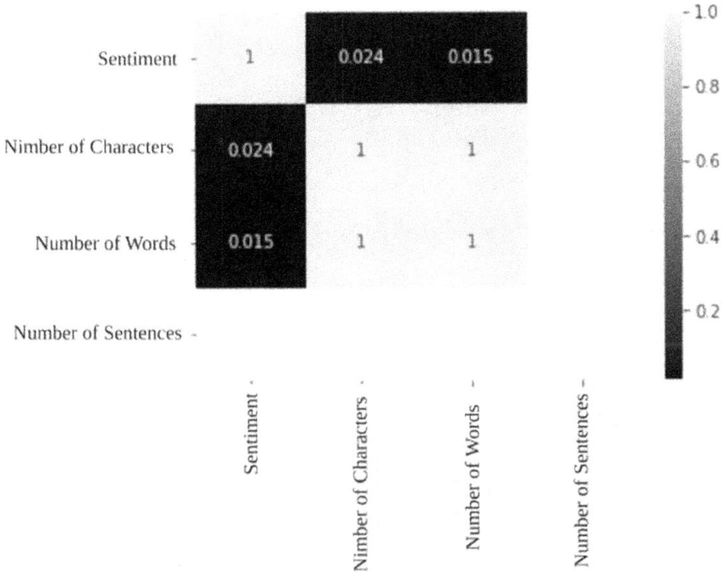

FIGURE 11.16   Correlation matrix for sentiment, number of characters, number of words, and number of sentences.

FIGURE 11.17   WordCloud for words in positive reviews.

FIGURE 11.18　Word Cloud for words in negative reviews.

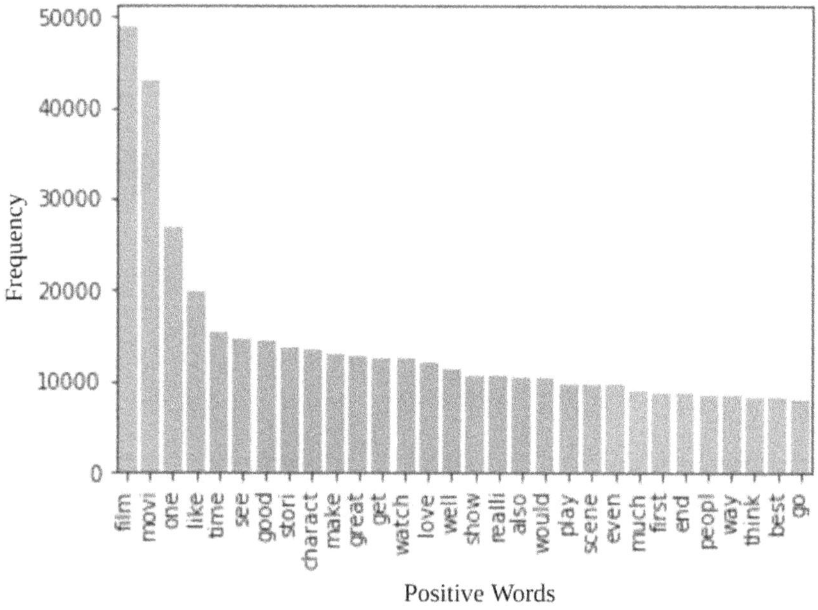

FIGURE 11.19　Top 30 words in positive reviews.

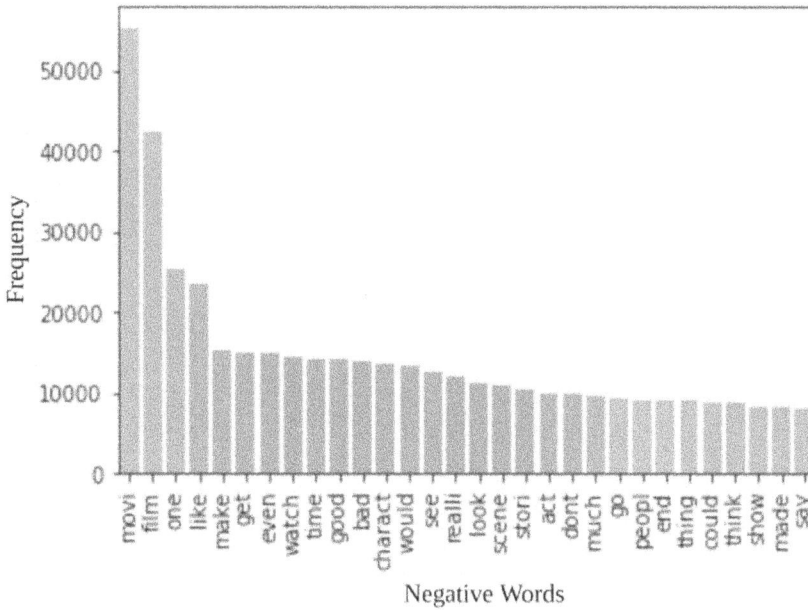

FIGURE 11.20   Top 30 words in negative reviews.

and negative reviews. The bar plot representation of the most frequently occurring positive and negative text can be used for business analytics.

11.3.2 DATASET GENERATION

A raw dataset is generated to generate datasets for order-wise, product-wise, and client-wise analysis. For every month in a year, orders are generated with random amounts of product and quantity per order. The client name for an order is decided randomly. All month's order sales data is aggregated and analyzed. Monthwise, city-wise, client-wise, and product-wise sales are plotted.

The client and salesperson chats are selected randomly from the movie reviews dataset and concatenated into one chat, the average sentiment is calculated by calculating the mean of sentiments of selected chats. The sentiment is also calculated using the stacking model, feed-forward neural network, and gated recurrent unit neural network models. These chats are fed into the dataset for every order. They represent the chats made by the client and salesperson before the order was made.

Order-wise data as shown in Table 11.8, is grouped by using the orderID. Every row contains the orderID, client name, month, hour, minute, day, products sold in that order, client's mean sentiment, salesperson's mean sentiment, client's and salesperson's sentiment predicted by feed-forward neural network model, gated recurrent unit model, stacking model. The total quantity of the products along with the total order amount. The sales column is the target column and can be predicted using regression techniques using other columns as feature columns.

Product-wise data as shown in Table 11.9, is the entry of every product sold in every order. Every row contains the product name, client name, month, orderID, order quantity, order total, client's average sentiment, salesperson's average sentiment, hour, minute of the time of sale, day of the sale, and the total product price. Client's and salesperson's sentiment before that order is predicted by the feed-forward neural network model, gated recurrent unit model, and stacking model. Product name and client name are the target columns and can be predicted using classification techniques. Product

TABLE 11.8
Order-Wise Sales Data

Order ID	Client Name	Month Num	Hour	Minute	Day	Prods	Qty Ordered	Sales	Cli Avg Senti	Sp Avg Senti	Cli Chats	Sp Chats	ANN cli senti	ANN sp senti	GRU cli senti	GRU sp senti	Cli Senti	Sp Senti
1001	Client1	1	20	51	4	prod14,prod16,prod2	5	2400	0.8	0.6	It's awesome! In Story….	OK, so in any Wile E. ….	0.998079	0.996931	0.902179	0.942251	0	0
1002	Client8	1	21	3	0	prod2,prod15,prod15	10	9500	0.4	0.4	Thistender beautifully …	Man were do I start,...	0.972840	0.998424	0.943390	0.383383	1	0
1003	Client0	1	18	22	4	prod2,prod9,prod1	10	3200	0.2	0.6	God this film was…	I saw this film ….	0.000107	0.003524	0.069862	0.199325	0	0

TABLE 11.9
Product-Wise Sales Data

Prod Name	Client Name	Month Num	Order ID	Order Qty	Order Total	Cli Avg Senti	Sp Avg Senti	Prod Price	Hour	Minute	Day	Cli Senti	Sp Senti	ANN Cli Senti	ANN Sp Senti	GRU Cli Senti	GRU Sp Senti
prod3	Client4	2	4191	15	6700	0.4	0.4	2100	21	33	3	0	0	0.473894	0.998727	0.183884	0.738530
prod13	Client0	2	4086	13	11000	0.0	0.6	200	22	53	6	0	0	0.000126	0.697719	0.406582	0.905176
prod6	Client5	9	12580	11	5700	0.4	0.4	1200	18	54	0	0	0	0.003496	0.873117	0.826856	0.927612
prod1	Client0	9	12933	27	22500	0.6	0.4	4900	22	16	1	0	0	0.433903	0.347175	0.677025	0.616245
prod2	Client5	8	12208	12	8500	0.8	0.4	3000	19	27	0	0	0	0.932095	0.99987	0.955717	0.919963

TABLE 11.10
Client-Wise Sales Data

ClientName	MonthNum	Day	Prods	QtyOrdered	Sales
Client0	4	2	prod3,prod12,prod1	9	4900
Client2	10	5	prod15,prod15	9	9000
Client2	6	4	prod1,prod4	2	1500
Client2	7	1	prod1.prod15	4	3700
Client0	4	0	prod10,prod1,prod10,prod1	21	16800

price is also a target column and can be predicted using regression techniques using other columns as feature columns.

Client-wise data as shown in Table 11.10, is generated by grouping the client names with the month, day of sale, and the products basket. Every row contains the client name, month, day, products, quantity ordered, and the total order amount of those products. The sales column is the target column and will be predicted using regression techniques.

11.4 PERFORMANCE EVALUATION

11.4.1 METRICS

R2_Score is the coefficient of determination, goodness of fit. It is 1 minus the ratio of the regression line by the mean line. If it is 0 means that the regression line and the mean line are the same. If it is 1, it indicates that the regression line has done no mistake and the model is very accurate. If it is −1, it indicates that the mean line is better performing than the regression line. *Accuracy* helps determine how many predictions are correct out of the total predictions. It depends on the data and problem to decide how much accuracy is good for the model. It does not indicate the type of error made by the model. It cannot be reliable when the data is imbalanced. *Precision* detects the type 1 error. It is higher when false positives are less in a model. *Recall* detects the type 2 error. It is higher when false negatives are less in a model. These can be used to compare two models with similar accuracies. *F1_Score* is the harmonic mean of precision and recall. Precision and recall have a tradeoff. F1_ Score stays on the lower side of the two. *MCC* is used for imbalanced datasets. It portrays the difference between the correctly classified and incorrectly classified points. It is 0 when the product of true positive, true negative, and false positive, false negative is equal. It is 1 when there are no false positives and false negatives. It is −1 when there are no true positives and true negatives.

11.4.2 PERFORMANCE ANALYSIS

The incoming client and salesperson conversations were preprocessed using techniques including stopword removal and stemming. Word2Vec representation was found to be an effective text representation for the conversations. Ensemble and deep learning models for text classification were trained. The conversation sentiments were integrated as features in the sales data. Sales prediction models were trained for order-wise, product-wise, and client-wise sales data.

The stacking model performs the best for ensemble text classification with an accuracy of 86.5%, MCC of 0.73, and F1_Score of 86.5%. The ANN performs the best for deep learning text classification with an accuracy of 90.8%, a precision of 88.1%, an MCC of 81.9%, and an F1_Score of 91%. Order-wise sales are predicted most accurately by XGBoost Regressor with an R2 Score of 0.63. Product-wise sales are predicted most accurately by the XGBoost model with an R2_Score of 0.45. Client name is predicted most accurately by the XGBoost classifier and bagging classifier with an accuracy of 0.61 and Matthews Correlation Coefficient of 0.57. Product name is predicted most accurately by the random forest model with an accuracy of 96.9% and Matthews correlation coefficient of 0.965. Client-wise sales are predicted most accurately by the XGBoost model with an R2_Score of 0.71.

11.4.3 Discussion

Sentiment of the client and salesperson includes how the conversation between them is flowing, it may go towards a positive end or a negative end. Since the text classification model predicts the overall sentiment, it takes into account the scenarios where sentiment is a key factor of conversation for instance flow of the conversation, the objections faced by the salesperson, and sentimental values towards the product or the buyer.

Order-wise sales predicted by decision tree regressor, on tuning gave the best results by taking 25% of the samples split, 100% of the features, maximum depth of 4, and using mean squared error for loss criterion. XGB regressor, on tuning gave the best results on taking 240 estimators, max depth as 10, and learning rate as 0.1. The most important features include the quantity of the order, products in the order, and sentiment of the salesperson predicted using ANN and GRU as shown in Figure 11.21.

Product-wise sales predicted by decision tree regressor, on tuning gave the best results by taking 25% of the sample split, 75% of max features, maximum depth of 8, and using mean squared error for loss criterion. Bagging regressor, on tuning gave the best results on taking max estimators as 100, 50% max samples, 100% max features, not bootstrapping features, and base estimator as decision tree regressor. XGB regressor on tuning gave the best results on taking 50 estimators, keeping the max depth to 10 and the learning rate at 0.1. The most important features include the total price of the order, the total quantity of the order, the client's sentiment predicted using GRU, and the salesperson's sentiment predicted using ANN as shown in Figure 11.22.

Client-wise sales predicted by decision tree regressor on tuning gave the best results by taking 25% of the sample split, 100% of max features, maximum depth of 4, and using mean absolute error for loss criterion. Bagging regressor on tuning gave the best results on taking max estimators

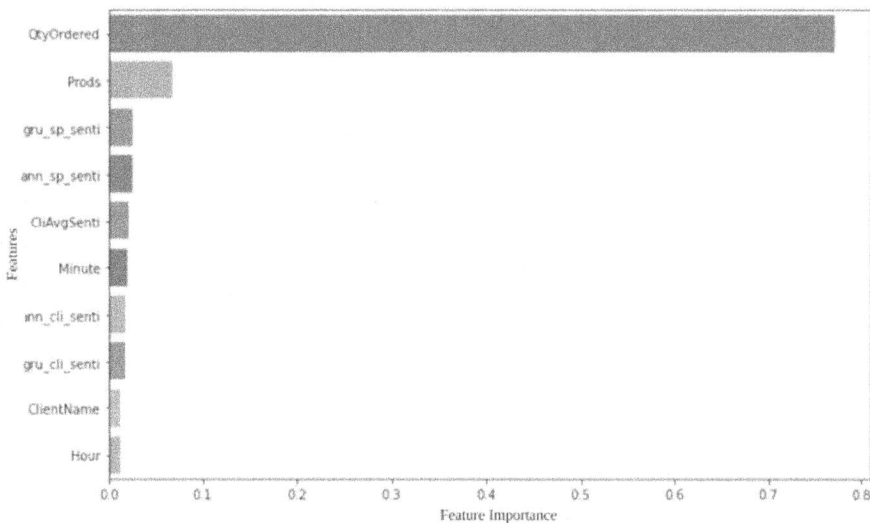

FIGURE 11.21 Feature importance for order-wise sales data.

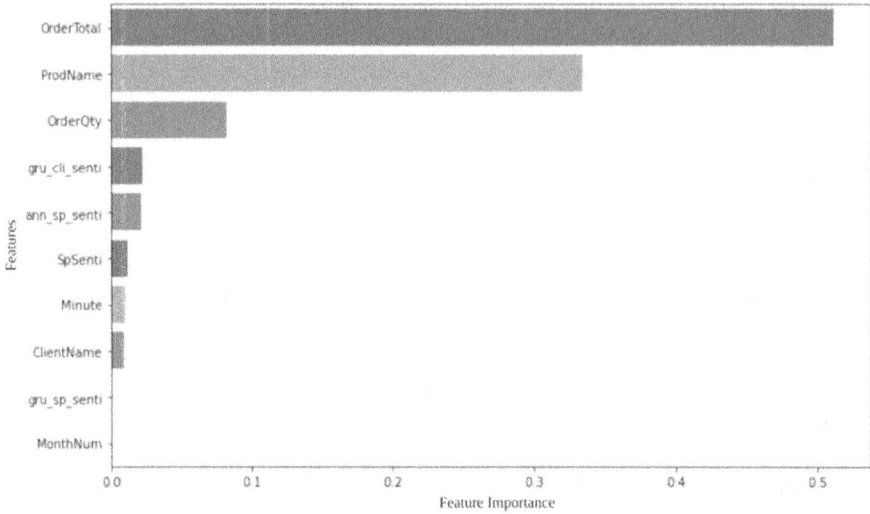

FIGURE 11.22   Feature importance for product-wise sales data.

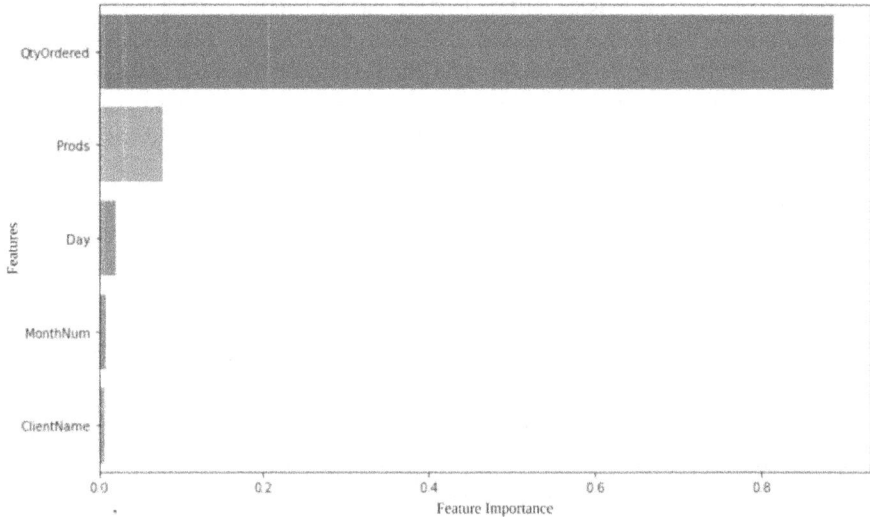

FIGURE 11.23   Feature importance for client-wise sales data.

as 100, 100% max samples, 100% max features, bootstrapping features and dataset, and base estimator as decision tree regressor. The XGBoost regressor on tuning gave the best results on taking 240 estimators, keeping the max depth to 4 and the learning rate at 0.9. The most important features include quantity ordered, products chosen, and the day on which the order was made as shown in Figure 11.23.

11.5   CONCLUSION

Stopword removal and stemming were performed to preprocess the text inputs. Word2Vec representation was used to convert the text to numerical vectors. The stacking ensemble model gave the best accuracy of 0.86 for binary text classification, ANN gave an accuracy of 0.9 for probabilistic text classification in a range of 0 to 1. The stacking model, ANN, and GRU model predictions for client and salesperson chats were integrated into order-wise, product-wise sales data. Order-wise, product-wise, and client-wise sales predictions were done best by XGB regressors. Products were predicted best by bagging and XGB classifiers and clients were predicted best by random forest classifier. Sentiments of client and salesperson conversations were among the most significant features for order-wise and product-wise sales prediction. The text classification and sales prediction pipelines built in the study will be helpful for integrating sentiments between client and salesperson conversations into the data corpus for sales prediction. The text classification model takes into account the flow of conversation, along with the objections faced by the salesperson, and sentimental values towards the product or the buyer. The text classification models are trained using the movie review dataset, which limits the models to predict sentiments for a particular field of work. The future scope would be to execute the pipelines by training the models with conversations data from different fields of work as well as adding more features to sales prediction by building the characteristics of the prospect.

ACKNOWLEDGMENTS

I would like to thank my project guide, Dr. T. Chellatamilan for his guidance through each stage of the progression of the study. His expertise was invaluable in formulating the research questions and methodology. Your insightful thought process and feedback, motivated me to take my work to a higher level.

REFERENCES

[1] Buitinck, L., Louppe, G., Blondel, M., Pedregosa, F., Mueller, A., Grisel, O., Niculae, V., Prettenhofer, P., Gramfort, A., Grobler, J., Layton, R., Vanderplas, J., Joly, A., Holt, B., & Varoquaux, G. (2013). *API Design for Machine Learning Software: Experiences from the Scikit-Learn Project.* https://arxiv.org/abs/1309.0238

[2] Maas, A. L., Daly, R. E., Pham, P. T., Huang, D., Ng, A. Y., & Potts, C. (2011). Learning word vectors for sentiment analysis. *49th Annual Meeting of the Association for Computational Linguistics: Human Language Technologies, Volume 1*, 142–150. https://doi.org/https://dl.acm.org/doi/10.5555/2002 472.2002491

[3] Gharatkar, S., Ingle, A., Naik, T., & Save, A. (2017). Review preprocessing using data cleaning and stemming technique. *2017 International Conference on Innovations in Information, Embedded and Communication Systems (ICIIECS).* https://doi.org/10.1109/iciiecs.2017.8276011

[4] Rui, Z., & Yutai, H. (2020). Research on short text classification based on Word2Vec microblog. *2020 International Conference on Computer Science and Management Technology (ICCSMT).* https://doi.org/10.1109/iccsmt51754.2020.00042

[5] Mikolov, T., Chen, K., Corrado, G., & Dean, J. (2013). Efficient Estimation of Word Representations in Vector Space. https://arxiv.org/abs/1301.3781

[6] Atallah, R., & Al-Mousa, A. (2019). Heart disease detection using machine learning majority voting ensemble method. *2019 2nd International Conference on New Trends in Computing Sciences (ICTCS).* https://doi.org/10.1109/ictcs.2019.8923053

[7] Bin Alam, Md. S., Patwary, M. J., & Hassan, M. (2021). Birth mode prediction using bagging ensemble classifier: A case study of Bangladesh. *2021 International Conference on Information and Communication Technology for Sustainable Development (ICICT4SD).* https://doi.org/10.1109/icict4sd50815.2021.9396909

[8] Yan, X., & Zhu, H. (2022). A novel robust support vector machine classifier with feature mapping. *Knowledge-Based Systems, 257*, 109928. https://doi.org/10.1016/j.knosys.2022.109928

[9] Arslan, H., & Arslan, H. (2021). A new COVID-19 detection method from human genome sequences using CPG island features and KNN classifier. *Engineering Science and Technology, an International Journal, 24*(4), 839–847. https://doi.org/10.1016/j.jestch.2020.12.026

[10] Singh Kushwah, J., Kumar, A., Patel, S., Soni, R., Gawande, A., & Gupta, S. (2022). Comparative study of regressor and classifier with decision tree using modern tools. *Materials Today: Proceedings, 56*, 3571–3576. https://doi.org/10.1016/j.matpr.2021.11.635

[11] Harshvardhan, G., Venkateswaran, N., & Padmapriya, N. (2016). Assessment of glaucoma with ocular thermal images using GLCM techniques and logistic regression classifier. *2016 International Conference on Wireless Communications, Signal Processing and Networking (WiSPNET)*. https://doi.org/10.1109/wispnet.2016.7566393

[12] Sperandei, S. (2014). Understanding logistic regression analysis. *Biochemia Medica*, 12–18. https://doi.org/10.11613/bm.2014.003

[13] Sibhi, K., Thanvir Ibrahim, S., Malik, A., & Praveen R, J. (2022). Career prediction using Naive Bayes. *2022 Third International Conference on Intelligent Computing Instrumentation and Control Technologies (ICICICT)*. https://doi.org/10.1109/icicict54557.2022.9917745

[14] Pavlyshenko, B. (2018). Using stacking approaches for machine learning models. *2018 IEEE Second International Conference on Data Stream Mining & Processing (DSMP)*. https://doi.org/10.1109/dsmp.2018.8478522

[15] Wang, T., Zhang, K., Thé, J., & Yu, H. (2022). Accurate prediction of band gap of materials using stacking machine learning model. *Computational Materials Science, 201*, 110899. https://doi.org/10.1016/j.commatsci.2021.110899

[16] Ong, A. K., Prasetyo, Y. T., Velasco, K. E., Abad, E. D., Buencille, A. L., Estorninos, E. M., Cahigas, M. M., Chuenyindee, T., Persada, S. F., Nadlifatin, R., & Sittiwatethanasiri, T. (2022). Utilization of random forest classifier and artificial neural network for predicting the acceptance of reopening decommissioned nuclear power plant. *Annals of Nuclear Energy, 175*, 109188. https://doi.org/10.1016/j.anucene.2022.109188

[17] Saabni, R. (2015). Ada-boosting extreme learning machines for handwritten digit and digit strings recognition. *2015 Fifth International Conference on Digital Information Processing and Communications (ICDIPC)*. https://doi.org/10.1109/icdipc.2015.7323034

[18] Dhananjay, B., Venkatesh, N. P., Bhardwaj, A., & Sivaraman, J. (2021). Cardiac signals classification based on extra trees model. *2021 8th International Conference on Signal Processing and Integrated Networks (SPIN)*. https://doi.org/10.1109/spin52536.2021.9565992

[19] Yao, S., Kronenburg, A., Shamooni, A., Stein, O. T., & Zhang, W. (2022). Gradient boosted decision trees for combustion chemistry integration. *Applications in Energy and Combustion Science, 11*, 100077. https://doi.org/10.1016/j.jaecs.2022.100077

[20] Shi, H., Wang, H., Huang, Y., Zhao, L., Qin, C., & Liu, C. (2019). A hierarchical method based on weighted extreme gradient boosting in ECG Heartbeat Classification. *Computer Methods and Programs in Biomedicine, 171*, 1–10. https://doi.org/10.1016/j.cmpb.2019.02.005

[21] Bian, Y., Ye, R., Zhang, J., & Yan, X. (2022). Customer preference identification from hotel online reviews: A neural network based fine-grained sentiment analysis. *Computers & Industrial Engineering, 172*, 108648. https://doi.org/10.1016/j.cie.2022.108648

[22] Hsu, J., & Klanecek, S. (2019, July 23). A Deep Dive into NLP with PyTorch. YouTube. https://youtu.be/4jROlXH9Nvc

[23] Lin, M., You, Y., Wang, W., & Wu, J. (2023). Battery health prognosis with gated recurrent unit neural networks and hidden Markov model considering uncertainty quantification. *Reliability Engineering & System Safety, 230*, 108978. https://doi.org/10.1016/j.ress.2022.108978

[24] Feng, X., Zhou, Y., Hua, T., Zou, Y., & Xiao, J. (2017). Contact temperature prediction of high voltage switchgear based on multiple linear regression model. *2017 32nd Youth Academic Annual Conference of Chinese Association of Automation (YAC)*. https://doi.org/10.1109/yac.2017.7967419

[25] Jha, R. S., Jha, N. N., & Lele, M. M. (2023). Stochastic gradient descent algorithm for the predictive modelling of grate combustion and boiler dynamics. *ISA Transactions, 136*, 571–589. https://doi.org/10.1016/j.isatra.2022.10.036

[26] Li, H., & Yamamoto, S. (2016). Polynomial regression based model-free predictive control for nonlinear systems. *2016 55th Annual Conference of the Society of Instrument and Control Engineers of Japan (SICE)*. https://doi.org/10.1109/sice.2016.7749264

[27] Fan, L., Chen, S., Li, Q., & Zhu, Z. (2015). Variable selection and model prediction based on Lasso, adaptive lasso and elastic net. *2015 4th International Conference on Computer Science and Network Technology (ICCSNT)*. https://doi.org/10.1109/iccsnt.2015.7490813

[28] Czajkowski, M., & Kretowski, M. (2016). The role of decision tree representation in regression problems – An evolutionary perspective. *Applied Soft Computing, 48*, 458–475. https://doi.org/10.1016/j.asoc.2016.07.007

[29] Gupta, S., & Prabha, C. (2021). Voting regression model for COVID-19 time series data analysis. *2021 3rd International Conference on Advances in Computing, Communication Control and Networking (ICAC3N)*. https://doi.org/10.1109/icac3n53548.2021.9725524

[30] Peng, L., Zheng, S., Zhong, Q., Chai, X., & Lin, J. (2023). A novel bagged tree ensemble regression method with multiple correlation coefficients to predict the train body vibrations using rail inspection data. *Mechanical Systems and Signal Processing, 182*, 109543. https://doi.org/10.1016/j.ymssp.2022.109543

[31] Kwak, S., Kim, J., Ding, H., Xu, X., Chen, R., Guo, J., & Fu, H. (2022). Machine learning prediction of the mechanical properties of γ-tial alloys produced using random forest regression model. *Journal of Materials Research and Technology, 18*, 520–530. https://doi.org/10.1016/j.jmrt.2022.02.108

[32] Wu, J., Guo, X., Fang, M., & Zhang, J. (2022). Short term return prediction of cryptocurrency based on XGBoost algorithm. *2022 International Conference on Big Data, Information and Computer Network (BDICN)*. https://doi.org/10.1109/bdicn55575.2022.00015

12 Sentimental Analysis on Amazon Book Reviews
A Deep Learning Approach

A. Vijayalakshmi, Koesha Sinha, and Debopriya Bose
Department of Statistics and Data Science, CHRIST (Deemed to be University), Bangalore, India

12.1 INTRODUCTION

Natural language processing (NLP), a component of artificial intelligence, is built to enable computers to recognize and infer important information from human language. NLP includes various computational techniques and algorithms to process, analyze, and operate natural language data, such as text, speech, and dialog. In the current world that is filled with data, NLP is used in various applications across different domains and is proved to improve in extracting information from data available. Within the realm of NLP, various tasks and techniques are employed to bridge the gap between human language and machine understanding. Text analysis techniques, such as part-of-speech tagging, named entity recognition, syntactic parsing, and semantic analysis, enable computers to derive meaning from textual data. Sentiment analysis allows for the classification of text as positive, negative, or neutral, providing insights into public opinion. Machine translation systems leverage NLP algorithms to automatically translate text between languages, fostering global communication and information exchange.

Through its many uses, natural language processing (NLP) has revolutionized a number of industries. In order to understand client sentiment and make wise decisions, organizations often use sentiment analysis, which entails extracting subjective information from text data. Machine translation technologies like Google Translate, Microsoft Translator, and DeepL are powered by NLP, enabling seamless language communication [1]. NLP algorithms are used by chatbots and virtual assistants to comprehend user inquiries and deliver automated responses, improving consumer experiences and lowering the human workload. Information extraction and text mining are made possible by NLP approaches, making it easier to glean insightful information from unstructured text for projects like trend analysis and market research. In addition, NLP-based question-answering systems can understand user questions and provide relevant information after retrieving them. Table 12.1 lists a few of the applications of NLP with the algorithms that can be used for the purpose [2–6].

12.2 SENTIMENTAL ANALYSIS

Sentimental analysis also known as opinion mining, is a widely used field of study within natural language processing (NLP) that focuses on automatically determining the sentiment or emotional tone of text data. With the proliferation of social media, online reviews, and other forms of user-generated content, the need to extract insights from vast amounts of opinionated text has become increasingly important. Opinion mining involves employing various computational techniques, including machine learning algorithms, statistical analysis, and linguistic processing, to analyze and classify text as positive, negative, or neutral based on sentiment. The term 'sentiment analysis' refers

DOI: 10.1201/9781003407959-12

TABLE 12.1
Applications of Natural Language Processing

Application	Algorithms/Techniques	Description
Sentiment Analysis	Naive Bayes, SVM, RNN, LSTM, Transformers	Analyzes text to determine the sentiment expressed, such as positive, negative, or neutral, to gauge customer opinions or feedback.
Machine Translation	SMT, NMT, Sequence-to-Sequence, Attention Mechanism	Translates text from one language to another, enabling communication and understanding across different languages.
Chatbots/Virtual Assistants	NLU, NER, Intent Recognition, Dialogue Management, RNN, Transformers	Provides automated conversational interfaces that as the ability to understand the user queries and respond to the queries in turn assisting with tasks and delivering information.
Information Extraction	NER, POS Tagging, Chunking, Relation Extraction, Named Entity Linking	Extracts structured information from unstructured text, identifying entities, relationships, and key facts for further analysis or organization.
Text Summarization	TextRank, LSA, Transformer-based models	Generates concise summaries of lengthy documents or articles, condensing important information while retaining key points.
Speech Recognition	HMM, DNN, RNN, CTC	Converts spoken language into written text, enabling voice commands, transcription services, and speech-to-text applications.
Named Entity Recognition	CRF, Bidirectional LSTM, Transformers	Identifies and classifies named entities like people, organizations, and locations in text for applications such as information retrieval or entity analysis.
Question Answering	Information Retrieval, Document Ranking, Passage Comprehension, NER, Question Classification, NLU	Understands user questions and provide relevant answers by retrieving and processing information from various sources.
Language Generation	RNN, Transformers, GANs	Generates coherent and contextually appropriate text, used in chatbots, content creation, and personalized responses.
Document Classification	Naive Bayes, SVM, Logistic Regression, Deep Learning	Automatically categorizes and organizes documents based on their content or topics, aiding in document management and information retrieval.

to the task of classifying texts or rather reviews to be positive or negative based on their content, in order to understand if the customers are liking a particular product or not [7]. For humans, this task is trivial as they can easily identify the nature of a particular text, but suppose one needs to classify thousands of such texts as positive or negative, then it would certainly require a lot of manpower and it will also be really time-consuming. Thus, as an alternative, we can easily perform sentiment analysis with the help of various existing techniques. When we are dealing with a huge amount of data, it is always beneficial to use neural networks or deep learning techniques.

Sentiment analysis has seen a significant boom in recent years. With the advent of social media, online reviews, and other user-generated content, there has been an explosion of text data that contains opinions, emotions, and sentiments expressed by individuals and communities. This has created a growing demand for automated methods to analyze and extract insights from this vast amount of data. With wide-ranging applications, ranging from businesses and marketing to social sciences and public opinion research sentimental analysis has gained its importance in every field. Companies can leverage sentiment analysis to gain insights into customer opinions about their products or services, monitor brand reputation, and inform marketing strategies. Researchers can use opinion mining to study public sentiment toward social issues, political events, or public policies [8]. Additionally, sentiment analysis can be applied in customer service, product development, market research, and other areas where understanding and analyzing human opinions and emotions from text data are valuable.

Opinion mining plays an important role in research and supports various application in the area of NLP. It is seen that there is increasing adoption of sentimental analysis in various industries and domains. Many businesses and organizations now rely on sentiment analysis to understand customer feedback, monitor brand reputation, inform marketing strategies, and gain competitive insights. Social scientists and researchers also use sentiment analysis to study public opinions on social issues, political events, and other topics of interest. Additionally, sentiment analysis has been applied in areas such as customer service, product development, market research, and more, to understand customer sentiments and preferences.

Recent advancements in various machine learning algorithms, deep learning techniques, as well as the access to enormous labeled datasets is a major contribution to the acceptance of opinion mining across various domains. These advancements have enabled the development of more accurate and sophisticated sentiment analysis models that can handle different languages, domains, and types of text data. Furthermore, the rapid pace of technological advancements, increased accessibility of text data through APIs and web scraping, and the growing need for data-driven decision-making have all contributed to the proliferation of opinion mining in recent years. Sentiment analysis has developed into a useful tool for deciphering and analyzing thoughts, emotions, and sentiments expressed in text as businesses and organizations strive to harness the power of text data, and its application is expanding across a range of industries and areas. Table 12.2 lists various domains where sentimental analysis can be used. The algorithms listed in the table are the commonly used algorithms and they can vary as per the problem requirements [7–10].

12.2.1 Types of Sentimental Analysis

The tasks powered by sentiment analysis can be classified into two levels: basic or core sentiment analysis (document-level, sentence-level, word-level, aspect-level sentiment analysis) and subcategory of the major sentiment analysis (concept-level, user-level, clause-level, sense-level sentiment analysis).

12.2.1.1 Document-Level Sentiment Analysis

A whole manuscript is used as the input in this kind of sentiment analysis approach, and the sentiment of the entire document is produced by combining the polarity of the sentences or words. This type of approach assumes or focuses only on one object or topic. So, if the document contains multiple objects or topics this approach may not be suitable for sentiment analysis. In order to build the document representation, [11] suggested a hierarchical network model that concentrates on key content. Modeling lengthy texts for establishing semantic linkages between sentences is a difficult situation in the case of document-level sentiment classification. This difficulty in linkage was addressed by introducing deep learning algorithms like long-short-term memory network (LSTM), RBM (restricted Boltzmann machine), and PNN (probabilistic neural network) in sentiment analysis [12].

12.2.1.2 Sentence-Level Sentiment Analysis

Combining the polarities of the words or phrases in a sentence helps in this category of sentimental analysis to uncover the sentiment that is hidden in the text. Sentence-level sentiment analysis can classify a sentence into subjective type or objective type. In case the sentence contains an opinion or judgment towards an entity, then it is classified as a subjective type. On the other hand, if the sentence contains just some facts and does not convey any kind of sentiment about some object, then

TABLE 12.2
Use Cases of Sentimental Analysis

Application	Description	Algorithms/Techniques
Social Media Monitoring	Analyzes sentiments expressed on social media platforms to track brand reputation, identify trends, and gather insights from customer feedback.	Naive Bayes, SVM, RNN, LSTM, Transformers
Customer Feedback Analysis	Analyzes sentiments in customer feedback, such as reviews and surveys, to understand customer satisfaction and identify areas for improvement.	Naive Bayes, SVM, RNN, LSTM, Transformers
Brand Monitoring	Monitors sentiments associated with a particular brand or product to assess customer perception and sentiment trends over time.	Naive Bayes, SVM, RNN, LSTM, Transformers
Market Research	Analyzes sentiments expressed in market research data to understand consumer opinions, preferences, and trends.	Naive Bayes, SVM, RNN, LSTM, Transformers
Reputation Management	Tracks and manages sentiments surrounding individuals, companies, or organizations to proactively address any negative sentiment and maintain a positive image.	Naive Bayes, SVM, RNN, LSTM, Transformers
Product Review Analysis	Analyzes sentiments in product reviews to evaluate customer satisfaction, identify product strengths and weaknesses, and make data-driven decisions for product improvements.	Naive Bayes, SVM, RNN, LSTM, Transformers
Political Sentiment Analysis	Evaluates sentiments expressed towards political figures, parties, or policies to gauge public opinion and sentiment trends.	Naive Bayes, SVM, RNN, LSTM, Transformers
Customer Service Analysis	Analyzes customer support interactions and sentiments to identify areas for improvement, enhance customer experiences, and measure customer satisfaction.	Naive Bayes, SVM, RNN, LSTM, Transformers
Brand Sentiment Analysis in Ads	Evaluates sentiments associated with brand advertisements to measure their impact, effectiveness, and customer response.	Naive Bayes, SVM, RNN, LSTM, Transformers
Financial Sentiment Analysis	Analyzes sentiments in financial news, reports, and social media discussions to assess market sentiment and predict trends in stock prices or investment decisions.	Naive Bayes, SVM, RNN, LSTM, Transformers

it is classified as an objective-type sentence. For example, "The work culture in this institution is very good" conveys a positive sentiment about the work culture in the institution and hence it is a subjective-type sentence with positive sentiment, whereas "I saw a group of people discussing over political influence in this city" does not convey any emotion or sentiment, rather it shares just a fact and hence this sentence is classified as an objective sentence. These types of objective sentences do not contribute to the polarity determination of a sentence and hence need to be filtered out. A model dubbed multi-level sentiment-enriched word embedding (MSWE), created by [13], uses a multi-layer perceptron (MLP) to model sentiment information at the word and tweet levels. In addition to learning sentiment-specific word embedding, the model also uses SVM to classify sentiment. But in general, deep learning-based approaches, like CNN, and LSTM are seen to give better accuracy [14].

12.2.1.3 Word-level Sentiment Analysis

When a text is considered for sentimental analysis, the impact of the polarity of a particular word is examined and explored in the whole text. This can be analyzed in two ways, corpus-based and dictionary-based. In corpus-based techniques, a word's co-occurrence patterns are used for sentiment analysis [15]. However, a big corpus is required to provide the statistical data required for determining a word's sentiment orientation, which is a basic need for any statistical model. The dictionary-based approach uses roots, synonyms, antonyms, and hierarchical structure to determine the sentiment of a particular word.

12.2.1.4   Aspect or Entity-Level Sentiment Analysis

In this case, before polarity is assigned in this sentiment analysis task, there is one more stage that must be completed. This phase is the identification of features, aspects, or entities, which is followed by the classification of the features as either positive or negative. This approach finds different aspects connected to a specific target entity, and the final polarity of the target is found by clubbing the polarities of each facet connected to it.

12.2.1.5   Concept-Level Sentiment Analysis

Sometimes simply extracting sentiments from individual words may result into wrong predictions. For example, "This machine has a long battery life" can convey a positive sentiment based on the positivity of the word "long," but "This way is so long" should convey negative sentiment based on the negativity of "long." So, only words are not sufficient for deciding upon the sentiment, a deep understanding of the meaning, context, and concept is required. This level of sentiment analysis is taken care of by concept-level sentiment analysis. Using online ontologies or semantic networks, concept-level sentiment analysis aims to communicate the semantic and affective information related to opinions [16]. This approach basically works by identifying features, assigning the right polarity to the words by considering different concepts and finally combining them to get the overall polarity.

12.2.1.6   User-Level Sentiment Analysis

User-level sentiment analysis considers the connection between the users of a certain social media platform. At the user-level analysis, reviews or comments of a certain influencer or reviewer can affect his or her followers and they may tend to have the same kind of opinion.

12.2.1.7   Clause-Level Sentiment Analysis

The ability to extract the polarity of two portions of a sentence separately by clauses is a strength of clause-level sentiment analysis. Clause-level sentiment analysis is concerned with the emotion related to each clause in light of its aspect, associated condition, domain, word relationships, etc. [17].

12.2.1.8   Sense-Level Sentiment Analysis

"I love going to work on holidays." In this sentence, though 'love' is an impactful word, the sentence is a pure example of sarcasm. Depending on how they are used in the sentence, the words that make up a sentence can have different meanings. Particularly when a term has more than one meaning, the context in which the word is used has a significant impact on the overall tone of the sentence or document. Therefore, resolving word sense ambiguity and conducting word sense disambiguation are essential components of creating an advanced sentiment analysis model [18].

12.3   RELATED WORKS

In recent years, NLP has experienced remarkable growth and innovation, making it possible for machines to comprehend, interpret, and generate human language. In order to give a comprehensive overview of contemporary research in the topic, this literature review will emphasize significant trends, methodologies, and difficulties.

NLP gives computers the ability to comprehend, decipher, and produce human language. One of the most important uses of NLP is sentiment analysis, which involves scanning through text data to extract sentiment and subjective information. To do sentiment analysis efficiently, a range of algorithms and techniques must be used. For sentimental analysis, traditional machine learning methods like naive Bayes and support vector machines (SVM) are widely used [19]. Deep learning techniques, such as recurrent neural networks (RNNs) with LSTM units, convolutional neural networks (CNNs), and transformer-based models like BERT (bidirectional encoder representations from transformers) and GPT (generative pre-trained transformer), have achieved cutting-edge performance in sentiment analysis [20].

Applying natural language processing techniques to extract and analyze the sentiment provided by customers in their reviews of books is a great advantage in assessing the popularity of books. Understanding customer attitudes towards particular books, authors, genres, and even particular elements like narrative, writing style, character development, and the pacing is crucially influenced by sentiment analysis of these evaluations. In this research work proposed by [21], paragraph vectors were employed to learn the syntactic and semantic relationship of a 'review text'. The review embeddings are further grouped and sorted and further given as input to the gated recurrent unit (GRU). SVM is used in training for sentiment classification. In [22] authors did a study on the sentiment analysis of Amazon.com customer product reviews. This study uses a dataset of product reviews to compare, train, and evaluate several machine learning algorithms. Comparisons were made between the multinomial naive Bayes (MNB), linear support vector machine (LSVM), and LSTM algorithm. The best performance was achieved by the LSTM with an accuracy of 90%. The authors concluded that LSTM networks are the most appropriate for binary sentiment analysis on product reviews. In [23] authors have compared K nearest neighbor, decision tree, naive Bayes, support vector machine, and random forest algorithms. They claim that random forest algorithm could analyze the sentiments of the text with an average accuracy of 90.15% and ranked highest when compared to other algorithms considered. Sentiments were classified on the basis of different emotions. After classification, the authors concluded the predominant emotions of authors on various books.

12.4 METHODOLOGY

The Amazon Kindle book reviews' dataset [24] is considered for this study. This is an unprocessed dataset available for sentiment analysis and there are 12,000 review texts in it. It includes every review text along with the relevant metadata in a single row. The book review dataset is preprocessed using methods like data cleaning, removal of URL tags, and removal of punctuation and stop words.

12.4.1 Preprocessing

Removal of URL

The dataset used for this study had URLs having prefix like 'http', 'www', and 'https', which does not contribute to sentiment analysis and hence they are removed.

Removal of punctuations

The dataset included various punctuations like full stop (.), comma (,), and brackets () and are removed since they do not convey sentiments. Apart from punctuation, special characters are also eliminated.

Removal of white spaces and numbers

All white spaces and numbers are removed as they do not contribute to the sentiment analysis

Stop word removal

Stop words are eliminated since they don't add much to the sentiment being represented in a text or have a significant meaning. Stop words are eliminated so that sentiment analysis algorithms can concentrate on terms in a text that are more significant and sentiment-rich, including adjectives, adverbs, and nouns that express feelings, views, or evaluations. Removing stop words can also help in decreasing noise and improves the effectiveness and accuracy of sentiment analysis algorithms.

By reducing the dimensionality of the text input, removing stop words can make the sentiment analysis process run more quickly and efficiently.

Convert to lowercase

We have defined a function using gensim.utils.simple_preprocess() to tokenize the reviews, but apart from tokenizing it is also converting all the characters in the reviews to lowercase characters, removing the tokens containing only two characters and also removing the accent marks from letters such that letters like 'é' will be converted to 'e'.

12.4.2 Classification Algorithms

12.4.2.1 Long-Short-Term Memory Networks

While working with short-term dependencies, RNN performs quite well, but it is unable to remember previous inputs for a long duration of time and apart from this limitation, RNN also has a problem of vanishing gradient. These limitations of RNN can be easily resolved with the help of LSTM networks, which are basically a variant of RNN. For performing sentiment analysis, it is really important to capture the entire text's sentiment, and for doing that, it is highly essential to consider the context of the entire text. Thus, LSTM is highly preferred for performing sentiment analysis as it is capable of capturing the long-term dependencies in textual data [22]. LSTM is also capable of ignoring useless data that is present in the text, which helps to solve the problem of vanishing gradients in the simple RNN.

The LSTM networks contain memory blocks known as cells in its hidden layers and each of these cells have an input gate, an output gate and a forget gate. The problem of vanishing gradient in RNN is solved in LSTM due to the presence of the forget gates, which is responsible for dropping the unnecessary information with the help of multiplication of a filter. The input gate is responsible for adding new information to the cell state and the output layer is responsible for determining which information is actually useful and according to give a suitable output.

12.4.2.2 Bidirectional LSTM

Bidirectional RNN is another variant of RNN, in which the input information is passed on through both directions, which enables it to make use of the information coming from both directions. For performing sentiment analysis, it can be really crucial to capture the context of the succeeding data along with the preceding and current data. Hence, bidirectional RNN can turn out to be really useful in this regard. It also helps to solve the problem of vanishing gradient in the simple RNN model by combining the data coming from both directions at each time step.

The bidirectional LSTM networks are designed by dividing LSTM neurons into two directions. One direction is that of the forward state and the other is the backward state. Figure 12.1 depicts the architectural configuration of LSTM and bidirectional LSTM networks. The advantage of this network is that the two directions that are included in this network allow the input data from both the past and future. This is different from LSTM. In LSTM, there is a delay in fetching information from the future. Bidirectional LSTM involves procedures that are used during backpropagation.

12.5 EXPERIMENTATION

In this chapter, we experiment with analyzing the sentiments of customers on Amazon Kindle book reviews. This kind of analysis is important in understanding customer opinions. Understanding the opinions/sentiments of customers on a book can give an understanding of what customers think about the book, whether they liked or disliked the book as well as what aspect of the book they were interested in and which aspect they did not like. This knowledge about the sentiments can help

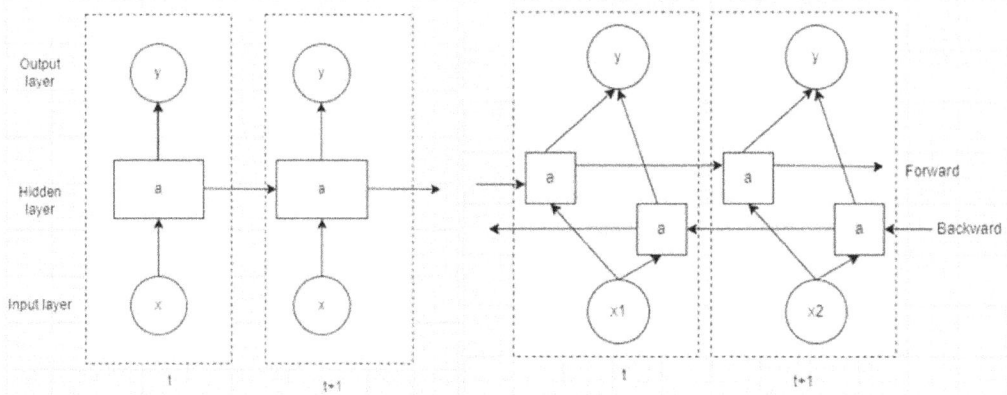

FIGURE 12.1 Architectural configuration of LSTM and bidirectional LSTM.

TABLE 12.3
Features in the Dataset

asin	This variable is representing the ID of the product, like B000FA64PK
helpful	This term explains the helpfulness rating of the book review – example: 2/3.
overall	This term is to signify the rating of the product.
reviewText	Text of the review (heading).
reviewTime	This term is used to represent the review timestamp(raw).
reviewerID	This term represents the ID of the reviewer, like A3SPTOKDG7WBLN
reviewerName	This variable is used in representing the reviewer's name
summary	This gives the review summary (description).
unixReviewTime	This represents the time stamp of review

in improving sales/make changes to future books. This kind of analysis can help in a great way in improving sales.

The dataset considered for this work is an unprocessed dataset of Amazon Kindle reviews. The dataset contains 12000 rows of review text and its corresponding metadata.

The dataset has 11 features/columns listed in Table 12.3.

Apart from these features, there are two more columns in this dataset, which are unnamed/insignificant.

The dataset includes much information about the book. In this work, our objective is to find the sentiments of customers on the book. Therefore, the "rating" and "reviewText" columns are the two that we mostly need to perform sentiment analysis. Thus, we are dropping all the other columns from this dataset. Further, we are converting all the ratings of values 1, 2 to negative review (denoted as 0), 3 to neutral review (denoted as 1) and 4, 5 to positive review (denoted as 2). This helps in categorizing the sentiments of the book review into a more systematic review.

For the purpose of preprocessing, a function is defined to remove the unnecessary punctuations, URL indicators such as 'https', 'www', etc., email address indicators like '@', and other characters like single quotes and new-line characters from our review text sentences, primarily with the help of 're' package. This is carried out since these characters do not contribute to finding sentiments in the review text. The second function is defined using gensim.utils.simple_preprocess() to tokenize the reviews, but apart from tokenizing, it is also converting all the characters in the reviews to lowercase

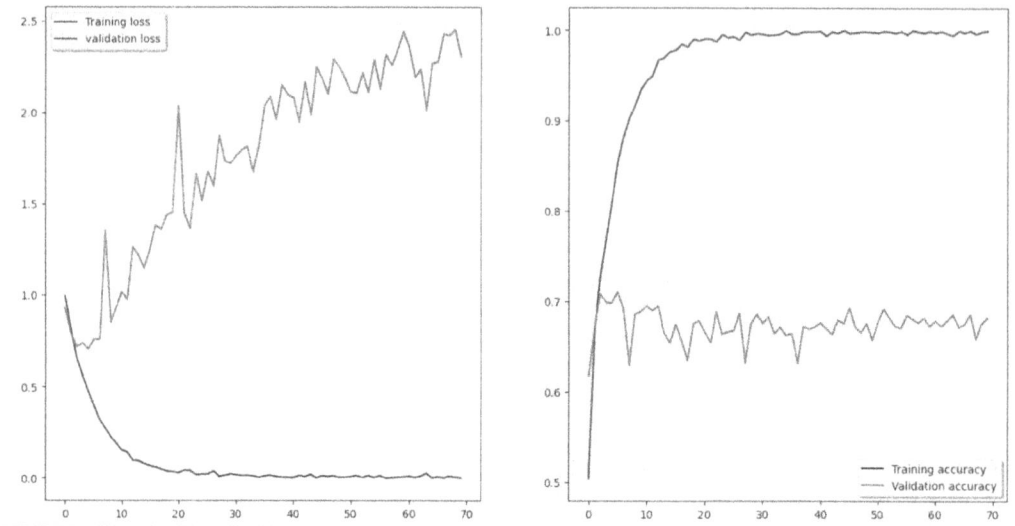

FIGURE 12.2 The curve representing training loss and validation loss (left) and training accuracy and validation accuracy (right) of the LSTM model.

characters, removing the tokens containing only two characters and it also removes the accent marks from letters such that letters like 'é' will be converted to 'e'. The created tokens of words are converted to sentences with the help of TreebankWordDetokenizer().detokenize() function before tokenizing it again into a sequence of integers.

All of the above functions are applied to the dataset in order to obtain the preprocessed data that can serve well for the purpose of sentiment analysis. After enhancing the quality of the reviews with the help of various functions, we tokenized the reviews into a sequence of integers, and then we padded it into sequences of length 200.

In the next step, we used Pandas get_dummies() function to perform one-hot encoding on the ratings. Here, we have used the bidirectional LSTM model to fit our dataset as it is capable of using a bidirectional propagation mechanism for capturing both historical and future data, which helps to get better results compared to LSTM as LSTM is unable to capture future data. We trained the model by using two-thirds, i.e., 66.67% of our dataset over 70 epochs and then we tested the model with the help of the remaining one-third, i.e., 33.33% of our dataset. We have used the softmax activation function as we are trying to perform multiclass classification over here and we have also used the RMSProp optimization technique, which stands for root mean squared propagation and it is basically a stochastic mini-batch learning method.

We acquire the training and validation loss after applying the model to the preprocessed dataset, as shown in Figures 12.2 and 12.3.

From Figures 12.2 and 12.3, we can clearly observe that the training loss is decreasing and the training accuracy is increasing with an increase in the number of epochs, which is certainly favorable. However, at the same time, we can also observe that the validation accuracy is constantly fluctuating and validation loss is increasing, instead of decreasing with an increase in the number of epochs, this may help to explain the model's inability to achieve even better performance.

The validation loss for the LSTM model in Figure 12.2 is increasing more rapidly compared to the validation loss for the bidirectional LSTM model in Figure 12.3 as the number of epochs is increasing and the validation accuracy for the LSTM model in Figure 12.2 also seems to be higher compared to the validation accuracy for the bidirectional LSTM model in Figure 12.3. Thus, from

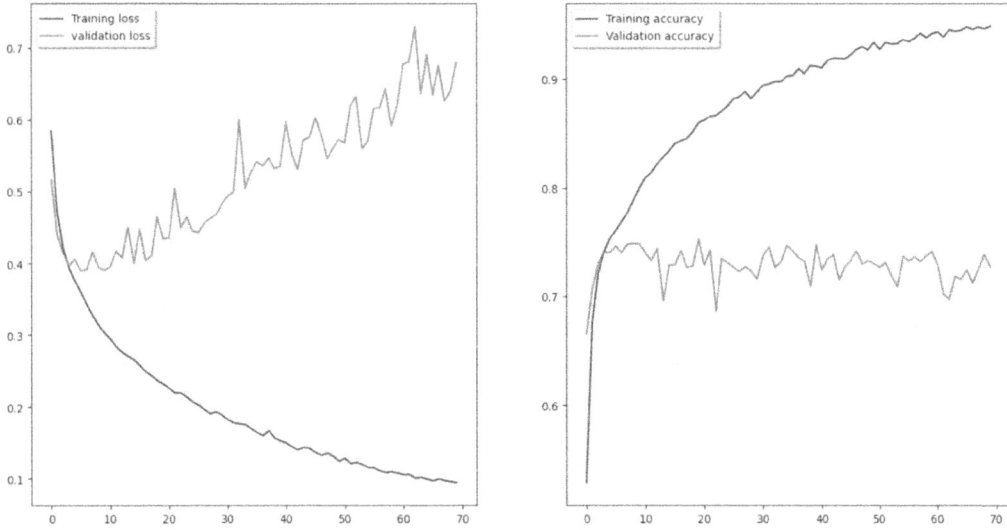

FIGURE 12.3 The curve representing training loss and validation loss (left) and training accuracy and validation accuracy (right) of the bidirectional LSTM model.

Figures 12.2 and 12.3, we can get a general idea that the bidirectional LSTM model is performing somewhat better than the LSTM model for this dataset.

In Table 12.2 and 12.3, '0' denotes negative reviews, '1' denotes neutral reviews, and '2' denotes positive reviews. While using a model for prediction, the predicted data can be placed into one of these four categories:

True positive: Both the actual and predicted values in this category are positive.
True negative: Both the actual value and the predicted value are negative in this category.
False positive: In this class, the predicted value is positive, but the actual value is negative.
False negative: The actual value falls positive and the predicted value is negative.

The most popular accuracy measures are as follows:
Accuracy: The percentage of right predictions to all of the model's predictions that were made is called accuracy.

$$\text{Accuracy} = \frac{\text{True positive} + \text{True Negative}}{\text{True Positive} + \text{False Positive} + \text{True Negative} + \text{False Negative}}$$

Precision: Precision is the ratio of the number of correct positive predictions to the total number of positive predictions, i.e., both true positive and false positive instances are considered.

$$\text{Precision} = \frac{\text{True Positive}}{\text{True Positive} + \text{False Positive}}$$

Recall: Recall is the ratio of the number of correct positive predictions to the total number of data points that are actually positive, i.e., both true positive and false negative instances are considered.

TABLE 12.4
Values of the Accuracy Measures Obtained for the LSTM Model

	Precision	Recall	F1-score	Support
0	0.65	0.79	0.71	319
1	0.29	0.22	0.25	140
2	0.79	0.74	0.76	528
accuracy			0.68	987
macro avg	0.58	0.58	0.58	987
weighted avg	0.68	0.68	0.68	987

$$\text{Recall} = \frac{\text{True Positive}}{\text{True Positive} + \text{False Negative}}$$

F1-score: F1-score mainly represents the harmonic mean of the precision and the recall. It can be really useful under certain circumstances as it takes both measures into consideration.

$$\text{F1 score} = 2 * \frac{\text{Recall*Precision}}{\text{Recall} + \text{Precision}}$$

Macro avg: The macro average value obtained for the precision, recall and F1-score is the actual mean of the precision, recall and F1-score values obtained for the three classes '0', '1', and '2' denoting negative, neutral, and positive reviews, respectively.

Weighted avg: For unbalanced data, it can be really useful to add some weights to the values obtained for the precision, recall, and F1-score measures as it will help us to get a more unbiased result. Hence, the weighted average value obtained for the precision, recall, and F1-score is basically the mean of the precision, recall, and F1-score values obtained for the three classes '0', '1', and '2' after adding the appropriate weights to the values.

Based on the evaluation models, the result obtained in represented in Tables 12.4 and 12.5.

From Table 12.4, we can observe that the model achieves an accuracy of 68% after fitting the LSTM model to the dataset. Since the dataset is imbalanced, so we are mainly going to focus on the weighted averages of precision, recall, and F1-score, which are all 68% as well.

From Table 12.5, we can observe that the model achieves an accuracy of 73% after fitting the bidirectional LSTM model to the dataset. Since the dataset is imbalanced, so we are mainly going to focus on the weighted averages of precision, recall and F1-score, which are 71, 73, and 71%, respectively.

Hence, after observing Tables 12.4 and 12.5, we can certainly claim that the bidirectional LSTM model is performing quite well in identifying the implied sentiments compared to the LSTM model for this dataset.

12.6   LIBRARIES AND PACKAGES USED

TensorFlow

TensorFlow is one of the most popular open-source deep learning libraries. It contains various functions that help in building the model for classifying the reviews as positive, negative, or neutral, based on the context of the review. The codes written for performing tasks using the functions in the

TABLE 12.5
Values of the Accuracy Measures Obtained for the Bidirectional LSTM Model

	Precision	Recall	F1-score	Support
0	0.76	0.75	0.76	1358
1	0.36	0.24	0.29	634
2	0.78	0.87	0.82	1968
accuracy			0.73	3960
macro avg	0.63	0.62	0.62	3960
weighted avg	0.71	0.73	0.71	3960

TensorFlow library are quite long in length as they are low-level and intricate in nature. It enables an individual to perform various complex operations that require a neural network to work [18].

Keras

Like TensorFlow, Keras is also an extremely popular open-source deep-learning library. It is much easier to learn and use and it requires fewer lines of codes compared to TensorFlow. Keras internally uses TensorFlow in the background [18]. For sentiment analysis, a suitable model can easily be constructed with the help of the functions in the Keras library.

NLTK (Natural Language Toolkit)

NLTK (Natural Language Toolkit) is an open-source Python library, which contains the functions that are primarily used to perform tasks concerning natural language processing such as sentiment analysis. This library contains various packages such as nltk.corpus, which contains several modules like stopwords, which just like the name suggests, is a collection of the most frequently used words in English, such as "the," "is," "and" and so on [19].

Gensim

Gensim is another open-source Python library, which is highly useful in natural language processing tasks such as sentiment analysis. It was developed with the intention of extracting semantic topics from documents. It contains several functions such as get_dummies(), which is used for one-hot encoding [20].

re

re is a built-in package in Python, which is really useful while working with regular expressions [21]. It can be primarily used for performing certain preprocessing tasks in sentiment analysis such as for removing unnecessary punctuations, URL indicators such as 'https', 'www', etc., email address indicators like '@', and so on.

12.7 SYSTEM CONFIGURATION

The proposed method is developed in Python and the experimental test is carried out on a computer with Intel(R) Core(TM) i5-1035G1 CPU @ 1.00GHz 1.20 GHz processor and Windows 10 operating system.

12.8 CONCLUSIONS AND FUTURE ENHANCEMENT

There is a large amount of unstructured data being generated every day from various sources like social media, e-commerce, and other online platforms. Sentiment analysis is gaining popularity because it provides a way to extract insights from these data and understand the opinion of people. These opinions can help in improving their products and services. In this chapter, we are analyzing the sentiments of customers based on the reviews from Amazon Kindle books. Bidirectional long-short-term memory (BiLSTM) is a type of RNN that can analyze sentiments particularly well for Amazon Kindle book reviews. This is particularly because BiLSTMs have the ability to process sequences of text in both forward and backward directions simultaneously, allowing them to capture more nuanced relationships between words and better understand the context in which they are used. In the case of sentiment analysis of Amazon Kindle book reviews, where reviews can often be lengthy and complex, a BiLSTM can provide a more nuanced understanding of the sentiment expressed in the text. This can help authors, publishers, and marketers better understand customer opinions and make more informed decisions about book promotion, marketing, and development.

Bidirectional LSTM is used to train the model, which is a powerful technique for gaining insights into reader opinions and preferences. BiLSTM's ability to process text sequences in both forward and backward directions allow for a more nuanced understanding of the sentiment expressed in the reviews. In the future, we can include additional features, such as metadata about the book or the reviewer, which could provide additional context for sentiment analysis.

REFERENCES

[1] Chowdhary, K. R. "Natural language processing." *Fundamentals of artificial intelligence* (2020): 603–649. Springer, New Delhi. https://doi.org/10.1007/978-81-322-3972-7_19

[2] Otter, Daniel W., Julian R. Medina, and Jugal K. Kalita. "A survey of the usages of deep learning for natural language processing." *IEEE transactions on neural networks and learning systems* 32.2 (2020): 604–624.

[3] Torfi, Amirsina, et al. "Natural language processing advancements by deep learning: A survey." arXiv preprint arXiv:2003.01200 4 (2020): 1–23.

[4] Goldberg, Yoav. "A primer on neural network models for natural language processing." *Journal of artificial intelligence research* 57 (2016): 345–420.

[5] Danilevsky, Marina, et al. "A survey of the state of explainable AI for natural language processing." arXiv preprint arXiv:2010.00711 1 (2020): 1–13.

[6] Khurana, Diksha, et al. "Natural language processing: State of the art, current trends and challenges." *Multimedia tools and applications* 82.3 (2023): 3713–3744.

[7] Medhat, Walaa, Ahmed Hassan, and Hoda Korashy. "Sentiment analysis algorithms and applications: A survey." *Ain Shams engineering journal* 5.4 (2014): 1093–1113.

[8] Wankhade, Mayur, Annavarapu Chandra Sekhara Rao, and Chaitanya Kulkarni. "A survey on sentiment analysis methods, applications, and challenges." *Artificial intelligence review* 55.7 (2022): 5731–5780.

[9] Mehta, Pooja, and Sharnil Pandya. "A review on sentiment analysis methodologies, practices and applications." *International journal of scientific and technology research* 9.2 (2020): 601–609.

[10] Shayaa, Shahid, et al. "Sentiment analysis of big data: Methods, applications, and open challenges." *IEEE Access* 6 (2018): 37807–37827.

[11] Yang, Zichao, et al. "Hierarchical attention networks for document classification." Proceedings of the 2016 conference of the North American chapter of the association for computational linguistics: human language technologies. 2016.

[12] Khan, Muhammad Taimoor, et al. "Sentiment analysis and the complex natural language." *Complex Adaptive Systems Modeling* 4 (2016): 1–19.

[13] Xiong, Shufeng, et al. "Towards Twitter sentiment classification by multi-level sentiment-enriched word embeddings." *Neurocomputing* 275 (2018): 2459–2466.

[14] Zhao, Wei, et al. "Weakly-supervised deep embedding for product review sentiment analysis." *IEEE transactions on knowledge and data engineering* 30.1 (2017): 185–197.

[15] Reyes, Antonio, and Paolo Rosso. "Making objective decisions from subjective data: Detecting irony in customer reviews." *Decision support systems* 53.4 (2012): 754–760.

[16] Cambria, Erik. "An introduction to concept-level sentiment analysis." Advances in soft computing and its applications: 12th Mexican international conference on artificial intelligence, MICAI 2013, Mexico City, Mexico, November 24–30, 2013, Proceedings, Part II 12. Springer Berlin Heidelberg, 2013.

[17] Bordoloi, Monali, and Saroj Kumar Biswas. "Sentiment analysis: A survey on design framework, applications and future scopes." *Artificial Intelligence Review* 56 (2023): 12505–12560.

[18] Wiebe, Janyce, and Rada Mihalcea. "Word sense and subjectivity." Proceedings of the 21st international conference on computational linguistics and 44th annual meeting of the association for computational linguistics. 2006.

[19] Chong, Wei Yen, Bhawani Selvaretnam, and Lay-Ki Soon. "Natural language processing for sentiment analysis: an exploratory analysis on tweets." 2014 4th international conference on artificial intelligence with applications in engineering and technology. IEEE, pp. 212–217, 2014.

[20] Dang, Nhan Cach, María N. Moreno-García, and Fernando De la Prieta. "Sentiment analysis based on deep learning: A comparative study." *Electronics* 9.3 (2020): 483.

[21] Shrestha, Nishit, and Fatma Nasoz. "Deep learning sentiment analysis of amazon.com reviews and ratings." International Journal on Soft Computing, Artificial Intelligence and Applications (IJSCAI), 8.1 (2019): 1–15.

[22] Güner, Levent, Emilie Coyne, and Jim Smit. "Sentiment analysis for Amazon. com reviews." Big Data in Media Technology (DM2583) KTH Royal Institute of Technology, Stockholm (2019), pp. 1–9. doi: 10.13140/RG.2.2.13939.37920

[23] Srujan, K. S., et al. "Classification of amazon book reviews based on sentiment analysis." Information systems design and intelligent applications: Proceedings of fourth international conference INDIA 2017. Springer Singapore, pp. 401–411, 2018. doi: 10.1007/978-981-10-7512-4_40.

[24] www.kaggle.com/datasets/meetnagadia/amazon-kindle-book-review-for-sentiment-analysis

13 A Deep LSTM Recurrent Learning Approach for Sentiment Analysis on Movie Reviews

G.R. Khanaghavalle[1], V. Rajalakshmi[1], R. Jayabhaduri[1], A. Kala[2], and P. Sharon Femi[2]

[1] Department of Computer Science and Engineering, Sri Venkateswara College of Engineering, Sriperumbudur, Chennai, Tamil Nadu, India
[2] Department of Information Technology, Sri Venkateswara College of Engineering, Sriperumbudur, Chennai, Tamil Nadu, India

13.1 INTRODUCTION

One of the most enjoyable ways to spend time in the present era is by watching movies, and many individuals enjoy sharing their thoughts and opinions about them. These expressions, known as movie reviews, provide a summary of a person's experience watching a movie. Movie reviews express the feelings of the audience and their expectations from the movie. This helps in making prior decisions about the movies. A well-written movie review can give readers an idea of what to expect from a movie, including its genre, tone, and overall quality. It offers an evaluation of a movie's artistic and technical merits. Movie reviews often spark discussion and debate about a movie's themes, messages, and cultural significance. Reviews can offer insight into how a movie fits into the larger cultural conversation and it offers an opportunity for viewers to express their views and opinions. It helps to promote a film by generating buzz and increasing awareness among potential audiences. Positive reviews can help draw in more viewers, while negative reviews can warn audiences away from a film that may not be worth their time or money.

Sentiment analysis for movie reviews can be useful in understanding audience reactions to a movie, identifying areas for improvement, and gauging the overall reception of a film. It helps in deciding whether a review is negative or positive. To perform analysis, a dataset of movie reviews has to be collected. This dataset can include reviews from various sources, such as social media, review websites, and blogs. The collected reviews need to be preprocessed by removing any unnecessary data such as special characters, punctuation, and stop words. Features or important words/phrases need to be extracted from the preprocessed text. This can be done using term frequency-inverse document frequency or bag-of-words methodologies. Finally, sentiment classification techniques such as machine learning, rule-based analysis, or deep learning can be implemented to classify the reviews. The reviews shall be categorized as positive, or negative based on the extracted features. Machine learning algorithms require a labeled dataset with the sentiment of reviews. The model is fed with features extracted from the text and predicts the sentiment of new text data obtained. The most commonly used ML algorithms include linear regression, support vector machine (SVM), random forest, and decision tree. Rule-based techniques utilize a predefined dictionary to categorize the sentiment of a text. Predefined dictionaries of words along with their associated sentiment scores

DOI: 10.1201/9781003407959-13

are used to classify the sentiment of text or sentences in lexicon-based techniques. The sentiment score is obtained by combining the sentiment score of words in the text.

Deep learning techniques involve building a neural architecture that automatically learns the representations of text data and makes predictions on new data. DNNs are predominantly implemented in the analysis of sentiment because of their ability to automatically extract patterns and features from large amounts of data, making them well-suited for processing text data. These algorithms can effectively analyze a huge volume of text data. A deep neural network can only understand numerical representations. A popular model for sentiment analysis is the long short-term memory (LSTM), which is suited for analyzing text data.

LSTM is a variant of recurrent neural network (RNN) that has proven its ability to process sequential data such as text [1]. A deep LSTM recurrent learning technique for the analysis of sentiment on the IMDB dataset can be a powerful tool for understanding audience reactions to films and identifying areas for improvement.

13.2 RELATED WORKS

In their research, Bodapati et al. [2] implemented RNN-LSTM networks to analyze the sentiment on movie reviews. The researchers treated the problem as a two-class sentiment classification problem, in which reviews were classified as either positive or negative. To address the variability in sentence length, the researchers employed sentence vectorization techniques. Additionally, the researchers examined the effect of hyperparameters, such as the number of layers, dropout rate, and activation functions, on the model performance. They analyzed the performance of the algorithms across different neural network configurations. The researchers conducted their experimental studies using the IMDB benchmark dataset.

Singh et al. [3] proposed a movie recommender system based on the idea of K-nearest neighbours and cosine similarity. Cosine similarity is based on a normalized score derived from the distance computation function. The efficiency of the recommender system is improved by applying the K-nearest neighbours algorithm.

Basiri et al. [4] implemented a deep learning model called the attention-based bidirectional CNN-RNN deep model (ABCDM), which employs GRU layers and two bi-LSTM layers. This model captures the features from the forward as well as from the backward directions to extract essential features from the past and present context. The emphasis in the output bidirectional layer is utilized using the attention mechanism. To extract local features and to reduce the features extracted, the model employs a pooling mechanism from the convolutional neural network. ABCDM is used for polarity detection in sentiment analysis.

Ullah et al. [5] developed a seven-layer convolutional neural network for analyzing the sentiment on the movie review dataset. The network contains an embedding layer that represents the vector sequence, and then two convolutional layers for feature extraction. The dimensions are reduced using a max-pooling layer. To address overfitting and enhance generalization error in the model, a dropout layer, and a dense layer are included. The final layer, a fully connected layer, is used for predicting between two classes in the binary classification namely positive and negative.

Gupta et al. [6] put forth an attention-based long short-term memory (Senti_ALSTM) for achieving the groundbreaking performance of sentiment analysis. This approach enhances neural networks by incorporating memory allocation and addresses the issue of vanishing gradients commonly associated with RNN. The attention mechanism in the LSTM model, considers the input sequence to remember the content and establish connections between input and output layers. The researchers conducted experiments with this model, and the benchmark data confirmed its efficacy over other existing models.

Jiang et al. [7] employed a hybrid model that relied on Word2Vec and LSTM to analyze the sentiment of movie reviews. The Word2Vec component was responsible for incorporating the contextual semantics of the text and generating corresponding text vectors, while the LSTM extracted semantic data to categorize the reviews as positive or negative. To gauge the classification capability of the Word2Vec model, the authors constructed models using the hash trick methods and the word index model. These benchmark models were then combined with various machine learning algorithms to create hash trick-based classifiers and word index-based classifiers. The experiments proved that the Word2Vec-LSTM performed the best.

Elfaik and Nfaoui [8] introduced a highly effective bidirectional LSTM network (bi-LSTM) for enhancing Arabic sentiment Analysis. Their approach employs forward-backward processing to capture contextual information from Arabic sentences. Results on widely used sentiment analysis datasets reveal that the LSTM model performs better than the DL models and traditional ML models, yielding significant improvements in accuracy.

Qamar et al. [9] explored the deep neural network that incorporates automated feature extraction and has the capability to handle large amounts of data. One of the main limitations of the current analysis of sentiment is that they rely on the domain they are trained on, resulting in reduced accuracy when applied to datasets from different domains. In order to overcome this limitation they proposed a model for analyzing sentiments, which is trained on datasets from diverse regions, including the US Presidential Election dataset, IMDB Movie Review Dataset, and Twitter US Airline Review dataset, making it domain-agnostic. The model is trained on deep learning models, namely CNN, GRU, CNN-GRU, LSTM, and CNN-LSTM.

Behera et al. introduced an approach that combines LSTM and CNN [1] for sentiment classification of movie reviews from different domains. CNNs are known for their effectiveness in local feature selection, while LSTMs excel in the analysis of lengthy sequential texts. The co-LSTM network is designed to be adaptable in analyzing large-scale social data, with scalability as a priority. In order to capture the relationship between word sequences in a review, an LSTM layer is employed. The distinctive architecture of LSTM, which includes memory at each network, appears to be capable of effectively capturing the long-term dependencies of words.

13.3   LEVELS OF SENTIMENT ANALYSIS

Analysis of sentiment can be performed at three levels as shown in Figure 13.1: sentence level, aspect/feature level, and document level.

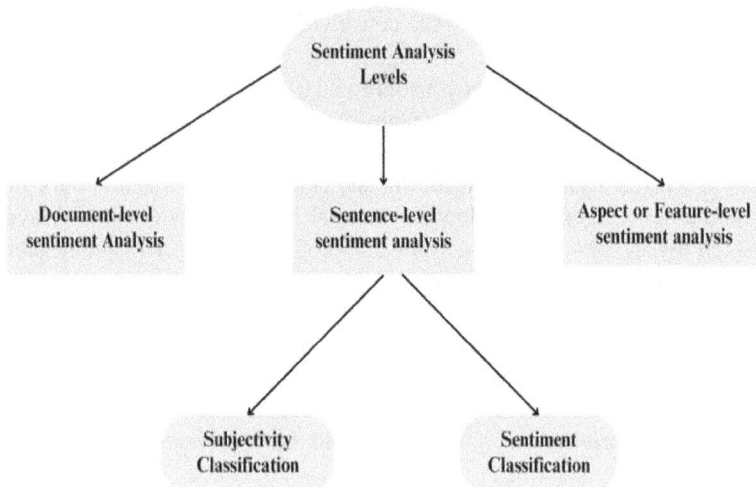

FIGURE 13.1   Levels of sentiment analysis.

13.3.1 Document-Based Sentiment Analysis

Document-based sentiment analysis stands as the basic form of analysis of sentiment. The basic unit of information comes from a single document that contains opinionated information about a single entity. The task is to analyze the entire document to classify whether the document holds a positive or negative feeling. Documents should be free from irrelevant information before processing. This type of analysis is not suitable if the document contains information about too many objects or entities. The supervised learning approach and unsupervised learning approach can be used for the classification of document-level analysis.

13.3.2 Sentence-Based Sentiment Analysis

The next level of analysis starts with the sentence. The sentence-based analysis is considered as a fine-tuned analysis of documents. Each sentence exhibits different opinions and each sentence is considered as a single unit of information.

These sentences are analyzed and the polarity of each sentence is calculated to decide an overall positive or negative opinion. Subjectivity and sentiment-based classification are the two sub-tasks of sentence-based analysis.

13.3.2.1 Subjectivity-Based Classification

A sentence shall be treated as an objective sentence or a subjective one. A subjective sentence consists of an opinion about an object or an entity. Whereas objective sentences are facts. It has no opinion about the object or entity. We bought a bike a month ago and its motor has tumbled off in a month. When this statement is considered, the first sentence does not convey any sentiment and it is a factual one. Hence this statement will not have a major impact in deciding the polarity of a sentence and it can be filtered out. The second sentence shows the opinion of the user, which is a subjective one and holds important information about the quality of the bike. Subjectivity analysis is helpful to mine opinion when a document contains information about many objects or entities.

13.3.2.2 Sentiment Classification

Sentiments can be classified into binary categories as positive and negative or multiclass categories. Sentiment analysis is carried out using two steps, which include sentiment polarity assignment and sentiment intensity assignment [10]. Sentiment polarity assignment is done by analyzing the sentence. The sentence is checked for positive and negative text. Based on the opinionated text in the sentence, the polarity is assigned. The sentiment of the user about the entity considered is defined in the sentiment polarity. Intensity describes the degree of polarity. Depending on the intensity of the sentiment (mild or strong) the polarity is decided.

13.3.2.3 Feature/Aspect-Based Sentiment Analysis

Feature-level analysis of sentiment is a fine-tuned sentiment analysis of the text [5]. Aspects refer to the attributes that contribute to the sentiment of a sentence. This kind of analysis allows businesses to analyze huge amounts of data to get opinions about the users or customers. It contributes to the particular aspect of sentiment in the context sentence. Consider the statement "Food served by the hotel was good but their service was dreadful". In this sentence the sentiment for the food is positive but for the restaurant it is negative. These types of analysis are challenging because rounding the context words around the target is difficult.

13.4 DATASET DESCRIPTION

For our work, we use the dataset in [11], which is publicly available in Kaggle. The word cloud for the IMDB dataset is shown in Figure 13.2. This dataset is curated for the analysis of sentiment on movie

FIGURE 13.2   Word cloud for the IMDB dataset.

reviews. All the reviews about the movies are gathered from IMDB (Internet Movie Database). The IMDB dataset contains a collection of 50,000 reviews about movies. The reviews are restricted to 30 per movie. The dataset contains a balanced ratio of reviews between positive and negative classes. The dataset belongs to a binary sentiment class. The reviews with a sentiment score of < 5 are given the value 0 or negative class. Whereas the review with a sentiment score > 7 is given the value 1 or positive class. Thus the dataset has 25,000 negative reviews and 25,000 positive reviews.

13.5 METHODOLOGY

13.5.1 DATA PREPROCESSING

IMDB dataset, which has 50,000 reviews annotated as positive and negative. The dataset is well-balanced since 50% of the reviews are positive and the remaining 50% are negative. The preliminary data preprocessing that we carried out is eliminating the stop words from the dataset so that we could focus on more important words that enables us to focus on the sentiment of the review. Tokenization process is carried out to create a bag of dictionary words that contain the most frequently used terms in movie reviews. The tokens created from movie reviews are then converted into a sequence of numbers [12]. Observations from Figure 13.3 shows that the movie review doesn't hold the same length of text. To overcome this issue, reviews with long sequences shall be truncated and reviews with short sequences shall be padded with null words, which then can be encoded to 0.

13.5.2 WORD EMBEDDING

The meaning of the word is extracted using the embedding process. Each word in a sentence is mapped to a vector in multidimensional space. In our work, embedding is carried out in two-dimensional

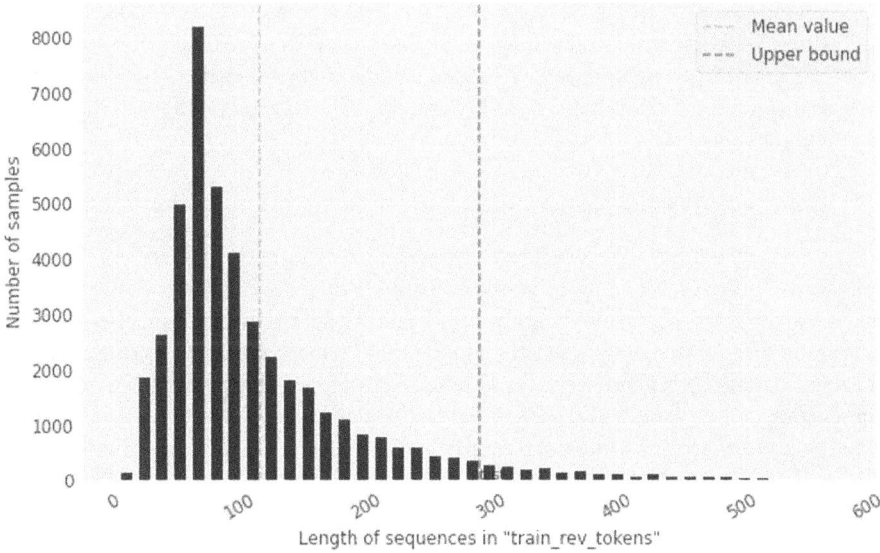

FIGURE 13.3   Length of sequences in dataset.

space. Each word is converted into a vector with coordinates (x, y) and then pointed to a plane. Through this process, words with similar meanings will be grouped together. This helps the neural network to learn from vectors in a similar way to learning from words. Word embedding solves two problems. When one-hot encoding is used, it causes performance degradation as it creates sparse vectors of high dimensions. On the other hand, embedding produces dense vectors of low dimensionality and the semantic relationship of words are projected in the distance and directions of the vectors.

31.5.3   CONVOLUTIONAL NEURAL NETWORK

CNN has its importance in image processing. Though it was intended for image recognition, this deep neural network can also be used for text classification. CNN [13] always learns independently where there is no relationship between the concerned input and desired output. A generalized CNN contains a convolutional layer, a max-pooling layer, a dropout layer, and an output layer. Word embedding carried out in the dataset provides an opportunity for the convolution layer [14] to extract features from it. Then the extracted features are fed into the pooling layer to convert into smaller dimensions. The max-pooling operation takes the highest value as a feature of a particular filter. The basic idea is to extract the highest value for one feature map. Generally, the model extracts multiple features using multiple feature maps. They form a penultimate layer, which is connected to a fully connected softmax layer. The softmax layer outputs a probability distribution over the labels.

13.5.4   RECURRENT NEURAL NETWORK

RNN learns from its previous experience. RNN works very well when we need to capture sequential pieces of information. RNN has an internal memory that enables it to remember current input with past sequences. Generally, inputs and outputs are independent in other neural networks. When we consider sentiment analysis the word sequence plays an important role in classifying the sentiment. So, it becomes important to remember the entire sequence of input in a neural network. RNN comes for rescue in this place.

RNN has a hidden layer that tries to remember some outline about the information. This hidden layer has a memory, which remembers all the calculated data. RNN provides the same weight and bias to all layers in order to standardize each hidden layer with the same characteristics. This process in turn reduces the number of parameters. RNN constructs only one hidden layer, which gets looped over many times as required.

13.5.5 LONG-SHORT TERM MEMORY

RNN has had many variations in recent days. One of the most common problems that occur in RNN is the exploding and vanishing gradient problem. To solve this, LSTM came into existence. LSTM works on the RNN but with a simple variation. When RNN works on sequential data it tries to remember all the information about the data. On the other hand, the LSTM recurrent network tries to forget irrelevant data by remembering the essential data alone. This process is achieved by a new activation function called "gates". Each LSTM unit computes a vector named internal cell state. This decides the information to be retained in previous LSTM units.

13.5.6 BIDIRECTIONAL LONG-SHORT TERM MEMORY

Bi-LSTM [15], a variation of RNN-LSTM where LSTM can be made to read the sequence both in the backward and forward directions. In RNN during the backpropagation stage, the gradients become too small to handle the long-term sequences. This causes an inability to update the weights in the model. LSTM solves this by adding a feed-forward connection in the architecture. The LSTM can also be stacked on top of CNN to move in forward and backward directions. This type of neural network is termed bidirectional LSTM.

13.6 EXPERIMENTAL RESULTS

We implemented the most common deep neural networks like RNN, RNN- LSTM, RNN- Bi-LSTM, and CNN on the IMDB dataset. We performed experiments on IMDB to capture the sentiment of the review using neural networks. The IMDB dataset has reviews of more than one sentence. The main purpose of this work is to explore the performance of deep neural networks towards the classification of sentiments in movie reviews. The word embedding technique applied to the dataset converts the reviews into numbers, which are fed into the different neural networks. The accuracy measures for different neural networks are shown in Table 13.1. The experiments on the IMDB dataset show that the bi-LSTM performs very well for sentiment classification as shown in Figure 13.4. We performed the same preprocessing techniques as discussed above on our dataset. Each model has a batch size of 256 with a learning rate of 0.01 and trained for 150 epochs. All the neural network models are trained in NVIDIA TITAN X GPU using the TensorFlow library.

TABLE 13.1
Deep Neural Network Sentiment Classification Results on IMDB Dataset

Model	Precision	Recall	F-Measure	Accuracy
CNN	80.32	79.68	80.00	80.00
LSTM	84.40	92.20	87.14	88.24
RNN	83.69	83.00	83.69	83.83
Bi-LSTM	**97.73**	**95.56**	**96.93**	**96.67**

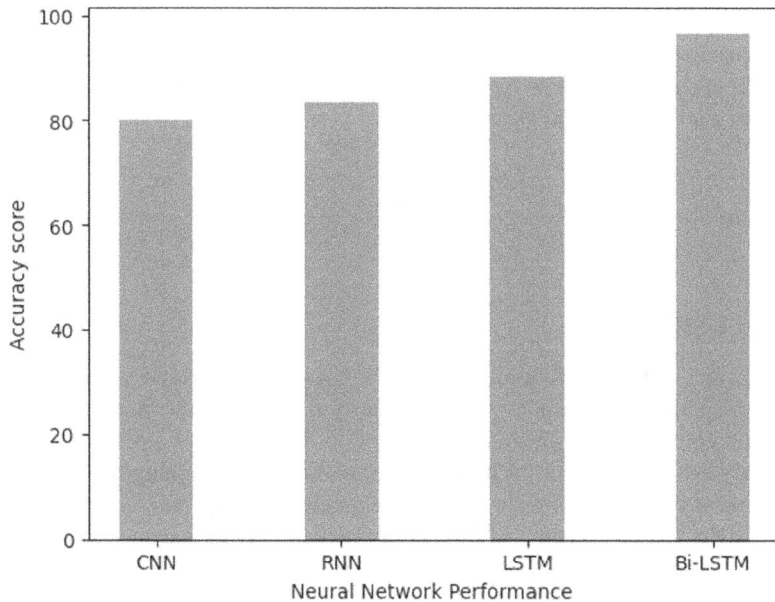

FIGURE 13.4 Accuracy comparison of neural networks on IMDB dataset.

13.6.1 EVALUATION METRICS

In order to evaluate the performance of deep neural networks we used precision, recall, F-measure, and accuracy scores.

13.6.1.1 Precision

Classifier's ability to return relevant instances is found using precision. It represents the quality of positive predictions made by the model. Precision denotes the ratio of true positives to the total number of positives.

$$precision = \frac{True\ Positives}{True\ Positives + False\ Positives} \tag{13.1}$$

13.6.1.2 Recall

Recall is also termed as aensitivity. The recall is the ratio of true positives to the total number of true positives and false negatives.

$$Recall = \frac{True\ Positives}{True\ Positives + False\ Negatives} \tag{13.2}$$

13.6.1.3 F-Measure

F-measure defines the quality of test accuracy of a model. It is calculated by combining precision and recall.

$$F - measure = 2 * \frac{1}{\dfrac{1}{precision} + \dfrac{1}{recall}} \tag{13.3}$$

13.6.1.4 Accuracy

Accuracy is a most common metric to evaluate the model based on classification. It is obtained by dividing the total correct predictions by the total number of inputs. When the dataset is balanced accuracy gives us a good indicator of our model.

$$Accuracy = \frac{True\ Positives + True\ Negative}{True\ Positives + True\ Negative + False\ Positives + False\ Negative} \quad (13.4)$$

13.7   CONCLUSION

The Internet provides a huge opportunity for people around the world to share their feelings and thoughts about each and everything they experience in life. This gives us a huge volume of data to explore. Sentiment analysis gives a great insight into the feeling of a person about a particular product. We analyzed the sentiment of movie reviews using deep learning models. CNN captures long-term dependency but also contains excessive convolutional layers. To overcome this problem we moved onto RNN and its variants. RNN is performing well for the sequential dataset. From our observations, we conclude that bi-LSTM gives a good accuracy score of 96% for the sentiment classification on the IMDB dataset.

REFERENCES

1. Behera, R. K., Jena, M., Rath, S. K. and Misra, S. (2021). Co-LSTM: Convolutional LSTM model for sentiment analysis in social big data. Information Processing & Management, 58(1), p. 102435.
2. Bodapati, J. D., Veeranjaneyulu, N. and Shareef, S. N. (2019). Sentiment analysis from movie reviews using LSTMs. *Ingénierie des Systèmes d Inf.*, *24*(1), pp. 125–129.
3. Singh, R. H., Maurya, S., Tripathi, T., Narula, T. and Srivastav, G. (2020). Movie recommendation system using cosine similarity and KNN. *International Journal of Engineering and Advanced Technology*, *9*(5), pp. 556–559.
4. Basiri, M. E., Nemati, S., Abdar, M., Cambria, E. and Acharya, U. R. (2021). ABCDM: An attention-based bidirectional CNN-RNN deep model for sentiment analysis. *Future Generation Computer Systems, 115*, pp. 279–294.
5. Ullah, K., Rashad, A., Khan, M., Ghadi, Y., Aljuaid, H. and Nawaz, Z. (2022). A deep neural network-based approach for sentiment analysis of movie reviews. *Complexity*, 5217491. https://doi.org/10.1155/2022/5217491
6. Gupta, C., Chawla, G., Rawlley, K., Bisht, K. and Sharma, M. (2021). Senti_ALSTM: Sentiment analysis of movie reviews using attention-based LSTM. In *Proceedings of 3rd International Conference on Computing Informatics and Networks: ICCIN 2020* (pp. 211–219). Springer Singapore.
7. Jiang, H., Hu, C. and Jiang, F. (2022, July). Text sentiment analysis of movie reviews based on Word2Vec-LSTM. In *2022 14th International Conference on Advanced Computational Intelligence (ICACI)* (pp. 129–134). IEEE.
8. Elfaik, H. and Nfaoui, E. H. (2020). Deep bidirectional LSTM network learning-based sentiment analysis for Arabic text. *Journal of Intelligent Systems*, *30*(1), pp. 395–412.
9. Qamar, M., Rao, H., Farooq, S. A. and Bhuyan, A. S. (2023, March). Sentiment analysis using deep learning: A domain independent approach. In *2023 Second International Conference on Electronics and Renewable Systems (ICEARS)* (pp. 1373–1380). IEEE.
10. Tian, L., Lai, C., and Moore, J. D. "Polarity and intensity: The two aspects of sentiment analysis." (2018). *arXiv preprint arXiv:1807.01466*.
11. Young, T., Hazarika, D., Poria, S. and Cambria, E. (2018). Recent trends in deep learning based natural language processing. *IEEE Computational Intelligence Magazine*, *13*(3), pp. 55–75.

12. Maas, A. L., Daly, R. E., Pham, P. T., Huang, D., Ng, A. Y. and Potts, C. (2011). Learning Word Vectors for Sentiment Analysis. *The 49th Annual Meeting of the Association for Computational Linguistics (ACL 2011),* pp. 142–150.
13. Kim, Y. (2019). Convolutional neural networks for sentence classification. arXiv 2014. arXiv preprint arXiv:1408.5882.
14. Zhang, L., Wang, S., and Liu, B. (2018). Deep learning for sentiment analysis: A survey. WIREs Data Mining Knowledge Discovery. 1253. https://doi.org/10.1002/widm.1253
15. Sangeetha, J. and Kumaran, U. (2023). A hybrid optimization algorithm using BiLSTM structure for sentiment analysis. *Measurement: Sensors, 25*, p. 100619.

14 Cognitive Intelligent Personal Learning Assistants for Enriching Personalized Learning

D. Ramalingam[1] and Mahalakshmi Dharmalingam[2]
[1] Majan University College, Muscat, Oman
[2] Department of Information Technology, AVC College of Engineering, Mayiladuthurai, Tamil Nadu, India

14.1 OVERVIEW OF MODERN LEARNING PEDAGOGY

The field of education has been evolving rapidly in recent years, with new technologies, pedagogical approaches, and research shaping the way we think about learning. One of the key trends in modern learning pedagogy is the shift towards a more student-centred, personalized approach to learning. This approach focuses on meeting the unique needs and learning styles of each individual student, rather than trying to fit all students into a one-size-fits-all mould. Learning theories are frameworks that help to explain how and why people learn. These theories are based on the latest research in cognitive psychology, neuroscience, and education, and provide a comprehensive understanding of the different factors that influence learning. The popularly accepted theories are listed here.

Constructivism is a theory of learning that emphasizes the active role of the learner in constructing their own understanding of the world. It suggests that people learn best when they are actively engaged in the process of constructing meaning from new information, rather than passively receiving information. Constructivism is based on the idea that learning is an active, dynamic process that involves building new knowledge from existing knowledge. For example, a math teacher uses a problem-based learning approach where students work in groups to solve real-world math problems. This approach aligns with constructivism as it allows students to actively construct their own understanding of math concepts by applying them to real-world situations. It is effective for students in middle and high school as it allows them to connect new concepts to their existing knowledge [1, 2].

Social constructivism is an extension of the constructivism theory and emphasizes the importance of social interactions and cultural context in learning. It argues that people learn best when they can interact with others and share their perspectives, as well as when they are exposed to diverse cultural perspectives [3]. For example, a history teacher creates a collaborative learning environment where students work in small groups to research and present a historical event from multiple perspectives. This approach aligns with social constructivism as it allows students to construct their understanding of history through interactions with their peers and exposure to diverse perspectives [4, 5].

Connectivism is a learning theory that emphasizes the importance of networks and connections in learning. It suggests that people learn best when they are able to connect new information to existing knowledge and experiences, as well as when they are able to access and connect to a diverse range of resources and networks. For example, a language teacher uses a blended learning approach where

DOI: 10.1201/9781003407959-14

students use online resources and networks to learn a new language. This approach aligns with connectivism as it allows students to connect new language concepts to their existing knowledge and experiences and access a diverse range of resources and networks. It is effective for students of all ages as it allows them to learn at their own pace and tailor the learning experience to their specific needs [6, 7].

Self-determination theory is a theory of motivation that emphasizes the importance of autonomy, competence, and relatedness in learning. It argues that people are more likely to engage in and persist with learning activities when they feel a sense of autonomy and competence, and when the learning activities are related to their interests and values. For example, an art teacher allows students to choose their own art projects that align with their interests and values. This approach aligns with self-determination theory as it allows students to feel a sense of autonomy and competence, which increases their motivation to learn. Additionally, the teacher provides regular feedback and encourages students to reflect on their learning process. This approach is effective for students of all ages, as it allows them to take ownership of their learning and fosters intrinsic motivation [8, 9].

Mind, brain, education science (MBE) is a new and interdisciplinary field that aims to understand how the brain learns, how different teaching methods affect the brain, and how to use this knowledge to improve education. MBE integrates insights from cognitive science, neuroscience, and education, and uses this knowledge to develop evidence-based teaching strategies that can improve student learning. A science teacher uses brain-based teaching strategies to increase student engagement and understanding. For example, the teacher uses a combination of visual, auditory, and kinaesthetic teaching methods to cater to different learning styles. Additionally, the teacher uses techniques such as spaced repetition and interleaving to optimize the brain's capacity to retain new information. This approach aligns with MBE as it uses knowledge of the brain's processes to inform teaching strategies and improve learning outcomes. It is effective for students of all ages as it allows them to learn in a way that is best suited to their individual needs [10, 11].

These are some of the modern learning theories that have gained popularity in the field of education. They provide a comprehensive understanding of the different factors that influence learning and help educators to develop effective teaching strategies. However, it's important to note that these theories are not mutually exclusive, and many educators may find it useful to use a combination of these theories to create a holistic learning environment.

14.2   LEARNING IN A POST-COVID ERA

The COVID-19 pandemic has had a significant impact on education, with many schools and universities shifting to online and remote learning. According to a report by the UNESCO Institute for Statistics, more than 1.5 billion students were affected by school closures worldwide, with many countries shifting to distance learning in order to curb the spread of the virus [12]. As a result of this shift, there is an increasing focus on developing effective online and remote learning strategies and technologies. In the post-COVID era, there is an increasing focus on developing educational systems that can quickly and easily adapt to changing circumstances, such as school closures or changes in student demographics. Additionally, there is also a growing emphasis on developing digital literacy skills, as well as critical thinking, problem-solving, and collaboration skills [13].

One of the latest trends in learning is the use of technology, such as artificial intelligence (AI) and machine learning (ML), to personalize the learning experience for students. This approach involves using data and analytics to understand the unique learning needs and preferences of each student, and then adapting the content and delivery of the material accordingly. Additionally, gamification and simulations are also gaining popularity in the education field [14, 15].

According to the Higher Education Academy (HEA) and the Quality Assurance Agency for Higher Education (QAA), learning happens when students are actively engaged in the process of constructing meaning from new information [16]. This means that students should be provided with opportunities to actively participate in their own learning, rather than simply receiving information passively. Furthermore, the HEA report suggests that learning is most effective when it is relevant to the student and when it is supported by guidance and feedback. The QAA report emphasizes the importance of creating an environment that supports student learning, including providing clear learning outcomes, effective assessment and feedback, and a positive learning culture. Therefore, learning happens when students are actively engaged in the process of constructing meaning from new information, in a relevant context and supported by guidance and feedback, and in an environment that fosters positive learning culture [17].

14.3　PERSONALIZED LEARNING

Personalized learning is a teaching method that tailors education to the individual needs, abilities, and interests of each student. The need for personalized learning support for students is becoming increasingly important as the education system faces new challenges, such as the increasing diversity of student populations and the rapid advancement of technology.

One of the main benefits of personalized learning is that it addresses the diverse needs of students. Every student is unique, with different learning styles, abilities, and interests. A one-size-fits-all approach to education is not effective for all students. Personalized learning allows teachers to differentiate instruction and provide each student with the support they need to succeed.

Personalized learning also helps to engage students in their own learning. When students can direct their own learning, they are more likely to be motivated and invested in their education. This is particularly important for students who are at risk of disengaging from school, such as those from disadvantaged backgrounds or with special needs.

In addition, personalized learning can help to close achievement gaps. Research has shown that students who receive personalized instruction are more likely to make significant gains in their academic performance. This is particularly important for students from disadvantaged backgrounds, who are more likely to struggle in school.

Furthermore, the rapid advancement of technology has made personalized learning more feasible than ever before. With the use of AI and other digital tools, teachers can now easily personalize instruction for each student and provide real-time feedback and support. Figure 14.1 shows the elements of personalized learning.

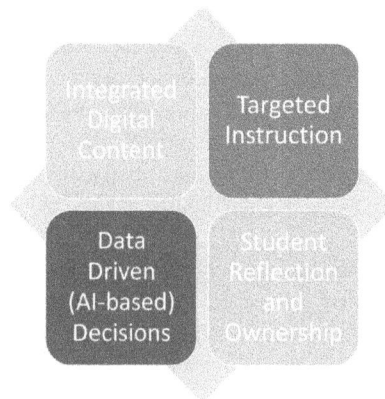

FIGURE 14.1　Elements of personalized learning.

There is a growing body of research that supports the need for personalized learning in the technological era.

14.4 INTRODUCTION TO CIPLAS AND THEIR CAPABILITIES

Cognitive intelligent personal learning assistants (CIPLAs) are a new class of technology that aim to enhance the way people learn and acquire new information. They use AI and ML techniques to understand the user's learning needs and preferences and provide personalized and interactive learning experiences.

One of the key capabilities of CIPLAs is their ability to adapt to the user's learning style. They can analyze the user's past performance and preferences and adjust the content and delivery of the material accordingly. For example, if a user prefers visual learning, the CIPLA may present information in the form of infographics or videos, while if a user prefers interactive learning, the CIPLA may provide interactive quizzes and games. This is supported by research, such as a study by Ramalingam that found that multimedia learning that includes text and images is more effective than text-only learning.

Another important capability of CIPLAs is their ability to provide real-time feedback and guidance. They can track the user's progress and provide instant feedback on their performance, as well as suggestions for improvement. This allows the user to quickly identify areas where they need to focus more attention and adjust their learning strategy as needed. Research supports that providing feedback is an important aspect of effective learning [18].

CIPLAs also could provide a wide range of learning materials, including text, audio, video, and interactive elements. They can be used to learn a variety of subjects, from language and mathematics to history and science. They can also be used to learn practical skills, such as coding, photography, and cooking. This is supported by research, such as that which argues multimedia learning is more effective than text-only learning [19].

In addition to providing learning materials and feedback, CIPLAs can also act as a personal tutor and coach. They can provide personalized guidance and advice, and help the user set and achieve learning goals. They can also track the user's progress over time and provide reports and analytics that can help the user understand their strengths and weaknesses. This is supported by research such as that which argues that providing guidance and support is an important aspect of effective learning [20].

Overall, CIPLAs have the potential to revolutionize the way people learn and acquire new information. By providing personalized and interactive learning experiences, real-time feedback, and a wide range of learning materials, CIPLAs can help users learn more effectively and efficiently. Additionally, by acting as a personal tutor and coach, CIPLAs can help users set and achieve their learning goals and improve their overall performance.

14.5 USE CASES OF CIPLAS

14.5.1 USE CASE OF CIPLAS IN EDUCATION

CIPLAs are a type of technology that can play a significant role in personalized learning. These assistants use AI and ML to adapt to the individual needs and learning styles of students. They can provide personalized feedback, support, and guidance to students as they learn.

CIPLAs can be used to provide personalized instruction, for example, by creating custom learning paths for students based on their abilities and progress. They can also provide personalized assessments and evaluate students' understanding of the material. Additionally, CIPLAs can be used to analyze students' performance data and provide insights to teachers to help them make data-driven decisions about instruction. Overall, CIPLAs can help to make learning more efficient, effective, and engaging for students by providing customized support and feedback, enabling

Sl No.	Name of the Bot
1	OpenAI's GPT-3
2	Gradescope
3	Cognii
4	Ivy Chatbot
5	Nuance's Dragon Speech Recognition
6	Dragon Speech Recognition
7	Century Tech

FIGURE 14.2   Popular AI-based conversational bots.

students to progress at their own pace, and providing teachers with valuable data to inform instruction. One example of a cognitive intelligent personal learning assistant is ALEKS (assessment and learning in knowledge spaces). It is a web-based education platform that uses AI to provide students with personalized learning paths. ALEKS uses adaptive assessments to determine students' knowledge levels and then creates a customized curriculum for each student based on their strengths and weaknesses [21]. Another example is Carnegie Learning's Cognitive Tutor. It is a software that uses AI to provide personalized instruction to students in math and science. The Cognitive Tutor tracks students' progress and adapts the instruction to the student's individual learning style and abilities [22].

There are several other examples of AI-based conversation bots in education. Here are a few examples:

OpenAI's GPT-3: GPT-3 is an AI-powered language model that can understand and respond to natural language input. It can be used to generate responses to student questions, provide explanations, and even create entire lesson plans. One of the primary uses of ChatGPT in education is as an intelligent personal learning assistant. It can provide personalized recommendations and resources to students based on their individual learning needs and preferences. For example, a student might ask ChatGPT for resources on a particular topic or for help understanding a concept, and ChatGPT can provide relevant articles, videos, or other resources to help the student learn. In addition to providing resources, ChatGPT can also be used to answer student questions. This can be particularly useful in large classes where it may be difficult for the instructor to answer every student question [23].

Students can ask ChatGPT for clarification on a topic or for help with a problem, and ChatGPT can provide a response that is tailored to the student's individual needs. Another way ChatGPT can be used in education is as a writing assistant. Students can input their writing into ChatGPT and receive feedback on their grammar, spelling, and syntax. ChatGPT can also provide suggestions for improving the clarity and organization of the writing. ChatGPT can also be used in language learning. Students can practice their language skills by conversing with ChatGPT in the target language. ChatGPT can provide feedback on grammar and pronunciation and can even provide suggestions for new vocabulary words to learn.

Gradescope: A digital platform that allows educators to grade, provide feedback, and manage assignments and exams in a more efficient way. It was created by a group of computer science professors at the University of California, Berkeley in 2014 and has since been adopted by many educational institutions. One of the primary features of Gradescope is its ability to handle a variety of assignment formats, including written assignments, coding assignments, and even multiple-choice exams. Students can submit their assignments through the platform, and instructors can then grade them online. This eliminates the need for physical paper submissions and allows instructors to grade assignments more quickly. Gradescope also includes tools to help instructors grade more

Top Ways to Use Chat GPT as a Student

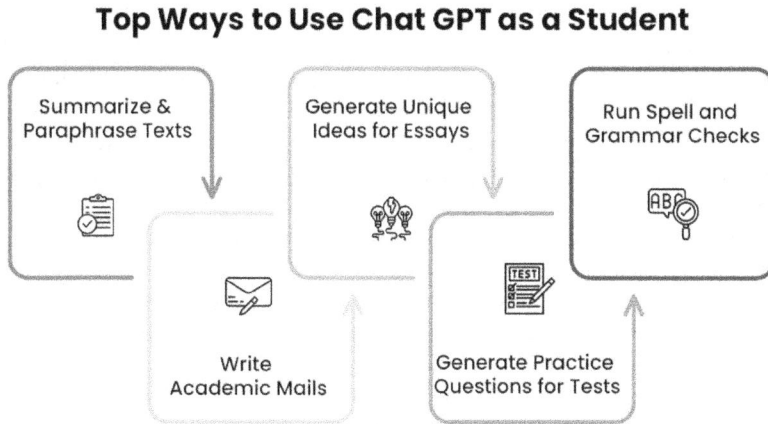

FIGURE 14.3 Uses of ChatGPT for learning.

accurately. For example, it provides a rubric creation tool that allows instructors to create detailed grading criteria for each assignment. This makes it easier to grade consistently and fairly across multiple students. In addition, Gradescope includes a variety of grading tools, including highlighters, commenting tools, and even AI-assisted grading. The AI-assisted grading feature uses machine learning algorithms to grade assignments automatically, reducing the amount of time instructors need to spend grading [24]. Gradescope also includes several features to help students understand their grades and receive feedback. For example, it allows instructors to provide feedback on specific parts of the assignment, and students can view their grades and feedback online. This makes it easier for students to understand where they need to improve and how they can do better on future assignments.

Cognii: A virtual tutor that uses natural language processing and machine learning to provide personalized feedback and coaching to students. Its flagship product is the Cognii Virtual Learning Assistant, which uses natural language processing and machine learning algorithms to provide personalized and interactive learning experiences for students [25]. One of the key features of Cognii is its ability to provide personalized feedback to students based on their individual learning needs and preferences. It can analyze student responses to open-ended questions and provide feedback that is specific to the student's writing style, grammar, and other writing factors. Additionally, Cognii can provide feedback on the overall quality of the student's response, including the accuracy of their ideas and the depth of their analysis. Cognii can also be used to assess student understanding of complex concepts. It can analyze student responses to multiple-choice questions and provide feedback on their thought process and reasoning. In this way, it can help students identify areas where they may be struggling and provide targeted feedback to help them improve their understanding of the material.

Another key feature of Cognii is its ability to adapt to the needs of different learners. It can recognize patterns in student responses and adjust its feedback and recommendations accordingly. For example, if a student consistently struggles with a particular concept, Cognii can provide additional resources and support to help the student master the material. Cognii also provides instructors with a variety of tools to help them monitor student progress and provide targeted support. Instructors can view detailed analytics on student performance and identify areas where students may be struggling. Additionally, Cognii can provide instructors with recommendations for additional resources and support to help their students succeed.

Ivy Chatbot: A virtual student advisor that uses natural language processing and machine learning to assist students with their academic and career goals. Its natural language processing capabilities and ability to engage in conversation make it as an ideal companion for students who are studying a subject and need assistance [26]. One of Ivy's most valuable features for learners is its ability to provide personalized learning experiences. Ivy can analyze a user's input and tailor its responses to their specific needs, making sure that they receive the information they require in a way that is accessible and easy to understand. This can be particularly useful for students who are struggling with a particular topic, as Ivy can identify the areas where they need additional help and provide targeted support.

In addition to providing personalized assistance, Ivy can also act as a virtual tutor, helping students to practice and reinforce their knowledge of a subject. It can provide interactive quizzes, flashcards, and other learning activities, enabling students to test their understanding and practice their skills. This can be especially useful for students who are learning a new language or studying a technical subject, as Ivy can provide immediate feedback on their progress and help them to improve their skills.

Nuance's Dragon Speech Recognition: A speech recognition software that allows users to control their computer and dictate text using their voice. Dragon uses advanced machine learning algorithms to analyze speech and accurately transcribe it into text, enabling users to dictate emails, documents, and other text-based content [27]. One of the key features of Dragon Speech Recognition is its accuracy. Nuance claims that Dragon is up to three times faster and more accurate than typing, with an accuracy rate of up to 99% [28]. This high level of accuracy is achieved through Dragon's advanced speech recognition algorithms, which are constantly learning and adapting to the user's voice and speaking patterns. As a result, Dragon can transcribe speech quickly and accurately, even in noisy environments or with users who have accents or speech impediments.

Another important feature of Dragon Speech Recognition is its customizability. Dragon can be trained to recognize individual user's voices and speaking patterns, improving its accuracy and responsiveness over time. Additionally, users can create custom commands and macros, enabling them to control their computer using voice commands for tasks such as opening applications, navigating the web, and controlling multimedia.

Dragon Speech Recognition also offers a range of productivity-enhancing features, such as the ability to transcribe audio recordings, integrate with popular productivity software such as Microsoft Office and Google Docs, and support for a wide range of languages and accents. These features make Dragon an ideal tool for professionals who need to transcribe large volumes of text, or for individuals with disabilities that make typing difficult or impossible.

Century Tech: A platform that uses data analytics and machine learning to personalize learning and improve student outcomes. Century Tech analyzes student performance data to identify areas where students need additional support, and provides tailored resources and activities to help them improve their understanding. In this essay, I will explore the features and capabilities of Century Tech, and explain how it can benefit learners in a range of contexts [29].

One of the key features of Century Tech is its ability to provide personalized learning experiences. Century Tech uses sophisticated machine learning algorithms to analyze student performance data, including their strengths and weaknesses, and provides tailored resources and activities to help them improve their understanding. This can be particularly useful for students who are struggling with a particular topic, as Century Tech can identify the areas where they need additional help and provide targeted support.

Another important feature of Century Tech is its ability to track student progress over time. Century Tech collects data on student performance and uses this data to create detailed progress reports, enabling teachers and students to monitor their progress and identify areas where they need additional support. This can help students to set learning goals, track their progress, and make informed decisions about their learning.

In addition to providing personalized learning experiences and progress tracking, Century Tech offers a range of interactive resources and activities to help students learn. These include interactive quizzes, multimedia resources, and other engaging content that can help students to improve their understanding and reinforce their knowledge.

Institutions can also create custom databases that include information such as exam dates, grading schemes, module details, and course information. This information can then be integrated into the bot's conversational flow, allowing students to easily access this information through the bot. AI-based bots in education are designed to be integrated with other systems, such as learning management systems (LMS) or student information systems (SIS).

To customize the database, institution can work with the vendors of the bot or hire a developer to work on this task. It is also important to keep the data private and secure, as the bot will handle sensitive information. Overall, integrating institution-specific information into an AI-based conversation bot in education can provide students with a more personalized and efficient way to access the information they need, and can also help institutions to manage and track student progress.

14.5.2 USE CASES OF CIPLAS IN PROFESSIONAL DEVELOPMENT

CIPLAs can be used to support professional development in a variety of ways. Here is an example of how a CIPLA might be used in a professional development context.

14.5.2.1 Career Advancement

A CIPLA can provide personalized career development plans for individual employees. By analyzing an employee's skills, interests, and career goals, the CIPLA can provide tailored training and development opportunities that align with the employee's career aspirations.

14.5.2.2 Compliance Training

CIPLAs can be used to provide interactive and engaging compliance training, such as workplace safety or anti-harassment training. They can also track employee progress and provide reminders when certifications or recertifications are needed.

14.5.2.3 Professional Certification

CIPLAs can be used to prepare employees for professional certifications, such as IT or project management certifications. They can provide interactive practice exams, study materials, and feedback on areas where the employee needs to improve.

14.5.2.4 Continuous Learning

CIPLAs can be used to provide continuous learning opportunities for employees, such as access to online courses, webinars, and industry events. An example of a company using CIPLA for professional development is a retail company that uses a CIPLA to train their sales associates. The CIPLA can be integrated with the company's CRM system to provide personalized training based on the sales associate's performance data. The CIPLA can provide interactive tutorials on topics such as product knowledge, customer service, and sales techniques. The CIPLA can also provide quizzes and assessments to test the sales associates' knowledge and provide feedback on areas where the sales associate needs to improve.

14.6 USE CASE DIAGRAM AND FRAMEWORK OF IMPLEMENTATION

The CIPLA's use case diagram (Figure 14.4) and its implementation framework (Figure 14.5) is a user-friendly, conversational AI agent that can be accessed through a mobile application or

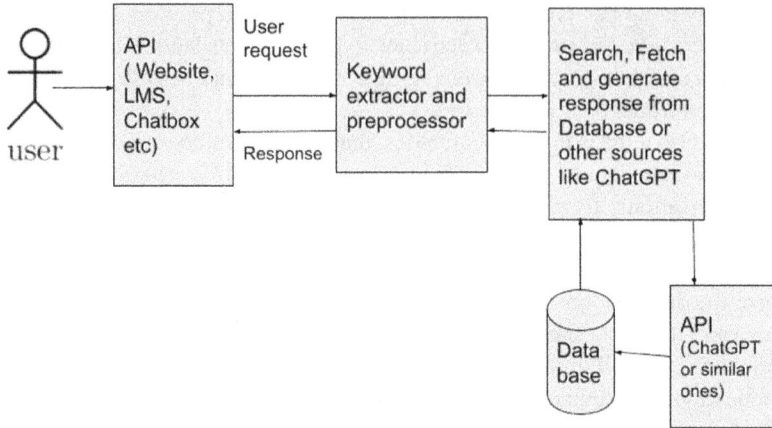

FIGURE 14.4   Use case diagram of the CIPLA.

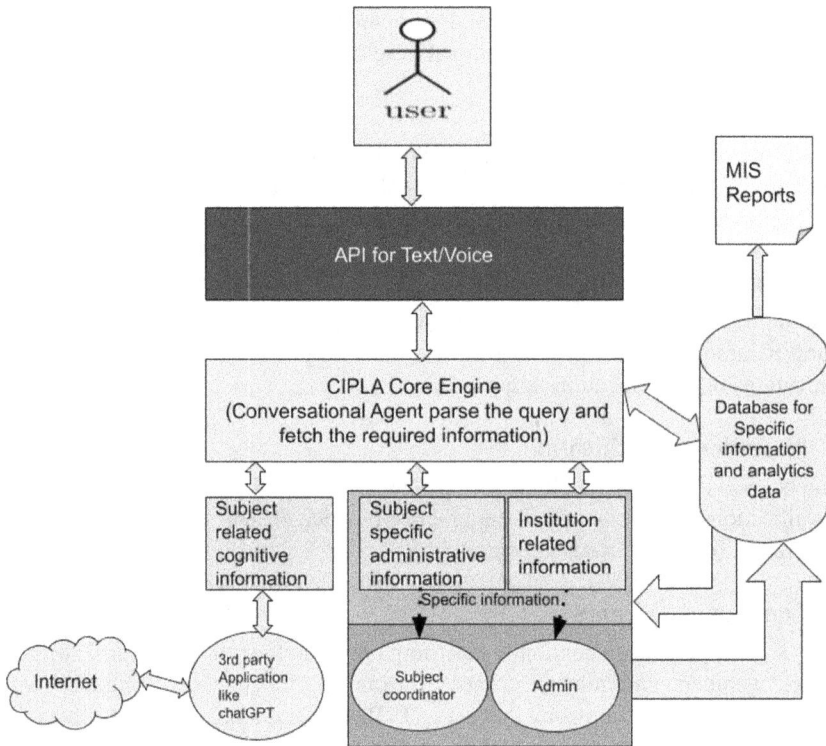

FIGURE 14.5   Framework for CIPLA.

website. The user can enter their query as text or voice message, which is then parsed by the CIPLA's core engine. The engine classifies the information into three categories – subject-related cognitive queries, institution-specific administrative information, and subject-related administrative information.

Subject-related cognitive queries are handed over to third-party applications like ChatGPT, which provides the required information. On the other hand, institution-specific and subject-related administrative information like class time, teacher availability, schedule of exams, academic calendar, availability of rooms, laboratories, books, location of rooms, buildings, opening hours of the library, and other facilities, email addresses of responsible persons, links to various forms, explanations of procedures, and regulations at the institution are updated by the respective subject coordinators and administrators into the database every semester or year, whichever is appropriate. These institution-specific information are then fetched from the database and given as a response to the user.

Every user query text or voice message parsed to text will be captured in the database for future analytics and system improvement. The CIPLA engine also recognizes and stores meta-information about the user and their past queries to provide a better experience.

The CIPLA framework has numerous benefits. It provides users with a personalized and interactive learning experience that is tailored to their needs. The AI agent's conversational interface makes learning more engaging and interactive, allowing users to ask questions and receive answers in real-time. Additionally, the system's ability to capture and analyze user data can provide valuable insights that can be used to improve the platform and enhance the learning experience further.

14.7 FUTURE DEVELOPMENTS AND ETHICAL IMPLICATIONS OF CIPLAS

The use of CIPLAs for professional development is expected to continue to grow in the future, as more companies look for ways to provide personalized and effective training to their employees. Some potential future developments for CIPLAs include the following:

- Use of AI: The use of AI and machine learning technologies in CIPLAs is expected to increase, allowing them to provide even more personalized and effective training.
- Virtual and augmented reality: CIPLAs may start using virtual and augmented reality technologies to provide immersive and interactive training experiences.
- Integration with other systems: CIPLAs may be integrated with other systems such as HR management systems, performance management systems, and recruitment systems to provide a more holistic approach to professional development.

However, with the increasing use of CIPLAs for professional development, there are also ethical implications that need to be considered. These include the following:

- Privacy: Ensuring that personal data collected by CIPLAs is handled and stored securely and in compliance with data protection regulations.
- Transparency: Ensuring that employees are aware of how their data is being collected, used, and shared by CIPLAs.
- Bias: Ensuring that CIPLAs are not biased in the training they provide or the career development opportunities they recommend, this is important to avoid discrimination.
- Job displacement: Ensuring that the use of CIPLAs does not lead to job displacement or loss of human skills.
- Lack of human touch: It is important to carefully consider the moral consequences of using conversational bots in place of human educators. Bots can respond quickly and consistently, but they lack the human touch, empathy, and nuanced knowledge that teachers bring to the classroom. Conversational bots should be used as a complement rather than a replacement for human teachers, who should still play an important part in education.
- Accuracy of the information: A significant ethical concern is the issue of conversational bots presenting inaccurate or subpar information. Users rely on these bots to offer accurate and trustworthy information when they communicate with them. If the bots frequently provide false or misleading information, it may have detrimental effects on the users.

14.8 CONCLUSION

In conclusion, CIPLAs have the potential to revolutionize professional development by providing personalized, engaging, and effective training experiences. CIPLAs can adapt to the learning style, strengths, and weaknesses of individual employees, providing tailored training and development opportunities that align with their career aspirations. With the integration of artificial intelligence and machine learning, virtual and augmented reality technologies, as well as the integration with other systems, CIPLAs are expected to continue to evolve and improve in the future. However, it is important to consider the ethical implications of CIPLAs, such as privacy, transparency, bias, and job displacement, and to ensure that they are used in a responsible and ethical manner. Overall, CIPLAs can play a vital role in modern learning pedagogy and can help organizations adapt to the changing learning landscape of the post-COVID era.

REFERENCES

[1] K. Reid-Martinez and L. D. Grooms, 'Constructivism in 21st Century Online Learning', in *Handbook of Research on Modern Educational Technologies, Applications, and Management*, IGI Global, 2021, pp. 730–743, doi: 10.4018/978-1-7998-3476-2.ch045

[2] A. Earl, V. A. Carbee, K. Becerra-Murillo, and A. M. Evans, 'The Four C's, Constructivism, and Digitizing Curriculum', in *Handbook of Research on Barriers for Teaching 21st-Century Competencies and the Impact of Digitalization*, IGI Global, 2021, pp. 65–86, doi: 10.4018/978-1-7998-6967-2.ch004

[3] V. Yeravdekar and R. Raman, 'A social constructivism approach to learning digital technologies for effective online teaching in Covid-19', *CARDIOMETRY*, no. 23, pp. 761–764, August 2022, doi: 10.18137/cardiometry.2022.23.761764

[4] Z. Jie, M. Puteh, and A. H. Sazalli, 'A social constructivism framing of mobile pedagogy in english language teaching in the digital era', *Indonesian Journal of Electrical Engineering and Computer Science*, vol. 20, no. 2, p. 830, November 2020, doi: 10.11591/ijeecs.v20.i2.pp830-836

[5] N. Morchid, 'The social constructivist response to educational technology', *International Journal of English Literature and Social Sciences*, vol. 5, no. 1, pp. 263–270, 2020, doi: 10.22161/ijels.51.46

[6] F. Corbett and E. Spinello, 'Connectivism and leadership: Harnessing a learning theory for the digital age to redefine leadership in the twenty-first century', Heliyon, vol. 6, no. 1, p. e03250, January 2020, doi: 10.1016/j.heliyon.2020.e03250

[7] N. M. Al-Mutairi, 'Connectivism learning theory to enhance higher education in the context of COVID-19 pandemic', *International Journal of Educational Sciences*, vol. 35, no. 1–3, September 2021, doi: 10.31901/24566322.2021/35.1-3.1197

[8] A. H. Al-Hoorie, W. L. Q. Oga-Baldwin, P. Hiver, and J. P. Vitta, 'Self-determination mini-theories in second language learning: A systematic review of three decades of research', *Language Teaching Research*, June 2022, doi: 10.1177/13621688221102686

[9] E. Deci, R. Vallerand, L. Pelletier, and R. Ryan, 'Motivation and education: The self-determination perspective', *Educational Psychologist,* 26(3–4), 325–346. https://doi.org/10.1207/s15326985ep2 603&4_6

[10] A. E. Egger, 'New resources for education researchers', *Journal of Geoscience Education*, vol. 67, no. 1, pp. 1–2, January 2019, doi: 10.1080/10899995.2018.1562270

[11] M. H. Hobbiss *et al.*, '"UNIFIED": Bridging the researcher–practitioner divide in mind, brain, and education', *Mind, Brain, and Education*, vol. 13, no. 4, pp. 298–312, January 2019, doi: 10.1111/mbe.12223

[12] UNESCO, '1.3 billion learners are still affected by school or university closures, as educational institutions start reopening around the world, says UNESCO', 2020. https://en.unesco.org/news/13-bill ion-learners-are-still-affected-school-university-closures-educational-institutions (accessed January 20, 2023).

[13] C. Sánchez-Cruzado, R. Santiago Campión, and M. T. Sánchez-Compaña, 'Teacher digital literacy: The indisputable challenge after COVID-19', *Sustainability*, vol. 13, no. 4, p. 1858, February 2021, doi: 10.3390/su13041858

[14] K.-Y. Tang, C.-Y. Chang, and G.-J. Hwang, 'Trends in artificial intelligence-supported e-learning: A systematic review and co-citation network analysis (1998–2019)', *Interactive Learning Environments*, vol. 31, no. 4, pp. 1–19, January 2021, doi: 10.1080/10494820.2021.1875001

[15] H. Munir, B. Vogel, and A. Jacobsson, 'Artificial intelligence and machine learning approaches in digital education: A systematic revision', *Information*, vol. 13, no. 4, p. 203, April 2022, doi: 10.3390/info13040203

[16] Advance HE, 'Student engagement through partnership | Advance HE', 2014. Accessed: January 20, 2023. [Online]. Available: www.advance-he.ac.uk/guidance/teaching-and-learning#priorities

[17] QAA UK, 'Learning and teaching', 2018. www.qaa.ac.uk/the-quality-code/advice-and-guidance/learning-and-teaching (accessed January 20, 2023).

[18] C. W. Okonkwo and A. Ade-Ibijola, 'Chatbots applications in education: A systematic review', *Computers and Education: Artificial Intelligence*, vol. 2, p. 100033, January 2021, doi: 10.1016/J.CAEAI.2021.100033

[19] K. Mageira, D. Pittou, A. Papasalouros, K. Kotis, P. Zangogianni, and A. Daradoumis, 'Educational AI chatbots for content and language integrated learning', *Applied Sciences*, vol. 12, no. 7, p. 3239, 2022.

[20] G.-J. Hwang, Y.-F. Tu, and K.-Y. Tang, 'AI in online-learning research: Visualizing and interpreting the journal publications from 1997 to 2019', *International Review of Research in Open and Distributed Learning*, vol. 23, no. 1, pp. 104–130, 2022.

[21] H. Harati, L. Sujo-Montes, C.-H. Tu, S. J. W. Armfield, and C.-J. Yen, 'Assessment and learning in knowledge spaces (ALEKS) adaptive system impact on students' perception and self-regulated learning skills', *Educational Science (Basel)*, vol. 11, no. 10, p. 603, 2021.

[22] K. R. Koeclinger, A. T. Corbett, and S. Ritter, 'Carnegie learning's cognitive tutor: Summary research results', *Cité en,* p. 126, 2000. https://ies.ed.gov/ncee/wwc/Docs/InterventionReports/wwc_cognitivetutor_062116.pdf

[23] M. A. AlAfnan, S. Dishari, M. Jovic, and K. Lomidze, 'ChatGPT as an educational tool: Opportunities, challenges, and recommendations for communication, business writing, and composition courses', *Journal of Artificial Intelligence and Technology*, vol. 3, no. 2, 2023, pp. 60–68. https://doi.org/10.37965/jait.2023.0184

[24] D. Hovemeyer, 'A framework for declarative autograders', in *Proceedings of the 54th ACM Technical Symposium on Computer Science Education*, vol. 2, 2022, p. 1282.

[25] K. G. Srinivasa, M. Kurni, and K. Saritha, 'Harnessing the power of AI to education', in *Learning, Teaching, and Assessment Methods for Contemporary Learners: Pedagogy for the Digital Generation*, Springer, 2022, pp. 311–342.

[26] M. Ehrenpreis and J. DeLooper, 'Implementing a chatbot on a library website', *Journal of Web Librarianship*, vol. 16, no. 2, pp. 120–142, 2022.

[27] L. Chan, L. Hogaboam, and R. Cao, 'Artificial intelligence in education', in *Applied Artificial Intelligence in Business: Concepts and Cases*, Springer, 2022, pp. 265–278.

[28] Nuance.com, 'Helping students reach their full potential Dragon ® speech recognition', 2009. Accessed: March 11, 2023. [Online]. Available: www.nuance.com/asset/en_us/collateral/dragon/whitepaper/wp-helping-students-reach-their-full-potential-en-us.pdf

[29] Century, 'CENTURY for Business & HE', 2023. https://integrations.century.tech/ (accessed January 27, 2023).

15 Natural Language Processing for Fake News Detection Using Hybrid Deep Learning Techniques

B. Valarmathi[1], Aditya Kocherlakota[2], Yuvraj Das[2], Aritam[2],
N. Srinivasa Gupta[3], and V. Mohanraj[4]

[1*] Department of Software and Systems Engineering, School of Computer Science Engineering and Information Systems, Vellore Institute of Technology, Vellore, Tamil Nadu, India
[2] Former B.Tech. (Information Technology) Student, Department of Information Technology, School of Computer Science Engineering and Information Systems, Vellore Institute of Technology, Vellore, Tamil Nadu, India
[3] Department of Manufacturing Engineering, School of Mechanical Engineering,
Vellore Institute of Technology, Vellore, Tamil Nadu, India
[4] Department of Information Technology, Sona College of Technology, Salem, Tamil Nadu, India

15.1 INTRODUCTION

In recent years, different web-based entertainment stages like Twitter, Facebook, Instagram, and so on have become exceptionally well known since they work with the simple procurement of data and give a fast stage to data sharing. The accessibility of unauthentic information via web-based entertainment stages has acquired gigantic consideration among specialists and become a problem area for sharing phony news. Counterfeit news has been a significant issue because of its enormous adverse consequence. It has expanded consideration among scientists, columnists, lawmakers, and the overall population. With regards to composing style, counterfeit news is composed or distributed with the plan to deceive individuals and to harm the picture of an organization, element, or individual, either for monetary or political advantages.

In this work, the information was gathered from the Information Security and Object Technology (ISOT) dataset, which contains 38529 exceptional passages. In the first place, we are eliminating the prevent words and accentuation from the articles in the dataset and afterwards tokenizing every one of the words by making them related with an exceptional number. From that point onward, brain organizations, for example, RNN, LSTM and GRU are utilized for characterization. A similar investigation has been finished to track down the classifier that gives the most noteworthy discovery precision.

15.2 LITERATURE SURVEY

They summed up the different artificial intelligence (AI) calculations that can recognize counterfeit news from news stories and other text-based materials [1]. It likewise specifies the solid sources

DOI: 10.1201/9781003407959-15

that can be utilized to gather the information and the few calculations, for example, random forest, logical regression, and decision tree classifier that can be utilized to sort the news things as one or the other phony or authentic. Credulous Bayes classifier [2] was likewise supposed to be great at grouping phony and veritable news things. Akshay Jain and Amey Kasbe gathered information from Facebook Posts and assessed their genuinity by arranging the information as autonomous and equivalent. They likewise utilized mixture models [3] to give better exactness of results. Nasir and his partners utilized a mix of brain networks to be specific CNN and RNN on the FA-KES and IOST dataset. They figured out how to accomplish a precision of ~100% exactness on the ISOT dataset, which contains 45000 articles and ~60% exactness on the FA-KES dataset, which contains 804 articles.

Here, a quick-fix method that requires minimal modification and nearly no human involvement was suggested by Carmel Mary Belinda and her colleagues [4]. Their model demonstrated its ability to effectively identify individuals experiencing depression from a large dataset, making it an invaluable tool in recognizing such cases. Additionally, a notable feature of this model was its capability to classify tweets written in English, Hindi, and Hinglish. By accommodating both English and Hindi languages, the architecture can be applied globally, with relevance to India, and across various platforms. This automation of the process contributes to addressing the increasing prevalence of depression.

Identification of fake news can likewise be stretched out to various mixed media designs by successfully using the data of a few modalities, for example, printed, visual, and social modalities [5]. They fostered a visual geometry group (VGG) model utilizing the CNN calculation to perceive pictures. They had the option to accomplish an exactness of more than 70% after the consummation of the main age. This gives a further degree for the identification of abnormalities after media reusing.

Ahmad et al. [6] utilized troupe methods utilizing LIWC (linguistic enquiry and word count) with different etymological capabilities to arrange news stories from various areas as evident or counterfeit. They had the option to accomplish the most noteworthy review, accuracy, and F1 score of 95.25%.

The relationship between partisan bias of publishers and the validity of news content [7] was used by Shu and his colleagues. Partisan bias refers to the perceived bias of the publisher in the manner in which news is reported and covered. They used the trirelationship embedding framework on the BuzzFeed Dataset to achieve an F1 score of >80%. They gave a principled method for demonstrating trirelationship among distributer, news, and significant client commitment at the same time. The news inert component implanting from news content was first acquainted and afterward showed how with model client social commitment and distributer hardliner independently. With RIST they accomplished a precision of around 61% and with Castillo 65%.

Mahesh et al. [8] suggested the AdaBoost Ensemble model, which depends on recognized feature patterns, for the prediction of heart disease. It could be contrasted with traditional data mining techniques in the diagnosis of cardiac illness. During the feature extraction stage, ensemble classification algorithms took the role of more conventional information extraction techniques. In this work, homogeneous classifiers and ensemble classifiers, created by mixing different methods were created by mixing different methods.

Oshikawa et al. [9] proposed the different ideas, for example, tokenization, lemmatization and different machine learning and deep learning methods had been utilized.

Mishra et al. [10] conducted an experiment using the naive Bayes classifier. This model provided a total accuracy of between 98–99%. Tacchini et al. [11] performed a cross-validation analysis between logistic regression and harmonic regression and had reported an accuracy of 99% and 99.4%, respectively. A three-phase method based on graphs, known as GTUT (graph mining methods over textual, user, and temporal data), was created by Reddy Gangireddy et al. [12] and addressed the task using a graph-based methodology. For the detection of unsupervised bogus news, the GTUT saw accuracy improvements of up to almost 80%, or more than 10 percentage points.

In order to improve learning, Kumar Kaliyar et al. proposed a model [13] that was combined with bidirectional encoder representations from transformers (BERT) and three parallel one-dimensional convolutional neural network (1D-CNN) blocks with various convolutional kernel-sized layers and varying filters. Kaur et al. [14] conducted a comparison of precision measurements using machine learning classifiers. Passive aggressive (93.3%), stochastic gradient descent (93.5%), and linear support vector machine [LinearSVC] (93%) outperform the rest on the three news articles using the term frequency-inverse document frequency (TF-IDF) and CV feature extraction technique.

By employing TF-IDF and linear discriminant analysis (LDA) on the dataset, Xu et al. [15] created a model that discriminated fake news using domain reputation and content understanding. The framework proposed by Elsaeed et al. [16] extracted the features from the dataset using TF-IDF and document to vector (Doc2Vec), and then the best features are extracted using the chi-square and analysis of variance (ANOVA) technique. The proposed architecture by Saleh et al. [17] was to identify fake news and they suggested an improved convolutional neural network model (OPCNN-FAKE). Using four benchmark datasets for fake news, they compared the performance of the OPCNN-FAKE with the two deep learning methods and the six normal machine learning (ML) methods.

Zhou and Zafarani [18] utilized datasets that were gathered from PolitiFact and BuzzFeed and their methodology earned them a precision of 92.9% and 83.5% individually.

Aphiwongsophon and Chongstitvatana [19] analyzed naïve Bayes (NB) versus brain organization and support vector machine (SVM). The consequence of review, F-measure, and precision with naïve Bayes were 96.10%, 97.90%, and 96.08% individually. Brain organization and support vector machine were identical outcomes with review, F-measure, and precisions were 99.90%, 99.80%, and 99.90% individually. Parikh and Atrey [20] directed a study that discusses pretty much every one of the various techniques accessible and furthermore all the different datasets.

Utilizing a number of innovative and cutting-edge classifiers, such as naive Bayes (NB), random forest (RF), K-nearest neighbors (KNN), extreme gradient boosting (XGBoost), and support vector machine with a radial basis function (RBF) segment, the discriminative ability of the previous features was studied by Reis et al. [21]. The XGB and RF classifiers, separately achieving 0.85 and 0.86 for area under the receiver operating characteristic (ROC) curve (AUC), achieved the best outcomes.

Monti et al. [22] pondered two interesting settings of fake news disclosure: uniform resource locator (URL) wise and overflowing astute, including comparative designing for the two settings. For URL-wise game plan, they used five randomized getting ready/test/endorsement parts. Generally, the arrangement, test, and endorsement sets contained 677, 226, and 226 URLs, independently, with 83.26% legitimate and 16.74% misdirecting marks (± 0.06% and 0.15% for planning and endorsement/test set separately). For flood-wise gathering they used a comparative split at first recognized for URL-wise request all wellsprings began by URL u were placed in a comparative wrinkle as u). Spills over containing under 6 tweets were discarded; the legitimization for the choice of this edge is animated underneath. Full overflow length (24 hour) was used for the two settings of this assessment. The planning, test, and endorsement sets contained on ordinary 3586, 1195, 1195 wellsprings, exclusively, with 81.73% substantial and 18.27% sham imprints (± 3.25% and 6.50%) for getting ready and endorsement/test set separately.

Six layers make up OPCNN-FAKE [17]: an inserting, a dropout, a pooling, a convolutional, a level, and a result layer, which outflanked LSTM and RNN comprising various layers. Jiang et al. [23] presented an original thought of stacking models with random forest with the best not set in stone by the remedied McNemmar test (99.94%). Li et al. [24] proposed a model in light of an auto-encoder to dissect the secret data and the interior relations, and afterward acknowledges unaided phony news discovery giving AUC score of 0.8305 for UFNDA_UPI. The paper's [25] main concern was identifying bogus COVID-19 information. Information was obtained from the WHO (World Health Organization), UNICEF, and the United Nations as well as health surveillance originating from a number of fact-verification websites. Elhadad et al. also offered a method for spotting false information. Reputable sources should be used to guarantee the validity of the data. Using the obtained verified true data, they implemented a classification model that uses machine learning techniques to identify false information. Ten machine learning algorithms and seven feature extraction methodologies were used to create a voting ensemble machine learning classifier. They used fivefold cross-validation to check the accuracy of the data that was collected and to assess the 12 performance metrics.

A model to distinguish was put up by Shahbazi et al. [26] by combining natural language processing, reinforcement learning (RL), and block chain. The proposed framework beat each usually utilized AI calculation by accomplishing the most reduced RMSE and other blunder scores. Here Shu and his associates utilized LIWC (linguistic inquiry and word count) [27] to group the clients between the people who every now and again post counterfeit news. They had the option to accomplish a precision of 78.2 % exactness for rhetorical structure theory (RST) and 83 % precision for LIWC.

15.3 METHODOLOGY

The proposed work is outlined in the list of steps below and it is shown in Figure 15.1.

Step 1: Selecting a dataset from Kaggle.

Step 2: Removal of stop words and Punctuations

The NLTK (Natural Language Toolkit) software is used to first eliminate stop words and punctuation from each article in the dataset. The sample document is shown in Figure 15.2.

Before the removal of stop words and punctuations, the number of words present in the sample document is 761. After removing the stop words and punctuations from the sample document, the number of words present in the sample document is 470. It is shown in Figure 15.3.

Step 3: Data preprocessing

Step 3.1: Vectorization of the data

In many natural language processing (NLP) applications, including named entity recognition (NER), semantic analysis, text classification, and others, vectorization is carried out using Word2Vec, an algorithm to generate word embeddings through distributed semantic representations.

The most common forms of deriving relationships between words are given below.

(1) **Bag of Words (BoW)**
BoW model is the easiest method of text encoding. Like the term itself, we can depict a sentence as a bag of vectors (an array of digits) linked to individual words.

(2) **Term Frequency – Inverse Document Frequency (TF-IDF)**
This strategy recognizes words that seem ordinarily and are familiar words yet gives data about the particular archive. One of the most utilized term frequency models was presented in Gomathy et al. [1]. TF-IDF formulae are displayed in situation 1 and condition 2.

Term frequency (TF) quantifies the occurrence rate of a term, T, within a document, D. It is calculated by dividing the number of times (N) the term 'T' appears in the document 'D' by the total number of terms present in the document.

$$TF = (N_{T,D}) / \text{number of terms in the document} \qquad (15.1)$$

where

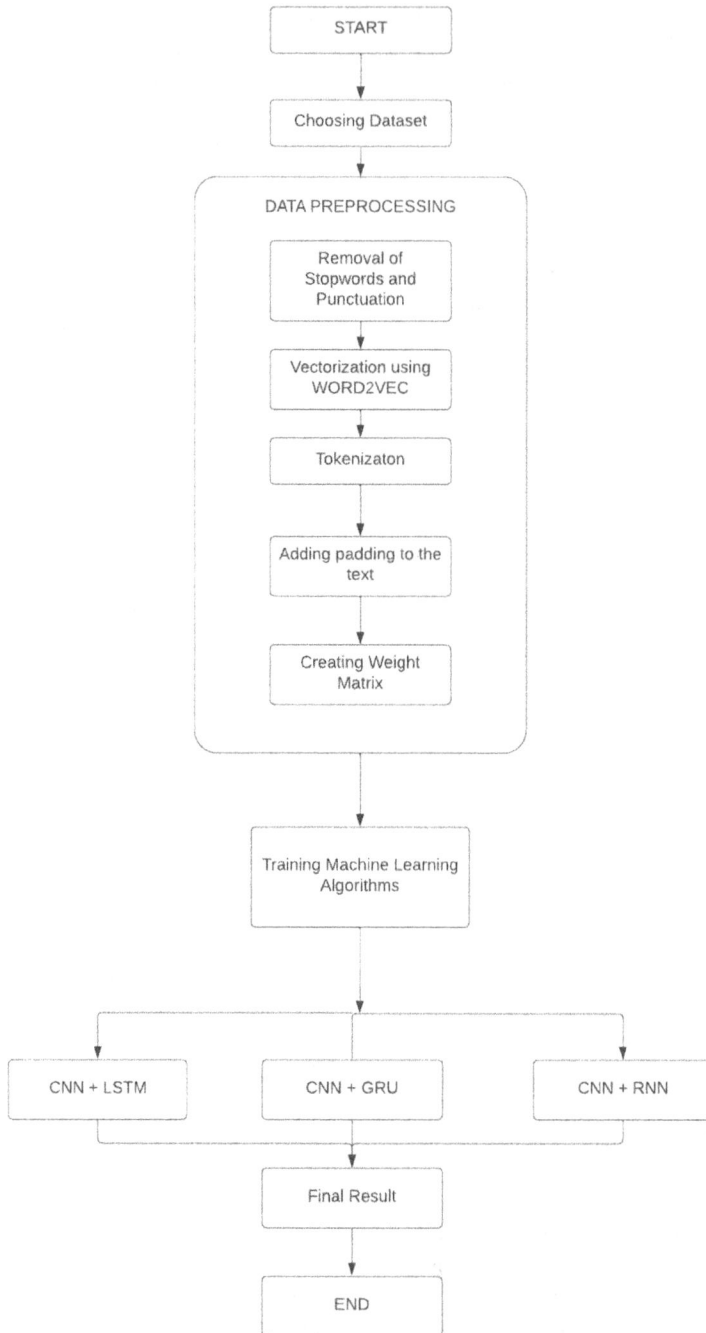

FIGURE 15.1   The proposed work's flow chart.

```
In [59]: print("no of words before removing stop words and punctuations:",len(data["text"][0].split(" ")))
         print("\n",data["text"][0])
```

no of words before removing stop words and punctuations: 761

As U.S. budget fight looms, Republicans flip their fiscal script The head of a conservative Republican faction in the U.S. Co
ngress, who voted this month for a huge expansion of the national debt to pay for tax cuts, called himself a "fiscal conservati
ve" on Sunday and urged budget restraint in 2018. In keeping with a sharp pivot under way among Republicans, U.S. Representativ
e Mark Meadows, speaking on CBS' "Face the Nation," drew a hard line on federal spending, which lawmakers are bracing to do bat
tle over in January. When they return from the holidays on Wednesday, lawmakers will begin trying to pass a federal budget in a
fight likely to be linked to other issues, such as immigration policy, even as the November congressional election campaigns ap
proach in which Republicans will seek to keep control of Congress. President Donald Trump and his Republicans want a big budget
increase in military spending, while Democrats also want proportional increases for non-defense "discretionary" spending on pro
grams that support education, scientific research, infrastructure, public health and environmental protection. "The (Trump) adm
inistration has already been willing to say: 'We're going to increase non-defense discretionary spending ... by about 7 percen
t,'" Meadows, chairman of the small but influential House Freedom Caucus, said on the program. "Now, Democrats are saying tha
t's not enough, we need to give the government a pay raise of 10 to 11 percent. For a fiscal conservative, I don't see where th
e rationale is. ... Eventually you run out of other people's money," he said. Meadows was among Republicans who voted in late D
ecember for their party's debt-financed tax overhaul, which is expected to balloon the federal budget deficit and add about $1.
5 trillion over 10 years to the $20 trillion national debt. "It's interesting to hear Mark talk about fiscal responsibility," D
emocratic U.S. Representative Joseph Crowley said on CBS. Crowley said the Republican tax bill would require the United States
to borrow $1.5 trillion, to be paid off by future generations, to finance tax cuts for corporations and the rich. "This is one
of the least ... fiscally responsible bills we've ever seen passed in the history of the House of Representatives. I think we'r
e going to be paying for this for many, many years to come," Crowley said. Republicans insist the tax package, the biggest U.S.
tax overhaul in more than 30 years, will boost the economy and job growth. House Speaker Paul Ryan, who also supported the tax
bill, recently went further than Meadows, making clear in a radio interview that welfare or "entitlement reform," as the party
often calls it, would be a top Republican priority in 2018. In Republican parlance, "entitlement" programs mean food stamps, ho
using assistance, Medicare and Medicaid health insurance for the elderly, poor and disabled, as well as other programs created
by Washington to assist the needy. Democrats seized on Ryan's early December remarks, saying they showed Republicans would try
to pay for their tax overhaul by seeking spending cuts for social programs. But the goals of House Republicans may have to take
a back seat to the Senate, where the votes of some Democrats will be needed to approve a budget and prevent a government shutdo
wn. Democrats will use their leverage in the Senate, which Republicans narrowly control, to defend both discretionary non-defen
se programs and social spending, while tackling the issue of the "Dreamers," people brought illegally to the country as childre
n. Trump in September put a March 2018 expiration date on the Deferred Action for Childhood Arrivals, or DACA, program, which p
rotects the young immigrants from deportation and provides them with work permits. The president has said in recent Twitter mes
sages he wants funding for his proposed Mexican border wall and other immigration law changes in exchange for agreeing to help
the Dreamers. Representative Debbie Dingell told CBS she did not favor linking that issue to other policy objectives, such as w
all funding. "We need to do DACA clean," she said. On Wednesday, Trump aides will meet with congressional leaders to discuss t
hose issues. That will be followed by a weekend of strategy sessions for Trump and Republican leaders on Jan. 6 and 7, the Whit
e House said. Trump was also scheduled to meet on Sunday with Florida Republican Governor Rick Scott, who wants more emergency
aid. The House has passed an $81 billion aid package after hurricanes in Florida, Texas and Puerto Rico, and wildfires in Calif
ornia. The package far exceeded the $44 billion requested by the Trump administration. The Senate has not yet voted on the aid.

FIGURE 15.2 The sample document – before removal of stop words and punctuations.

```
In [62]: print("No. of words after removing stop words and punctuations:", len(X[0]))
         for x in X[0]:
             print(x,end=" ")
```

No. of words after removing stop words and punctuations: 470
As budget fight looms Republicans flip fiscal script The head conservative Republican faction Congress voted month huge expansi
on national debt pay tax cuts called fiscal conservative Sunday urged budget restraint 2018 In keeping sharp pivot way among Re
publicans Representative Mark Meadows speaking CBS Face Nation drew hard line federal spending lawmakers bracing battle January
When return holidays Wednesday lawmakers begin trying pass federal budget fight likely linked issues immigration policy even No
vember congressional election campaigns approach Republicans seek keep control Congress President Donald Trump Republicans want
big budget increase military spending Democrats also want proportional increases non defense discretionary spending programs su
pport education scientific research infrastructure public health environmental protection The Trump administration already will
ing say We going increase non defense discretionary spending percent Meadows chairman small influential House Freedom Caucus sa
id program Now Democrats saying enough need give government pay raise 10 11 percent For fiscal conservative see rationale Event
ually run people money said Meadows among Republicans voted late December party debt financed tax overhaul expected balloon fed
eral budget deficit add trillion 10 years 20 trillion national debt It interesting hear Mark talk fiscal responsibility Democra
tic Representative Joseph Crowley said CBS Crowley said Republican tax bill would require United States borrow trillion paid fu
ture generations finance tax cuts corporations rich This one least fiscally responsible bills ever seen passed history House Re
presentatives think going paying many many years come Crowley said Republicans insist tax package biggest tax overhaul 30 years
boost economy job growth House Speaker Paul Ryan also supported tax bill recently went Meadows making clear radio interview wel
fare entitlement reform party often calls would top Republican priority 2018 In Republican parlance entitlement programs mean f
ood stamps housing assistance Medicare Medicaid health insurance elderly poor disabled well programs created Washington assist
needy Democrats seized Ryan early December remarks saying showed Republicans would try pay tax overhaul seeking spending cuts s
ocial programs But goals House Republicans may take back seat Senate votes Democrats needed approve budget prevent government s
hutdown Democrats use leverage Senate Republicans narrowly control defend discretionary non defense programs social spending ta
ckling issue Dreamers people brought illegally country children Trump September put March 2018 expiration date Deferred Action
Childhood Arrivals DACA program protects young immigrants deportation provides work permits The president said recent Twitter m
essages wants funding proposed Mexican border wall immigration law changes exchange agreeing help Dreamers Representative Debbi
e Dingell told CBS favor linking issue policy objectives wall funding We need DACA clean said On Wednesday Trump aides meet con
gressional leaders discuss issues That followed weekend strategy sessions Trump Republican leaders Jan White House said Trump a
lso scheduled meet Sunday Florida Republican Governor Rick Scott wants emergency aid The House passed 81 billion aid package hu
rricanes Florida Texas Puerto Rico wildfires California The package far exceeded 44 billion requested Trump administration The
Senate yet voted aid

FIGURE 15.3 The sample document – after removing stop words and punctuations.

$N_{T,D}$ = number of times (N) the term 'T' appears in the document 'D'

$$Z_{i,j} = df_{i,j} \times \log(N / tf_i) \tag{15.2}$$

where,

$df_{i,j}$= count of occurrences of the term 'i' in the document
tf_i= count of documents containing the term 'i'
N = total count of documents

Word2Vec endeavors to decide a part of the issues associated with the BoW approach and they are high perspective vectors and words totally insignificant to each other. Using a mind network with several layers, Word2Vec endeavors to get to know the associations among words and integrates them into a lower-layered vector space. To do this, Word2Vec takes words against various words that envelop them in the data corpus, getting a piece of the significance in the word gathering. Two novel techniques were made. They are continuous bunch of words (CBoW) and skip-gram.

Mutual information (MI) describes relationships with respect to uncertainty. MI between two quantities is a measure of how much knowledge of one quantity reduces uncertainty over another. The MI formula is shown in equation (15.3).

$$MI(A; B) = \sum_{a \in A} \sum_{b \in B} p(a,b) \times \log \frac{p(a,b)}{p(a)p(b)} \tag{15.3}$$

where A and B are two independent variables. They represent all the bi-grams in the corpus such that b comes after a.

Pointwise mutual information (PMI) is the measure of dependence between concrete occurrences of a and b. The PMI formula is shown in equation (15.4).

$$PMI(a : b) = \log\left(\frac{p(a,b)}{p(a)p(b)}\right) \tag{15.4}$$

PMI($a : b$) will have a high value, while it will have a value of 0 if a and b are completely independent.

For a comparison between all the bi-grams, we need to include the upper bound of all the values. Hence we need to certain bi-grams above a certain threshold. This is achieved through normalized pointwise mutual information (NPMI). It is shown in the equation (15.5).

$$NPMI(a : b) = \frac{\log\left(\frac{p(a,b)}{p(a)p(b)}\right)}{-\log p(a,b)} \tag{15.5}$$

The FBI refers to the Federal Bureau of Investigation. 'Federal' means U.S. national government. Departments or divisions of the government are also referred to as 'Bureaus' in this context. 'Investigative' means gathering data and proof to identify and stop offences. Figure 15.4 displays a list of words that were found to be most comparable to the Federal Bureau of Investigation (FBI).

The list of words are found to be most similar to the word 'Iran' is shown in Figure 15.5.

Step 3.2: Tokenizing the text into numbers

Tokenizing the text means that we represent every word as a number. The word index property of the tokenizer maintains the correspondence between the initial word and the number. Tokenizer performs fundamental processing, like changing lowercase words. For example, in Figure 15.6, the lists of words are mapped into a unique number.

```
In [27]:  w2v_model.wv.most_similar("FBI")

Out[27]:  [('bureau', 0.687788724899292),
           ('Comey', 0.6846204400062561),
           ('CIA', 0.6338177919387817),
           ('investigators', 0.6336326003074646),
           ('DOJ', 0.6027809977531433),
           ('Mueller', 0.6020801663398743),
           ('investigation', 0.6015434265136719),
           ('investigations', 0.5699388980865479),
           ('investigated', 0.5599686503410339),
           ('Investigation', 0.5591317415237427)]
```

FIGURE 15.4   The list of words are found to be most similar to the Federal Bureau of Investigation (FBI).

```
In [62]:  1  w2v_model.wv.most_similar("Iran")

Out[62]:  [('Tehran', 0.8931469917297363),
           ('Iranian', 0.7581676840782166),
           ('Hezbollah', 0.6662694215774536),
           ('Riyadh', 0.6413218975067139),
           ('nuclear', 0.6356945037841797),
           ('disarmament', 0.5988152027130127),
           ('Syria', 0.5944088101387024),
           ('IranDeal', 0.5905274748802185),
           ('Doha', 0.5897599458694458),
           ('destabilizing', 0.5897470712661743)]
```

FIGURE 15.5   The list of words are found to be most similar to the word 'Iran'.

```
In [79]:  1  #Lets check few word to numerical replesentation
          2  #Mapping is preserved in dictionary -> word_index property of instance
          3  word_index = tokenizer.word_index
          4  for word, num in word_index.items():
          5      print(f"{word} -> {num}")
          6      if num == 10:
          7          break

trump -> 1
said -> 2
the -> 3
president -> 4
would -> 5
people -> 6
one -> 7
state -> 8
new -> 9
it -> 10
```

FIGURE 15.6   The words are allotted with unique numbers (token).

Step 3.3: Padding all the news articles with less than 700 words to 700 with '0'.

In order to establish uniformity across all of the articles in the dataset, padding is applied to news articles that are less than 700 words long, and long words are truncated. After padding, all news articles have 700 words in numerical form. If any news article has less than 700 words they are padded with 0.

Step 3.4: Embedding the words to create a word matrix.

Neural network embeddings are helpful on the grounds that they can decrease the dimensionality of unmitigated factors and seriously address classifications in the changed space. It assists us with recognizing connections between words

```
[[ 0.          0.          0.          ... 0.          0.
   0.         ]
 [ 0.24346469  0.93695128  1.13295805 ... 0.1420199   0.21923901
   1.09807146]
 [-1.29763162  0.64046514  1.94017291 ... 0.97805142 -0.76303488
   0.34911418]
 ...
 [ 0.          0.          0.          ... 0.          0.
   0.         ]
 [ 0.          0.          0.          ... 0.          0.
   0.         ]
 [ 0.          0.          0.          ... 0.          0.
   0.         ]]
```

FIGURE 15.7   The embedded weight matrix is created.

In Figure 15.7, each row represents a word vector. Each vector uniquely identifies a word and its relationship with other words.

Step 4: Training the machine learning algorithms

Different algorithms like CNN, RNN, GRU & LSTM and their CNN+ RNN, CNN+GRU, and CNN+LSTM hybrid models are trained as well as the standalone counterparts to compare their accuracy. The training is done with permutations of major activation functions and optimizers.

In convolutional neural network (CNN), we mainly use the convolution of the kernel. A small matrix of numbers (referred to as a kernel or filter) is used in this process, which then runs the text from our input through the filter values. Equation (15.6) carries out the subsequent feature mapping.

$$L[x,y] = (k*l)[x,y] = \sum_a \ \sum_b \ h[a,b]k[x-a,y-b] \tag{15.6}$$

f = input
Kernel = h
Index of rows and columns = a and b, respectively.

We take each worth from the Kernel and duplicate them two by two with comparing values from the picture. At long last, we summarize all that and put the outcome perfectly located in the result highlight map.

RNN is modified to function with time series information or information that includes a succession of some sort or another. The typical feedforward brain networks are just implied for information focuses, which are free of one another anyway when we have an information grouping where the information at one point relies upon the past information point the brain network must be changed to incorporate the conditions between the two arrangements of information. RNN utilizes memory to store the conditions of the past contributions to create the following result of the arrangement.

Suppose we have an activation function h in the unfolded network the following operation takes place and it is shown in equation (15.7).

$$f_{t+1} = h\left(x_t, f_t, w_x, w_f, b_f\right) = h\left(w_x x_t + w_f f_t + b_f\right) \tag{15.7}$$

The output y at time t is computed as follows and it is shown in equation (15.8).

$$y_t = h\left(f_t, w_y\right) = h\left(w_y, f_t + b_y\right) \tag{15.8}$$

x_t inputs at time step t

w_x, w_y = input weights

w_f = weights attached to the hidden layer

A variant of the recurrent neural network known as the gated recurring unit (GRU) allows the gating of the hidden layer.

For a given time step t, suppose that the input is a minibatch $Y_t \in K^{n \times d} Y_t \in K^{n \times d}$ (number of examples = n, number of inputs = d) and the hidden state of the previous state is $H_{t-1} \in K^{n \times h} H_{t-1} \in K^{n \times h}$ (no of hidden units = h), then the reset gate $S_t S_t$ and the update gate $U_t U_t$ is given as follows and they are given in equation (15.9) and equation (15.10).

$$S_t = \sigma\left(X_t W_{xr} + H_{t-1} W_{hr} + b_r\right) \tag{15.9}$$

$$U_t = \sigma\left(X_t W_{xz} + H_{t-1} W_{hz} + b_z\right) \tag{15.10}$$

where $W_{xr}, W_{xz} \in K^{d \times h}$ and $W_{hr}, W_{hz} \in K^{h \times h}$ and b_r, b_z are biases.

Below is a representation of candidate hidden state equation (15.11).

$$H_{t-1} = \tanh\left(X_t W_{xh} + (R_t \odot H_{t-1}) W_{hh} + b_h\right) \tag{15.11}$$

where

$W_{xr}, W_{xz}, W_{hr}, W_{hz}$ are weight parameters

H_{t-1} represents the hidden state at the time '$t-1$'

R_t represents the reset gate

The following drawbacks of the recurrent neural network are addressed by the long-short term memory (LSTM) version.

- The long term dependency problem
- Vanishing gradient and exploding gradient

The cell state of the LSTM provides memory so that it can remember the previous state of the input. Input, forget, and output gates are the three gates of LSTM. They are given in equation (15.12), equation (15.13), and equation (15.14).

$$I_t = \sigma\left(V_i\left[h_{t-1}, a_t\right] + b_i\right) I_t = \sigma\left(V_i\left[h_{t-1}, a_t\right] + b_i\right) \tag{15.12}$$

$$G_t = \sigma\left(V_g\left[h_{t-1}, a_t\right] + b_g\right) G_t = \sigma\left(V_g\left[h_{t-1}, a_t\right] + b_g\right) \tag{15.13}$$

$$Y_t = \sigma\left(V_y\left[h_{t-1}, a_t\right] + b_y\right) Y_t = \sigma\left(V_y\left[h_{t-1}, a_t\right] + b_y\right) \tag{15.14}$$

I_t = input gate

G_t = forget gate

Y_t = output gate

σ = sigmoid function

By computing a weighted total and additionally adding bias to it, activation work determines whether a neuron should be turned on or off. The goal of the initiation effort is to introduce non-linearity into a neuron's output.

The different activation functions are available and few of them are given below.

(a) A non-linear function is a **sigmoid function**. The range of x's number is -2 to 2. The values of the output are very significant. This implies that substantial differences in output value would result from even minor changes in x. It can be found in equation (15.10). It is typically applied in the binary classification's output layer. Equation (15.15) illustrates it.

$$\text{sigmoid function} = \frac{1}{1 + e^{-x}} \tag{15.15}$$

Hyperbolic tangent (tanh) function is non-linear in nature. Values are between -1 and $+1$. Because its values range from -1 to 1, which is how the hidden layer of a neural network emerges to 0 or very near to it, it is typically used in the hidden layers of a neural network. Due to the mean being closer to 0, this aids in centering the data. It is given in equation (15.16).

$$\text{tanh function} = \frac{e^x - e^{-x}}{e^x + e^{-x}} \tag{15.16}$$

(b) Rectified linear unit is referred to as the rectified linear activation function (**ReLU**). The most common activation method is this one. Most neural network implementations take place in hidden levels. It is a non-linear function, and because only a small number of neurons are active at once, the network is sparse, which makes calculation simple and effective. Compared to tanh and sigmoid functions, it performs quicker and requires less computing power. It can be found in equation (15.17).

$$\text{relu function} = \max(0, x) \tag{15.17}$$

(c) When handling multiple classes, the **softmax** function is employed. The outputs for each class are constrained to fall between 0 and 1, and it would also split by the sum of the outputs. It is typically applied to the classifier's output layer.

The learning rate and weights of the neural network are two parameters that optimizers alter to minimize the loss. The different optimizers currently in use are given below.

(i) Adam (adaptive moment estimation) operates on first- and second-order momenta. The reason we chose Adam is because we don't want to go so fast just because we can skip the minimum, we want to lower the speed a little for a thorough search. They are shown in equation (15.18) and equation (15.19).

$$\hat{l}_t = \frac{l_t}{1 - \beta_1^t} \hat{l}_t = \frac{l_t}{1 - \beta_1^t} \tag{15.18}$$

$$\widehat{S}_t = \frac{s_t}{1 - \beta_2^t} \widehat{S}_t = \frac{s_t}{1 - \beta_2^t} \tag{15.19}$$

l_t = mean of the gradients
s_t = uncentered variance of the gradients

(ii) Stochastic gradient descent (SGD) frequently updates the parameters in the model. In this instance, the model's parameters are modified following the computation of the loss for each training sample. It is given in equation (15.20).

$$\theta=\theta-\alpha\cdot\nabla J(\theta;k(i);l(i)) \qquad\qquad (15.20)$$

where

$k(i),l(i)\}$ are the training examples

α = learning rate

J = cost function

Θ = parameter to be updated

15.4   EXPERIMENTAL RESULTS

This dataset is hosted on Kaggle and is highly applauded by the community for its usability. It is employed in the creation of a machine learning program to determine when an article might be false news. The dataset itself is divided into two parts, fake (Fake.csv) and true (True.csv). Fake.csv contains all the fake news and True.csv contains all the real news. All the fake news is contained in Fake.csv, whereas all the actual news is found in True.csv.

Each row consists of four columns and they are given below.

- **Title** – This column includes all the titles of the news articles.
- **Text** – This column contains all the contents of the articles.
- **Subject** – The subject of the article. This is further divided into three categories, which are news, politics, and other.
- **Date** – Date at which the article was published.

Figures 15.8 and 15.9 depict the sample dataset of politics for the false and accurate reviews.

Numbers of subjects in the Fake.csv dataset are news, politics, Government news, United States (US) news, left news, and Middle East. It is shown in Figure 15.10. The count of the number of records in each subject is also displayed along with the total number of records of all the subjects combined. The details are given below. The subject 'politicsNews' has 6841 records, 'News' has 9050 records, 'left-news' has 4459 records, 'Government News' has 1570 records, 'US_News' has 783 records, and 'Middle-east' has 778 records within the dataset. The total records are 23481.

Number of subjects in the True.csv dataset are politicsNews and worldnews. It is shown in Figure 15.11. The count of the number of records in each subject is also displayed along with the total number of records of all the subjects combined. The details are given below. The subject 'politicsNews' has 11271 records and 'worldnews' has 10145 records within the dataset. The total records are 21416.

	A	B	C	D
1	title	text	subject	date
2	Donald Trump Sends Ou	Donald Trump just couldn t wis	News	December 31, 2017
3	Drunk Bragging Trump S	House Intelligence Committee	News	December 31, 2017
4	Sheriff David Clarke Bec	On Friday, it was revealed that	News	December 30, 2017
5	Trump Is So Obsessed H	On Christmas day, Donald Trum	News	December 29, 2017
6	Pope Francis Just Called	Pope Francis used his annual Ch	News	December 25, 2017
7	Racist Alabama Cops Br	The number of cases of cops br	News	December 25, 2017
8	Fresh Off The Golf Cour	Donald Trump spent a good poi	News	December 23, 2017
9	Trump Said Some INSAN	In the wake of yet another cou	News	December 23, 2017
10	Former CIA Director Sla	Many people have raised the al	News	December 22, 2017
11	WATCH: Brand-New Pro	Just when you might have thou	News	December 21, 2017
12	Papa Johnâ€™s Founde	A centerpiece of Donald Trump	News	December 21, 2017

FIGURE 15.8   The sample dataset of politics for the fake reviews.

	A	B	C	D
1	title	text	subject	date
2	As U.S. budget fight looms, Republicans flip their fiscal script	WASHINGTON (Reuters) - The head of a conservative Republican faction in th	politicsNews	December 31, 2017
3	U.S. military to accept transgender recruits on Monday: Pent	WASHINGTON (Reuters) - Transgender people will be allowed for the first tim	politicsNews	December 29, 2017
4	Senior U.S. Republican senator: 'Let Mr. Mueller do his job'	WASHINGTON (Reuters) - The special counsel investigation of links between t	politicsNews	December 31, 2017
5	FBI Russia probe helped by Australian diplomat tip-off: NYT	WASHINGTON (Reuters) - Trump campaign adviser George Papadopoulos tol	politicsNews	December 30, 2017
6	Trump wants Postal Service to charge 'much more' for Amaz	SEATTLE/WASHINGTON (Reuters) - President Donald Trump called on the U.S	politicsNews	December 29, 2017
7	White House, Congress prepare for talks on spending, immig	WEST PALM BEACH, Fla./WASHINGTON (Reuters) - The White House said on	politicsNews	December 29, 2017
8	Trump says Russia probe will be fair, but timeline unclear: N	WEST PALM BEACH, Fla (Reuters) - President Donald Trump said on Thursday	politicsNews	December 29, 2017
9	Factbox: Trump on Twitter (Dec 29) - Approval rating, Amazc	The following statementsÂ were posted to the verified Twitter accounts of U	politicsNews	December 29, 2017
10	Trump on Twitter (Dec 28) - Global Warming	The following statementsÂ were posted to the verified Twitter accounts of U	politicsNews	December 29, 2017
11	Alabama official to certify Senator-elect Jones today despite	WASHINGTON (Reuters) - Alabama Secretary of State John Merrill said he wil	politicsNews	December 28, 2017
12	Jones certified U.S. Senate winner despite Moore challenge	(Reuters) - Alabama officials on Thursday certified Democrat Doug Jones the	politicsNews	December 28, 2017
13	New York governor questions the constitutionality of federa	NEW YORK/WASHINGTON (Reuters) - The new U.S. tax code targets high-tax	politicsNews	December 28, 2017
14	Factbox: Trump on Twitter (Dec 28) - Vanity Fair, Hillary Clint	The following statementsÂ were posted to the verified Twitter accounts of U	politicsNews	December 28, 2017
15	Trump on Twitter (Dec 27) - Trump, Iraq, Syria	The following statementsÂ were posted to the verified Twitter accounts of U	politicsNews	December 28, 2017

FIGURE 15.9   The sample dataset of politics for the true reviews.

```
In [19]:    1  plt.figure(figsize=(8,5))
            2  sns.countplot("subject", data=fake)
            3  plt.show()
```

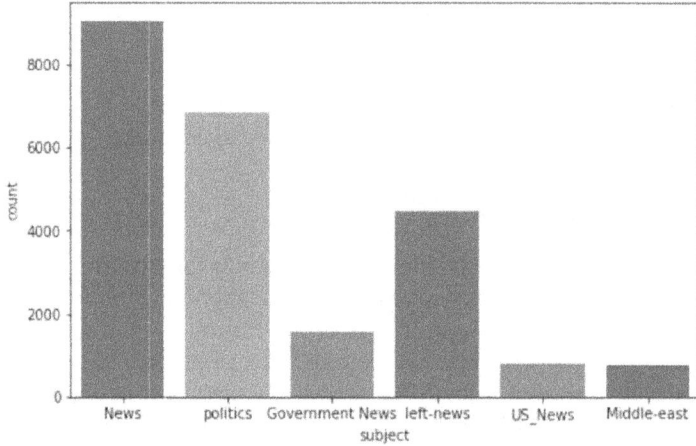

FIGURE 15.10   Number of subjects in the Fake.csv dataset.

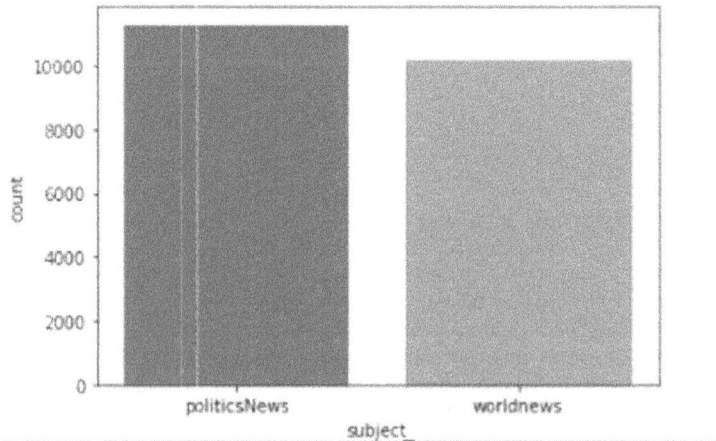

FIGURE 15.11   Number of subjects in the True.csv dataset.

TABLE 15.1

The Accuracy for CNN, GRU, LSTM and RNN Algorithms with 'adam' & 'sgd' Optimizers and 'sigmoid', 'relu', 'softmax', and 'tanh' Activation Functions

| Algorithm | ADAM | | | | SGD | | | |
	relu	sigmoid	softmax	tanh	relu	sigmoid	softmax	tanh
CNN	0.5300	0.9200	0.4700	0.9200	0.4700	0.9200	0.4700	0.5200
GRU	0.5250	0.5290	0.9780	0.9810	0.5170	0.8970	0.4730	0.4740
LSTM	0.4700	**0.9900**	0.4700	0.9700	0.5200	0.9500	0.4700	0.9000
RNN	0.8840	0.9200	0.4800	0.8990	0.4780	0.8940	0.4800	0.4730

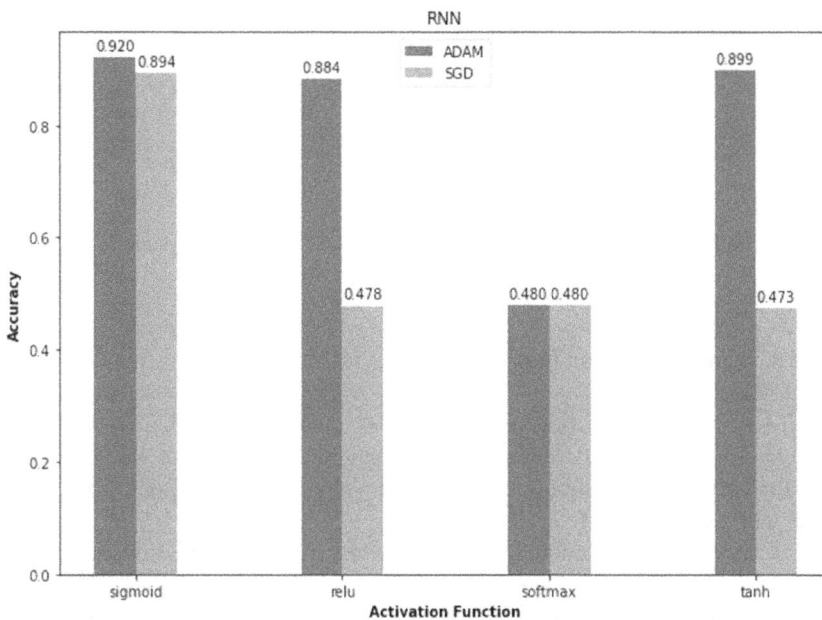

FIGURE 15.12   The RNN-based prediction model's accuracy when utilizing various activation functions and optimizers.

Keeping the preprocessing of the data in all cases the same we get the following results. In the proposed study, graphs between the test data and the predicted data were also drawn using CNN + RNN, CNN + LSTM, and CNN + GRU for classification purposes. The proposed work is contrasted with present algorithms like CNN, GRU, LSTM and RNN algorithms with 'adam' & 'sgd' optimizers and 'sigmoid', 'relu', 'softmax', and 'tanh' activation functions. The metrics calculated in each algorithm is the accuracy.

The accuracy for CNN, GRU, LSTM, and RNN algorithms with 'adam' and 'sgd' optimizers and 'sigmoid', 'relu', 'softmax', and 'tanh' activation functions are shown in Table 15.1. The accuracy of different algorithms with different optimizers and 'sigmoid', 'relu', 'softmax', and 'tanh' activation functions, LSTM algorithm with 'adam' optimizer and 'sigmoid' activation function is performing very well and gives the accuracy of 99%.

Figure 15.12 shows the RNN model's accuracy when using different activation functions such as 'sigmoid', 'relu', 'softmax', and 'tanh' and pairing them with optimizers 'ADAM' and 'SGD'. In

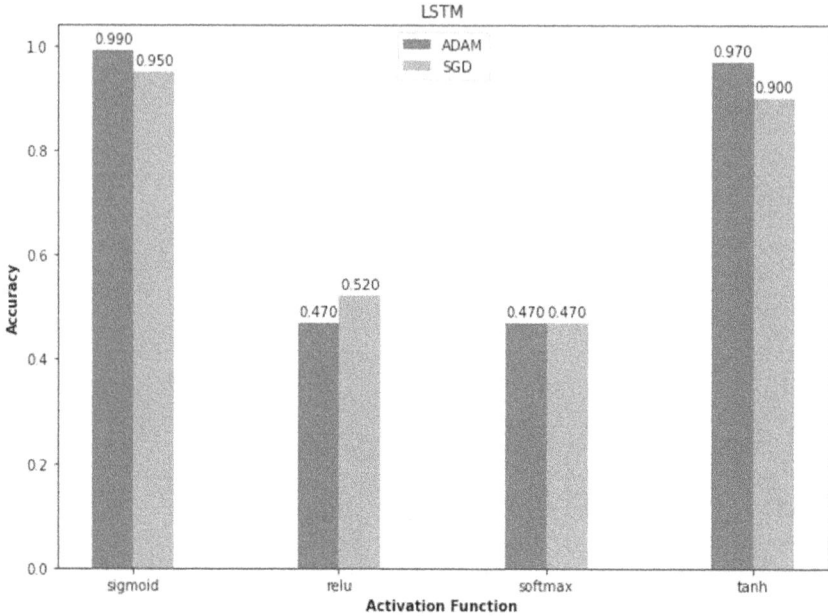

FIGURE 15.13 The LSTM-based prediction model's accuracy when utilizing various activation functions and optimizers.

the RNN model, sigmoid activation function with 'ADAM' optimizer performs very well and gives an accuracy of 92%.

Figure 15.13 displays the accuracy of the LSTM model while utilizing various activation functions such as 'sigmoid', 'relu', 'softmax', and 'tanh' and pairing them with optimizers 'adam' and 'sgd'. In the LSTM model, sigmoid activation function with 'adam' optimizer performs very well and gives an accuracy of 99%.

Figure 15.14 shows the GRU model's accuracy when using various activation functions such as 'sigmoid', 'relu', 'softmax', and 'tanh' and pairs them with optimizers 'adam' and 'sgd'. In the CNN + RNN hybrid model, sigmoid activation function with 'adam' optimizer performs very well and gives an accuracy of 97.8%.

Figure 15.15 shows the accuracy of the hybrid model deployed consisting of 'RNN + CNN' with the various activation functions such as 'sigmoid', 'relu', 'softmax' and 'tanh' and pairing them with optimizers 'adam' and 'sgd'. In the 'RNN + CNN' hybrid model, sigmoid activation function with 'sgd' optimizer performs very well and gives an accuracy of 93.6%.

Figure 15.16 shows the accuracy of the hybrid model deployed consisting of 'LSTM + CNN' with the various activation functions such as 'sigmoid', 'relu', 'softmax', and 'tanh' and pairing them with optimizers 'adam' and 'sgd'. In the 'LSTM + CNN' hybrid model, sigmoid activation function with 'adam' optimizer performs very well and gives an accuracy of 97.6%.

Figure 15.17 shows the accuracy of the hybrid model deployed consisting of 'GRU + CNN' with the various activation functions such as 'sigmoid', 'relu', 'softmax', and 'tanh' and pairing them with optimizers 'adam' and 'sgd'. In 'GRU + CNN' hybrid model, sigmoid activation function with 'sgd' optimizer performs very well and gives an accuracy of 97.8%.

Accuracy for 'CNN+GRU', 'CNN+LSTM', and 'CNN+RNN' hybrid algorithms in combination with embedding matrix extracted from using Word2Vec with 'adam' and 'sgd' optimizers and 'sigmoid', 'relu', 'softmax', and 'tanh' activation functions are shown in Table 15.2. The precision of different algorithms with various optimizers and 'sigmoid', 'relu', 'softmax', and 'tanh'

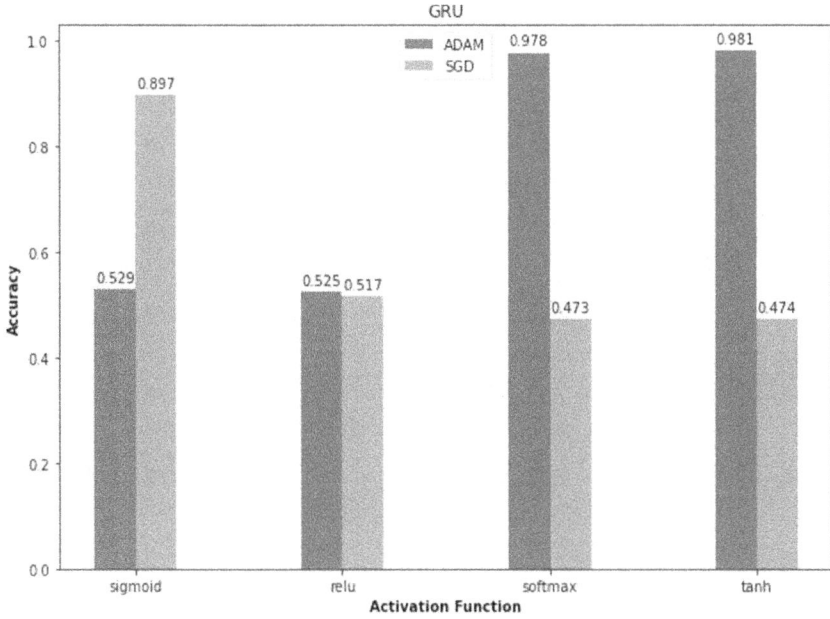

FIGURE 15.14   The GRU-based prediction model's accuracy when utilizing various activation functions and optimizers.

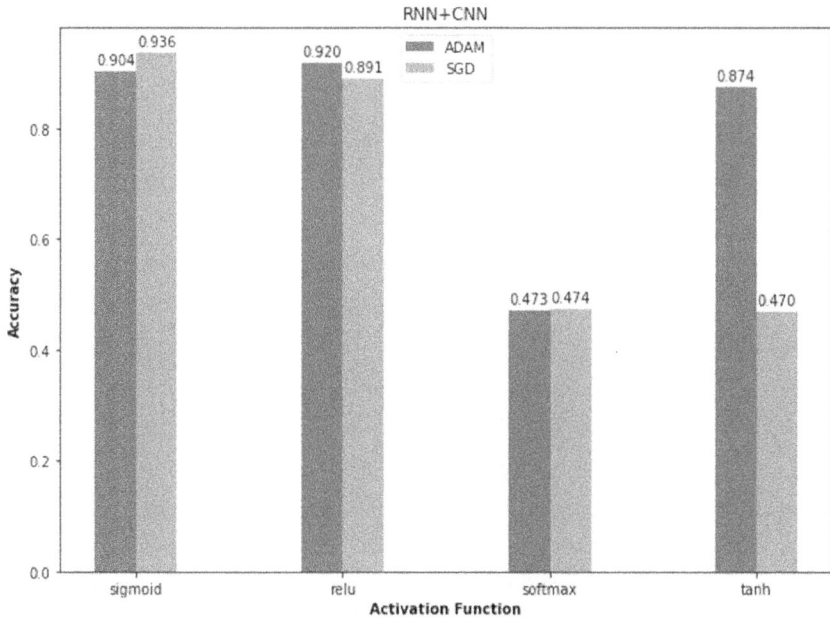

FIGURE 15.15   The RNN + CNN – based prediction model's accuracy when utilizing various activation functions and optimizers.

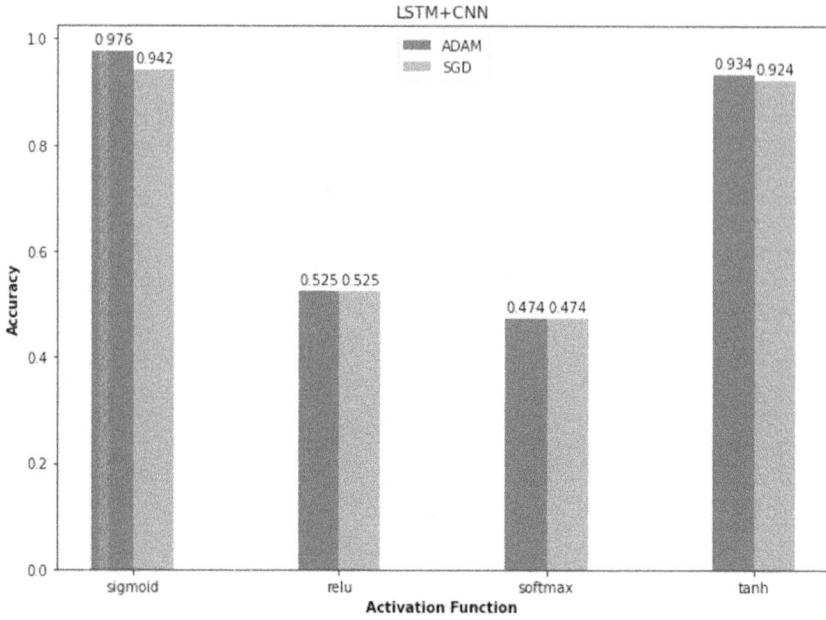

FIGURE 15.16   The LSTM + CNN – based prediction model's accuracy when utilizing various activation functions and optimizers.

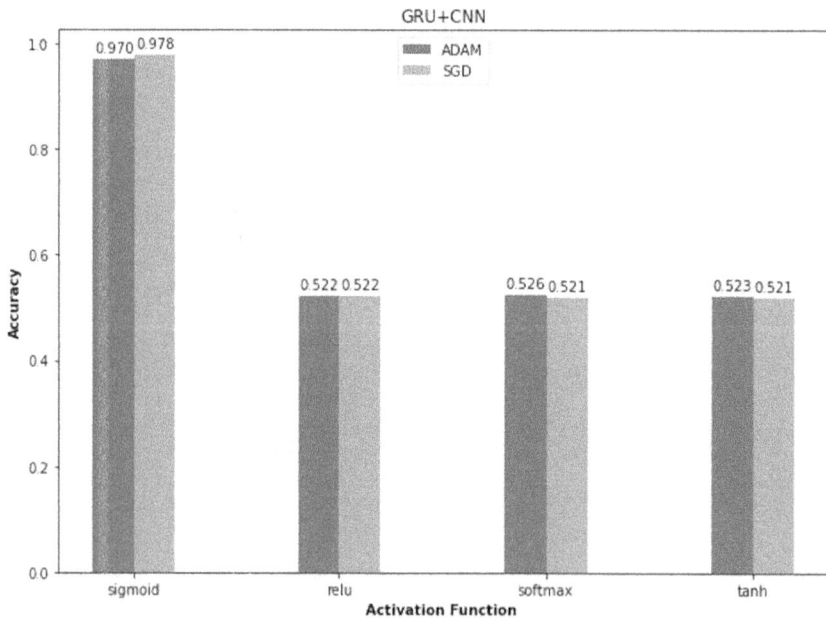

FIGURE 15.17   The GRU + CNN – based prediction model's accuracy when utilizing various activation functions and optimizers.

TABLE 15.2
Accuracy for CNN+GRU, CNN+LSTM, and CNN+RNN Algorithms in Combination with Embedding Matrix Extracted from Using Word2Vec with 'adam' and 'sgd' Optimizers and 'sigmoid', 'relu', 'softmax' and 'tanh' Activation Functions

Algorithm	ADAM				SGD			
	relu	sigmoid	softmax	tanh	relu	sigmoid	softmax	tanh
CNN+GRU	0.5220	0.9700	0.5260	0.5230	0.5220	**0.9780**	0.5210	0.5210
CNN+LSTM	0.5250	0.9760	0.4740	0.9340	0.5250	0.9420	0.4740	0.9240
CNN+RNN	0.9200	0.9040	0.4730	0.8740	0.8910	0.9360	0.4740	0.4700

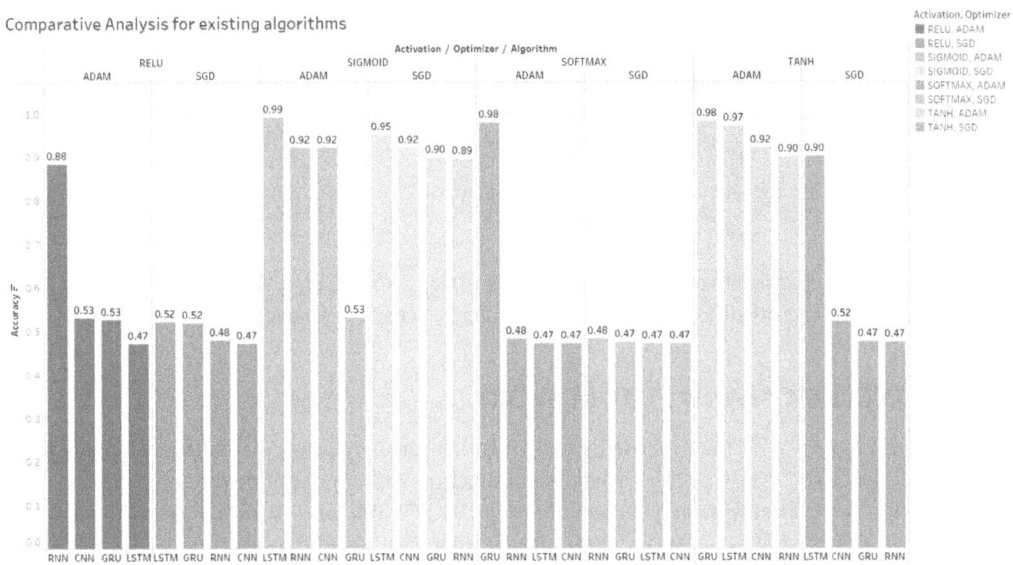

FIGURE 15.18 Accuracy comparisons for the CNN, GRU, LSTM, and RNN algorithms using various optimizers and activation functions.

activation functions, 'CNN+GRU' hybrid algorithm with 'sgd' optimizer and 'sigmoid' activation function performs very well and gives an accuracy of 97.8%.

The accuracy comparisons for the CNN, GRU, LSTM and RNN algorithms with 'adam' and 'sgd' optimizers and 'sigmoid', 'relu', 'softmax', and 'tanh' activation functions is shown in Figure 15.18. Among these four existing algorithms, the LSTM algorithm with 'adam' optimizer and 'sigmoid' activation function performs very well and gives an accuracy of 99%.

The comparative analysis of the accuracy for 'CNN + GRU', 'CNN + LSTM' and 'CNN + RNN' hybrid algorithms with 'adam' and 'sgd' optimizers and 'sigmoid', 'relu', 'softmax', and 'tanh' activation functions is shown in Figure 15.19. First an initial CNN layer is used, to which a layer of another algorithm is added. Among these four existing algorithms, the 'CNN+GRU' hybrid algorithm with 'sgd' optimizer and SIGMOID activation function performs very well and gives an accuracy of 97.8%.

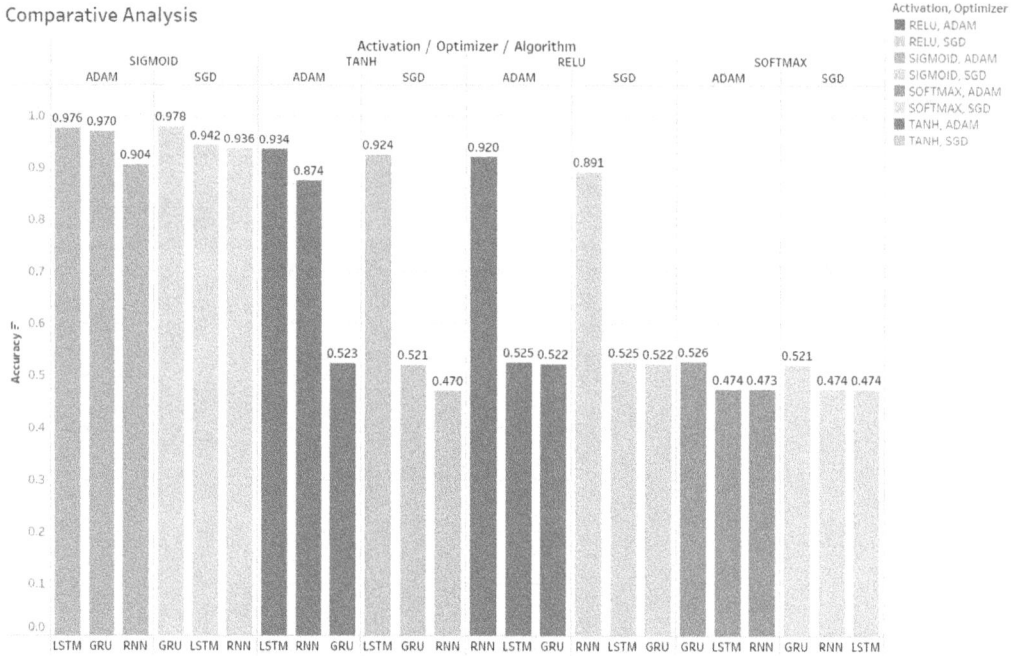

FIGURE 15.19 The comparative analysis of the accuracy for 'CNN + GRU', 'CNN + LSTM' and 'CNN + RNN' hybrid algorithms with different optimizers and different activation functions.

15.5 CONCLUSION AND FUTURE WORK

As seen from Figure 15.17, the proposed work finds that for the purpose of 'fake news prediction', the ensemble model of 'CNN + GRU' when in conjunction with 'sgd' optimizer and 'sigmoid' activation function works best with an accuracy of around 97.8%. Whereas 'softmax' works poorly with all the models with the lowest accuracy across the board regardless of optimizer or model, the lowest being 47.4%. We can also see that LSTM seems to work with more consistency when compared to GRU and RNN with all the optimizers and the activation functions used.

Sequential neural networks can be used to encode the required information using a variety of current and practical approaches. The utilization of a deep neural network employing bidirectional training has shown remarkable efficacy and precision when it comes to detecting fake news. A few papers have explored this idea and have come up with multiple conclusions. There has been no comprehensive paper that compares and contrasts the different approaches available. Hence the proposed work aims to perform an exhaustive comparative analysis of these different approaches and figures out the best outcome possible.

For future work, we find that an interesting domain would be to employ similar approaches for real-time fake news detection in videos and audio files. We can also survey how different embedding techniques affect the accuracy of the ensemble models.

REFERENCES

1. Gomathy, C. K. (2021). The fake news detection using machine learning algorithms. *International Research Journal of Engineering and Technology (IRJET)*, 8(10), pp. 37–40.
2. Jain, A. and Kasbe, A. (2018). Fake news detection. 2018 IEEE International Students' Conference on Electrical, Electronics and Computer Science (SCEECS), Bhopal, India, pp. 1–5, doi: 10.1109/SCEECS.2018.8546944

3. Nasir, J. A., Khan, O. S., and Varlamis, I. (2021). Fake news detection: A hybrid CNN-RNN based deep learning approach. *International Journal of Information Management Data Insights*, 1(1), p. 100007, ISSN 2667-0968.

4. Carmel Mary Belinda M J, Ravikumar S, Muhammad Arif, Dhilip Kumar V, Antony Kumar K, and Arulkumaran G. (2022). Linguistic analysis of Hindi-English mixed tweets for depression detection, Hindawi, *Journal of Mathematics*, 2022(3225920), pp. 1–7, https://doi.org/10.1155/2022/3225920.

5. Cao, J., Qi, P., Sheng, Q., Yang, T., Guo, J., and Li, J. (2020). Exploring the Role of Visual Content in Fake News Detection. In: Shu, K., Wang, S., Lee, D., Liu, H. (eds) *Disinformation, Misinformation, and Fake News in Social Media. Lecture Notes in Social Networks*. Springer, Cham. https://doi.org/10.1007/978-3-030-42699-6_8

6. Shu, K., Zhou, X., Wang, S., Zafarani, R., and Liu, H. (2019). The role of user profiles for fake news detection. In Proceedings of the 2019 IEEE/ACM International Conference on Advances in Social Networks Analysis and Mining. Association for Computing Machinery, New York, NY, USA, 436–439. https://doi.org/10.1145/3341161.3342927

7. Ahmad, I., Yousaf, M., Yousaf, S., and Ahmad, M. O. (2020). Fake news detection using machine learning ensemble methods. *Complexity*, 2020(8885861), pp. 1–11. doi: 10.1155/2020/8885861

8. Shu, K., Wang, S, and Liu, H. (2017). Fake News Detection on Social Media: A Data Mining Perspective. *SIGKDD Explor.* 19(1), pp. 22–36.

9. Mahesh, T. R., Kumar, V. D., Kumar, V. V., Asghar, J., Geman, O., Arulkumaran, G., and Arun, N. (2022). AdaBoost ensemble methods using K-fold cross validation for survivability with the early detection of heart disease. Hindawi, *Computational Intelligence and Neuroscience*, pp. 1–10, https://doi.org/10.1155/2022/9005278

10. Monti, F., Frasca, F., Eynard, D., Mannion, D., and Bronstein, M. M. (2019). Fake news detection on social media using geometric deep learning. arXiv.org, 10-Feb-2019. [Online]. Available: https://arxiv.org/abs/1902.06673.

11. Mishra, V., Verma, M., and Sharma, A. (2021). Fake news detection with naive Bayes classifier. *International Journal of Innovative Research in Technology*, 30-May-2021. [Online]. Available: https://ijirt.org/Article?manuscript=151486.

12. Tacchini, E., Ballarin, G., Della Vedova, M., Moret, S., and Alfaro, L. (2017). Some like it hoax: Automated fake news detection in social networks. CEUR Workshop Proceedings, 1960, pp. 1–15. arXiv:1704.07506

13. Gangireddy, S., Deepak, P., Long, C., and Chakraborty, T. (2020). Unsupervised fake news detection: A graph-based approach. In Proceedings of the 31st ACM Conference on Hypertext and Social Media (HT '20). Association for Computing Machinery, New York, NY, USA, pp. 75–83. https://doi.org/10.1145/3372923.3404783

14. Kaliyar, R., Goswami, A., and Narang, P. (2021). FakeBERT: Fake news detection in social media with a BERT-based deep learning approach. *Multimedia Tools and Applications*. 80. 10.1007/s11042-020-10183-2

15. Kaur, S., Kumar, P., and Kumaraguru, P. Automating Fake News Detection System using Multi-level Voting Model. *Soft Comput* 24, pp. 9049–9069 (2020). https://doi.org/10.1007/s00500-019-04436-y

16. Xu, K., Wang, F., Wang, H., and Yang, B. (February 2020). Detecting fake news over online social media via domain reputations and content understanding, in *Tsinghua Science and Technology*, 25(1), pp. 20–27, doi: 10.26599/TST.2018.9010139

17. Elsaeed, E., El-Daydamony, E., Elmogy, M. M., Atwan, A., and Ouda, O. (2021). Detecting fake news in social media using voting classifier. *IEEE Access*, 9, pp. 161909–161925. pp. 1–1. 10.1109/ACCESS.2021.3132022

18. Oshikawa, R., Qian, J., and Yang Wang, W. (2020). A survey on natural language processing for fake news detection. Proceedings of the 12th Conference on Language Resources and Evaluation (LREC 2020), pp. 6086–6093.

19. Zhou, X. and Zafarani, R. (2019). Network-based fake news detection: A pattern-driven approach, arXiv.org, 10-Jun-2019. [Online]. Available: https://arxiv.org/abs/1906.04210

20. Aphiwongsophon, S., and Chongstitvatana, P. (2018). Detecting fake news with machine learning method. 2018 15th International Conference on Electrical Engineering/Electronics, Computer, Telecommunications and Information Technology (ECTI-CON), 528–531.

21. Parikh, S. B., and Atrey, P. K. (2018). Media-rich fake news detection: A survey. 2018 IEEE Conference on Multimedia Information Processing and Retrieval (MIPR), pp. 436–441, doi: 10.1109/MIPR.2018.00093

22. Reis, J. C. S., Correia, A., Murai, F., Veloso, A., Benevenuto, F., and Cambria, E. (2019). Supervised learning for fake news detection. *IEEE Intelligent Systems, 34,* 76–81.

23. Saleh, H., Alharbi, A., and Alsamhi, S. H. (2021). OPCNN-FAKE: Optimized convolutional neural network for fake news detection, in IEEE Access, 9, pp. 129471–129489, doi: 10.1109/ACCESS.2021.3112806

24. Jiang, T., Li, J. P., Haq, A. U., Saboor, A., and Ali, A. (2021). A novel stacking approach for accurate detection of fake news, in *IEEE Access,* 9, pp. 22626–22639, doi: 10.1109/ACCESS.2021.3056079

25. Li, D., Guo, H., Wang Z., and Zheng, Z. (2021). Unsupervised fake news detection based on autoencoder. in IEEE Access, 9, pp. 29356–29365, doi: 10.1109/ACCESS.2021.3058809

26. Elhadad, M. K., Li, K. F., and Gebali, F. (2020). Detecting misleading information on COVID-19. in IEEE Access, 8, pp. 165201–165215, doi: 10.1109/ACCESS.2020.3022867.

27. Shahbazi, Z. and Byun, Y.-C. (2021). Fake media detection based on natural language processing and blockchain approaches. in IEEE Access, 9, pp. 128442–128453, doi: 10.1109/ACCESS.2021.3112607.

LIST OF ABBREVIATIONS

1d-CNN	One-dimensional Convolutional Neural Network
Adam	Adaptive Moment
AI	Artificial intelligence
ANOVA	Analysis of Variance
AUC	Area Under the ROC Curve
BERT	Bidirectional Encoder Representations from Transformers
BoW	Bag of Words
CBoW	Continuous bunch of words
CNN	Convolutional Neural Network
CNN+GRU	Convolutional Neural Network + Gated Recurrent Unit
CNN+LSTM	Convolutional Neural Network + Long Short-Term Memory
CNN+RNN	Convolutional Neural Network + Recurrent Neural Networks
CV	Computer Vision Library
Doc2Vec	Document to Vector
FBI	Federal Bureau of Investigation
GRU	Gated Recurrent Unit
GTUT	Graph mining methods over Textual, User and Temporal data
ISOT	Information Security and Object Technology
KNN	K-Nearest Neighbours
LDA	Linear Discriminant Analysis
LinearSVC	Linear Support Vector Machine
LIWC	Linguistic Inquiry and Word Count
LSTM	Long Short-Term Memory
MI	Mutual information
NB	Naïve Bayes
NER	Named Entity Recognition
NLP	Natural Language Processing
NLTK	Natural Language Toolkit
NPMI	Normalized Pointwise Mutual Information
PMI	Pointwise Mutual Information
RBF	Radial basis function
ReLU	Rectified Linear Activation function
RF	Random Forest

RL	Reinforcement Learning
RNN	Recurrent Neural Network
ROC	Receiver Operating Characteristic
RST	Rhetorical Structure Theory
SGD	Stochastic gradient descent
SVM	Support Vector Machine
Tanh	Hyperbolic Tangent
TF-IDF	Term Frequency-Inverse Document Frequency (TF-IDF)
URL	Uniform Resource Locator
US	United States
VGG	Visual Geometry Group
Word2Vec	Word to Vector
XGBoost	Extreme Gradient Boosting

16 A Comparative Analysis of Deep Learning Models for Fake News Detection and Popularity Prediction of Articles

Jayanthi Devaraj
Department of Information Technology, Sri Venkateswara College of Engineering, Sriperumbudur, Chennai, Tamilnadu, India

16.1 INTRODUCTION

Fake news detection is very important in this digital era since there is an increasing number of fake news spreads with various intentions. The reason for spreading fake news may be to damage any entity, which leads to the wrong way. Also, for commercial and political reasons, the spreading of fake news happens in real life.

Figure 16.1 depicts how fake news is spreading faster than real news, which imposes serious threats to society. The false news headline is released and the spreading of false news with false headlines increases 10 times more than the real news. So it is a serious concern to implement prediction algorithms for fake news detection to retain humanity worldwide.

In recent years, the leading misinformation can be detected automatically using natural language processing algorithms and artificial intelligence. Due to the advancement in technology, fake news patterns are recognized easily for separating fake and real news. There exist many models and classifiers, capable of extracting the textual features using methods like character vector, N-gram, count vectorization, term and inverse term-document frequency, etc. The models accurately categorize the type of news by training and testing the data. 80% to 100% accuracy can be achieved by the models depending on the functional layers and the metrics used [1].

Deep learning (DL) models can automatically extract the textual features and various classifications of data are available such as news content-based data, which deals with only text and images. The models can extract visual and textual features and can handle multimodal data. Next, there are social context-based methods that deal with identifying the propagation pattern and finding out user credibility where articles are tested and analyzed with other factors like the publisher, owner, and the posts. The last category is the knowledge-based approach where the attention mechanisms and the graph-based knowledge are used for prediction [2]. A study of deep learning and machine learning models is carried out using different datasets. For the Fakenews dataset, the ensemble modeling of machine learning models achieves better performance. Models like convolutional neural network (CNN), and bidirectional long-short term memory (BiLSTM) are used for capturing long-term sequences in the data and discovering the hidden and salient features for detecting fake news. Still, advanced techniques like integration of deep CNN with the bidirectional encoder representations from the transformers (BERT) achieve good results compared to the other hybrid models [3].

DOI: 10.1201/9781003407959-16

FIGURE 16.1 Fake news [Appendix-1].

Even there is existing literature that uses only the title for classifying the fake news data without considering the other aspects of the content. On the world news dataset and the political dataset, support vector machine (SVM), random forest classification, and other ensemble models are used for detecting news with a title field [4]. Fake or distrustful reviewers can be detected automatically using deep learning models. Real-time Yelp datasets are used for capturing the behavioral patterns from the text. A hybrid output of behavioral sensitive features with the attention mechanisms and transformer models effectively classifies whether the reviewer is fake or genuine [5]. The LSTM model can find the long-term dependencies in the dataset and is widely used for time series analysis. The fact-checking dataset is used by the authors for fake news identification by training the BiLSTM model on the data. A good classification accuracy of 99.8% is achieved by the model by evaluating the various performance metrics [6].

There are a variety of models presently used by researchers for making accurate predictions in real-time. In order to alert the people, and to give warning about the increasing number of fake news to the government, it is necessary to predict the fake news in real-time accurately. This work involves the development of a new hybrid DL model for classifying news articles.

The remaining sections are organized as follows. Section 16.2 defines the literature works being carried out; Section 16.3 elaborates on the proposed study with a detailed explanation of modules and architecture diagrams. Section 16.4 demonstrates the dataset used, and the detailed analysis of the data; Section 16.5 shows the implementation results and discussion section; Section 16.6 concludes and presents future scope of improvement.

16.2 RELATED WORKS

There are existing works being carried out by researchers in detecting fake news. Some of the recent literature studies on this work are highlighted in this section.

Due to the increasing number of social media networks in this digital world, a huge volume of data is being generated every second. Quantitative argumentation-based automated explainable decision-making system (QA-AXDS) is used to identify news types, which are fully data-driven approaches where machine learning models are employed for prediction [7]. The preprocessing methods like count vectorizer, N-gram technique, and term frequency and inverse term frequency (TF-IFD) vectorizer along with models like SVM, logistic regression, random forest, and naïve Bayes are used for the detection of fake news [8,9,10]. The multi-lingual and cross-language prediction [11, 12] and ensemble models [13, 14] along with features related to contextual learning are done. Feature propagation with graph and network-based features is used. The model with a stance network and Recurrent Neural Network (RNN) model for encoding patterns is used and 94% of accuracy is achieved [15]. In most of the related works, textual features [16] are extracted, and text summarization techniques [17] and sentiment analysis [18] are done along with machine learning [19] or other DL algorithms [20] for improving the detection rate. Knowledgeable prompt learning [21] is adopted, which incurs the process of injecting the knowledge using PolitiFact and Gossip datasets and an average of 94% accuracy rate is achieved. Various other techniques and models like stacked models and progressive fusion networks are developed to improve the fake news detection rate [22, 23]. Also, the popularity prediction of news articles can be very helpful to the users to predict the priority and to help people to distinguish if the news article published is fake or real [24]. Real-time fake news detection is done by combining the events and extracted topics together and parallel processing of real-time streaming data is carried out. This technique improves memory management and a new knowledge base is constructed [25].

Fake news propagation affects the social, political, and economic environments and imposes various impacts on people in making wrong decisions. Ensemble models, optimization algorithms, diverse classifiers, natural language processing techniques, machine learning, and deep learning models are widely used for the detection of fake news. CNN and LSTM provides an accurate detection rate when compared with the other models. The variants of the CNN model are used in the recent literature on fake news detection.

Table 16.1 highlights current studies on detection techniques with datasets and models used with significant outcomes and limitations.

16.2.1 RESEARCH GAP AND MOTIVATION

In most of the studies, the raw features in the dataset are used directly, and feature extraction and feature representation is carried out on the raw features. The dependencies in the data cannot be captured well where the behavioral features are missed out in the prediction. Although the existing algorithms predict fake news, models are trained only on past data but still it is a difficult task to detect fake news on the evolving events. Also, it is very expensive and difficult to fine-tune the models for better performance. Hybrid deep learning models can improve the detection rate when compared to the other models.

The main research contribution is mentioned below.

1. Data preprocessing techniques like word embedding are employed before training the model.
2. Classification of news articles using a hybrid model (BERT + BiLSTM) is carried out to find the article class.
3. To check the robustness of the proposed work, various evaluation metrics such as accuracy, precision, recall, F1 score, etc. are used.
4. Popularity prediction of the news articles is done.
5. A comparative analysis between the proposed hybrid models with other models is carried out.

16.3 PROPOSED METHODOLOGY

This section deals with the methodology for discovering fake news. This section describes the architecture diagram of the proposed system. Figure 16.2 shows the detailed architecture diagram.

The study contains modules like extracting the news titles from the dataset and performing data preprocessing where the unwanted characters are eliminated. The preprocessed data is split into training and testing phases and fed to the models to train the data and to categorize the test input as fake or genuine. The dataset is trained using models like BiLSTM, and BERT models separately and the novel hybrid model BERT-BiLSTM is built to improve the classification accuracy.

TABLE 16.1
Literature Study on Fake News Detection

Ref No.	Dataset Used	Model Used	Significant Outcome	Limitations
[26]	The dataset includes fake news data with 23,502 articles and real news data with 21,417 articles.	(BERT) and transfer learning model is used.	99% of accuracy is achieved using the pretrained model and the news dataset is used for training at the last layer.	More expensive because of more weight updates with increasing training time.
[27]	The Fakenews dataset from Kaggle is used.	Bidirectional LSTM and recurrent neural networks (RNN) are used.	98.2% accuracy is obtained using the BiLSTM-RNN model.	An attention mechanism is not used and RNN does not capture the long-term dependencies in the data.
[28]	The dataset is collected from United Daily News, Liberty Times, Apple Daily, and China Times. Headlines and content of the news are trained.	BERT and artificial neural network (ANN) model is used.	The accuracy value of the BERT model is 91.2%. The accuracy value of ANN is 82.75%.	A hybrid model with more attributes can be used for improving the performance of the model.
[11]	The news dataset contains attributes like id, text, label, and title. Another news dataset with URL, body, and headline details is used.	Comparative analysis between CNN, recurrent neural network, unidirectional LSTM+RNN, and bidirectional LSTM+RNN is carried out.	The developed hybrid bidirectional model achieved 99% training accuracy, 89% validation accuracy, and 91% of test accuracy.	Fewer features are considered and the attention mechanism is not used for improving the performance.
[13]	The fake news dataset is used for classifying real and fake news. Data is collected from 1 Jan 2019 to 31 May 2021.	Artificial neural networks and BERT models are used for classification.	BERT model achieved a 91.2% hit rate and ANN with 65.5%.	More training time and expense. Other performance metrics can be used to further evaluate the model accurately.
[29]	English news articles are used as the source scenarios and French, Spanish, and Chinese are used as the target languages.	Pretrained BERT model is used with the semantic features. Cross-language prediction is done.	An average rank of 1.32 is obtained for fine-tuned and 2.15 for the re-trained model.	Advert law object is used and only text classification tasks are performed and the flaw detection has to be improved.
[30]	Real or fake news dataset with harmful and non-harmful categories in each type is considered for prediction.	The BERT model along with text sentiment analysis is employed.	70% of the accuracy value and 72.8% of the F1 score value is obtained.	Dataset size should be increased and only identifying harmful news is more focused than identifying fake news.

(Continued)

TABLE 16.1 (Continued)
Literature Study on Fake News Detection

Ref No.	Dataset Used	Model Used	Significant Outcome	Limitations
[31]	ISOT, a Fake news detection dataset is used.	CNN model is used for the detection of fake data. A cooperative technique is adopted where the user feedback is obtained and fed to the CNN model for predicting the ranking of news articles.	98% accuracy is achieved using the CNN model.	The body content and the usage of other special symbols and characters in the news articles are not considered.
[32]	A fake news detection dataset is used.	A genetic algorithm for selecting the neural network along with preprocessing techniques is used.	An accuracy score of 74.2% is achieved and 89% of the detection rate is obtained.	A hybrid deep learning model is not employed for improving the performance.
[33]	Four datasets are used. ISOT FN, Fake news detection data, and real or fake data.	CNN, ResNet, and Bidirectional LSTM models are used.	The bidirectional LSTM model along with Glove or fastText word embedding techniques provide good results than the other models. 94% to 98% of accuracy is obtained.	The number of records in two datasets is less. Data preprocessing is not done in all the datasets.
[34]	Fake News (FNs) and Fake News Challenge Datasets (FNC) from Kaggle are used.	CNN, concatenated CNN, gated recurrent unit (GRU), and LSTM models are used for comparative analysis.	A good accuracy of 99.6% is attained for concatenated CNN when compared with other models.	The text preprocessing and encoding techniques can be improved.
[35]	Twitter, Weibo, Politi, and Gossip datasets are used.	Bidirectional cross-model fusion (BCMF) is used.	Embedding the input data, a fusion of text and image data processing is carried out for prediction.	85% to 92% of accuracy is obtained for four different datasets. BCMF model performs better than the other.
[36]	Twitter and Weibo datasets are used to discover fake news.	BERT domain adaptation neural network (BDANN) is used for extracting the textual features, the BERT model is used and for images, the visual geometry group (VGG-19) model is used for training.	83% accuracy is achieved for the BDANN model for Twitter data and 84% accuracy for Weibo data.	Post-processing techniques are not employed for better performance.

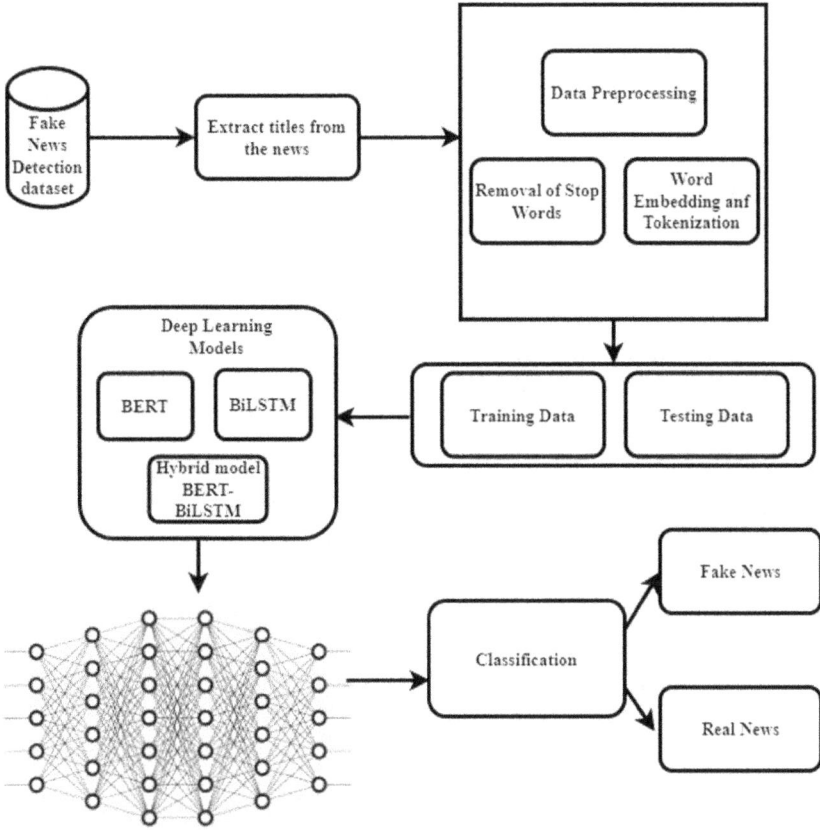

FIGURE 16.2 Architecture diagram of hybrid model.

TABLE 16.2
Dataset Categories

	News	Politics
Fake	9051	14,452
Real	10,145	11,273

16.3.1 DATASET DESCRIPTION

The dataset with fake and real categories is collected, which contains data related to various world news and politics. The entire number of records is 44,921 and used for training and testing. In the world news subset of data, 9051 records are labeled as fake data, and 10,145 records are labeled as real data. For politics subset data, 14,452 records are labeled as fake, and 11,273 records are labeled as real. 80% is taken to train data and 20% is to test data. News titles are extracted from the dataset and the dataset contains fields like news title, text, subject, and the target variable class (0 or 1). The different categories of data considered are shown in Table 16.2.

16.3.2 Data Preprocessing

Data cleaning and preprocessing is an important step that has a major impact on the model performance. Preprocessing is done and the steps for the preprocessing task are mentioned below.

1. Removal of stop words and other special characters.
2. In the LSTM model, word embedding is used for preprocessing, which assigns a unique index value to every word. Word embedding is represented as a d-dimensional vector for every index. The maximum embedding dimension value is set as 40. The vector space is generated by combining the words with similar semantic contexts together.
3. In the BERT model, the Bert tokenizer is used for generating the token vectors and the input masks fed to train the data and to perform classification.
4. In the BERT LSTM hybrid model, the BERT tokenization result is provided as input to the LSTM layer.

16.3.3 Stacked Bidirectional LSTM Model

The LSTM model is capable of learning the sequence of words present in the sentences of the dataset. The dataset consists of fields like title, text, subject, and class variables with 0 or 1 values. The total number of records is nearly 45,000. 80% is used for training and 20% for testing, where nearly 35,000 records are used for training and 10,000 records for testing. The sentences are passed to the tokenizer where the index for the words is generated and the total number of words is found as 1, 43,756.

The index value is generated for each title for further processing and is represented in Table 16.3. The layered LSTM is depicted in Figure 16.3.

The LSTM model retains the long-term dependencies in the dataset. LSTM model can acquire data and sequence of words from the sentences. A stacked bidirectional LSTM model is used and in each LSTM layer, the order of the word from sentences is preserved with the help of a memory gate. LSTM model basically has input, memory, forget, and output gates. The memory gate is used to retain the context of the words and preserved them in the same order from all the sentences. The first layer receives the embedded vector as the input and is fed to the next BiLSTM layer followed by two layers of dropout and dense layers. Drop out value is set to 0.2, which helps to remove the neurons, which do not have an impact on the training. The number of trainable and non-trainable parameters is depicted below in Table 16.4.

The embedded vectors are generated from the news titles and fed to the stacked BiLSTM layer with dropout and dense layers.

The value of epochs is given as 3 with 1094 steps in each epoch. Model validation is done on 10% of test data and a validation accuracy value of 99.72% is obtained with a loss value of 0.01. The training accuracy value is 99.2% with 0.043 as a loss value.

TABLE 16.3
Data Transformation Step

Index	Title	Text	Class
0	Donald Trump sends out embarrassing new year.	Entire body content	0
1	Drunk bragging Trump.	Whole text of the news title	0

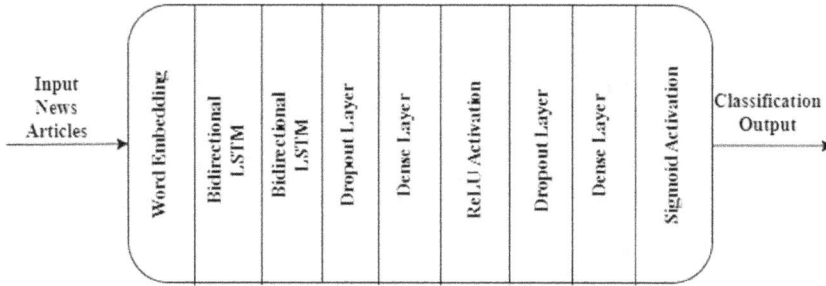

FIGURE 16.3 LSTM architectural layers.

TABLE 16.4
Trainable and Non-Trainable Parameters of the BiLSTM Model

Layers	No. of Parameters
Embedding layer	400000
Bidirectional layer	18688
Bidirectional_1 layer	24831
dropout	0
dense	1560
dropout_1	0
dense_1	25
Total parameters: 445,105	
Trainable parameters: 445,105	
Non-trainable parameters: 0	
None	

16.3.4 BERT MODEL

BERT is an attention model, which is widely used in numerous natural language processing tasks. Data is trained using the bidirectional transformer where the training of data takes place from either left to right or training from right side to left side and vice-versa. The context of the data is analyzed in a deeper sense using BERT. The transformer encoder in BERT can read the whole sequence of words at the same time simultaneously. The context is retained and relevant to all the neighboring words in the dataset. The sentence-level classification is done using the BERT model. There are two types of BERT models used. One is the BERT- Base model with the stacked layers consisting of 12 encoders, hidden units as 768, and attention heads as 12 with 110M parameters. The other one is the BERT large model with 24 encoders with an increased number of attention heads and hidden units. BERT tokenizer generates the vector, which is fed as input to the other model.

Figure 16.4 displays the frequency of real data occurrence in both subsets. Figure 16.5 depicts the fake data count in all the subsets of data. Figures 16.6 and 16.7 show the real and fake data word cloud maps, respectively. The frequently occurring word in the text content is displayed in the word cloud map. It does not consider the commonly occurring words like "the", "for", "an", "a" etc.

The architectural layers of BERT model is denoted in Figure 16.8.

In the BERT model, the input data is read by the encoder and the prediction is done at the decoder part. The pretrained model is used for converting data into a format appropriate for training. The attention mechanism and input masks are fed into the model. The hidden state is used to obtain the full sentence and there are two input layers followed by the BERT layer with the base model and

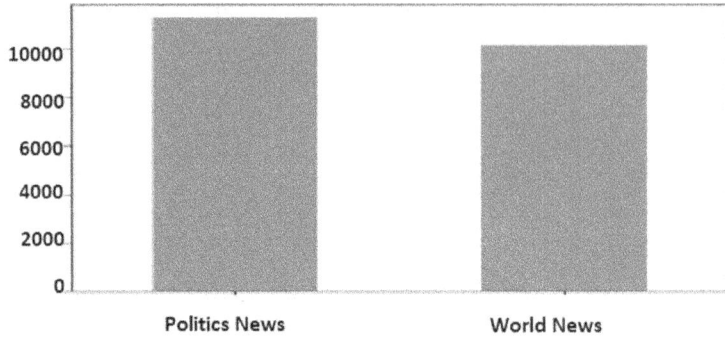

FIGURE 16.4   Real data frequency.

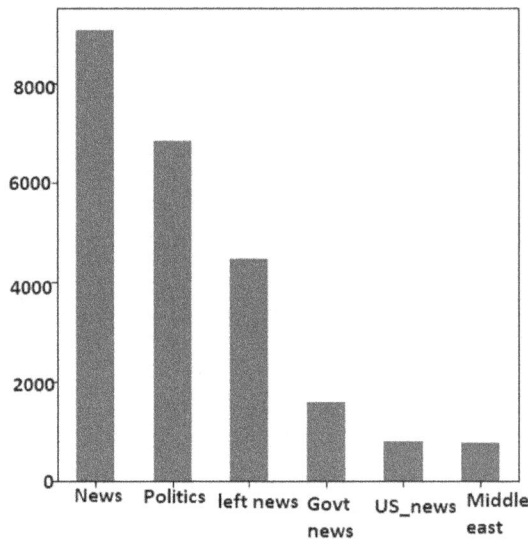

FIGURE 16.5   Fake data frequency.

cross attentions, dropout, and dense layers. The trainable and non-trainable parameters count and the training at different layers of the BERT model are represented in Table 16.5.

Epoch's value was set as 3 with 957 steps in each epoch. The BERT model achieved 99.8% of training accuracy with a loss value of 0.040 and validation accuracy of 99% with a loss value of 0.004.

16.3.5   HYBRID MODEL (BERT-BiLSTM)

The classification accuracy can be enhanced by developing a hybrid model of BiLSTM and BERT. The architecture diagram of the proposed hybrid model is represented in Figure 16.9.

The input sentences are preprocessed and the preprocessed data is fed into the BERT tokenizer, which generates the input mask values and input ids. The attention mask and ids are fed into the BERT model and embedded vectors are generated. The hidden size of each token is 768 since the base model of BERT is considered for training. The hidden state value is fed to the BiLSTM model followed by the batch normalization layer. The LSTM layer retains the contextual data from all the

FIGURE 16.6 Real-data word cloud map.

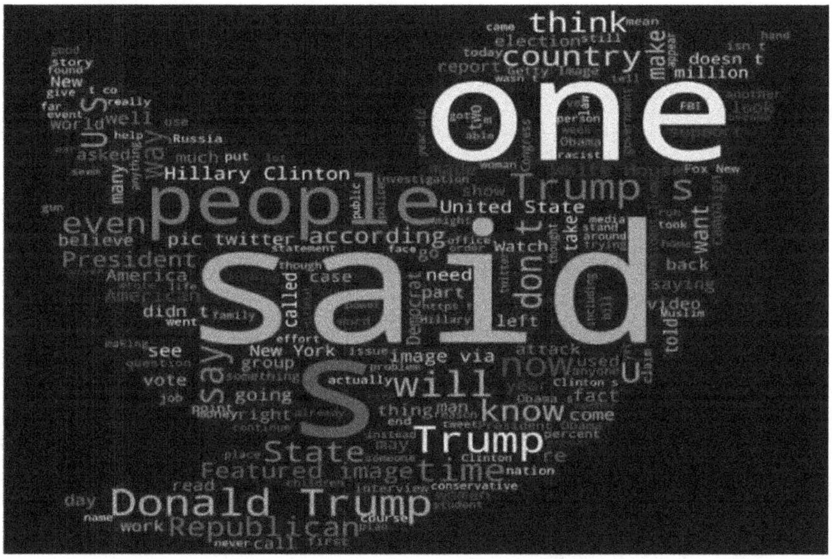

FIGURE 16.7 Fake-data word cloud map.

sentences. The semantics of the data is retained by the LSTM model for better classification. The standard deviation, mean of activation values are determined by the batch normalization layer. For standardizing the inputs and for faster training, a batch normalization layer is used. In the hybrid model, the dropout value is chosen as 0.5, and randomly the network architecture can be modified. The output size of the feed-forward layer is chosen as 2, which detect the test value as either fake or real. The BERT-BiLSTM model achieved 99.97% of training accuracy with a loss value of 0.030 and validation accuracy of 99.9% with a loss value of 0.003.

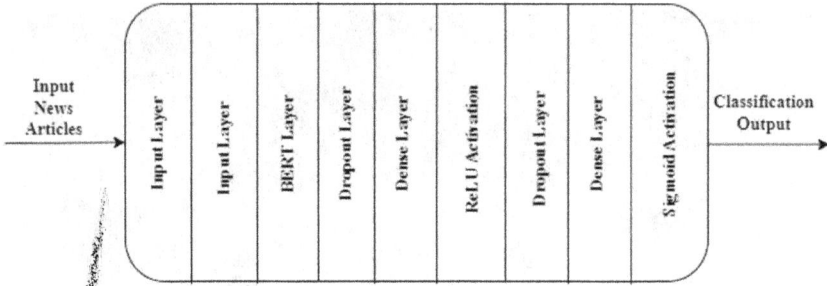

FIGURE 16.8   BERT architectural layers.

TABLE 16.5
Trainable and Non-Trainable Parameters of the BERT Model

Type of Layer	Parameters	Connected state
Input layer 1	0	[]
Input layer 2	0	[]
Bert layer	109472240	
dropout_1 layer	0	bert_model
Dense_1 layer	49216	dropout_1
dropout_2 layer	0	Dense_1
dense_2 layer	65	dropout_2

Total count of parameters: 109,531,521
Count of trainable parameters: 109,531,521
Count of non-trainable parameters: 0

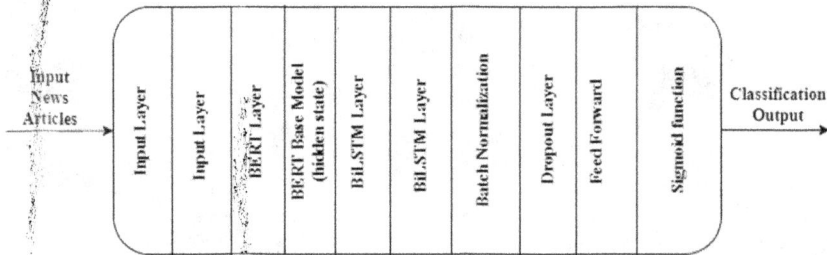

FIGURE 16.9   Architectural layers of BERT-BiLSTM hybrid model.

16.3.6 POPULARITY PREDICTION OF NEWS ARTICLES

The popularity of news articles is predicted using a regression model. The probabilities are computed by extracting the sentiment of the contextual content. The different classes like positive, negative, and neutral are predicted using a multi-linear regression model. There can be a case, where any new article can be generated and can be spread over the Internet. There are various factors to be considered such as computing the time variance among the release period of the article and the very first tweet for an article. The re-tweets after a specified duration of time can be computed and the number of followers during the particular time period can be considered for popularity prediction. The architecture diagram of the popularity prediction is shown in Figure 16.10.

Figure 16.10 indicates finding the popularity of articles, where the user comments are taken and passed to linear regression. The sentimental analysis is used along with this and a sample output is used for finalizing the articles having positive, negative, or neutral comments. The final output of linear regression shows the lifespan of the article.

16.3.6.1 Polarity

TextBlob package is used to perform text processing tasks with an interface. Polarity and subjectivity are the two categories of outputs obtained from text blob. The value of the polarity is a float value ranging between −1.0 to +1.0 where the neutral class is indicated as 0.0, a positive class with positive sentiment values is indicated as +1.0 and the negative class with negative sentiment values is indicated as −1.0.

Polarity mapping for the popularity prediction details with subjectivity and objectivity. Subjectivity expresses the various subjects from the dataset like emotions, personal feelings, different views and beliefs, users' opinions, etc. whereas objectivity describes only the factual information. The values choice is from [0.0 to 1.0], 0.0 indicates the sentences in the dataset are very subjective and the value of 1.0 indicates the data is very objective.

16.4 PERFORMANCE EVALUATION

In order to assess the proposed model, different metrics like accuracy, precision, recall, and F1 score are used. These metrics are evaluated based on the confusion matrix table with actual and predicted values obtained by the classification task. Table 16.6 shows the confusion matrix table used for computation.

16.4.1 ACCURACY (A)

The accuracy measure is denoted as the ratio between the correct classifications of news type to the entire number of data points.

$$Accuracy(A) = \frac{TP + TN}{TP + TN + FP + FN}$$

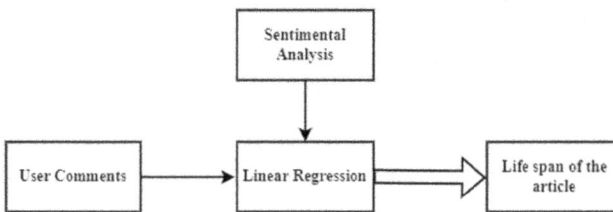

FIGURE 16.10 Architecture diagram for popularity prediction.

TABLE 16.6
Confusion Matrix Table

	Positive values – predicted	Negative values – predicted
Positive values – actual	True Positive – TP	False Negative – FN
Negative values – actual	False Positive – FP	True Negative – TN

16.4.2 Precision (P)

Precision is denoted as the ratio between the number of positive predictions made to the total number of positive data points.

$$Precision(P) = \frac{TP}{TP + FP}$$

16.4.3 Recall (R)

The recall is denoted as the ratio between the positive value counts to the sum of true positive values and false negative values.

$$Recall(R) = \frac{TP}{TP + FN}$$

16.4.4 F1 Score (F1)

The F1 score is denoted as the balance or harmonic mean between the recall and precision values.

$$F1\ score(F1) = \frac{2*(Precision*Recall)}{Precision + Recall} = \frac{2*TP}{2*TP + FP + FN}$$

16.5 RESULTS AND ANALYSIS

For the proposed study, the fake news detection dataset is used with real and fake records. The total count of different categories of datasets is mentioned in Table 16.7.

16.5.1 BiLSTM Model

The fake news dataset trained with the BiLSTM model includes stacked layers with dropout and dense layers. Figures 16.11 and 16.12 show the accuracy of the model and the loss values, respectively. The model achieves a validation accuracy of 99.72% with a loss value of 0.01. The training accuracy value is 99.2% with 0.043 as a loss value.

16.5.2 BERT Model

The BERT model, which is a transformer-based encoder decoder model, performs better than the BiLSTM. BERT tokenization process and training at BERT layer by passing the attention mask and

---

TABLE 16.7
News Articles Count in the Dataset

	News	Politics
Fake	9051	14,452
Real	10,145	11,273

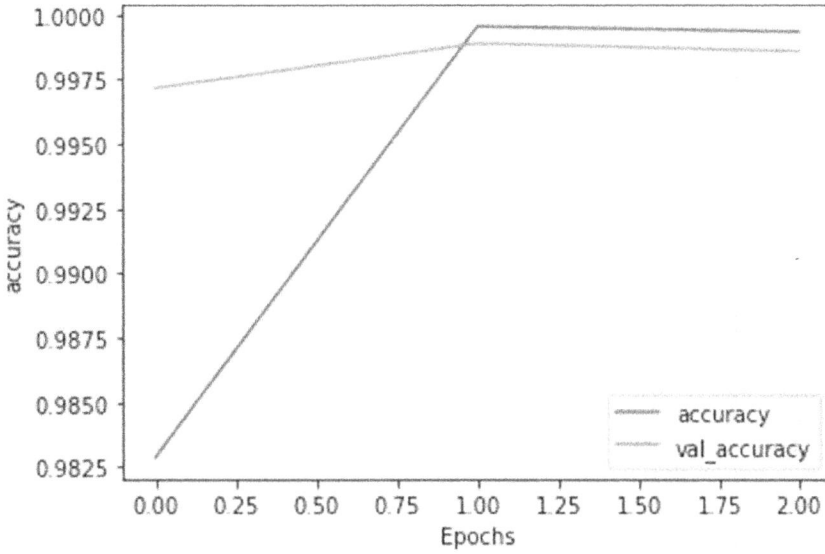

FIGURE 16.11 Accuracy graph of BiLSTM.

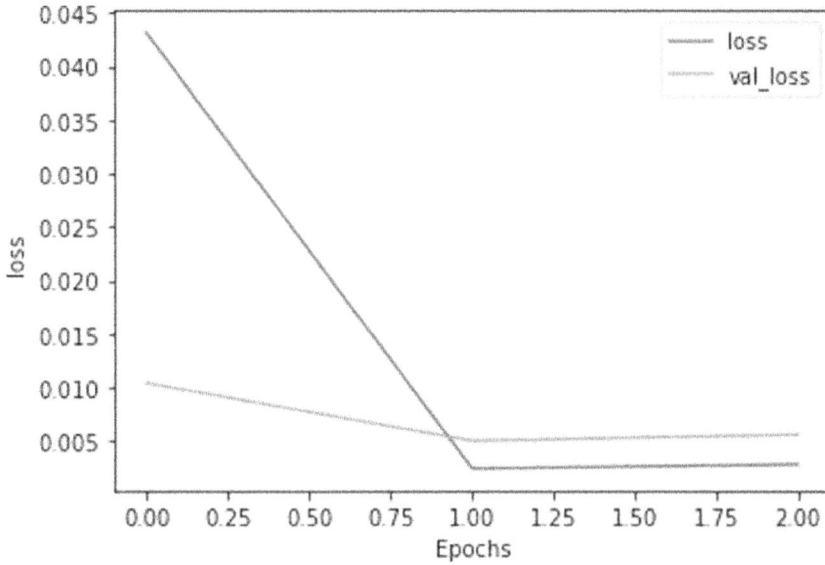

FIGURE 16.12 Loss graph of BiLSTM.

input ids improves the performance. Figure 16.13 shows the accuracy plot of BERT and Figure 16.14 depicts loss graph of the model. BERT model achieved 99.7% of training accuracy with loss value of 0.040 and validation accuracy of 99.8% with loss value of 0.004. The generation of vector embedding and the input mask is fed to the classifier for training.

16.5.3 HYBRID BERT-BiLSTM MODEL

The novel hybrid model shows a slight improvement over the performance of the other two models. Figures 16.15 and 16.16 display the accuracy plot and loss graph of the hybrid model. BERT model

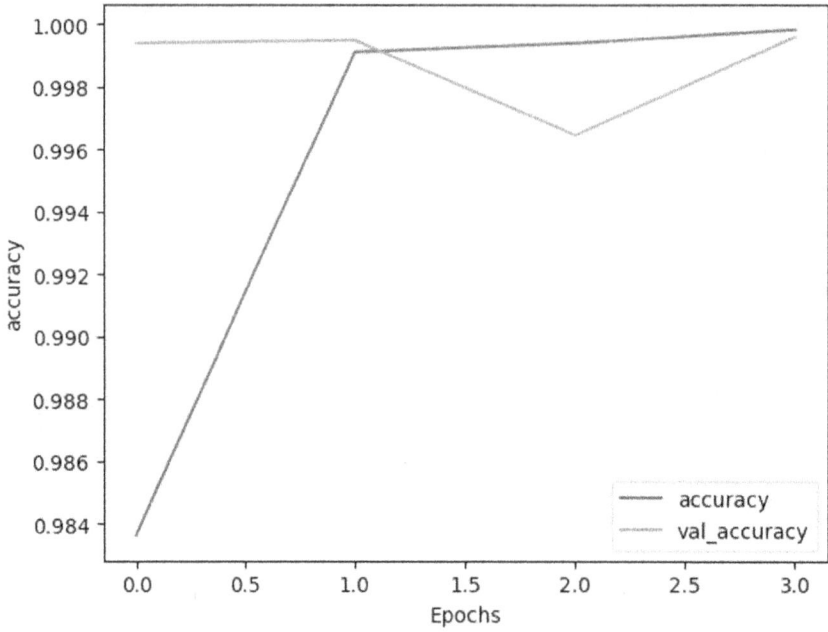

FIGURE 16.13 Accuracy plot of BERT model.

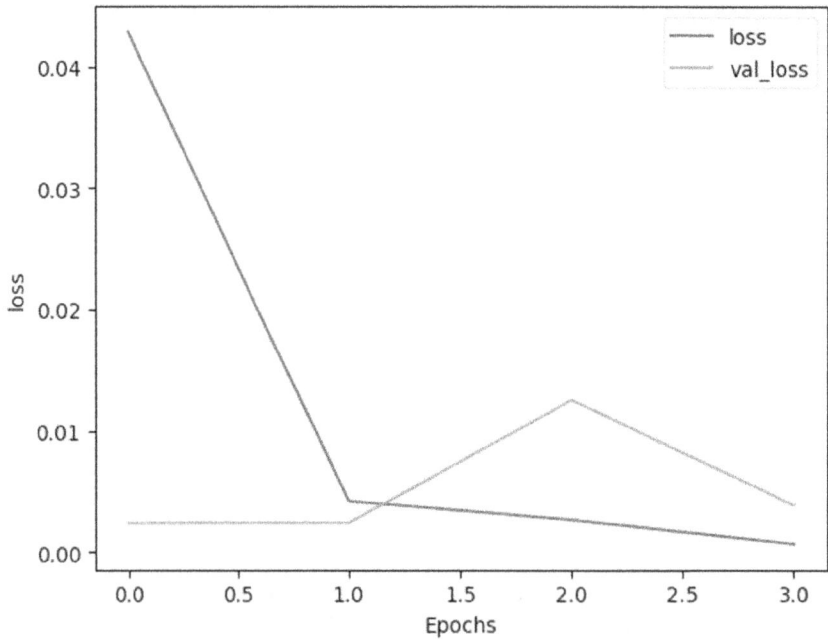

FIGURE 16.14 Loss graph of BERT.

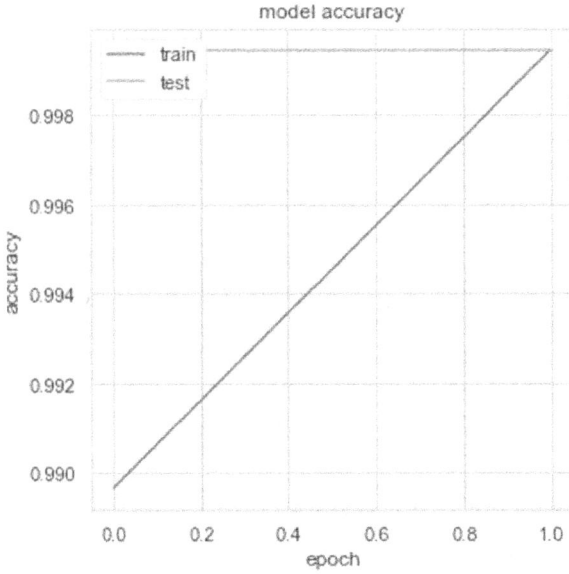

FIGURE 16.15   Accuracy diagram of BERT-BiLSTM model.

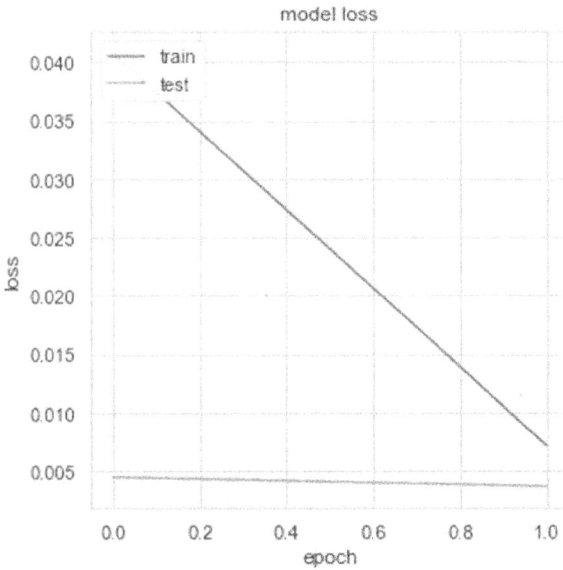

FIGURE 16.16   Loss graph for BERT-BiLSTM.

tokenization output is given to the BiLSTM layers, which capture the semantics of the sentences with long-distance relationships. Hence, the hybrid model is enriched with 99.97% of training accuracy with loss value of 0.030 and validation accuracy of 99.9% with loss value of 0.003. The false positive and true negative values obtained by confusion matrix are very less.

The output size of final feed forward layer in the classification is set as 2. The inclusion of batch normalization layer makes training process to converge faster and improves the training efficiency. Dropout layer value set as 0.5 is used for changing the network randomly.

16.5.4 COMPARATIVE ANALYSIS OF MODEL PERFORMANCE

Table 16.8 shows the comparative analysis of results obtained for various models used in the study.

16.5.5 POPULARITY PREDICTION OF NEWS ARTICLES

Figure 16.17 shows the sentiment analysis and the popularity prediction of articles based on the news title. The different classes identified are the positive class, negative class, and neutral class. A multi-linear regression model is used for finding the correlation between the sentimental features. The probability values are predicted using the polarity method, which separates the contextual and the semantic features. The temporal features in the data can also be extracted to find out the correlation between the news article publishing time and the views at specified interval periods. The very important characteristic of online news is polarity prediction since the mining of contextual text and extraction of subjective and objective information is a relatively important task for popularity prediction.

The news titles are extracted from the data and a multiple linear regression model is applied which captures the relationship between multiple features by generating a feature matrix and the response vector. The lifespan of the article can be predicted based on user comments and tweets. Figure 16.18 shows the output of linear regression analysis for the popularity prediction of news articles with news headlines or news titles. The evaluation index used are mean squared error (MSE), mean absolute error (MAE), root mean squared error (RMSE), and R^2 value.

TABLE 16.8
Performance Analysis of Deep Learning Models

Model	Training Accuracy	Model Loss Value	Precision	Recall	F1 Score
BiLSTM	99.2%	0.043	99.17%	99.18%	99.1%
BERT	99.7%	0.040	99.64%	99.61%	99.6%
Proposed BERT-BiLSTM (word embedding + fine-tuning parameters)	99.97%	0.030	99.96%	99.95%	99.93%

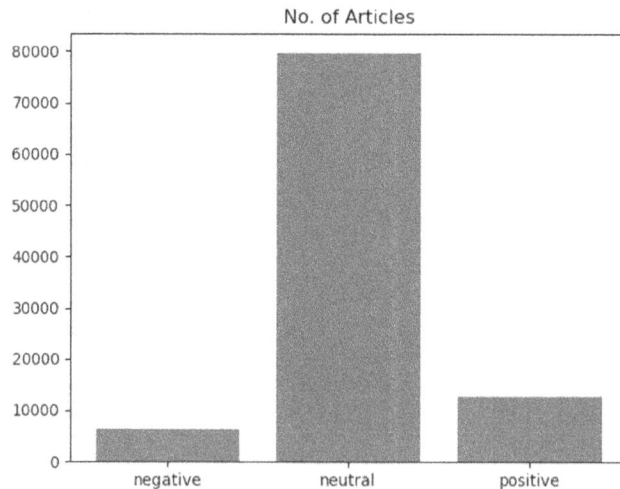

FIGURE 16.17 Bar graph for sentiment analysis of news headlines.

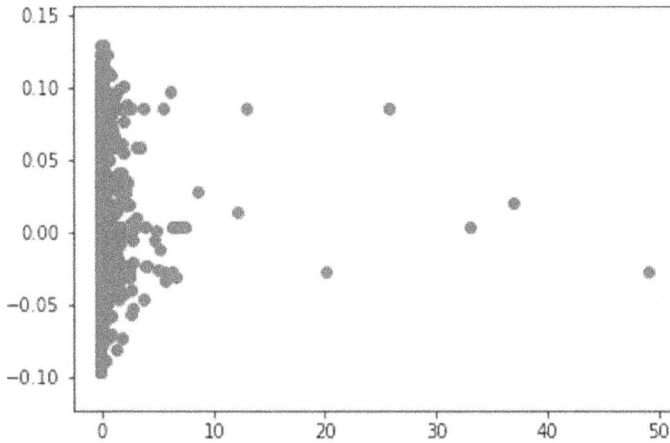

FIGURE 16.18 Output for popularity prediction.

TABLE 16.9
Analysis of Popularity Prediction of Articles

Mean Squared Error	Mean Absolute Error	Root Mean Square Error	R-Squared
0.02	0.014	0.029	0.921

Table 16.9 shows the experimental results obtained for the prediction of popularity for news articles.

16.6 CONCLUSION AND FUTURE SCOPE

A relative analysis of various models is done for identifying fake news. Fake news datasets with nearly 45,000 records are trained using the BiLSTM model, BERT model, and the novel hybrid BERT-BiLSTM model. Compared to BiLSTM and BERT models, a slight improvement in classification accuracy is achieved using the hybrid model. Since there are a lot of online networks and people rely on the data being posted on social media, fake news detection is important in the current era. BERT model is used to identify the textual context using the BERT tokenization process and LSTM is used for maintaining the semantics in the sentence for a longer period of time. The BERT-BiLSTM model achieved 99.97% of training accuracy and 99.8% of validation accuracy. It performs better compared to the single models.

Although deep learning models provide good accuracy and can extend with new features, there are still many challenges involved in detecting fake news. Since the models are trained with the past history of data, the new information arrives only in the evolving events; there will be fewer data or posts for evolving events. It is necessary to build a model, which can detect and identify misinformation with a limited volume of data. The textual context, writing style, various classes, and vocabulary may differ for each type of article. It is also a challenging task to perform multi-lingual detection. The future scope is to build a model and to train the datasets of different languages and to perceive fake news in texts. Also, popularity prediction of news article study can be extended using the LSTM model where the word embedding feature extraction technique can be used to improve the prediction based on the user comments.

REFERENCES

1. Abdulrahman A. and Baykara M. Fake news detection using machine learning and deep learning algorithms. International Conference on Advanced Science and Engineering (ICOASE), Duhok, Iraq, pp. 18–23, doi: 10.1109/ICOASE51841.2020.9436605, 2020.
2. Hu L., Wei S., Zhao Z. and Wu B. Deep learning for fake news detection: A comprehensive survey. AI Open, 2022, vol. 3, pp.. 133–155.doi. https://doi.org/10.1016/j.aiopen.2022.09.001
3. Alghamdi J., Lin Y. and Luo S. A comparative study of machine learning and deep learning techniques for fake news detection. *Information*, vol. 13, no. 12, p. 576. https://doi.org/10.3390/info13120 576, 2022.
4. Barbara P., Piotr S. and Jan K. Rapid detection of fake news based on machine learning methods. *Procedia Computer Science*, vol. 192, pp. 2893–2902, 2021.
5. Zhang D., Li W., Niu B. and Wu C. A deep learning approach for detecting fake reviewers: Exploiting reviewing behavior and textual information. *Decision Support Systems*, vol. 166, p. 113911, 2023.
6. Jiang T., Li J. P., Haq A. U. and Saboor A. Fake news detection using deep recurrent neural networks. 2020 17th International Computer Conference on Wavelet Active Media Technology and Information Processing (ICCWAMTIP), Chengdu, China, pp. 205–208, doi: 10.1109/ICCWAMTIP51612.2020.9317325, 2020.
7. Chi H. and Liao B. A quantitative argumentation-based Automated eXplainable Decision System for fake news detection on social media. *Knowledge-Based Systems*, vol. 242, p. 108378, 2022.
8. Santhosh Kumar A., Kalpana P., Sharan Athithya K. and Sri Ajay Sundar V. Fake news detection on social media using machine learning, *Journal of Physics: Conference Series*, vol. 1916, International Conference on Computing, Communication, Electrical and Biomedical Systems (ICCCEBS), 2021. doi: 10.1088/1742-6596/1916/1/012235
9. Seetharaman R., Tharun M., Sreeja Mole S. S. and Anandan K. Analysis of fake news detection using machine learning. *Materials Today: Proceedings*, vol. 51, Part 8, pp. 2218–2223, 2022.
10. Senthil Raja M. and Arun Raj L. Fake news detection on social networks using machine learning techniques. vol. 62, pp. 4821–4827. https://doi.org/10.1016/j.matpr.2022.03.351, 2022.
11. Choudhary A. and Arora A. Linguistic feature based learning model for fake news detection and classification. *Expert Systems with Applications*, vol. 169, p. 114171, 2021.
12. Kishwar A. and Zafar A. Fake news detection on Pakistani news using machine learning and deep learning. *Expert Systems with Applications*, vol. 211, p. 118558, 2023.
13. Das S. D., Basak, A., and Dutta, S. A heuristic-driven uncertainty based ensemble framework for fake news detection in tweets and news articles. *Neurocomputing*, vol. 491, pp. 607–620, 2022.
14. Koloski B., Perdih T. S., Robnik-Šikonja M., Pollak S. and Škrlj B. Knowledge graph informed fake news classification via heterogeneous representation ensembles. *Neurocomputing*, vol. 496, pp. 208–226, 2022.
15. Davoudi M., Moosavi M. R. and Sadreddini M. H. DSS: A hybrid deep model for fake news detection using propagation tree and stance network. *Expert Systems with Applications,* vol. 198, p. 116635, 2022.
16. Dimitrios P. K., Paraskevas K. and Christos T. Exploiting textual information for fake news detection. *International Journal of Neural Systems*, vol. 32, no. 12, p. 2250058, 2022.
17. Hartl P. and Kruschwitz U. Applying automatic text summarization for fake news detection. In Proceedings of the Thirteenth Language Resources and Evaluation Conference, pp. 2702–2713, Marseille, France. European Language Resources Association, 2022.
18. Ahmadian A., and Ciano T. Covid-19 fake news sentiment analysis. *Computers and Electrical Engineering,* vol. 101, p. 107967, 2022.
19. Lahby M., Aqil S., Yafooz W. M., and Abakarim Y. Online fake news detection using machine learning techniques: A systematic mapping study. Combating Fake News with Computational Intelligence Techniques, pp. 3–37, Springer International Publishing, 2022. doi: 10.1007/978-3-030-90087-8_1
20. G, S.K. Deep learning for fake news detection. In: *Data Science for Fake News: The Information Retrieval Series*, vol. 42, pp. 71–100. Springer, Cham. https://doi.org/10.1007/978-3-030-62696-9_4, 2021.
21. Jiang G., Liu S., Zhao Y., Sun Y. and Zhang M. Fake news detection via knowledgeable prompt learning. *Information Processing & Management,* vol. 59, no. 5, p. 103029.

22. Jiang T., Li J. P., Haq A. U., Saboor A. and Ali A. A novel stacking approach for accurate detection of fake news. in *IEEE Access*, vol. 9, pp. 22626–22639, doi: 10.1109/ACCESS.2021.3056079, 2021.

23. Jing J., Wu H., Sun J., Fang X. and Zhang H. Multimodal fake news detection via progressive fusion networks. *Information Processing & Management*, vol. 60, no. 1, p. 103120, 2023.

24. Kong S., Ye F. and Feng L. Predicting future retweet counts in a microblog. *Journal of Computational Information Systems*, vol. 10, no. 4, pp. 1393–1404, 2014.

25. Zhang C., Gupta A., Qin X. and Zhou Y. A computational approach for real-time detection of fake news. *Expert Systems with Applications*, vol. 221, p. 119656, 2023.

26. Aljawarneh S. A. and Swedat S. A. Fake news detection using enhanced BERT. in IEEE Transactions on Computational Social Systems, pp. 1–8, doi: 10.1109/TCSS.2022.3223786, 2022.

27. Bahad P., Saxena P. and Kamal R. Fake news detection using bi-directional LSTM-recurrent neural network. *Procedia Computer Science*, vol. 165, pp. 74–82, 2019.

28. Chiang T. H., Liao C. S., and Wang, W. C. Investigating the difference of fake news source credibility recognition between ANN and BERT algorithms in artificial intelligence. *Applied Sciences*, vol. 12, no. 15, p. 7725, 2022.

29. Li M., Zhou H., Hou J., Wang P. and Gao E. Is cross-linguistic advert flaw detection in Wikipedia feasible? A multilingual-BERT-based transfer learning approach. *Knowledge-Based Systems*, vol. 252, p. 109330, 2022.

30. Lin S. Y., Kung Y. C. and Leu F. Y. Predictive intelligence in harmful news identification by BERT-based ensemble learning model with text sentiment analysis. *Information Processing & Management*, vol. 59, no. 2, p. 102872, 2022.

31. Mallick C., Mishra S. and Senapati M. R. A cooperative deep learning model for fake news detection in online social networks. Journal of Ambient Intelligence Humanized Computing, vol. 14, no. 4, pp. 4451–4460. doi: 10.1007/s12652-023-04562-4, 2023.

32. Okunoye O. B. and Ibor A. E. Hybrid fake news detection technique with genetic search and deep learning. *Computers and Electrical Engineering*, vol. 103, p. 108344, 2022.

33. Sastrawan I. K., Bayupati I. P. A. and Arsa D. M. S. Detection of fake news using deep learning CNN–RNN based methods. *ICT Express*, vol. 8, no. 3, pp. 396–408, 2022.

34. Sedik A., Abohany A. A., Sallam K. M., Munasinghe K. and Medhat T. Deep fake news detection system based on fake news detection on social media using machine learning concatenated and recurrent modalities. *Expert Systems with Applications,* vol. 208, p. 117953, 2022.

35. Yu C., Ma Y., An L. and Li G. BCMF: A bidirectional cross-modal fusion model for fake news detection. *Information Processing & Management*, vol. 59, no. 5, p. 103063, 2022.

36. Zhang T., Wang D., Chen H., Zeng Z., Guo W., Miao C. and Cui L. BDANN: BERT-based domain adaptation neural network for multi-modal fake news detection. Proceedings of the 2020 International Joint Conference on Neural Networks (IJCNN), pp. 1–8. doi:10.1109/IJCNN48605.2020.9206973, 2020.

APPENDIX

1. Figure 1. Fake news, adapted from "The conversation", global edition, by Fatemeh Torabi Asr, August 14, 2019, https://theconversation.com/the-language-gives-it-away-how-an-algorithm-can-help-us-detect-fake-news-120199.

17 Internet of Things (IoT)-Based Smart Maternity Healthcare Services

P. Vinothiyalakshmi[1], V. Pallavi[2], N. Rajganesh[1], and V. Adityavignesh[1]

[1] Department of Computer Science and Engineering, Sri Venkateswara College of Engineering, India

[2] Graduate Student in Computer Science (Artificial Intelligence), USC Viterbi School of Engineering, University of Southern California, Los Angeles, California, USA

17.1 INTRODUCTION

Healthcare in rural areas is not accessible with the advanced technology equipment needed to monitor injuries and routine checkups. Particularly, women undergo many body changes like hormone fluctuation, anxiety, depression, breathlessness, heart burn, and many other problems. These problems have to be monitored in a regular basis to maintain healthy lifestyle. Studies suggest that there is high probability for the women in next generation for high-rise pregnancy due to polycystic ovarian syndrome, thyroid disorders, anxiety, depression, indulging more of fast foods, improper sleep–wake cycle and lack of physical exercise. In order to avoid these conditions, proper healthcare has to be provided to the general public. Research shows that 1 in 10 women of child bearing age suffer from polycystic ovarian syndrome. Regular time-to-time checkups will help in reducing congenital disorders and fetal mortality rate. The overall length of the pregnancy ranges from 37–42 weeks. During this period, there is a possibility of various complications due to preeclampsia, gestational diabetes, depression and anxiety, high blood pressure, and pregnancy loss/miscarriage. Baby can gain more weight due to high sugar level of the mother and high blood pressure can result in gestational hypertension. If the baby's weight is less than the standard weight, then there are more possibilities for the occurrence of preterm baby. All these complications can only be prevented and cured by having advanced medical equipment. Traveling a long distance during the time of pregnancy is not advised by a physician. So, it would be a great advantage for them if there is an accessibility of monitoring from home or having regular healthcare checkups in nearby known primary healthcare centers and then communicating with the best specialist in the world via the internet. Various sensors are available to measure heart beat rate, body temperature, respiration, SPO2, ECG, which is transferred to the specialist at a distant hospital. The specialist interprets the data they received and decides whether the patient's condition is normal or not.

17.2 LITERATURE SURVEY

In this section, the review of existing works is done on the concept of various healthcare services is as follows: Fox et al. [1] proposed a framework to demonstrate the time and effort of labor and delivery of midwifes and nurses who get involved in monitoring the heartbeat of the fetus by means of external monitoring. Sharma et al. [2] proposes efficient fetal state health monitoring using novel enhanced binary bat algorithm, which addresses various issues of hypnoxic status of the fetuses by

DOI: 10.1201/9781003407959-17

using random forest classifier. Watson et al. [3] gathers insight about the experience of the midwifes and the doctors in using telemetry for monitoring and measuring the recordings. Sarhaddi et al. [4] presented Internet-of-Things (IoT)-based technology in providing ubiquitary maternal health monitoring during the phase of pregnancy. They integrated the presented model with the current healthcare system. Nafissa et al. [5] proposed an approach in which fetal ECG signal (fECG) is extracted and then R peaks are detected to deduce FHR. Tahmina et al. [6] demonstrated a clinical study between the fetal stimulation in the evaluation of fetal well-being at the time of labor and the no stimulation among women with a singleton pregnancy. Frenken et al. [7] presents a retrospective study on the association between the uterine contraction frequency and the fetal scalp pH in women. Wang et al. [8] describe a solution to all the challenges being faced in today's medical industry for ultrasound fetal head edge detection. Christoph et al. [9] address various issues such as constitutionally fetus with small size, congenital infections, various chromosomal abnormalities, or any genetic conditions. Ashley et al. [10] hypothesized a sheep model of hypoglycemia for monitoring maternal undernutrition deficiency. Benjamin et al. [11] proposed that the findings support investigation into an indirect method of effect relevant to the pyrethroids on fetal development in mammals. Kamel et al. [12] presented the correlation between transabdominal and transperineal assessment of fetal head position, and to research the shape of fetus at different stages of pregnancy and head positions. Kumar et al. [13] described a project that simplifies the monitoring of the pregnant women's fetal condition by using various deep learning models. Hema et al. [14] proposed a framework, which detects the fetal heartrate by the instrument called as fetal digital stethoscope sensor with the microcontrollers. Daniel et al. [15] demonstrated a transabdominal fetal oximetry (TFO) system that ensures in utero metrics of fetal SpO2. Radek et al. [16] focused on the three types of noise (ambient noise, Gaussian noise, and movement artifacts of the mother and the fetus) having different amplitudes for filtering of an fPCG signal. Castaldo et al. [17] reviewed the existing literature investigating, in healthy subjects, the associations between acute mental stress and short-term heart rate variability (HRV) measures in time, frequency, and non-linear domain to provide the reliable information about the trends and the pivot values of HRV measures during mental stress. Mehrabadi et al. [18] evaluated the sleep parameters of the Oura ring along with the Samsung Gear Sport watch in comparison with a medically approved actigraphy device in a midterm everyday setting, where users engage in their daily routines. Shaffer et al. [19] reviewed the mechanisms that generate 24 hours, short-term, and ultra-short-term HRV, the importance of HRV, and its implications for health and performance. Malik et al. [20] focused on the oscillation in the interval between consecutive heart beats as well as the oscillations between consecutive instantaneous heart rates. Pecchia et al. [21] reviewed the various rigorous methods to assess the validity of ultra-short HRV features in a control situation and identified the reliable ultra-short HRV features. Choi et al. [22] analyzed the variability in the time and frequency domain from pulse rate obtained through PPG with the electrocardiogram using sampling frequency. Amiri et al. [23] presented the Internet-of-Things (IoT)-based system to provide ubiquitous maternal health monitoring during pregnancy and postpartum. Gopalan et al. [24] provided a comprehensive survey of recent energy-efficient medium access control (MAC) protocols for wireless body area networks (WBANs) and discussed about the comparison of the various approaches pursued. Aijaz et al. [25] focused the emerging standardization efforts and the latest developments on protocols for cognitive M2M networks and its communications from a protocol stack perspective. Anzanpour et al. [26] presented a remote monitoring and diagnostic system in a holistic perspective of patients and their health conditions. They explained the concept of self-awareness used in various parts of the system such as information collection through wearable sensors, confidence assessment of the sensory data, the knowledge base of the patient's health situation, and automation of reasoning about the health situation. Sasai et al. [27] proposed an algorithm for heartbeat interval error compensation and incorporated a low-noise readout circuit to improve the signal-to-noise ratio (SNR). Clawson et al. [28] conducted the iterative inductive and deductive analyses of personal health-tracking technologies and identified the health motivations and rationales. Fantinelli et al. [29] discussed about the telemedicine for gestational diabetes mellitus (GDM) by considering the role of psychological dimensions such as empowerment/self-efficacy, engagement and satisfaction. Lau et al. [30] discussed the best available evidence to evaluate the efficacy of internet-based self-monitoring interventions in improving maternal and neonatal outcomes among perinatal diabetic women.

17.3 MATERNAL AND FETAL MONITORING

Maternal and fetal monitors bring forth details about fetal heart rate, uterine contractions, maternal electro cardiogram, and maternal blood pressure levels. Ultrasound transducer technology is the technique used to record the condition of the fetus in the prenatal stage. A ray of light hits the maternal abdomen recording the fetal conditions in ultrasound transducer technique. Motion in the heart will result in the Doppler shift. Difference in the frequency of the light beam detects the heart rate. Uterine activity is determined in the same way. Both heart rate and uterine contractions help in a complete understanding of the fetus.

17.3.1 TYPES OF MATERNAL MONITORING SERVICES

With the help of IOT-based devices, long-term maternity care services can be monitored. A few types are as follows:

A. SLEEP MONITORING

The third trimester is an ideal period in which pregnant women experience various sleep disorders. Anxiety, hormonal fluctuations, risk of preterm birth, gestational hyperglycemia, and various mood disorders are some of the conditions, which are associated with the sleep disorders. Generally, average sleep time decreases slowly during pregnancy, but there are several conditions where severe sleep disorders will cause early postpartum period. Proper sleep–wake cycle has to be monitored regularly for a pregnant woman to avoid sleep complications. Individualized sleep information will give us knowledge about the circadian rhythm, which in turn helps in maintaining the healthy lifestyle.

B. PHYSICAL ACTIVITY MONITORING

Physical activity is indispensable for the wellbeing of healthy human beings. Various complications caused due to high blood pressure, gestational diabetes, preeclampsia, polycystic ovarian syndrome, and depression [31] can be prevented by adopting physical activity in lifestyle. Observational research shows that women who exercise regularly have many advantages such as a decreased number of caesarean births and postpartum recovery time. Women who experience postpartum depression after pregnancy also find physical activity to be a predominant factor beneficial to them.

C. STRESS MONITORING

Due to various hormonal changes, women generally experience high level of stress and anxiety. When it comes to pregnant women, stress levels will affect the health of the fetus as well. Depression is the main outcome resulted when stress levels are not managed. A high level of unmanaged stress level will result in complicated pregnancy disorders such as hypertensive disorders and preterm birth. Advance research indicates that unmanaged stress levels during pregnancy will have a long-term impact on the fetus and the pregnant women. So, stress monitoring should be given an importance during maternity care. Practicing yoga, meditation, listening to favorite music, exercise, going on a selfcare spa, seeking the help of the therapy by professionals when needed are some of the techniques involved in stress management.

17.4 IOT-BASED MATERNAL HEALTH MONITORING SYSTEM USING CLOUD COMPUTING

This section includes IoT-based system designed for smart maternity healthcare services using cloud computing technology used during pregnancy and postpartum. The presented architecture is depicted in Figure 17.1. This system allows monitoring the maternal healthcare continuously and integrates both internal and external health conditions of the mothers, thus collecting data from the mothers providing the comprehensive view of maternal and fetal health. The monitoring consists (Figure 17.2) of four layers: perception layer, gateway layer, cloud layer,

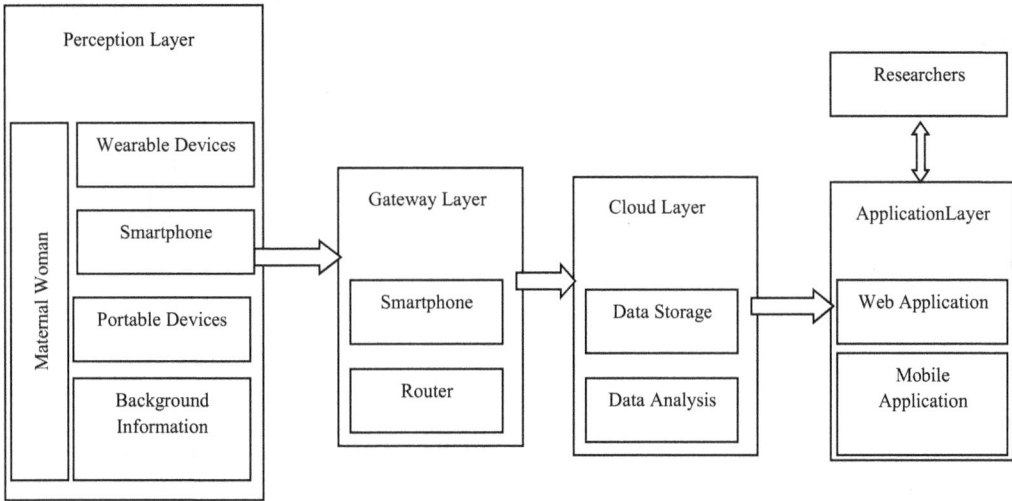

FIGURE 17.1 IoT-based maternal health monitoring system using cloud computing.

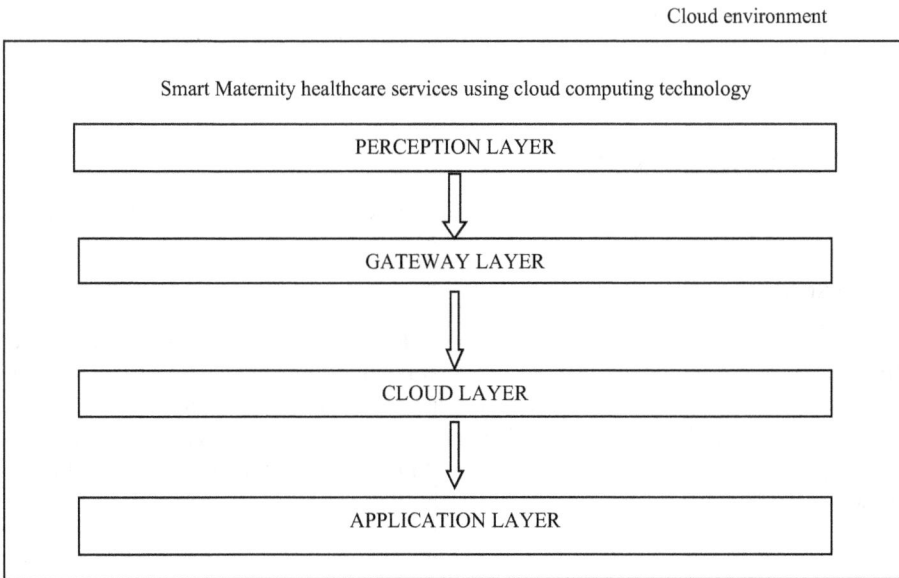

FIGURE 17.2 Layers of the monitoring system.

and application layer. Perception layer collects the physiological data from the pregnant women, data which has been collected are sent to the cloud layer through routers from the gateway layer. Cloud layer mainly focuses on storing and analyzing the fetched data for visualization purpose. Either in the form of web application or mobile application, the application layer visualizes the data monitored.

A. PERCEPTION LAYER

Perception layer is the first layer of monitoring in which the physical condition of women is administered using various types of sensors, smartphones, portable devices, and background monitoring system. Photoplethysmogram, accelerometer, and gyroscope are some of the sensors, which provide information about maternal health conditions. Wearable devices, which are worn by the pregnant women have to be small, free from attracting harmful rays, and easy to be wear with enhanced levels of scalability and feasibility. Day-to-day physical activity can be monitored using smartphones.

B. GATEWAY LAYER

Gateway layer interconnects both perception and the cloud layer. All the recorded physiological information from the perception layer is sent to the cloud layer for storing and accessing through various routers or smart devices. Data compression and data mining are a few functions being performed in advanced gateway layers.

C. CLOUD LAYER

Through the gateway layer, the cloud layer receives the recorded information from the perception layer. It facilitates reliable centralized storage to store a large volume of data, which could be accessed from anywhere at any time. It even forms various data science techniques to keep up with the trends and the anomalies of the collected data. A cloud server is used as a server part in a client–server prototype to demonstrate the data for the end users.

D. APPLICATION LAYER

All the data stored in the cloud are integrated and visualized to the end users either in the form of web application or mobile application. This layer acts as a dashboard where health professionals and mothers communicate in a real-time environment. Application layer is now used for self-monitoring of their own health data, which is an ideal way of self-management.

17.5   FETAL MONITORING SENSORS

The clinical condition of the mother and the fetus are sensed with the help of the sensors. The various types of sensors are the following:

A. ECG SENSOR

ECG levels of the mother and fetus are monitored by ECG sensors. Fetal heart rate combined with the mother heart rate is a predominant parameter for monitoring the health of the women. Fetal abnormalities and fetal distress can be detected using an ECG sensor. Invasive scalp electrode method and non-invasive abdominal electrode method are the two methods for obtaining fetal ECG. The frequency magnitude of the mother's ECG level in the abdomen is about two–ten times that of the fetal ECG level.

B. OXYGEN SENSOR

An oxygen sensor is an electronic device that indicates the fraction of oxygen (O_2) in the gas or liquid being examined. When a pregnant woman express symptoms, such as getting fatigue or feeling drowsy, it may be an indication of having low blood oxygen level. Hence, monitoring oxygen level can be done using an oximeter or other sensors. Preeclampsia is one of the causes of schizophrenia, which is a serious mental disorder where people experience disorganized speech, hallucinations, etc.

C. TEMPERATURE SENSOR

Temperature is one of the data to be monitored for constant maintenance as the change in temperature will be the first symptom for any complex condition. Novel intra-body sensor displays data collection that is required in knowing the relationship between temperature variations and fetal–maternal health conditions, such as detection of pregnancy contractions and complications and pre-term labor prevention.

D. BLOOD PRESSURE SENSOR

The mother's blood pressure is calculated to determine her clinical condition, both systolic and diastolic pressures, and to alert when it deviates from the standard threshold values. It is the non-invasive sensor designed to measure human blood pressure. High blood pressure has many complications, such as decreased blood flow to the placenta, which reduces the nutrition supply to the baby. All these complex conditions can be reduced by constant blood pressure monitoring.

17.6 CASE STUDY

According to the World Health Organization, 98 percent of all maternal deaths occur in the developing countries. Approximately 861 women died from childbirth or pregnancy in 2020. A major number of these fatalities would have been prevented through advanced and proper medical care. Those who die or suffer unnecessary complications often do so as a result of high blood pressure, severe bleeding, infections, blood clots, or obstructed labor. There are a various number of ways in which distributed mobile technologies can aid in improving maternal care and helps in reducing infant mortality. They can improvise and advance training for medical workers by providing access to precise and up-to-date information regarding the health complications and medication as well as the current ideas on handling specific problems. When the training includes solution to all complex problems, frontline health workers will be able to handle diverse illnesses and have an idea of where to go when they require additional healthcare information. Mobile Midwife and Text4Baby are the technologies that includes innovative maternal healthcare programs in the developing nation. The program focuses on promoting healthy pregnancies by motivating women to access prenatal care, debunking information on myths about pregnancy and childbirth and teaching lectures on topics such as postpartum depression, exercise, breastfeeding, immunization, delivery, and good nutrition. Women also receive tips on preventing malaria, pneumonia and managing pregnancy-related costs and consequences. Pradhan Mantri Matru Vandana Yojana, Mathrushree Scheme, Muthulakshmi Maternity Benefit Scheme, KCR Kit and Amma Odi Scheme are some of the major healthcare schemes providing facilities to maternal women. Mobile Ultrasound Patrol program in Morocco makes use of portable ultrasound machines and 3G smartphones to enhance diagnostic times for maternal women. Implemented in cooperation with Qualcomm's Wireless Reach initiative, this program is integrated in all rural clinics throughout Morocco, accessing healthcare professionals with utilities including devices that are wirelessly integrated with the maternal health specialists in advanced hospital clinics. Having current medical knowledge of the healthcare professionals in detecting high-risk pregnancies and the reduced number of home deliveries, which is the riskiest kind of delivery in the developing nations. When looking at the front line of accommodating care, clinical professionals are the first and often only link to healthcare services. The lack of quality and proper maternity management, infant and child care has a devastating impact by causing the increase in mortality rates in developing countries. Most unwanted deaths in developing nations could be prevented through access to proper healthcare and management, adequate healthy nutrition and the presence of a skilled birth professional attendant during delivery time.

The proposed model results are more reliable and accurate estimates compared with the conventional models where the deletion methods are unfit for real-time decision making. Also, the existing traditional imputation methods, machine learning-based methods and model-based methods are underestimate the variability of the missing heart rate values, delivering estimates with high error rates. With the ceaseless improvement of the national medical system, health monitoring integrated with cloud computing and Internet of Things has become a major consideration. This study mainly focuses on the development of a medical health monitoring IoT system based on cloud computing. The sensor terminal present in the wearable device is used to measure physiological indicators, such as blood pressure, electrocardiogram, heart rate, and other physiological indicators and the gateway terminal is served as a connector between the sensor terminal to receive physiological indicators and forward them to the cloud platform for the purpose of storage, data analysis, and prediction.

17.7 CONCLUSION

Each and every pregnancy has a unique story. Maternal healthcare monitoring is essential to provide the health and wellbeing of the pregnant women and the fetus, as many health problems occur during pregnancy, which has an effect throughout their lifetime. Although there exists many IoT-based maternal healthcare monitoring systems, few were limited to particular health problems, short-term data collection or self-management questionnaires. Ending preventable maternal mortality must remain at the top of the medical agenda. Simultaneously, surviving pregnancy and childbirth should not be the indicator of successful maternal healthcare. It is critical and demanding to expand efforts and resources reducing maternal injury. The proposed study involves less staffing, reduced cost price monitoring thus making it affordable for all class of patients. By connecting the respective pregnant women in home with the gynecologist department in the hospital, complications caused due to traveling would be prevented and provides the secure environment for both the mother and the fetus. In future, we will try to transfer the proposed model to other scenarios and in other hospitals. With effective implementation of the module, hospitals can extract data and track the development of the fetal growth in the maternal body. Thus, it results in the keen surveillance of the maternal health in detecting, observing, and delineating the results of the maternal body functions. During the case of emergency, maternal healthcare monitoring comes to the rescue.

REFERENCES

[1] Fox A., Glasofer A., Long D., Time and effort by labor nurses to achieve and maintain a continuous recording of the fetal heart rate via external monitoring, *Nursing for Women's Health*, February 2022, 26(1), 44–50, https://doi.org/10.1016/j.nwh.2021.12.001

[2] Sharma P., Sharma K., Fetal state health monitoring using novel Enhanced Binary Bat Algorithm, *Computers and Electrical Engineering*, July 2022, 101, 108035, https://doi.org/10.1016/j.compeleceng.2022.108035.

[3] Watson K., Mills T. A., Lavender T., Experiences and outcomes on the use of telemetry to monitor the fetal heart during labour: Findings from a mixed methods study, *Women and Birth*, May 2022, 35(3), e243–e252, https://doi.org/10.1016/j.wombi.2021.06.004

[4] Sarhaddi F., Azimi I., Long-term IoT-based maternal monitoring: System design and evaluation, MDPI, *Sensors*, 2021, 21(7), 2281, https://doi.org/10.3390/s21072281

[5] Dia N., Fontecave-Jallon J., Fetal heart rate estimation by non-invasive single abdominal electrocardiography in real clinical conditions, *Biomedical Signal Processing and Control*, January 2022, 71, Part B, 103187, https://doi.org/10.1016/j.bspc.2021.103187

[6] Sultana Tahmina M. S., Mary Daniel M. D., Manual fetal stimulation during intrapartum fetal surveillance: A randomized controlled trial, *American Journal of Obstetrics & Gynecology MFM*, March 2022, 4(2), 100574, https://doi.org/10.1016/j.ajogmf.2022.100574

[7] Frenken M. W. E., van der Woude D. A. A., The association between uterine contraction frequency and fetal scalp pH in women with suspicious or pathological fetal heart rate tracings: A retrospective

study, *European Journal of Obstetrics & Gynecology and Reproductive Biology*, April 2022, 271, 1–6, https://doi.org/10.1016/j.ejogrb.2022.01.023.

[8] Wang X., Wang W., Automatic measurement of fetal head circumference using a novel GCN-assisted deep convolutional network, *Computers in Biology and Medicine*, June 2022, 145, 105515, https://doi.org/10.1016/j.compbiomed.2022.105515.

[9] Lees, C. C., Romero, R., Stampalija, T., Dall'Asta, A., DeVore, G. R., Prefumo, F., ... & Hecher, K. The diagnosis and management of suspected fetal growth restriction: an evidence-based approach, *American Journal of Obstetrics and Gynecology*, March 2022, 226(3), 366–378, https://doi.org/10.1016/j.ajog.2021.11.1357

[10] Meakin A. S., Darby J. R. T., Maternal-placental-fetal drug metabolism is altered by late gestation undernutrition in the pregnant ewe, *Life Sciences*, 1 June 2022, 298, 120521, https://doi.org/10.1016/j.lfs.2022.120521

[11] Elser B. A., Simonsen D., Maternal and fetal tissue distribution of α-cypermethrin and permethrin in pregnant CD-1 mice, *Environmental Advances*, July 2022, 8, 100239, https://doi.org/10.1016/j.envadv.2022.100239

[12] Rasha Kamel M. D., Sherif Negm M. D., Fetal head descent assessed by transabdominal ultrasound: A prospective observational study, *American Journal of Obstetrics and Gynecology*, January 2022, 226(1), 112.e1–112.e10, https://doi.org/10.1016/j.ajog.2021.07.030

[13] Vinoth Kumar P., Balaji M., An IoT based solution for smart maternal and fetal health monitoring with one dimensional CNN, *International Journal of Science Engineering and Technology*, 2021, 9(3): 1–7.

[14] Ansari S., Ansari M. B., Smart health monitoring system for pregnant women, *International Journal of Engineering and Advanced Technology* (IJEAT), 2020, 9(4): 923–926, 10.35940/ijeat.D7114.049420

[15] Fong D. D., Yamashiro K. J., Design and in vivo evaluation of a non-invasive transabdominal fetal pulse oximeter, *IEEE Transactions on Biomedical Engineering*, January 2021, 68(1), 256–266, https://doi.org/10.1109/TBME.2020.3000977

[16] Martinek R., Barnova K., Passive fetal monitoring by advanced signal processing methods in fetal phonocardiography, *IEEE Access*, 8, 2020, 221942–221962, 10.1109/ACCESS.2020.3043496.

[17] Castaldo R., Melillo P., Bracale U., Caserta M., Triassi M., Pecchia L., Acute mental stress assessment via short term HRV analysis in healthy adults: A systematic review with meta-analysis. *Biomedical Signal Processing Control* 2015, 18, 370–377, https://doi.org/10.1016/j.bspc.2015.02.012

[18] Mehrabadi M. A., Azimi I., Sarhaddi F., Axelin, A., Niela-Vilén H., Myllyntausta S., Stenholm S., Dutt N., Liljeberg P., Rahmani A. M., Sleep tracking of a commercially available smart ring and smartwatch against medical-grade actigraphy in everyday settings: Instrument validation study. *JMIR MHealth UHealth*, 2020, 8, e20465, https://doi.org/10.2196/20465

[19] Shaffer F., Ginsberg J., An overview of heart rate variability metrics and norms. *Frontiers Public Health*, 2017, 5, 258, https://doi.org/10.3389/fpubh.2017.00258

[20] Malik M., Bigger J. T., Camm A. J., Kleiger R. E., Malliani A., Moss A. J., Schwartz P. J., Heart rate variability: Standards of measurement, physiological interpretation, and clinical use. *European Heart Journal*, 1996, 17, 354–381, https://doi.org/10.1093/oxfordjournals.eurheartj.a014868

[21] Pecchia L., Castaldo R., Montesinos L., Melillo P., Are ultra-short heart rate variability features good surrogates of short-term ones? State-of-the-art review and recommendations. *Healthcare Technology Letters*, 2018, 5, 94–100, https://doi.org/10.1049/htl.2017.0090

[22] Choi A., Shin H., Photoplethysmography sampling frequency: Pilot assessment of how low can we go to analyze pulse rate variability with reliability? *Physiological Measurement*, 2017, 38, 586, https://doi.org/10.1088/1361-6579/aa5efa

[23] Amiri D., Anzanpour A., Azimi I., Levorato M., Rahmani A. M., Liljeberg P., Dutt N., Edge-assisted sensor control in healthcare IoT. In Proceedings of the 2018 IEEE Global Communications Conference (GLOBECOM), Abu Dhabi, United Arab Emirates, 9–13 December 2018, pp. 1–6, https://doi.org/10.3390%2Fs21072281

[24] Gopalan S. A., Park J. T., Energy-efficient MAC protocols for wireless body area networks: Survey. In Proceedings of the International Congress on Ultra Modern Telecommunications and Control Systems, Moscow, Russia, 18–20 October 2010, 739–744, https://doi.org/10.1109/ICUMT.2010.5676554

[25] Aijaz A., Aghvami A. H., Cognitive machine-to-machine communications for internet-of-things: A protocol stack perspective, *IEEE Internet Things Journal*, 2015, 2, 103–112, https://doi.org/10.1109/JIOT.2015.2390775

[26] Anzanpour A., Azimi I., Götzinger M., Rahmani A. M., TaheriNejad N., Liljeberg P., Jantsch A., Dutt N., Self-awareness in remote health monitoring systems using wearable electronics. In Proceedings of the Conference on Design, Automation & Test in Europe, Lausanne, Switzerland, 27–31 March 2017, 1056–1061, https://doi.org/10.23919/DATE.2017.7927146

[27] Sasai K., Izumi S., Watanabe K., Yano Y., Kawaguchi H., Yoshimoto M., A low-power photoplethysmography sensor using correlated double sampling and reference readout circuit. In Proceedings of the 2019 IEEE SENSORS, Montreal, QC, Canada, 27–30 October 2019, 1–4, https://doi.org/10.1109/tbcas.2019.2956948

[28] Clawson J., Pater J. A., Miller A. D., Mynatt E. D., Mamykina L., No longer wearing: Investigating the abandonment of personal health-tracking technologies on craigslist. In Proceedings of the 2015 ACM International Joint Conference on Pervasive and Ubiquitous Computing, Osaka, Japan, 7 September 2015, 647–658, http://dx.doi.org/10.1145/2750858.2807554.

[29] Fantinelli S., Marchetti D., Verrocchio M. C., Franzago M., Fulcheri M., Vitacolonna E., Assessment of psychological dimensions in telemedicine care for gestational diabetes mellitus: A systematic review of qualitative and quantitative studies. *Frontiers in Psychology*, 2019, 10, 153, https://doi.org/10.3389/fpsyg.2019.00153.

[30] Lau Y., Htun T. P., Wong S. N., Tam W. S. W., Klainin-Yobas P., Efficacy of internet-based self-monitoring interventions on maternal and neonatal outcomes in perinatal diabetic women: A systematic review and meta-analysis. *Journal Medical Internet Re*search, 2016, 18, e220, https://doi.org/10.2196/jmir.6153.

[31] Nagarajan R., Thirunavukarasu R., A neuro-fuzzy based healthcare framework for disease analysis and prediction. *Multimedia Tools and Applications*, 2022, 81(8), 11737–11753.

18 A Real-Time Automated Face Recognition and Detection System for Competitive Examination

Rajalakshmi Gurusamy[1] and B. Ben Sujitha[2]
[1] Department of Information Technology, Sethu Institute of Technology, Madurai, Tamilnadu, India
[2] Department of Computer Science & Engineering, Noorul Islam Centre for Higher Education, Kumaracoil, Kanyakumari, Tamil Nadu, India

18.1 INTRODUCTION

Over the years, standardized testing has been the most common approach to selecting recruits for all services and occupations. The functioning of today's increasingly electronically interconnected information society relies on accurate automated personal identification. Traditional methods of automated personal identification that rely on something already know, such as personal identification numbers (PINs), ID cards, keys, etc., are no longer considered reliable enough to meet the security requirements of electronic transactions.

The main problem with the examination system is malpractice. This has been identified due to the lack of a reliable identity verification system for both offline and online exams. To overcome the above problems, researchers turned to using artificial intelligence and using biometrics. Biometrics is technologies that uniquely identify individuals based on their physiological or behavioral characteristics. Because it is based on personal identification, it can inherently distinguish between authorized persons and fraudsters [1].

Facial recognition is an important topic in AI applications worldwide, as it is essential for human–computer interaction and widely used in surveillance systems, enterprise surveillance, and other industries [2–5]. The effectiveness of facial recognition applications is increased by combining artificial intelligence and physical device acceleration [6]. Fingerprint recognition is easier, but facial recognition seems to be more reliable and harder to counterfeit.

Accurate facial recognition systems in security, surveillance, and personal identification are enabled by machine learning and deep learning with CNNs. For example, the method that combines local Gabor filters, principal component analysis (PCA), and linear discriminant analysis (LDA) to extract discriminative features, reduce dimensionality, and improve classification performance [7].

Emotion, lighting, location, and noise affect the effectiveness of facial recognition systems. To address changes in lighting, position, and facial expressions, state-of-the-art facial recognition methods combine near-infrared and visible-light photographs with deep neural networks, potentially improving performance in real-world applications [8]. Combining CNN and particle swarm optimization (PSO), this technique accurately identifies and eliminates noise while preserving key features, making it a significant advancement in image denoising [9].

The use of facial recognition brings various benefits contingent on the situation [10–14]. For example, automated facial recognition technology is used in airport border controls to speed up the

DOI: 10.1201/9781003407959-18

screening process, authenticate travelers' identities, and retain employees. In order to reduce manual crime, businesses are also utilizing facial recognition technology to regulate access to rapid check-ins for authorized personnel.

The objective of the research is to authenticate a person's identity using facial recognition, and to distinguish between authorized persons and impostors. The research involves developing a facial biometric system to help eliminate identity theft during trials. A deep convolutional neural network (DCNN) is a deep learning (DL) technique that differs from a regular convolutional neural network (CNN) in that it typically uses more than five hidden layers to extract more features and make predictions, which improves the accuracy.

In order to analyze an image's RGB components all at once, DCNN employs a 3D NN. As a result, the network can evaluate not only individual pixels, but how those pixels are related in space, thus improving image recognition and classification. This significantly reduces the number of interconnected neurons required for image processing compared to traditional neural networks. A classifier is trained using the images as input to the DCNN. Instead of matrix multiplication, the network uses a unique mathematical operation known as convolution. Convolution operations help the network; identifying patterns and spatial relationships in image data.

The organizational structure of the research is as follows. Previous studies on face recognition and detection are described in Section 18.2. A recommended DCNN structure is provided in Section 18.3. Section 18.4 provides a detailed analysis of the experimental results obtained from the proposed method. Finally, Section 18.5 summarizes the main findings of the study and discusses implications for future research.

18.2   RELATED WORKS

DNNs are widely used in machine vision applications such as face recognition, object detection, and image classification. In order to improve face detection in computer vision, Sun et al. introduced a fast region-based CNN (RCNN) [15] method that combines techniques like region proposals, supervised pre-training, structured output layers, and an extended fully convolutional network. By eliminating geometric variances, Zhou et al. [16] presented a Grid face approach to enhance recognition performance. To predict the position of faces and shapes from coarse to fine, Zhang et al. [17] developed a cascade architecture with meticulously designed three-level deep convolutional network.

An innovative deep neural network (DNN)-based feature learning approach by Zhang et al. [18] was focused. This methodology effectively removes the spatial complexity of parameters associated with traditional approaches and isolates the most robust high-level features that enable models to recognize facial expressions from different angles.

Kim et al. [19] proposes a new CNN architecture with two feature extractors. A content-based extractor that recognizes facial components and a trace-based extractor that detects signs of forgery such as painting, transformation, and splicing. Extracted content and feature traces help identify real faces in media. It is a more extensive form of the separate restricted convolution for grayscale images presented in [20], and the authors also suggested inter restricted convolution of colored input images. This architecture uses a Siamese structure to process multiple iterations of the same face image using two identical networks (left and right).

Capturing global and local facial features improves recognition accuracy and system robustness through the integration of discrete wavelet transform (DWT) and machine learning [21]. Jain et al. [22] use the Viola–Jones method along with PCA and LDA to offer a thorough examination of face detection and identification systems. The system becomes more effective, accurate, and robust as a result of the integration of various strategies.

Saranke et al. [23] proposed facial emotion recognition and classification based on deep neural networks. Incorporating CNN enables accurate and efficient recognition of emotions from facial expressions. The CK+ and FER2013 datasets and other publicly available facial expression databases are used for experimental evaluation of the proposed methodology. CNNs can be used to enhance images that have undergone smoothing by restoring lost details and improving overall image quality [24].

CNNs aid in 2D image representation, enabling feature positioning and translation for image processing. Applications such as object detection, semantic segmentation, and image classification have shown promising results by modifying traditional CNNs. Convolutional neural networks will continue to be developed to increase their effectiveness and efficiency as this research area expands. The designs of LeNet [25], AlexNet [26], ConvvNet [27], and GoogleNet [28] are just a few examples of designs produced with traditional CNN models. This architecture can be used to accurately segment images by using large datasets to train neural networks based on patterns and features.

ResNets were trained in emotion recognition using supervised and unsupervised learning methods. The fer2013 dataset was utilized by the researchers to characterize facial expressions automatically using ResNets [29–31]. To accurately and efficiently analyze facial emotions, Storey et al. [32] created a deep learning technique, EmotionoNet, which combines face identification and facial landmark localization. Deep learning has shown promise in speech recognition and image recognition [33–35], but there are still problems such as the need for large amounts of labeled data. To replace manual feature-based methods [36, 37], the proposed system leverages deep learning algorithms to recognize human emotions based on real-time facial expressions.

A literature review also found that neural networks outperform traditional methods in performance and accuracy when processing complex datasets. The characteristics of feature maps and the DCNN architecture determines the accuracy of face detection and recognition. Certain image features, such as edges, curves, and lines are mapped using these feature maps. The recommended approach is a combination of grayscale co-occurrence matrix (GLCM) and region proposal network (RPN) approaches. The proposed RPN+GLCM method was able to identify and detect faces with 98.65% accuracy.

18.3 PROPOSED METHODOLOGY

The proposed DCNN-based face detection and verification technology incorporating the security strength of examination and the accuracy. DCNN performs the following steps to accurately detect the spoofers in competitive environment and functional modules of DCNN are depicted in Figure 18.1.

18.3.1 FACIAL IMAGE ACQUISITION

A webcam is essential during the competitive examination to record video. To maximize image variation, participants can turn their face when the image is taken. The video is divided into 20–30 sequence of image frames for further processing.

18.3.2 PREPROCESSING

It is essential to format all input images before starting the training process. Preprocessing includes RGB to grayscale conversion, resizing, denoising, and binarization to simplify the image analysis. Grayscale conversion reduces data processing, resizing ensures consistency, denoising removes noise, and binarization converts grayscale images into binary ones, simplifying analysis and highlighting specific features.

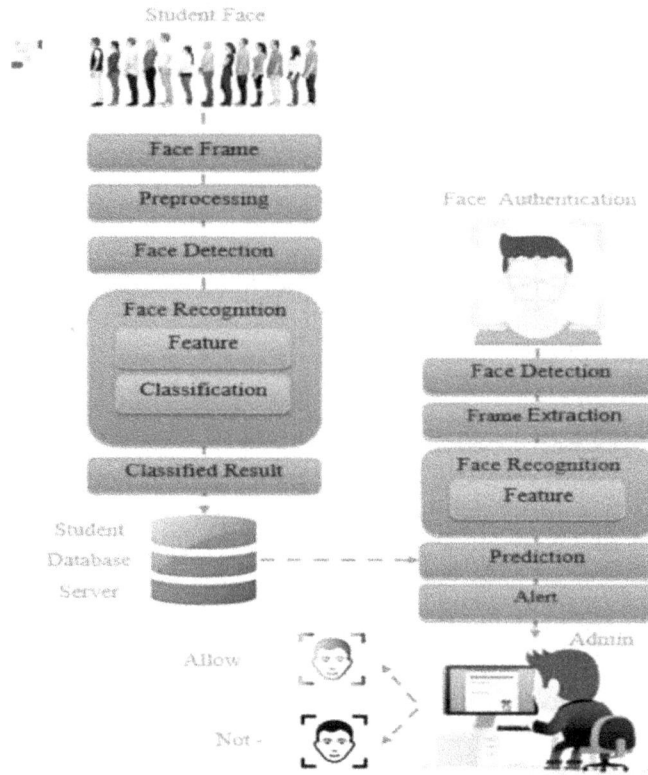

FIGURE 18.1 Functional modules of DCNN.

Resize all input images (width, height, and number of RGB channels) as needed (224, 224, and 3). After removing unwanted noise and smoothly scaling the image, Gaussian filtering is performed. It smooths out irregularities and sharp edges, improving facial feature detection and matching accuracy. Binarization is a technique to reduce image size by taking input grayscale images and converting them to binary images.

18.3.3 Face Detection

Face detection is a more effective method for segmenting and identifying faces based on region proposal networks (RPN). It can be utilized to create regions of interest (RoI) on a map of features by dragging a window over anchors with various scales and aspect ratios. Figure 18.3 shows the DCNN architecture.

18.3.3.1 RPN

The RPN is a complete CNN that predicts object borders and object-free scores concurrently at all locations. RPN excels in producing excellent regional proposals. Each feature (point) on the feature map, which is an output of CNN, which the system uses, is referred to as an anchor point. Overlay the image with nine anchor boxes (all sizes and proportions) per anchor point. These anchor boxes are centered on image locations that correspond to the anchor points of the feature map shown in Figure 18.2.

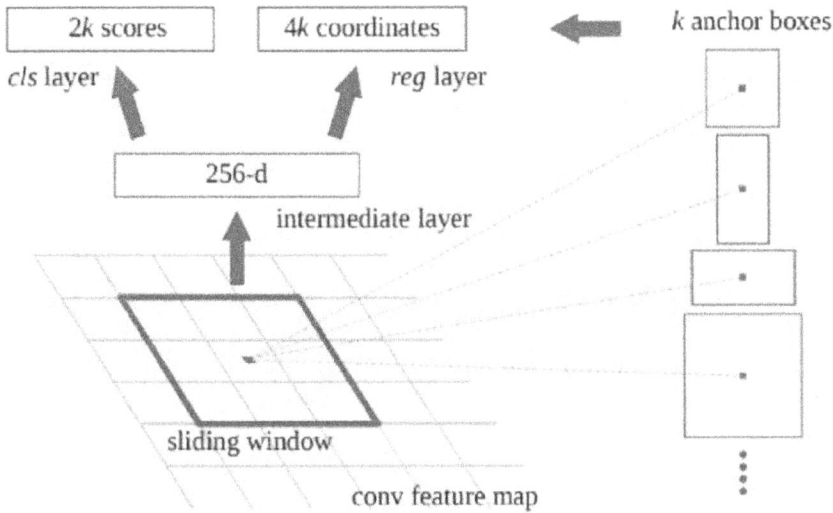

FIGURE 18.2 RPN convolutional feature map.

Source: [38].

Training RPN

Training RPN based on the foreground and background of the anchor boxes.

Algorithm 1: Training RPN
Input: Location of feature map
Output: Mini-batch of relevant anchor point for training
Required Parameters: 1. Foreground: Anchor box have an object.
                               2. Background: Anchor box does not have an object.
                               3. Initialize intersection over union (IoU) = 0.7.

1   Set the required parameters:
2   for all input image:
3   Assign labels to each anchor box (-1,0,1):
4   if (IoU>0.7)
5   Assign label 1: Foreground
6   End if
7   Else if (IoU<0.7)
8   Assign label -1: Background
9   End if
10 Else
11 Assign label 0: Ignore anchor box
12 End if
13 End for
14 Return mini-batch of anchor box.

RPN mini-batches were formed using back-propagation and stochastic gradient descent (SGD).

18.3.3.2 Gray-Level Co-Occurrence Matrix (GLCM)

In order to assess how frequently a specific pixel combination emerges in an image in a specific direction Θ and at a specific distance d, it investigates the spatial correlation among pixels. Four GLCMs (M) with $d = 1$ and $\Theta = 0, 30, 60,$ and 120 are created after each image is digitized with 16 grayscale (0 to 15).

Five features are recovered from each GLCM. So each image has 20 attributes. Before being provided to the classifier, each feature is normalized to have a value that ranges from zero to one, and that each model received the identical set of data.

Restored functionality can be divided into three categories. The first category characterizes the gray-level intensity of the region. The second category characterizes the characteristics defining the shape of the region.

Twenty-two grayscale dependency matrix (GLDM) functions, 16 grayscale range length matrix (GRLM) functions, and 14 GLCM functions make up the three types of texture functions. The texture of the area is determined by these attributes.

18.3.4 Face Identification

The face detection module receives the image of the face captured by the camera. This module identifies regions of images with faces and extracts facial features such as jaws, eyes, noses, and mouths. The feature extraction module takes a face image as input and identifies important features that are used for post-face recognition classification using a RPN. After the facial features are extracted, the DCNN classification method is used to classify the face using the extracted features. The algorithm recognizes patterns in facial features and uses a training approach to classify facial features based on those patterns depicted in Figure 18.3. The facial recognition module then receives the extracted features and compares them with facial models previously stored in the database. If a match is found, the individual's identity is established.

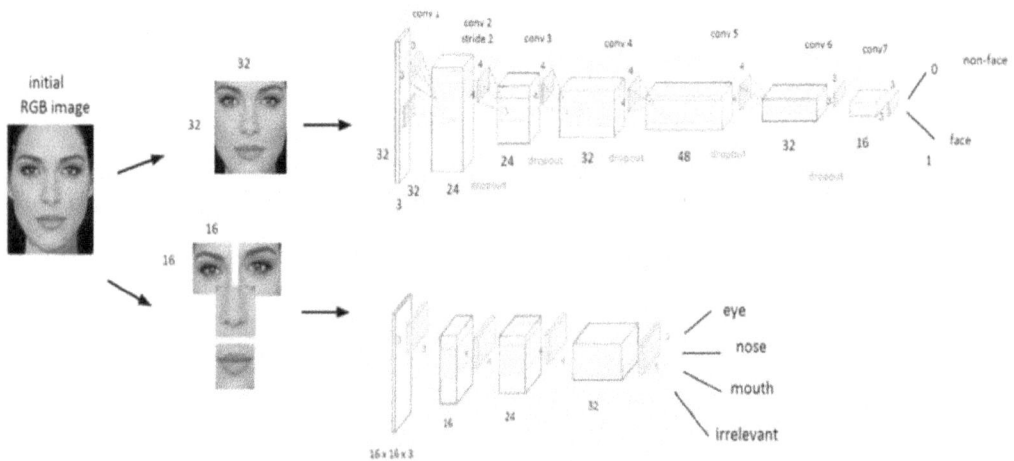

FIGURE 18.3   DCNN architecture.

Source: [39].

Training DCNN

Algorithm 2: Training DCNN

Input:	train(x),train(y): Training set
	test(x),test(y): Test set
Output:	Updated weight ($\Delta\hat{w}$) and bias ($\Delta\bar{b}$) of DCNN
Required Parameters:	1. Learning rate (η)
	2. Target error (ε): $c\varepsilon < \varepsilon$, training finished.
	// $c\varepsilon$ = current training_error.
	3. Max_time (€).
	4. Number of training samples (n)
Parameters Initialization:	1. Weight ($\hat{W}$) and bias ($\bar{b}$) set a random numbers.
	2. $t = 1$; // t = current time.
	3. $\mu(t) = 1$; //$\mu(t)$ = mean square error at time t.

1   Begin:
Set the required parameters and initialization
While t<€ and $\mu(t)$> ε
2   *for* all input training set image:
3   Calculate forward propagation as follows:

$$z_{i,j} = \text{sum}\left(\hat{W}_{m,n} * X\left[i+m, j+n\right]\right) + \bar{b}$$

Where, z = feature map output, X = feature map input.
Calculate soft-max activation function (a) as follows:

$$a = f\left(z_{i,j}\right)$$

Calculate loss function as follows:

$$\mu(t) = -\text{sum}\left(Y * \log\left(a\right)\right)$$

Where, Y = predicted output.
End for
4   Calculate backward propagation as follows:

$$\Delta z_{i,j} = \frac{\Delta\mu(t)}{\Delta a} * f'\left(z_{i,j}\right)$$

$$\Delta\hat{w} = \left(\Delta z_{i,j} * X^T\right)/n$$

$$\Delta\bar{b} = \text{sum}\left(\Delta z_{i,j}\right)/n$$

$$\Delta a = \hat{W}^T * \Delta z_{i,j}$$

Where, X^T = Transpose of input matrix, $\hat{W}^T$ = Transpose of weight matrix
5   Update the weight and bias as follows:

$$\hat{W} = \hat{W} - \eta * \Delta\hat{w}$$

$$\bar{b} = \bar{b} - \eta * \Delta\bar{b}$$

6   *t++*
7   *End while*
8   *End.*

18.3.5 PREDICTION

The matching method in this module is important for assessing the accuracy of the classification results. This task is performed using trained sensitive results and testing on sensitive data collected on live cameras. The difference between the two datasets is computed using the eigenvectors and this difference is used to predict the accuracy of the classification results. This step is essential for producing accurate and reliable results in the classification process.

18.4   RESULTS AND DISCUSSION

18.4.1   DATASET DESCRIPTION

We combined real video footage with computer-generated deep shadow videos to develop our own deep shadow dataset. First, we obtained videos among a variety sources (internet, TV shows, movies, etc.) that meet the requirements of a minimum face area size of 224 × 224 and a minimum resolution of 500p.

We then combine RPN and GLCM to exchange two IDs using DCNN architecture. All faces were acquired and allied by pre-trained face point recognition NN. In addition, we trained RPN and GLCM using facial photographs from the collected dataset to create fake deep movies. In stratified sampling, there are 45 training movies, 15 validation movies, and 15 test movies in the dataset. 50 facial images were obtained from each video.

18.4.2   DATASET DESCRIPTION

Probabilistic neural networks with linear discriminant analysis (PNN+LDA), principal component analysis (PCA), deep neural networks (DNN), support vector machines (SVM), residual networks (ResNet), and multi-layer perceptron (MLP) are used to compare the overall effectiveness of the proposed model. The comparison's outcomes demonstrate that the suggested model exceeds all other fundamental models in terms of precision. Specifically, the proposed model achieves a higher accuracy rate compared to PCA, PNN+LDA, SVM, MLP, DNN, and ResNet (shown in Table 18.1).

18.4.3   EXPERIMENTAL RESULTS

The recommended DCNN model combines the RPN and GLCM techniques. DCNN models should perform better on image classification tasks using the RPN+GLCM technique. This is because RPN extracts region proposals and GLCM analyzes texture information, making the model more efficiently identifies objects in the image. The performance of the model was evaluated by capturing real-time pictures shown in Figure 18.4. The findings indicate that the suggested DCNN outperforms the F1 score and accuracy of the conventional system.

The performance metrics of the model is depicted as follows: True positive ($True_{Pos}$): In this case, the student's face was correctly recognized and associated with a name by the DCNN algorithm. False negative ($False_{Neg}$): In this case, the DCNN algorithm was unable to recognize the student's face and match it to the name.

False positive ($False_{Pos}$): In this case, the DCNN algorithm incorrectly determined that the student's face belonged to someone else. A true negative ($True_{Neg}$) is produced when the DCNN algorithm correctly recognizes a face that does not belong to any of the enrolled students. The results show that the proposed method is more accurate than the others.

TABLE 18.1
Performance Comparison of Base Models

Techniques Used	F1 Score	Accuracy %
PCA [40]	0.72	87.13
PNN+LDA[41]	0.84	83.8
SVM [42]	0.86	94
MLP[43]	0.82	94.5
DNN [18]	0.73	80.1
ResNet [44]	0.832	84.83
Proposed	**0.99**	**98.65**

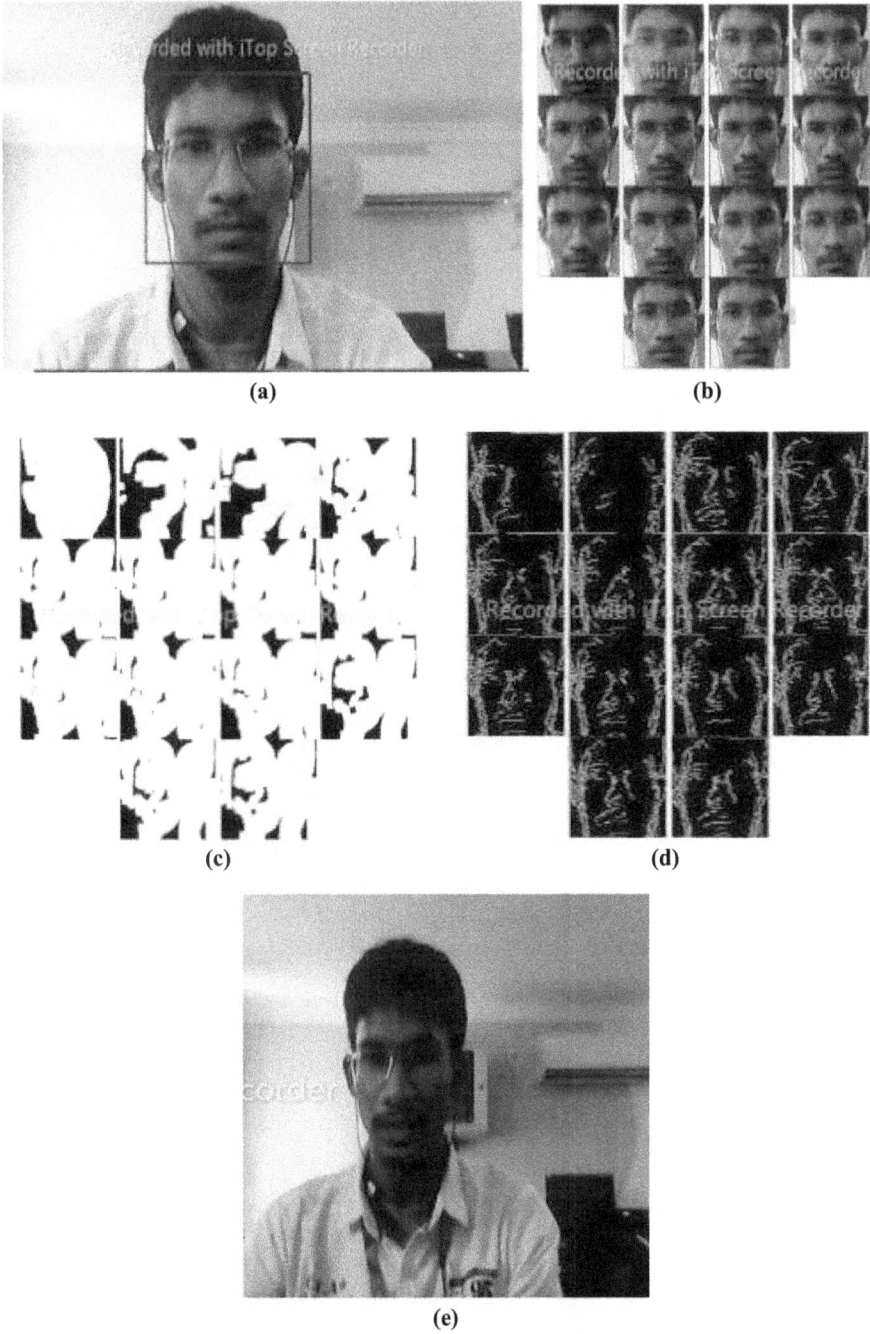

FIGURE 18.4 Model performance evaluation: (a) capturing image; (b) preprocessing; (c) segmentation; (d) feature extraction; and (e) prediction.

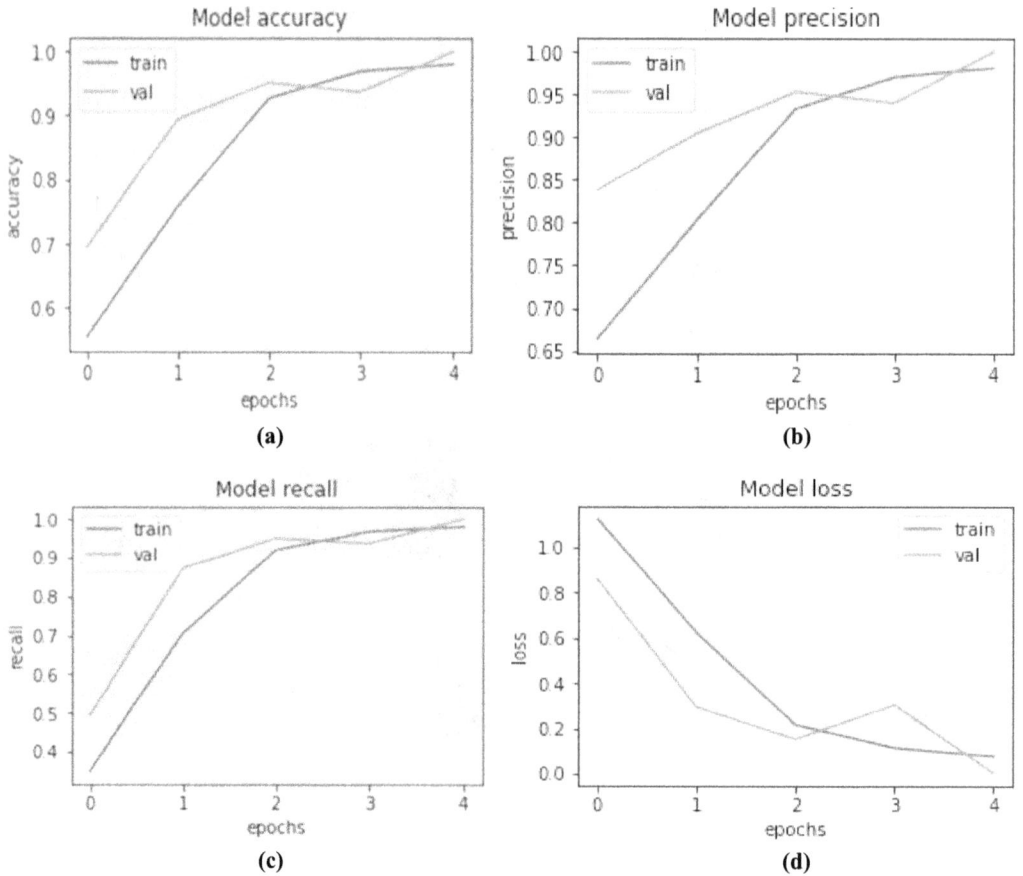

FIGURE 18.5 Performance metric analysis: (a) accuracy; (b) precision; (c) recall; and (d) F1 score.

The model performance measures are expressed as follows and the results are shown in Figure 18.5.

Accuracy refers to the ability of DCNN model to accurately recognize human faces and it is defined as follows:

$$Accuracy = \frac{\left(True_{Pos} + True_{Neg}\right)}{\left(True_{Pos} + True_{Neg} + False_{Pos} + False_{Neg}\right)} \tag{18.1}$$

Precision is an essential performance indicator for face recognition technology. It refers to the DCNN system's ability to accurately recognize faces and compare them to a database of known people. It is defined as follows:

$$Precision(p) = \frac{True_{Pos}}{\left(True_{Pos} + False_{Pos}\right)} \tag{18.2}$$

Recall rates, which assess the system's ability to accurately identify faces and minimize false positives and false negatives. It improves the accuracy and reliability of the model. It is defined as follows:

$$Recall(r) = \frac{True_{Pos}}{\left(True_{Pos} + False_{Neg}\right)} \tag{18.3}$$

The F1 score combines p and r to provide a more comprehensive assessment of DCNN performance. F1 score for facial recognition can be used to assess how accurately the DCNN system recognizes people based on their facial features.

$$F1\,Score = 2 * \left\{\frac{(p*r)}{(p+r)}\right\} \tag{18.4}$$

18.5 CONCLUSIONS

We used the GLCM approach to extract texture features from images and feed them into the RPN for face detection to achieve this high accuracy. The proposed GLCM + RPN approach was able to recognize and detect faces with 98.65% accuracy. The recommended method has uses in a variety of scenarios such as online testing, job interviews, and secure access control. The proposed classifier outperformed other state-of-the-art classifiers in recognition accuracy. In the future, we planned to use DCNN's facial recognition algorithms to verify voter identities so that only eligible voters can vote. This supports the integrity of the voting process and helps prevent fraud. In addition, technology can provide real-time updates on election results, making it easier for citizens to participate in democracy.

REFERENCES

1. Minaee, S., Abdolrashidi, A., Su, H., Bennamoun, M. and Zhang, D. Biometrics recognition using deep learning: A survey. Artificial Intelligence Review, vol. 3, pp. 1–49, 2023.
2. Wu, F., Bao, L., Chen, Y., Ling, Y., Song, Y., Li, S., Ngan, K. N. and Liu, W. Mvf-net: Multi-view 3d face morphable model regression. In Proceedings of the IEEE/CVF Conference on Computer Vision and Pattern Recognition, pp. 959–968, 2019.
3. Shao, X., Lyu, J., Xing, J., Zhang, L., Li, X., Zhou, X. and Shi, Y. 3D face shape regression from 2D videos with multi-reconstruction and mesh retrieval. In Proceedings of the IEEE/CVF International Conference on Computer Vision Workshops, pp. 1–6. 2019.
4. Dou, P., and Kakadiaris, I. A. Multi-view 3D face reconstruction with deep recurrent neural networks. Image and Vision Computing, vol. 80, pp. 80–91. 2018.
5. Zhou, H., Chen, P. and Shen, W. A multi-view face recognition system based on cascade face detector and improved Dlib. In MIPPR 2017: Pattern Recognition and Computer Vision, vol. 10609, pp. 37–42. Proceedings of SPIE, 2018.
6. Renuka, B., Sivaranjani, B., Maha Lakshmi, A., and Muthukumaran, N. Automatic enemy detecting defense robot by using face detection technique. Asian Journal of Applied Science and Technology, 2, no. 2, pp. 495–501, 2018.
7. Pumlumchiak, T., and Vittayakorn, S. Facial expression recognition using local Gabor filters and PCA plus LDA. In 2017 9th International Conference on Information Technology and Electrical Engineering (ICITEE), pp. 1–6, 2017.
8. Guo, K., Wu, S., and Xu, Y. Face recognition using both visible light image and near-infrared image and a deep network. CAAI Transactions on Intelligence Technology, 2(1), 39–47, 2017.
9. Khaw, H. Y., Soon, F. C., Chuah, J. H., and Chow, C. O. High-density impulse noise detection and removal using deep convolutional neural network with particle swarm optimisation. IET Image Processing, 13(2), 365–374, 2019.

10. Rodrigues, A. S. F., Lopes, J. C., Lopes, R. P., and Teixeira, L. F. Classification of facial expressions under partial occlusion for VR games. In Optimization, Learning Algorithms and Applications: Second International Conference, OL2A 2022, Póvoa de Varzim, Portugal, October 24–25, 2022, Proceedings, pp. 804–819. Cham: Springer International Publishing, 2023.

11. Zhu, X., Li, Z., and Sun, J. Expression recognition method combining convolutional features and transformer. Mathematical Foundations of Computing, 6, no. 2, pp. 203–217, 2023.

12. Gurusamy, R., and Seenivasan, S. R. DGSLSTM: Deep gated stacked long short-term memory neural network for traffic flow forecasting of transportation networks on big data environment. Big Data, 12, pp. 1–14. DOI: 10.1089/big.2021.0013, 2022.

13. Liu, S., Gao, P., Li, Y., Fu, W., and Ding, W. Multi-modal fusion network with complementarity and importance for emotion recognition. Information Sciences, 619, pp. 679–694, 2023.

14. Majidpour, J., Jameel, S. K., and Qadir, J. A. Face identification system based on synthesizing realistic image using edge-aided GANs. The Computer Journal, 66, no. 1, pp. 61–69, 2023.

15. Sun, X., Wu, P., and Hoi, S. C. Face detection using deep learning: An improved faster RCNN approach. Neurocomputing, 299, pp. 42–50, 2018.

16. Zhou, E., Cao, Z., and Sun, J. Gridface: Face rectification via learning local homography transformations. In Proceedings of the European Conference on Computer Vision (ECCV), pp. 3–19, 2018.

17. Zhang, K., Zhang, Z., Li, Z., and Qiao, Y. Joint face detection and alignment using multitask cascaded convolutional networks. IEEE Signal Processing Letters, 23(10), pp. 1499–1503, 2016.

18. Zhang, T., Zheng, W., Cui, Z., Zong, Y., Yan, J., and Yan, K. A deep neural network-driven feature learning method for multi-view facial expression recognition. IEEE Transactions on Multimedia, 18(12), pp. 2528–2536, 2016.

19. Kim, E., and Cho, S. Exposing fake faces through deep neural networks combining content and trace feature extractors. IEEE Access, 9, pp. 123493–123503, 2021.

20. Bayar, B., and Stamm, M. C. Constrained convolutional neural networks: A new approach towards general purpose image manipulation detection. IEEE Transactions on Information Forensics and Security, 13(11), pp. 2691–2706, 2018.

21. Tabassum, F., Islam, M. I., Khan, R. T., and Amin, M. R. Human face recognition with combination of DWT and machine learning. Journal of King Saud University-Computer and Information Sciences, 34(3), 546–556, 2022.

22. Jain, U., Choudhary, K., Gupta, S., and Privadarsini, M. J. P. Analysis of face detection and recognition algorithms using Viola Jones algorithm with PCA and LDA. In 2018 2nd International Conference on Trends in Electronics and Informatics (ICOEI), pp. 945–950, IEEE, 2018.

23. Salunke, V. V., and Patil, C. G. A new approach for automatic face emotion recognition and classification based on deep networks. In 2017 International Conference on Computing, Communication, Control and Automation (ICCUBEA), pp. 1–5, IEEE, 2017.

24. Yang, B., Sun, X., Cao, E., Hu, W., and Chen, X. Convolutional neural network for smooth filtering detection. IET Image Processing, 12(8), 1432–1438, 2018.

25. Venkata Kranthi, B., and Surekha, B. Real-time facial recognition using deep learning and local binary patterns. In Proceedings of International Ethical Hacking Conference 2018: eHaCON 2018, Kolkata, India, pp. 331–347, Springer Singapore, 2019.

26. Sekaran, S. A. R., Lee, C. P., and Lim, K. M. Facial emotion recognition using transfer learning of AlexNet. In 2021 9th International Conference on Information and Communication Technology (ICoICT), pp. 170–174, IEEE, 2021.

27. Onyema, E. M., Shukla, P. K., Dalal, S., Mathur, M. N., Zakariah, M., and Tiwari, B. Enhancement of patient facial recognition through deep learning algorithm: ConvNet. Journal of Healthcare Engineering, 2021, pp. 1–8, 2021.

28. Khan, S., Javed, M. H., Ahmed, E., Shah, S. A., and Ali, S. U. Facial recognition using convolutional neural networks and implementation on smart glasses. In 2019 International Conference on Information Science and Communication Technology (ICISCT), pp. 1–6, IEEE, 2019.

29. Khan, H. H., Malik, M. N., Alotaibi, Y., Alsufyani, A., and Alghamdi, S. Crowd sourced requirements rngineering challenges and solutions: A software industry perspective. Computer Systems Science & Engineering, 39(2), pp. 221–236, 2021.

30. Devries, T., Biswaranjan, K., and Taylor, G. W. Multi-task learning of facial landmarks and expression. In 2014 Canadian Conference on Computer and Robot Vision, pp. 98–103, IEEE, 2014.
31. Benitez-Quiroz, C. F., Srinivasan, R., Feng, Q., Wang, Y., and Martinez, A. M. Emotionet challenge: Recognition of facial expressions of emotion in the wild. Computer Vision and Pattern Recognition, 17(3), pp.1–8, 2017.
32. Storey, G., Bouridane, A., and Jiang, R. Integrated deep model for face detection and landmark localization from "in the wild" images. IEEE Access, 6, pp. 74442–74452, 2018.
33. Jorge-Martinez, D., Butt, S. A., Onyema, E. M., Chakraborty, C., Shaheen, Q., De-La-Hoz-Franco, E., and Ariza-Colpas, P. Artificial intelligence-based Kubernetes container for scheduling nodes of energy composition. International Journal of System Assurance Engineering and Management, 12(5), pp. 1–9, 2021.
34. Afriyie, R. K., Asante, M., and Onyema, E. M. Implementing morpheme-based compression security mechanism in distributed systems. International Journal of Innovative Research and Development, 9(2), 157–162, 2020.
35. Gurusamy, R., Mahalakshmi, M., Divya Bharathi, M., and Harishmaashree, G. Incident detection using LSTM and SAE. International Journal of Scientific Research in Computer Science, Engineering and Information Technology (IJSRCSEIT), ISSN: 2456-3307, 9(8), pp. 138–145, 2023.
36. Ranjan, R., Patel, V. M., Chellappa, R., and Castillo, C. D. U.S. Patent No. 10,860,837. Washington, DC: U.S. Patent and Trademark Office. 2020.
37. Liu, Y., Qin, B., Li, R., Li, X., Huang, A., Liu, H., ... and Liu, M. Motion-robust multimodal heart rate estimation using BCG fused remote-PPG with deep facial ROI tracker and pose constrained Kalman filter. IEEE Transactions on Instrumentation and Measurement, 70, pp. 1–15, 2021.
38. Wang, C., and Peng, Z. Design and implementation of an object detection system using faster R-CNN. In 2019 International Conference on Robots & Intelligent System (ICRIS), pp. 204–206, IEEE, 2019.
39. Triantafyllidou, D., and Tefas, A. A fast deep convolutional neural network for face detection in big visual data. In Advances in Big Data: Proceedings of the 2nd INNS Conference on Big Data, October 23–25, 2016, Thessaloniki, Greece 2, pp. 61–70, Springer International Publishing, 2017.
40. Sultoni, S., and Abdullah, A. G. Real time facial recognition using principal component analysis (PCA) and EmguCV. In IOP Conference Series: Materials Science and Engineering, 384(1), p. 012079. IOP Publishing, 2018.
41. Ouyang, A., Liu, Y., Pei, S., Peng, X., He, M., and Wang, Q. A hybrid improved kernel LDA and PNN algorithm for efficient face recognition. Neurocomputing, 393, pp. 214–222, 2020.
42. VenkateswarLal, P., Nitta, G. R., and Prasad, A. Ensemble of texture and shape descriptors using support vector machine classification for face recognition. Journal of Ambient Intelligence and Humanized Computing, 10(11), pp. 1–8, 2019.
43. Luaibi, M. K., and Mohammed, F. G. Facial recognition based on DWT–HOG–PCA features with MLP classifier. Journal of Southwest Jiaotong University, 54(6), pp. 1–12, 2019.
44. He, K., Zhang, X., Ren, S., and Sun, J. Deep residual learning for image recognition. in Proceeding IEEE Conference on Computer Vision Pattern Recognition (CVPR), pp. 770–778, 2016.

19 Medical Image Analysis with Vision Transformers for Downstream Tasks and Clinical Report Generation

Evans Kotei[1] and Ramkumar Thirunavukarasu[2]
[1] Department of Computer Science, Kumasi Technical University, Kumasi, Ghana
[2] School of Information Technology & Engineerig, Vellore Institute of Technology, Vellore, Tamil Nadu, India

19.1 INTRODUCTION

Medical imaging technologies have developed to the point that they are now crucial in the medical domain. Human bias, time constraints, and a range of explanations limit analysts' investigation of this imagery, which results in the illusion [1–3]. An abundance of information [4] essential for medical diagnosis and effective therapy is in medical images. About 90% of medical image data are from modalities like pathology, X-ray radiography, CT, MRI, and ultrasound. Interpreting AI-based analysis in scientific experiments is hampered by several challenges [4–7]. In recent years, algorithms for deep learning have had considerable success in automatically analyzing medical images. CNN techniques are the most widely used architecture for image feature extraction [8–11]. The strength of CNN lies in weight sharing, wherein comparable objects that occur in various parts of an image can share weights, hence decreasing the number of parameters. The CNN architecture consists of layers of convolution, a layer for pooling, plus one or more connected layers. Figure 19.1 is a condensed CNN framework.

CNN has made tremendous strides in tasks like identifying and categorizing diseases. Detecting the patient's lesion, identifying the location of interest from organs in medical imaging, and registering several images in one point for comparing or consolidating the details they contain are all examples of image processing techniques. CNNs are very good at performing feature extraction tasks even though they cannot encode the relative locations of different features. The deeper layers in a CNN can only see what the top levels have passed on to them. In doing so, they eliminate the characteristics' overall context. The model efficacy gets improved by adding more filters, but computation is high [12]. Researchers have proposed several architectural adjustments as an effective remedy throughout time and resulting in attention strategies [13]. Researchers are utilizing this design for diverse computer vision applications in response to the transformer network's success in natural language processing (NLP) tasks. When extracting fine-grained features from sequences of image patches, the authors in [14] developed an update to the transformer called vision transformer (ViT). In ViT, the entire image gets sliced into patches of global attention, which focuses on the image's overall prominent elements by eliminating long-range dependencies within the feature maps. Incorporating global context into the visual characteristics while maintaining computing efficiency makes the network superior. The attention mechanism produces a slight edge over traditional CNN models in extracting deep feature representations [15–17]. The chapter focuses on vision

DOI: 10.1201/9781003407959-19

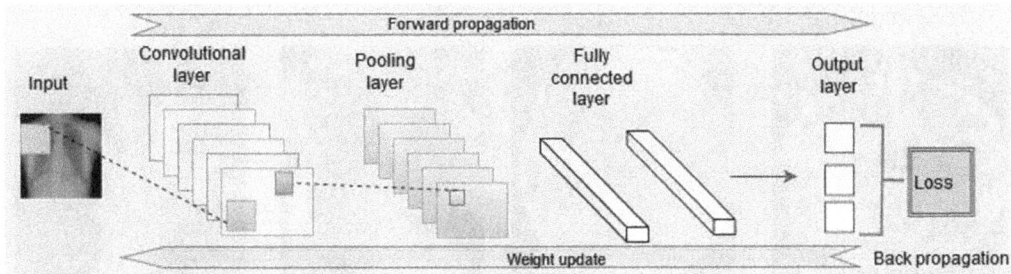

FIGURE 19.1 Simplified CNN framework.

transformers to overcome challenges to automatic disease diagnosis utilizing medical imaging modalities. The readers of this book who work in the medical and disciplines of computer vision make up our target audience.

19.2 OVERVIEW OF MEDICAL IMAGE MODALITIES

Understanding medical image modalities would help computer vision practitioners build fundamental domain knowledge and use it to diagnose and cure various illnesses. The techniques to acquire a medical image are sophisticated compared to natural images. The acquisition techniques may include physical phenomena such as electromagnetic waves, light, sound, nuclear electromagnetic resonance, and radiation to create medical imaging of the human body's external or internal organs.

Tools for capturing images, known as image modalities, have advanced because of their importance in modern healthcare delivery. These modalities are critical for monitoring patients' progress throughout therapy or the progression of disease conditions. As they optimize diagnosing accuracy, these modalities are vital for public health and preventative actions. These clinical images can record many bodily parts, including the arms, legs, brain, chest, eyes, and heart. Figure 19.2 contains examples of images generated based on some image modalities.

19.2.1 X-Ray Image Modality

X-ray is radiation similar to visible light but has higher energy to pass through opaque bodies to capture internal body parts or organs like fractures, luxation, and bone and lung diseases [18]. The conventional x-ray machine has a detector and an image generator to produce images, as depicted in Figure 19.3. There are two methods for creating x-ray images, film-screen radiography, and digital radiography. In contrast, digital radiography transforms multiple signals to produce an image [19]. Bone fractures [20], COVID-19 [21], an enlarged heart [22], lung disorders [23], arthritis [24], tooth decay [25] are just a few of the ailments that are often diagnosed with x-ray modality.

19.2.2 Computed Tomography Image Modality

Computed tomography is a computerized x-ray imagery procedure in which thin x-rays are swiftly alternated around an individual to produce image slices. These slices, known as tomographic images, contain much information compared to conventional x-rays. The computer automatically stacks multiple successive slices to create a three-dimensional (3D) image of the patient, making it simpler to discern between the patient's traits and any potential anomalies. A variety of organs and body parts are diagnosed with CT scans, including the heart [26], brain tumors [27], clots that

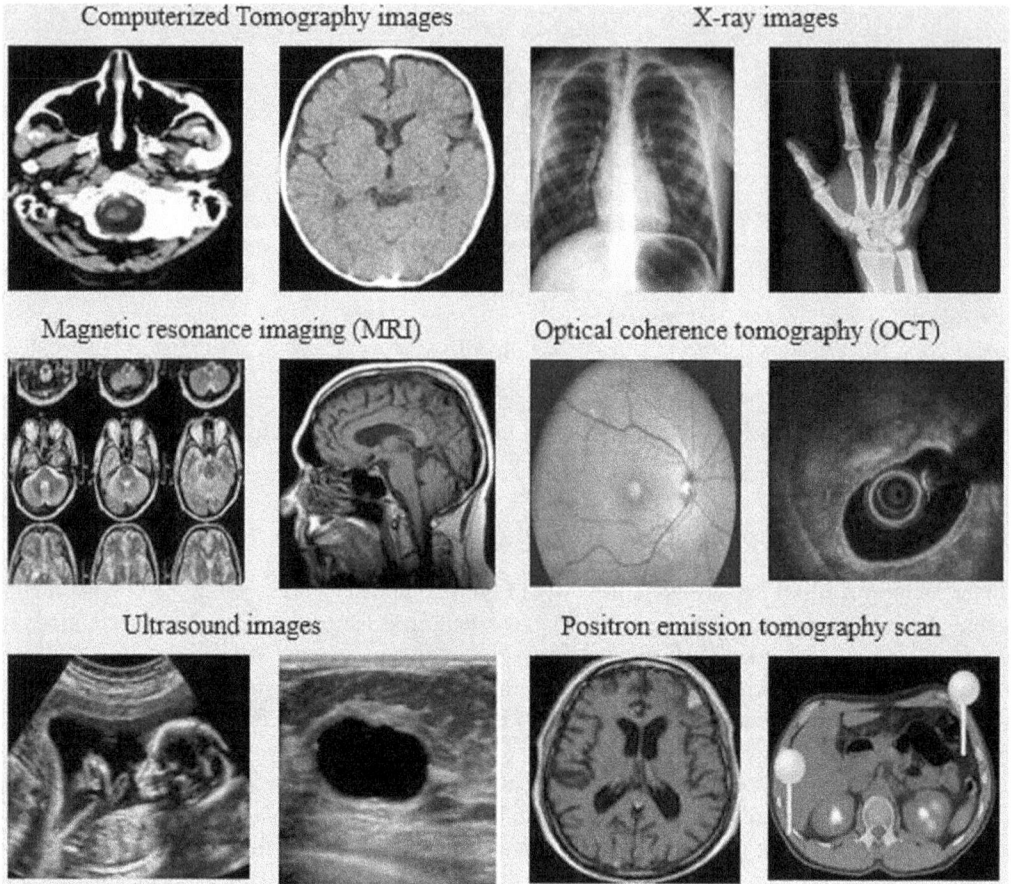

FIGURE 19.2 A list of the medical imaging technologies used by vision transformers for aided diagnosis.

cause strokes, bleeding, and other disorders [28]. Lung CT scans help detect cancer [29], tumors, and blood clots [30].

19.2.3 MAGNETIC RESONANCE IMAGING (MRI)

MRI is a quick method that creates precise images of inside body components with an MRI machine which can be either closed-bore or open, which forms the two primary varieties. Open MRI machines may offer more comfort during imaging because there is no confined environment, even though closed-bore MRI machines produce the highest-quality images. With an open MRI scanner, you typically lie between two flat magnets positioned above and under you. The space on both sides of the machine reduces the claustrophobia that individuals feel with closed-bore MRI scanners. Open MRI machines do not provide as sharp images as closed-bore MRI equipment. With close-bore MRIs, the head-to-ceiling distance is quite limited. Some people may experience anxiety and pain even though these MRI scanners produce the highest-quality images. In the advent of multiple images throughout therapy, particularly in the brain, MRI scanning is the best option because it uses no radiation. MRI is most suited for catching the body's soft tissues, although it is more expensive than x-rays and CT scans. MRI is a diagnosing tool for aneurysms, brain tumors, and distinguishing between white and gray matter in the brain and other body areas. Other applications for MRI include the spinal cord, nerves, muscles, and ligaments [31–33].

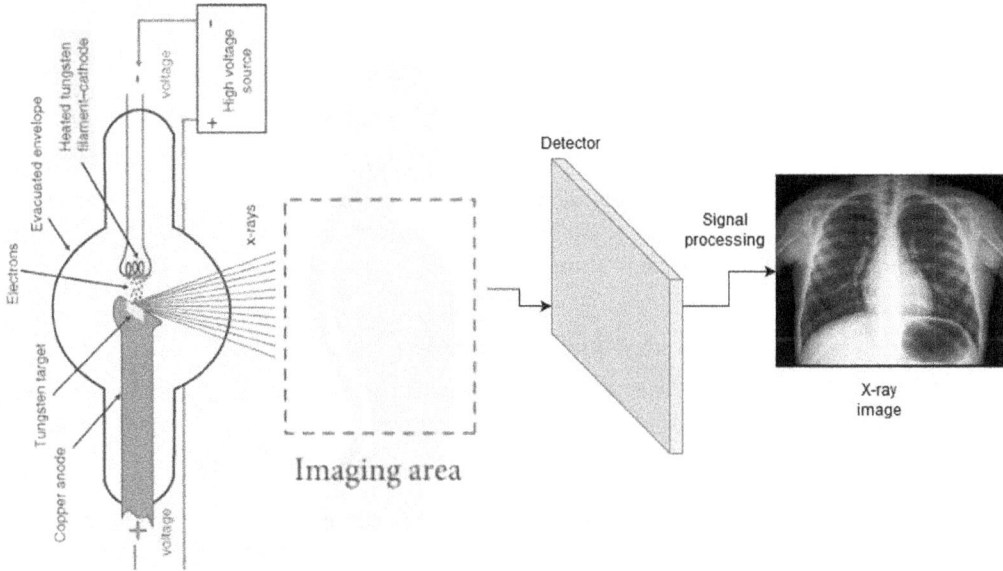

FIGURE 19.3 A schematic of an imaging system using x-rays.

19.2.4 ULTRASOUND IMAGE MODALITY

Ultrasound imaging uses sound waves to view the body's internal organs. Instantaneous ultrasound imaging shows the movement of the internal organs of the body as well as blood circulation. The test gets conducted by placing a transducer (probe) on the skin or into a body opening, where a thin coating gel is applied to the skin to enable ultrasonic waves to travel through it and enter the body. Through the body's structures reflecting the waves, ultrasound creates pictures. The strength and length of the waves moving through the body offer the data necessary to build images. Ultrasound modality captures abnormalities in the following organs: heart, eyes, brain, thyroid, blood vessels, abdominal organs, skin, and muscles. A heartbeat or blood flow through arteries can be captured in 3D by ultrasound [34, 35].

19.2.5 OPTICAL COHERENCE TOMOGRAPHY (OCT)

OCT is an imaging method that represents your retina. The non-invasive method produces an image by measuring how much dim red light reflects the retina and optic nerve. The modality can measure the dimension of the retina and optic nerve. During procedures known as cardiac catheterization, experts use OCT to capture photographs of the blood vessels for further analysis. OCT imaging aid physicians in cancer, dentistry, dermatology, gastrointestinal, and gastroenterology diagnosis. In addition, OCT can be helpful in the identification of eye problems, macular wrinkles, macular oedema, central serous retinopathy, diabetes mellitus, and macular holes [36, 37].

19.2.6 HISTOPATHOLOGY OR WHOLE-SLIDE IMAGING (WSI) MODALITY

WSI is a digitized high-resolution image from tiny tissue specimens captured on a glass slide from a biopsy. These images are taken by montaging several high-resolution tiles or strips to form a complete picture of a histological segment [38]. Slide management systems enable the efficient storage, access, analysis, and sharing of samples converted digital files. WSI gets employed in many disease diagnostics, prognoses, and therapies, including survival prediction, detection of tissue phenotypes [39], automatic grade categorization [40, 41] and segmentation of micro vessels [42].

19.2.7 POSITRON EMISSION TOMOGRAPHY–COMPUTED TOMOGRAPHY (PET–CT) MODALITY

The PET scan is a type of nuclear medicine imaging procedure that employs radioactive elements known as radiotracers to detect, evaluate, and treat bodily abnormalities. These illnesses range from gastrointestinal, endocrine, or neurological problems to cancer and heart disease. Also, they may indicate how well a patient responds to treatment [43–45].

19.3 COMPUTER VISION IN MEDICAL DOMAIN

A convolutional neural network is the go-to deep neural network in computer vision, spanning multiple disciplines. CNN systems proposed for automatic disease diagnosis aid doctors in identifying diseases effectively. A simple computer-aided diagnosing system based on computer vision is in Figure 19.4.

The transformer [13] is an attention-based model to tackle several deep learning problems, including text categorization, language processing, and question-answering [46]. This new network serves as the foundation for several cutting-edge models, including text-to-text transformers, robustly optimized BERT pre-training, generative pre-trained transformers, and bidirectional encoder representations from transformers (BERT) [47–49]. The self-attention technique that can learn sequences from an element serves as the foundational design of the transformer framework. The transformer handle has the power to handle models with high capacities, such as BERT-large [47], which has 340 million parameters, and GPT-3, which has 175 billion parameters. Another illustration is the functionality of the 1.6 trillion parameter expert switch transformer [50]. Transformer networks are now being focused on for computer vision jobs because of the performance transformations demonstrated in natural language processing. Transformer gets applied in computer vision problems like text-image synthesis, object identification, image segmentation, video understanding, visual question answering, and classification [51–53]. The network is on two fundamental

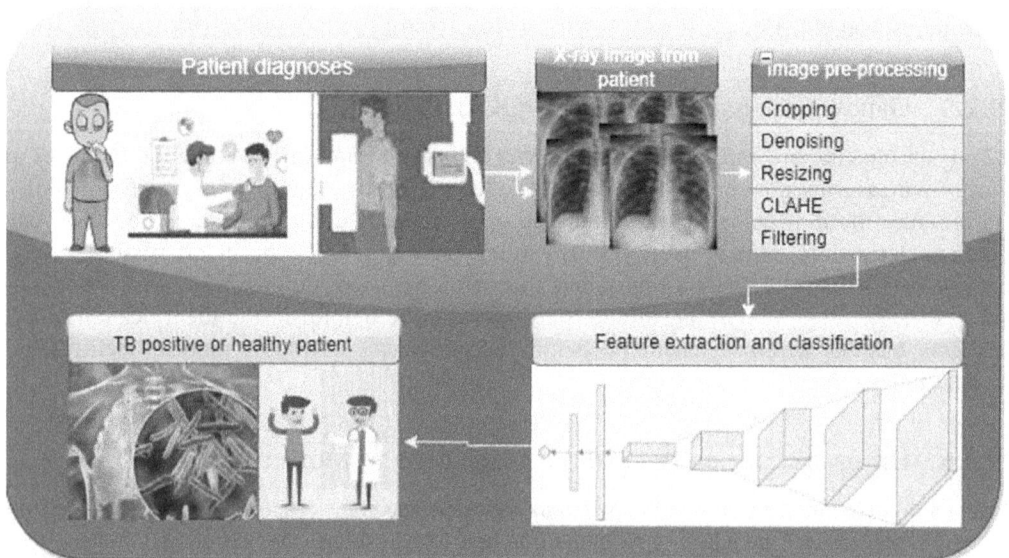

FIGURE 19.4 Computer vision-based diagnosing system.

ideas: self-attention and self-supervised learning, which entails pretraining on a large amount of unlabeled data and fine-tuning on a relatively small amount of labeled data [47, 49].

19.3.1 MEDICAL IMAGE ANALYSIS USING A VISION TRANSFORMER NETWORK

The transformer-based network's clinical use in medical image analysis includes categorization, segmentation, and detection. Tables 19.3 and 19.4 provide an overview of the transformer network suggested for classifying medical images. CNN networks excel at identifying local elements in images but struggle with identifying distant links. A DL method that can extract long-range associations in the features from these multimodal images is the vision transformer network. The vision transformer has demonstrated cutting-edge results in downstream tasks. The prominent ones are discussed below.

19.3.2 MEDICAL IMAGE CLASSIFICATION USING A PURE VISION TRANSFORMER

Direct transformers are those whose fundamental architecture does not dramatically change from the original transformer architecture. In [54] is a lightweight vision transformer framework (POCFormer) for COVID-19 identification.

The open-sourced lung ultrasonography for COVID-19 [55] dataset got used in the experiment, and the model did well, as evidenced by its accuracy of 93.9%. A comprehensive YOLOv3 network was created by the authors of [56] using ViT to classify bone femurs. The region of interest extracted with the YOLOv3 network gets forwarded to the ViT network for categorization and visualization using attention maps. The precision, recall, and F1-score values of the model were 0.77 (CI 0.64–0.90), 0.76 (CI 0.62–0.91), and 0.77 (CI 0.64–0.89), respectively. An innovative screening method for pancreatic disease classification based on 450 pancreatic ductal adenocarcinomas (PDACs) and 394 non-PDACs with the ViT network [57]. For identifying anomalies, the suggested model obtains a sensitivity of 95.2% and a specificity of 95.8%.

19.3.3 CNN AND ViT COMBINED FOR MEDICAL IMAGE ANALYSIS

Even though pure ViTs have shown excellent results in medical image analysis, substantial research into merging ViTs with CNNs extracts complicated data distributions better with improved performance. To evaluate hepatocellular carcinoma (HCC), the authors in [58] suggested a multi-function transformer regression network (mfTrans-Net) (CEMRI). TransMed [59], model got proposed for classifying multi-modal medical images. For feature extraction and establishing long-distance relationships between the modalities, the TransMed framework uses CNN and a transformer network to learn local and global information for prediction based on the MRNet datasets. The model attained 10.1% and 1.9% more accurate results. ScATNet, which combines self-attention and CNN, is a framework for classifying melanocytic skin lesions in digital WSIs [17]. The CNN component uses the image to learn patch-wise representations. The model was trained on the skin biopsy dataset [60], and when compared to other cutting-edge models developed for comparable tasks, the model fared well. A summary of model performance evaluation and employed datasets for experimentation are in Table 19.1.

A variant of the ViT network 128 was proposed for TB diagnosis from x-ray images. The suggested model combines the original vision transformer with EfficientNet. The model got trained on two combined datasets (the COVID-19 dataset and the RSNA Pneumonia Detection Challenge datasets) [61]. The suggested framework is significant for TB classification based on chest x-ray radiographs since it achieves an accuracy of 97.72% with an AUC of 100%. Females commonly die from cervical cancer, which may be automatically detected and treated in its early stages. The authors in [62] presented a transfer learning and token-to-token vision transformer (T2T-ViT) framework for

TABLE 19.1
An Overview of the Performance of the Vision Transformer Network for Disease Diagnosis

Reference	Disease	Organ/ Body Part	Image Modality	Dataset	Feature Extraction	Evaluation Metrics (%)
He et al. [15]	Brain age estimation	Brain	MRI	BGSP [65], OASIS-3 [66] DLBS [67], CMI [68]	ViT CNN	MAE = 6.85±0.65
Park et al. [16]	COVID-19	Lung	X-ray	CheXpert [69], Brixia [70], NIH ChestX-ray8 [71] datasets	ViT	AUC = 94.7
Wu et al. [17]	Melanocytic skin lesions	Skin	Skin biopsy images	Skin biopsy dataset [60]	ViT CNN	ACC = 64 AUC = 79
Liu and Yin [72]	COVID-19	Lung	X-ray	COVID-19 chest x-ray dataset [73]	Vision Outlooker [74]	ACC = 99.7
Perera et al. [54]	COVID-19	Lung	Ultrasound	Open-sourced lung ultrasound COVID-19 [55]	ViT and Linformer	ACC = 93.9
Tanzi et al. [56]	Bone fracture	Femur	X-ray	Femur images	ViT	PREC = 0.77 (CI 0.64-0.90), Recall = 0.76
Xia et al. [57]	Pancreatic disease screening	Pancreas	CT	Single-phase pancreas CTs [75] and abdominal CTs [76]	ViT	SEN = 95.2 SPEC = 95.8
Zhao et al. [58]	Liver cancer	Liver	MRI	Private datasets of 4000 images with 138 HCC subjects.	CNN and ViT	MAE = 2.35±0.25
Dai et al. [59]	Parotid gland tumors and knee injury	Head and Neck	MRI	Parotid gland tumor (PGT) dataset and the MRNet dataset [77].	Resnet, Transformer	Average accuracy = 10.1% and 1.9%
Zhao et al. [62]	Cervical cancer	Cervix	Cervical cancer smear cell images	Liquid-based cytology Pap smear dataset [78], SIPAKMeD [63], Herlev dataset [64]	ViT, CNN	ACC = 98.79±0.82, SEN = 98.61±0.62

cervical cancer diagnosis. The model was trained on the SIPAKMeD [63], Herlev [64] and liquid-based cytology Pap smear dataset. The classification accuracy rates for the model were 98.79%, 99.58%, and 99.88%. The novelty and limitations identified in these models are in Table 19.2.

A deep learning network built on a transformer-like network was presented for diagnosing COVID-19 [72]. The model's backbone is on the Vision Outlooker (VOLO) network [74]. The model was trained and verified using the suggested VOLO model customized with a transfer learning approach to utilize a larger x-ray dataset, producing outstanding performance.

19.4   VISION TRANSFORMER FOR MEDICAL IMAGE SEGMENTATION

The ViT with different variants has performed well in medical image segmentation tasks. An overview of the proposed models for image segmentation, datasets used, and focus is in Table 19.3. With medical image segmentation, U-Net is the network to use. The CNN-based architecture's weakness is that it cannot struggle with long-range dependencies. Transnet, a transformer and U-Net model [79], was suggested to remedy the drawback. In the TransClaw U-Net network [80], a

TABLE 19.2

An Overview of the Novelties, Benefits, and Restrictions Based on ViT Medical Image Analysis

Reference	Proposed method	Novelty	Advantages	Limitations
He et al. [15]	Global local transformer	Proposed a transformer- and CNN-based two-pathway network for estimating brain age.	The proposed model employs an attention method to capture the brain age information on tumor free regions of a patient.	Limited training and testing dataset
Park et al. [16]	ViT	Transformer network trained to discover low-level anomalous CXR results in the built-in large-scale dataset to incorporate feature corpus suited for high-level illness categorization.	With high generalization performance on unknown datasets, the suggested technique outperforms both baseline vision transformer and SOTA models.	Enhanced performance is possible if more training dataset is utilized
Wu et al. [17]	Scale-Aware Transformer Network (ScATNet)	Proposed an end-to-end system based on self-attention for identifying whole slide images (WSIs) at various input sizes.	The technique outperformed other cutting-edge WSI classification techniques by a wide margin.	The suggested model's overall performance got affected by the short sample size (115 WSIs) utilized to assess it.
Liu and Yin [72]	Vision Outlooker (VOLO) [74]	One of the pioneering research projects on using VOLO for image classification tasks and confirming the model's generalizability	Comparing the model performance to modern models for COVID-19 screening, it performs well.	Pre-processing methods might further enhance the performance.
Perera et al. [54]	POCFormer	An innovative ViT network design for COVID-19 diagnosis that uses ultrasound clips	Lightweight and compatible with cutting-edge technology	ViTs' performance depends on a larger dataset, the suggested model got trained on a smaller dataset, which affected the model's performance.
Tanzi et al. [56]	ViT	Free classification pipeline from CNN using the ViT network	Produced a satisfactory categorization for bone fractures.	Unbalanced class data that may have an impact on the model's performance
Xia et al. [57]	Anatomy-aware Hybrid Transformer	For pancreatic cancer detection based on non-contrast CT images through a transformer network has been proposed.	The suggested model outperforms the mean radiologists by a wide margin, obtaining excellent sensitivity and specificity on a large dataset.	The suggested system needs testing on unseen datasets for generalizability.
Zhao et al. [58]	Multi-function Trans-former regression network (mfTrans-Net)	CNN with the mfTransNet architecture, which offers a solution for multi-modality images.	Integrating multi-phase CEMRI allows for automatic multi-index quantification of HCC, which offers a quick and accurate prognosis.	Extraction of precise HCC lesion characteristics is challenging because of interference information from other anatomical structures shown in the MRI.

(Continued)

TABLE 19.2 (Continued)
An Overview of the Novelties, Benefits, and Restrictions Based on ViT Medical Image Analysis

Reference	Proposed method	Novelty	Advantages	Limitations
Dai et al. [59]	TransMed	A cutting-edge multi-modal image synthesis technique for knowledge extraction based on different modalities.	TransMed combines the advantages of CNN and Transformer to efficiently extract modest knowledge from images and create long-range links between modalities.	Because the ViT structure lacks an inductive self-attention mechanism, it performs worse when working with smaller datasets, such as medical images.
Zhao et al. [62]	CCG-taming transformers (T2T-ViT-24)	Transformer-based image generation model cervical cancer.	The model produces accurate findings, making it appropriate for cervical cancer screening.	Due to overlapping cervical cancer cells, the model performs poorly in classification.

TABLE 19.3
Overview of Models for Medical Visual Segmentation Employing Vision Transformers

Reference	Proposed Model	Organ	Modality	Dataset	Focus
Chang et al. [80]	TRANSCLAW U-NET	Abdomen	CT images	Synapse multi-organ data [86]	Combining CNNs and the transformer mechanism for image segmentation.
You et al. [81]	Class-aware transformer (CASTformer)	Aorta, Liver Kidney, Spleen, Pancreas	CT images	Synapse multi-organ CT data	The innovative adversarial transformer CASTformer is presented for 2D image segmentation.
Sha et al. [82]	Transformer-Unet	Pancreas	CT	CT82	Combine the transformer and U-Net networks to segment the area of the pancreas that is of interest.
Zhang et al. [83]	TransFuse	Tissue	Camera Colonoscopy	Kvasir [87], CVC-ClinicDB [88], CVC-ColonDB [89], EndoScene [90], ETIS [91]	Segmentation with CNN and transformer network
Karimi et al. [84]	Transformer network	Brain	MRI and CT	dHCP dataset [85]	Explore the potential of self-attention-based DL networks for 3D medical image segmentation.

combined CNN and a transformer capture brief spatial characteristics to preserve the original resolution and encrypts the patches to pull global information between sequences. The TransClaw U-Net outperforms other cutting-edge models. It achieves DSC accuracy ratings of 0.6% and HD accuracy values of 5.3%, respectively. For 2D medical image segmentation, an aggressive transformer of the class-aware transformer module (CASTformer) type is proposed [81]. The approach uses generative adversarial networks (GANs) with pyramid structures to learn detailed local and global multi-scale spatial representations. The class-aware transformer module gradually learns areas, including the structural data and underlying anatomical properties. The experiment's Synapse multi-organ CT dataset was used, and the model received Dice scores of 82.55% and 73.82%, or around a 5.88%, respectively.

Theoretically, combining the transformer and U-net would be more inventive than using either technique alone. Based on the CT82 dataset, a framework combining a transformer and U-net is proposed [82] for pancreatic segmentation. The authors' approach substituted end-to-end training for the whole model for the conventional pre-trained and fine-tuning routine. The suggested model did well, scoring 0.8301 0.7966 on the mIoU scale. For polyp segmentation, the TransFuse transformer model got proposed [83]. The process got accomplished by combining the extracted characteristics from the two modules using controlled skip-connection. The model works well by producing a cutting-edge polyp segmentation result. A convolution-free architecture that utilizes an attention mechanism for 3D image embedding was suggested in [84] based on a publicly accessible dHCP dataset [85] for segmentation. The models' performance was superior to other outstanding CNN applications proposed for similar tasks.

Notwithstanding the complexity of the transformer network on segmentation tasks, its performance to CNN models proposed for segmentation is outstanding.

19.4.1 REGISTRATION

Transformer networks are a good tool for registering medical images because they extract spatial connections between and within images. The creation of this correlation between moving and stationary images is known as image registration. A CNN and transformer-based architecture called TransUNetViT-V-Net performed volumetric brain MRI registration [92]. The model learned associations amid points in both moving and static descriptions based on 260 T1-weighted brain MRI images. The suggested ViT-V-Net network outperformed existing cutting-edge image registration techniques with a dice score of 0.7260.130.

Recent research has shown interest in deformable registration, a crucial problem in medical image processing. VoxelMorph is a suggested medical image registration model in [93] that conceptualizes registration as a function based on CNN to pin an input data pair for precise alignment. Appearance matching gets registered by maximizing image intensities in an unsupervised scenario according to the objective functions chosen for the model. The authors of [94] propose a framework for image registration and segmentation. The concept uses a contrastive patch-based unsupervised learning method to achieve complete alignment and feature representation. In its operation, PC-SwinMorph creates non-overlap patches from data input [95] to an encoder to produce patch representations. The technology is new and needs further research to fully realize its potential in image registration, notwithstanding transformer networks' potential in medical diagnosis.

19.4.2 DETECTION

Object detection in computer vision is identifying an item's position in an image and classifying it into a category. Two distinct network types for object identification (two-stage networks and one-stage networks) are employed. The two-stage networks are built on the region proposal algorithms, R-CNN [96] and faster R-CNN [97]. The alternative method operates directly on images; examples are YOLO [98] and SSD [99]. There is a compromise in accuracy and speed between the two architectures. While one-stage networks are faster, two-stage networks can achieve more accuracy. So, selecting these models for object detection relies on the task and the dataset. Biomedical applications frequently need the detection and segmentation of object instances. For example, identification of lesions, finding tumors in WSIs and obtaining bacteria information from microscopy. Modern deep-learning techniques that effectively address the scenario include attention-based transformers. The cell-DETR model got recommended to identify information from biological data

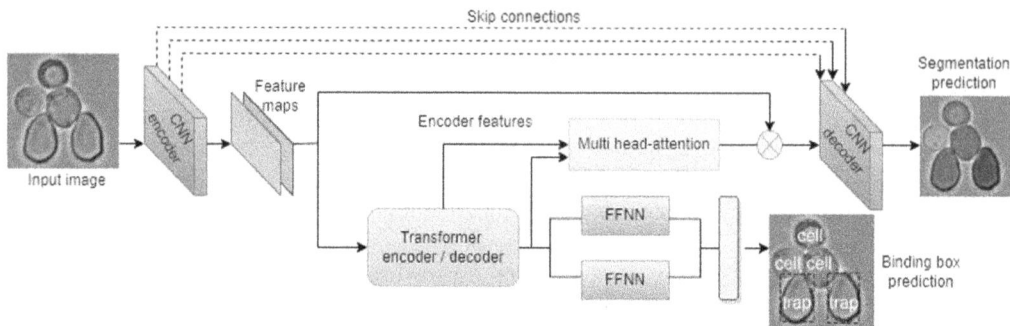

FIGURE 19.5   How the Cell-DETR network segments and detects objects.

Source: [100].

[100]. Figure 19.5 shows the suggested model architecture. The model accurately identifies each cell instance and produces segmentation maps to measure the fluorescence and cell shape.

19.5 CLINICAL REPORT GENERATION

Clinicians write thorough reports of their assessments after seeing radiological images like chest x-ray radiology, CT scans, MRIs, etc. For radiologists, this traditional approach of report drafting is laborious, time-consuming, and prone to mistakes [101]. Deep learning has made enormous strides toward producing medical reports automatically [102]. Automated report production can help clinical professionals make decisions quickly and accurately as an illustration, if an input image goes through the DL model, the output would be a comprehensive report that would include information on whether or not a tumor is present, its location, size, and other characteristics. The use of these models on medical data is image captioning. Modern models for NLP tasks, including language translation, sentiment categorization, and captioning, are produced by transformers for computer vision issues. One significant benefit is a societal influence in auto-report production in medical imaging using caption-generating models. Based on this, the TrMRG got developed to provide radioscopy reports without CNNs, which inherit spatial bias. The Indiana University Chest X-ray dataset (IU) [103] got employed for model training and evaluation [104]. An image-to-text-to-text producing system for medical image reporting based on the UI dataset got proposed based on curriculum learning [105].

With a transformer architecture, global notions are first generated from the image and then transformed into finer, more coherent language. To generate a fine-grained semantic report for radiologists, a Medical Concepts Generation Network (MCGN) was proposed [106]. This network uses RadGraph for feature extraction and classification. Using the most recent and comprehensive MIMIC-CXR dataset [107], the model got trained. For the creation of ophthalmic reports (ORG), a cross-modal clinical graph transformer (CGT) is presented [108]. CGT initially reconstructs a portion of the clinical graph from a collection of ophthalmic pictures before injecting the triples into the visual characteristics. The encoded cross-modal characteristics get decoded using a transformer decoder to forecast reports. Table 19.4 shows the various datasets employed for the report generation. Several tests on the FFA-IR benchmark dataset [109] show that the proposed CGT surpasses prior benchmarking techniques and attains cutting-edge results.

The authors in [110] suggested a deep learning model for radiologist reports generation using the IU-Xray dataset. The proposed framework in Figure 19.6 shows the model's creation, which consists of three phases. Fine-tune a pretrained Chexnet model to forecast certain tags from the images in phase 1. In phase 2, the pre-trained embeddings of the tag predictions aid in calculating weighted semantic features. Phase 3 produces comprehensive medical reports by training the GPT2

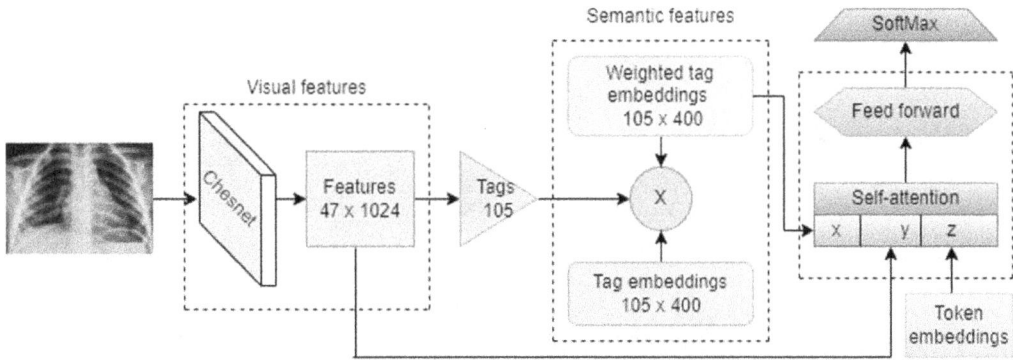

FIGURE 19.6   Proposed CDGPT2 model architecture.

Source: [110].

TABLE 19.4
Transformer-Based Report Radiological Report Generation Models with Dataset Sources

Model	Dataset	Metrics
TrMRG [104]	Indiana University Chest X-Ray dataset (IU) [103]	BLEU, METEOR, ROUGE-L and CIDEr
Progressive Transformer [105]	Indiana University Chest X-Ray dataset (IU) [103]	BLEU, ROUGE-L, METEOR
Medical Concepts Generation Network (MCGN) [106]	MIMIC-CXR dataset [107]	BLEU, ROUGE-L, METEOR, CIDEr
Cross-modal clinical Graph Transformer (CGT) [108]	FFA-IR benchmark dataset [109]	BLEU, ROUGE-L, METEOR, CIDEr
CDGPT2 [110]	MIMIC-CXR dataset [107]	BLEU, ROUGE-L, METEOR, CIDEr

model. A pre-trained transformer generated the medical reports with semantic similarity metrics based on visual and semantic data. Some of the predictions from the proposed CDGPT2 [111] model are in Figure 19.7.

19.6   DISCUSSION

Due to its superior performance over CNNs, vision transformers (ViT) is a revolutionary approach to computer vision tasks. In developing applications that can guarantee a quick and accurate diagnosis, CNNs have been tried and tested. Yet, the attention mechanism in vision transformers is necessary for the medical industry, where an exact output is crucial. ViT models outperform CNNs in terms of accuracy since they examine the image's whole context and its interpretability through the attention module. Quite recently, several proposals to explore the efficacy of vision transformers have emerged. These techniques excelled at different visual identification tasks, such as classifying, detecting, segmenting anatomical structures, and creating clinical reports. However, the full-strength vision transformer has not yet been reached and requires further research. Transformer models are still in the learning phase as they attempt to learn how to produce whole paragraphs of medical reports from sparse datasets. Most medical reports are made up of repeated and identical sentences of a descriptive character and do not explain abnormalities and diseases well. The assessment measures for the BLEU score are biased in favor of the imbalanced datasets. In the case

Normal	Accurate	Missing details	False

Ground truth

Negative. Normal cardio mediastinal silhouette is in size and contour. No focal consolidation, pneumothorax or large pleural effusion. normal	Stable cardiomegaly with mild pulmonary interstitial edema. Unchanged cardiomegaly. Negative for pneumothorax or focal consolidation. No large effusion. Mildly prominent interstitial opacities	Minimal left basilar atelectasis versus infiltrate. Low lung volumes. Normal cardio mediastinal contours. Low lung volumes with minimal left basilar opacities. No pneumothorax or pleural effusion.	Prominent hiatal hernia. Left basilar opacity compatible pleural effusion and atelectasis. Right pleural effusion. No pulmonary edema/ overt chf identified. Stable senescent mediastinal contour

Predictions

No evidence of active disease. The heart size and pulmonary vascularity appear within normal limits. The lungs are free of focal airspace disease. No pleural effusion or pneumothorax is seen	Cardiomegaly and mild interstitial pulmonary edema. Moderate cardiomegaly bibasilar and perihilar interstitial opacities. No pneumothorax	No acute cardiopulmonary abnormality. The lungs are clear, and without focal air space opacity. The cardio mediastinal silhouette is normal in size and contour. There is no mediastinal contour, large pleural effusion, pneumothorax, or focal airspace consolidation	Interval left subclavian central venous catheter with tip approximating the high svc. No evidence of pneumothorax. Generalized heart size and mediastinal contours appear within normal limits. Atherosclerotic changes of the aorta. Moderate degenerative changes of the thoracic spinae

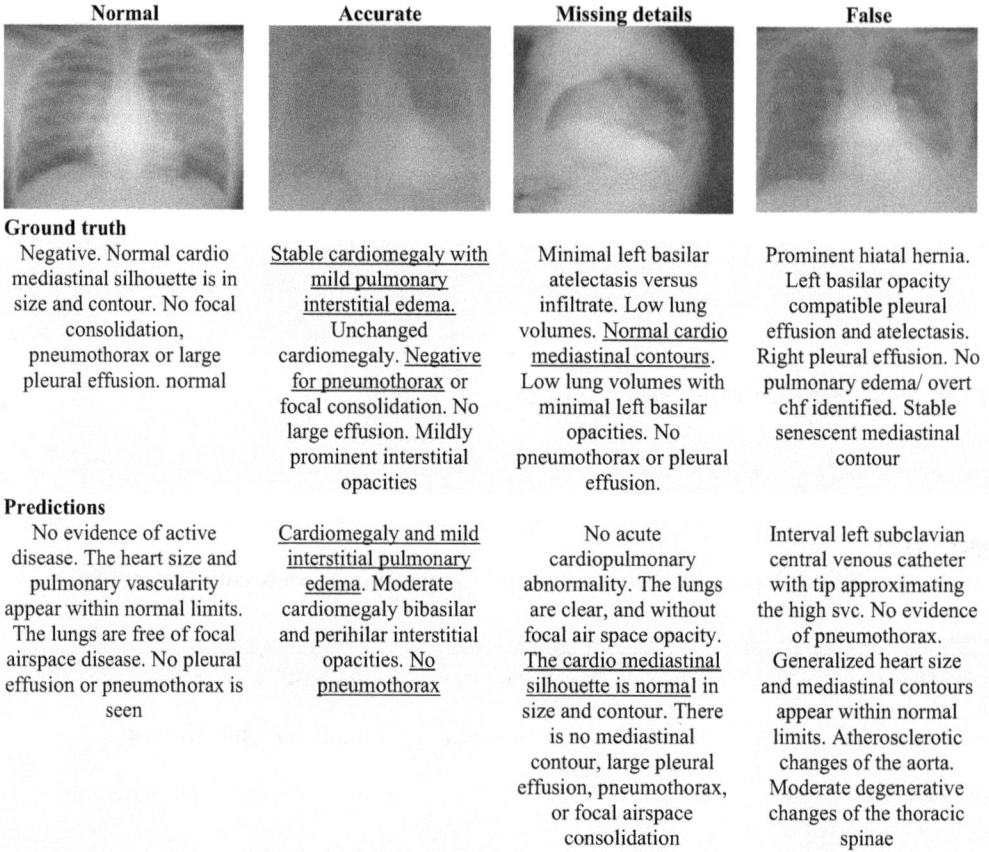

FIGURE 19.7 Examples of CDGPT2 model predictions. The underlined passages represent correct predictions in identifying anomalies and characterizing them with actual reports. In the prediction, the red text indicates any incorrect or missing information.

of medical reports, a model may get high accuracy score on these evaluation criteria by presenting only normal as the vast majority of the data consists of normal scenarios. Because of this, suggested solutions employed additional metrics, such as METEOR, ROUGE-L, and CIDEr, to effectively quantify the performance of the models to get around the BLEU scores' constraint.

19.6.1 Open Challenges and Future Directions

Thus far, we've discussed how transformers (particularly vision transformers) are used in medical image processing and analyzed cutting-edge models. There is still scope for enhancement in many areas to develop a more useable and medically accurate system by utilizing transformers, even though their efficiency is demonstrated in the previous sections by presenting their concepts and critically examining the essential facets. As a result, we highlight the difficulties and potential possibilities to aid researchers in understanding the constraints and creating more practical autonomous medical systems based on transformers.

19.6.1.1 Model Explainability and Visualization

Researchers are working to incorporate XAI methods into building transformer-based models to promote a more trustworthy and understandable system in various fields, including medical analysis.

This work is motivated by recent advancements in XAI (explainable artificial intelligence) and the introduction of algorithms that try to provide interpretable prediction in DL-based systems. Existing techniques concentrate on areas that help the model forecast by using attention maps. ViTs can deliver attention maps that highlight the essential linkages between the input and prediction regions. Nevertheless, the difficulty of quantitative instabilities and the ambiguity of the attention maps cause erroneous token associations, making interpretable ViTs an emerging research topic in computer vision, particularly for medical image analysis.

19.6.1.2   Enhanced Relational Representation

Building medical analytic systems requires an efficient and appropriate representation space. For tasks such as natural language processing (NLP), object identification, and speech recognition, transformers have shown their effectiveness in obtaining global information and catching long-term dependencies. On the other hand, CNNs excel in sifting through visual data to derive local context. To employ both local and global information simultaneously in clinical applications, numerous strategies stack transformers and CNNs (e.g., medical report generation). A multi-scaled feature representation is used, which improves performance in computer vision tasks since recent research has shown that the single-scale representation of ViTs limits improvement in dense prediction tasks. A future study might include applying this theory to ViTs in the medical field to improve clinically practical solutions.

19.6.1.3   Complex and Computationally Expensive

Larger datasets, which need enormous computational capacity for training, are necessary for the transformer network to be resilient and efficient. Extending ViTs for pretraining in new tasks and datasets is costly and burdensome. As a result, transfer learning got applied to pre-train models to solve the issue of capturing domain-specific characteristics for generalization and improved performance. Eventually, it is better to focus on building efficient transformer systems with small datasets while preserving optimal performance and robustness.

19.6.1.4   Medical Video Analysis

Extending ViT architectures to video identification challenges has drawn more and more attention from the visual field. However, due to the lack of available methods, video-based medical analysis is still in its infancy and is subject to further study.

19.7   CONCLUSION

The chapter analyzes visual transformer-based solutions proposed for downstream tasks such as classification, detection, segmentation, registration, and the creation of clinical reports. Prominent medical image modalities and their generation for analysis have been explained in this chapter. The chapter also presents information on datasets, categories, the suggested model, and assessment metrics to aid and lead the way to researchers and scholars new to medical computer vision applications. Last but not least, we list the main issues and explore several future study options to help improve research on medical image analysis for downstream tasks.

REFERENCES

[1]   S. K. Zhou *et al.*, "A Review of deep learning in medical imaging: Imaging traits, technology trends, case studies with progress highlights, and future promises," in *Proceedings of the IEEE*, vol. 109, no. 5, pp. 820–838, 2021. doi: 10.1109/JPROC.2021.3054390

[2]   I. Castiglioni *et al.*, "AI applications to medical images: From machine learning to deep learning," *Phys. Medica.*, vol. 83, March, pp. 9–24, 2021. doi: 10.1016/j.ejmp.2021.02.006

[3] E. Kotei, and R. Thirunavukarasu "Computational techniques for the automated detection of myco-bacterium tuberculosis from digitized sputum smear microscopic images: A systematic review," *Prog. Biophys. Mol. Biol.*, vol. 171, pp. 4–16, 2022. doi: 10.1016/j.pbiomolbio.2022.03.004

[4] K. Latha Bhaskaran, R. S. Osei, E. Kotei, E. Y. Agbezuge, C. Ankora, and E. D. Ganaa, "A survey on big data in pharmacology, toxicology and pharmaceutics," *Big Data Cogn. Comput.*, vol. 6, no. 4, 2022. doi: 10.3390/bdcc6040161

[5] G. A. Kaissis, M. R. Makowski, D. Rückert, and R. F. Braren, "Secure, privacy-preserving and federated machine learning in medical imaging," *Nat. Mach. Intell.*, vol. 2, no. 6, pp. 305–311, 2020. doi: 10.1038/s42256-020-0186-1

[6] J. M. Johnson and T. M. Khoshgoftaar, "Survey on deep learning with class imbalance," *J. Big Data.*, vol. 6, no. 1, 2019. doi: 10.1186/s40537-019-0192-5

[7] N. Tajbakhsh, L. Jeyaseelan, Q. Li, J. N. Chiang, Z. Wu, and X. Ding, "Embracing imperfect datasets: A review of deep learning solutions for medical image segmentation," *Med. Image Anal.*, vol. 63, no. 2019, 2020. doi: 10.1016/j.media.2020.101693

[8] P. Khosravi, E. Kazemi, M. Imielinski, O. Elemento, and I. Hajirasouliha, "Deep convolutional neural networks enable discrimination of heterogeneous digital pathology images," *EBioMedicine*, vol. 27, pp. 317–328, 2018. doi: 10.1016/j.ebiom.2017.12.026

[9] S. Dabeer, M. M. Khan, and S. Islam, "Cancer diagnosis in histopathological image: CNN based approach," *Informatics Med. Unlocked*, vol. 16, May, pp. 100231, 2019. doi: 10.1016/j.imu.2019.100231

[10] M. Porumb, S. Stranges, A. Pescapè, and L. Pecchia, "Precision medicine and artificial intelligence: A pilot study on deep learning for hypoglycemic events detection based on ECG," *Sci. Rep.*, vol. 10, no. 1, pp. 1–16, 2020. doi: 10.1038/s41598-019-56927-5

[11] E. Kotei, and R. Thirunavukarasu, "Ensemble technique coupled with deep transfer learning frame-work for automatic detection of tuberculosis from chest," *Healthcare*, vol. 10, no. 2335, 2022. doi: doi.org/10.3390/healthcare10112335

[12] E. Kotei, and R. Thirunavukarasu, "A Systematic review of transformer-based pre-trained language models through self-supervised learning," *Information*, vol. 14, no. 187, 2023. doi: doi.org/10.3390/info14030187

[13] A. Vaswani *et al.*, "Attention is all you need," *Adv. Neural Inf. Process. Syst.*, vol. 30, pp. 5999–6009, 2017.

[14] A. Dosovitskiy *et al.*, "An image is worth 16x16 words: Transformers for image recognition at scale," 2021. arXiv preprint arXiv:2010.11929

[15] S. He, P. E. Grant, and Y. Ou, "Global-Local transformer for brain age estimation," *IEEE Trans. Med. Imaging*, vol. 41, no. 1, pp. 213–224, 2022, doi: 10.1109/TMI.2021.3108910

[16] S. Park *et al.*, "Vision transformer for COVID-19 CXR diagnosis using chest x-ray feature corpus," *arXiv*, March 2021, pp. 1–10, 2021.

[17] W. Wu *et al.*, "Scale-Aware Transformers for diagnosing melanocytic lesions," *IEEE Access*, vol. 9, pp. 163526–163541, 2021. doi: 10.1109/ACCESS.2021.3132958

[18] M. Barani, M. Mukhtar, A. Rahdar, S. Sargazi, S. Pandey, and M. Kang, "Recent advances in nanotechnology-based diagnosis and treatments of human osteosarcoma," *Biosensors*, vol. 11, no. 2, pp. 1–24, 2021. doi: 10.3390/bios11020055

[19] X. Ou *et al.*, "Recent development in X-Ray imaging technology: Future and challenges," *Research*, vol. 2021, pp. 1–18, 2021. doi: 10.34133/2021/9892152

[20] M. M. Rahman, L. Dürselen, and A. M. Seitz, "Automatic segmentation of knee menisci – A systematic review," *Artif. Intell. Med.*, vol. 105, December 2018, pp. 101849, 2020. doi: 10.1016/j.artmed.2020.101849

[21] C. Sitaula, and M. B. Hossain, "Attention-based VGG-16 model for COVID-19 chest X-ray image classification," *Appl. Intell.*, vol. 51, no. 5, pp. 2850–2863, 2021. doi: 10.1007/s10489-020-02055-x

[22] D. Liu, S. Lu, L. Zhang, and Y. Liu, "Anomaly detection in chest X-rays based on dual-attention mechanism and multi-scale feature fusion," *Symmetry (Basel).*, vol. 15, no. 3, pp. 668, 2023. doi: 10.3390/sym15030668

[23] A. Souid, N. Sakli, and H. Sakli, "Classification and predictions of lung diseases from chest x-rays using mobilenet v2," *Appl. Sci.*, vol. 11, no. 6, 2021. doi: 10.3390/app11062751

[24] N. Bayramoglu, M. T. Nieminen, and S. Saarakkala, "Machine learning based texture analysis of patella from X-rays for detecting patellofemoral osteoarthritis," *Int. J. Med. Inform.*, vol. 157, October 2021, pp. 104627, 2022. doi: 10.1016/j.ijmedinf.2021.104627

[25] A. E. Yüksel *et al.*, "Dental enumeration and multiple treatment detection on panoramic X-rays using deep learning," *Sci. Rep.*, vol. 11, no. 1, pp. 1–10, 2021. doi: 10.1038/s41598-021-90386-1

[26] R. Arnaout, L. Curran, Y. Zhao, J. C. Levine, E. Chinn, and A. J. Moon-Grady, "An ensemble of neural networks provides expert-level prenatal detection of complex congenital heart disease," *Nat. Med.*, vol. 27, no. 5, pp. 882–891, 2021. doi: 10.1038/s41591-021-01342-5

[27] A. T. Alouani, and T. Elfouly, "Traumatic brain injury (TBI) detection: Past, present, and future," *Biomedicines,* vol. 10, no. 2472, 2022.

[28] M. Woźniak, J. Siłka, and M. Wieczorek, "Deep neural network correlation learning mechanism for CT brain tumor detection," *Neural Comput. Appl.*, vol. 0123456789, 2021. doi: 10.1007/s00521-021-05841-x

[29] J. Chamberlin *et al.*, "Automated detection of lung nodules and coronary artery calcium using artificial intelligence on low-dose CT scans for lung cancer screening: Accuracy and prognostic value," *BMC Med.*, vol. 19, no. 1, pp. 1–14, 2021. doi: 10.1186/s12916-021-01928-3

[30] J. Akilandeswari, G. Jothi, A. Naveenkumar, R. S. Sabeenian, P. Iyyanar, and M. E. Paramasivam, "Detecting pulmonary embolism using deep neural networks," *Int. J. Performability Eng.*, vol. 17, no. 3, pp. 322–332, 2021. doi: 10.23940/ijpe.21.03.p8.322332

[31] Z. Merali, J. Z. Wang, J. H. Badhiwala, C. D. Witiw, J. R. Wilson, and M. G. Fehlings, "A deep learning model for detection of cervical spinal cord compression in MRI scans," *Sci. Rep.*, vol. 11, no. 1, pp. 1–11, 2021. doi: 10.1038/s41598-021-89848-3

[32] M. Yang *et al.*, "A deep learning model for diagnosing dystrophinopathies on thigh muscle MRI images," *BMC Neurol.*, vol. 21, no. 13, pp. 1–9, 2021.

[33] M. J. Awan, M. S. M. Rahim, N. Salim, M. A. Mohammed, B. Garcia-Zapirain, and K. H. Abdulkareem, "Efficient detection of knee anterior cruciate ligament from magnetic resonance imaging using deep learning approach," *Diagnostics*, vol. 11, no. 1, 2021, doi: 10.3390/diagnostics11010105

[34] Y. Guo, L. Bi, E. Ahn, D. Feng, Q. Wang, and J. Kim, "A spatiotemporal volumetric interpolation network for 4D dynamic medical image," in *Proceedings of the IEEE Computer Society Conference on Computer Vision and Pattern Recognition*, 2020, pp. 4725–4734, doi: 10.1109/CVPR42600.2020.00478

[35] Z. Wang, G. Li, J. Zhou, and P. O. Ogunbona, "Optical flow networks for heartbeat estimation in 4D ultrasound images," *ACM Int. Conf. Proceeding Ser.*, November 2022, pp. 127–131, 2021. doi: 10.1145/3467707.3467725

[36] S. Asano *et al.*, "Predicting the central 10 degrees visual field in glaucoma by applying a deep learning algorithm to optical coherence tomography images," *Sci. Rep.*, vol. 11, no. 1, pp. 1–10, 2021. doi: 10.1038/s41598-020-79494-6

[37] E. Parra-Mora, A. Cazanas-Gordon, R. Proenca, and L. A. Da Silva Cruz, "Epiretinal membrane detection in optical coherence tomography retinal images using deep learning," *IEEE Access,* vol. 9, July, pp. 99201–99219, 2021. doi: 10.1109/ACCESS.2021.3095655

[38] M. Shaban *et al.*, "A novel digital score for abundance of tumour infiltrating lymphocytes predicts disease free survival in oral squamous cell carcinoma," *Sci. Rep.*, vol. 9, no. 1, pp. 1–13, 2019. doi: 10.1038/s41598-019-49710-z

[39] S. Javed *et al.*, "Cellular community detection for tissue phenotyping in colorectal cancer histology images," *Med. Image Anal.*, vol. 63, 2020. doi: 10.1016/j.media.2020.101696

[40] Y. Zhou, S. Graham, N. Alemi Koohbanani, M. Shaban, P. A. Heng, and N. Rajpoot, "CGC-net: Cell graph convolutional network for grading of colorectal cancer histology images," in *Proceedings – 2019 International Conference on Computer Vision Workshop, ICCVW 2019*, 2019, pp. 388–398. doi: 10.1109/ICCVW.2019.00050

[41] M. Shaban *et al.*, "Context-Aware convolutional neural network for grading of colorectal cancer histology images," *IEEE Trans. Med. Imaging*, vol. 39, no. 7, pp. 2395–2405, 2020. doi: 10.1109/TMI.2020.2971006

[42] S. Wazir, and M. M. Fraz, "HistoSeg: Quick attention with multi-loss function for multi-structure segmentation in digital histology images," *2022 12th Int. Conf. Pattern Recognit. Syst. ICPRS 2022*, 2022. doi: 10.1109/ICPRS54038.2022.9854067

[43] H. Guo, J. Wu, Z. Xie, I. W. K. Tham, L. Zhou, and J. Yan, "Investigation of small lung lesion detection for lung cancer screening in low dose FDG PET imaging by deep neural networks," *Front. Public Heal.*, vol. 10, 2022. doi: 10.3389/fpubh.2022.1047714

[44] B. H. Kann *et al.*, "Pretreatment identification of head and neck cancer nodal metastasis and extranodal extension using deep learning neural networks," *Sci. Rep.*, vol. 8, no. 1, pp. 1–11, 2018. doi: 10.1038/s41598-018-32441-y

[45] K. Etminani *et al.*, "A 3D deep learning model to predict the diagnosis of dementia with Lewy bodies, Alzheimer's disease, and mild cognitive impairment using brain 18F-FDG PET," *Eur. J. Nucl. Med. Mol. Imaging*, vol. 49, no. 2, pp. 563–584, 2022. doi: 10.1007/s00259-021-05483-0

[46] M. Ott, S. Edunov, D. Grangier, and M. Auli, "Scaling neural machine translation," in WMT 2018 – 3rd Conference on Machine Translation, Proceedings of the Conference, 2018, vol. 1, pp. 1–9. doi: 10.18653/v1/w18-6301

[47] J. Devlin, M. W. Chang, K. Lee, and K. Toutanova, "BERT: Pre-training of deep bidirectional transformers for language understanding," in NAACL HLT 2019 – 2019 Conference of the North American Chapter of the Association for Computational Linguistics: Human Language Technologies – Proceedings of the Conference, 2019, vol. 1, pp. 4171–4186.

[48] T. B. Brown *et al.*, "Language models are few-shot learners," in *Advances in Neural Information Processing Systems*, 2020, vol. 2020-Decem.

[49] Y. Liu *et al.*, "RoBERTa: A robustly optimized BERT pretraining approach," *arXiv preprint arXiv:1907.11692* , 2019.

[50] W. Fedus, B. Zoph, and N. Shazeer, "Switch Transformers: Scaling to trillion parameter models with simple and efficient sparsity," *J. Mach. Learn. Res.*, vol. 23, pp. 1–40, 2022.

[51] H. Touvron, M. Cord, M. Douze, F. Massa, A. Sablayrolles, and H. Jégou, "Training data-efficient image transformers & distillation through attention," *Proc. – Int. Conf. machine learning. PMLR*, pp. 10347–10357, 2021.

[52] N. Carion, F. Massa, G. Synnaeve, N. Usunier, A. Kirillov, and S. Zagoruyko, "End-to-end object detection with transformers," *Proc. – Int. Conf. computer vision. CHAM*, pp. 213–229, 2020.

[53] X. Zhu, W. Su, L. Lu, B. Li, X. Wang, and J. Dai, "Deformable DETR: Deformable transformers for end-to-end object detection," in ICLR 2021, 2020, pp. 1–16.

[54] S. Perera, S. Adhikari, and A. Yilmaz, "POCFormer: A lightweight transformer architecture for detection of covid-19 using point of care ultrasound," *Proc. – IEEE Int. Conf. on image processing. ICIP*, pp. 195–199, 2021.

[55] J. Born *et al.*, "POCOVID-Net: Automatic detection of COVID-19 from a new lung ultrasound imaging dataset (POCUS)," arXiv, 2021.

[56] L. Tanzi, A. Audisio, G. Cirrincione, A. Aprato, and E. Vezzetti, "Vision transformer for femur fracture classification," *Injury*, vol. 53, no. 7, pp. 2625–2634, 2022. doi: 10.1016/j.injury.2022.04.013

[57] Y. Xia *et al.*, "Effective pancreatic cancer screening on non-contrast CT scans via anatomy-aware transformers," *arXiv*, vol. 12905 LNCS, pp. 259–269, 2021. doi: 10.1007/978-3-030-87240-3_25

[58] J. Zhao *et al.*, "mfTrans-Net: Quantitative measurement of hepatocellular carcinoma via multifunction transformer regression network," *Lect. Notes Comput. Sci. (including Subser. Lect. Notes Artif. Intell. Lect. Notes Bioinformatics)*, vol. 12905 LNCS, September, pp. 75–84, 2021. doi: 10.1007/978-3-030-87240-3_8

[59] Y. Dai, Y. Gao, and F. Liu, "Transmed: Transformers advance multi-modal medical image classification," *Diagnostics*, vol. 11, no. 8, pp. 1–15, 2021. doi: 10.3390/diagnostics11081384

[60] J. G. Elmore *et al.*, "Diagnostic concordance among pathologists interpreting breast biopsy specimens," *JAMA*, vol. 313, no. 11, pp. 1122–1132, 2015. doi: 10.1001/jama.2015.1405.Diagnostic

[61] S. Rajaraman, and S. K. Antani, "Modality-specific deep learning model ensembles toward improving TB detection in chest radiographs," *IEEE Access*, vol. 8, pp. 27318–27326, 2020. doi: 10.1109/ACCESS.2020.2971257

[62] C. Zhao, R. Shuai, L. Ma, W. Liu, and M. Wu, Improving cervical cancer classification with imbalanced datasets combining taming transformers with T2T-ViT, vol. 81, no. 17. 2022.

[63] M. E. Plissiti, P. Dimitrakopoulos, G. Sfikas, C. Nikou, O. Krikoni, and A. Charchanti, "Sipakmed: A new dataset for feature and image based classification of normal and pathological cervical cells in pap

smear images," *Proc. – Int. Conf. Image Process. ICIP*, December, pp. 3144–3148, 2018. doi: 10.1109/ICIP.2018.8451588

[64] J. Jantzen, J. Norup, G. Dounias, and B. Bjerregaard, "Pap-smear benchmark data for pattern classification," Proc. NiSIS 2005, Albufeira, Port., January 2005, pp. 1–9, 2005.

[65] A. J. Holmes *et al.*, "Brain genomics superstruct project initial data release with structural, functional, and behavioral measures," *Sci. Data*, vol. 2, pp. 1–16, 2015. doi: 10.1038/sdata.2015.31

[66] D. LaMontagne, P.J. Benzinger, T.LS. Morris, J.C. Keefe, S. Hornbeck, R. Xiong, C. Grant, E. Hassenstab, J. Moulder, K. Vlassenko, "OASIS-3: Longitudinal neuroimaging, clinical, and cognitive dataset for normal aging and alzheimer disease," *medRxiv Prepr.*, 2016. doi: doi.org/10.1101/2019.12.13.19014902

[67] J. Park *et al.*, "Neural broadening or neural attenuation? Investigating age-related dedifferentiation in the face network in a large lifespan sample," *J. Neurosci.*, vol. 32, no. 6, pp. 2154–2158, 2012. doi: 10.1523/JNEUROSCI.4494-11.2012

[68] L. M. Alexander *et al.*, "Data Descriptor: An open resource for transdiagnostic research in pediatric mental health and learning disorders," *Sci. Data*, vol. 4, pp. 1–26, 2017. doi: 10.1038/sdata.2017.181

[69] J. Irvin *et al.*, "CheXpert: A large chest radiograph dataset with uncertainty labels and expert comparison," 33rd AAAI Conf. Artif. Intell. AAAI 2019, 31st Innov. Appl. Artif. Intell. Conf. IAAI 2019 9th AAAI Symp. Educ. Adv. Artif. Intell. EAAI 2019, pp. 590–597, 2019. doi: 10.1609/aaai.v33i01.3301590

[70] A. Signoroni *et al.*, "BS-Net: Learning COVID-19 pneumonia severity on a large chest X-ray dataset," *Med. Image Anal.*, vol. 71, pp. 102046, 2021. doi: 10.1016/j.media.2021.102046

[71] X. Wang, Y. Peng, L. Lu, Z. Lu, M. Bagheri, and R. M. Summers, "ChestX-ray8: Hospital-scale chest X-ray database and benchmarks on weakly-supervised classification and localization of common thorax diseases," in Proceedings – 30th IEEE Conference on Computer Vision and Pattern Recognition, CVPR 2017, 2017, pp. 3462–3471. doi: 10.1109/CVPR.2017.369

[72] C. Liu, and Q. Yin, "Automatic diagnosis of COVID-19 using a tailored transformer-like network," *J. Phys. Conf. Ser.*, vol. 2010, no. 1, 2021. doi: 10.1088/1742-6596/2010/1/012175

[73] M. E. H. Chowdhury *et al.*, "Can AI help in screening viral and COVID-19 pneumonia?," *IEEE Access*, vol. 8, pp. 132665–132676, 2020. doi: 10.1109/ACCESS.2020.3010287

[74] L. Yuan, Q. Hou, Z. Jiang, J. Feng, and S. Yan, "VOLO: Vision outlooker for visual recognition," IEEE Trans. Pattern Anal. Mach. Intell., pp. 1–13, 2022, doi: 10.1109/tpami.2022.3206108

[75] A. L. Simpson *et al.*, "A large annotated medical image dataset for the development and evaluation of segmentation algorithms," *arXiv preprint arXiv:1902.09063*, 2019.

[76] E. Gibson *et al.*, "Automatic multi-organ segmentation on abdominal CT with dense V-Networks," *IEEE Trans. Med. Imaging*, vol. 37, no. 8, pp. 1822–1834, 2018. doi: 10.1109/TMI.2018.2806309

[77] N. Bien *et al.*, "Deep-learning-assisted diagnosis for knee magnetic resonance imaging: Development and retrospective validation of MRNet," *PLoS Med.*, vol. 15, no. 11, pp. 1–19, 2018. doi: 10.1371/journal.pmed.1002699

[78] E. Hussain, L. B. Mahanta, H. Borah, and C. R. Das, "Liquid based-cytology Pap smear dataset for automated multi-class diagnosis of pre-cancerous and cervical cancer lesions," *Data Br.*, vol. 30, 2020. doi: 10.1016/j.dib.2020.105589

[79] J. Chen *et al.*, "TransUNet: Transformers make strong encoders for medical image segmentation," *arXiv preprint arXiv:2102.04306*, pp. 1–13, 2021.

[80] Y. Chang, H. Menghan, Z. Guangtao, and Z. Xiao-Ping, "TransClaw U-Net: Claw U-Net with transformers for medical image segmentation," *arXiv preprint arXiv:2107.05188*. 2021.

[81] C. You *et al.*, "Class-aware generative adversarial transformers for medical image segmentation," in 36th Conference on Neural Information Processing Systems (NeurIPS 2022), 2022, no. NeurIPS.

[82] Y. Sha, Y. Zhang, X. Ji, and L. Hu, "Transformer-Unet: Raw image processing with Unet," *arXiv preprint arXiv:2109.08417*, pp. 1–13, 2021.

[83] Y. Zhang, H. Liu, and Q. Hu, "TransFuse: Fusing transformers and CNNs for medical image segmentation," *Lect. Notes Comput. Sci. (including Subser. Lect. Notes Artif. Intell. Lect. Notes Bioinformatics)*, vol. 12901 LNCS, no. 1, pp. 14–24, 2021. doi: 10.1007/978-3-030-87193-2_2

[84] D. Karimi, S. D. Vasylechko, and A. Gholipour, "Convolution-free medical image segmentation using transformers," *Lect. Notes Comput. Sci. (including Subser. Lect. Notes Artif. Intell. Lect. Notes Bioinformatics)*, vol. 12901 LNCS, pp. 78–88, 2021. doi: 10.1007/978-3-030-87193-2_8

[85]   M. Bastiani *et al.*, "Automated processing pipeline for neonatal diffusion MRI in the developing human connectome project," *Neuroimage*, vol. 185, May 2018, pp. 750–763, 2019. doi: 10.1016/j.neuroimage.2018.05.064

[86]   Wikipedia, "Calvin Klein (fashion house)," Wikimedia Foundation, Inc.,. [Online]. Available: https://en.wikipedia.org/wiki/Calvin_Klein_(fashion_house) [Accessed: 17-Apr-2023]

[87]   D. Jha *et al.*, "Kvasir-SEG: A segmented polyp dataset," *arXiv*, vol. 11962 LNCS, pp. 451–462, 2020. doi: 10.1007/978-3-030-37734-2_37

[88]   J. Bernal, F. J. Sánchez, G. Fernández-Esparrach, D. Gil, C. Rodríguez, and F. Vilariño, "WM-DOVA maps for accurate polyp highlighting in colonoscopy: Validation vs. saliency maps from physicians," *Comput. Med. Imaging Graph.*, vol. 43, pp. 99–111, 2015. doi: 10.1016/j.compmedimag.2015.02.007

[89]   N. Tajbakhsh, S. R. Gurudu, and J. Liang, "Automated polyp detection in colonoscopy videos using shape and context information," *IEEE Trans. Med. Imaging,* vol. 35, no. 2, pp. 630–644, 2016. doi: 10.1109/TMI.2015.2487997

[90]   D. Vázquez *et al.*, "A benchmark for endoluminal scene segmentation of colonoscopy images," *J. Healthc. Eng.*, vol. 2017, 2017. doi: 10.1155/2017/4037190

[91]   J. Silva, A. Histace, O. Romain, X. Dray, and B. Granado, "Toward embedded detection of polyps in WCE images for early diagnosis of colorectal cancer," *Int. J. Comput. Assist. Radiol. Surg.*, vol. 9, no. 2, pp. 283–293, 2014. doi: 10.1007/s11548-013-0926-3

[92]   J. Chen, Y. He, E. C. Frey, Y. Li, and Y. Du, "ViT-V-Net: Vision transformer for unsupervised volumetric medical image registration," *arXiv preprint arXiv:2104.06468*, pp. 1–9, 2021.

[93]   G. Balakrishnan, A. Zhao, M. R. Sabuncu, J. Guttag, and A. V. Dalca, "VoxelMorph: A learning framework for deformable medical image registration," *IEEE Trans. Med. Imaging*, vol. 38, no. 8, pp. 1788–1800, 2019. doi: 10.1109/TMI.2019.2897538

[94]   L. Liu, Z. Huang, P. Liò, C.-B. Schönlieb, and A. I. Aviles-Rivero, "PC-SwinMorph: Patch representation for unsupervised medical image registration and segmentation," *arXiv preprint arXiv:2203.05684*, pp. 1–10, 2022.

[95]   D. N. Kennedy, C. Haselgrove, S. M. Hodge, P. S. Rane, N. Makris, and J. A. Frazier, "CANDIShare: A resource for pediatric neuroimaging data," *Neuroinformatics*, vol. 10, no. 3, pp. 319–322, 2012. doi: 10.1007/s12021-011-9133-y1080, 2008. doi: 10.1016/j.neuroimage.2007.09.031. Construction

[96]   P. Doll, R. Girshick, and F. Ai, "Mask R-CNN," *Proc. – IEEE Int. Conf. on computer vision.. ICCV*, pp. 2961–2969, 2017.

[97]   S. Ren, K. He, R. Girshick, and J. Sun, "Faster R-CNN: Towards real-time object detection with region proposal networks," *IEEE Trans. Pattern Anal. Mach. Intell.*, vol. 39, no. 6, pp. 1137–1149, 2017. doi: 10.1109/TPAMI.2016.2577031

[98]   Y. Su, Q. Liu, W. Xie, and P. Hu, "YOLO-LOGO: A transformer-based YOLO segmentation model for breast mass detection and segmentation in digital mammograms," *Comput. Methods Programs Biomed.*, vol. 221, pp. 106903, 2022. doi: 10.1016/j.cmpb.2022.106903.

[99]   Z. Tian, C. Shen, H. Chen, and T. He, "FCOS: Fully convolutional one-stage object detection," in Proceedings of the IEEE International Conference on Computer Vision, 2019, vol. 2019, October, pp. 9626–9635. doi: 10.1109/ICCV.2019.00972

[100] T. Prangemeier, C. Reich, and H. Koeppl, "Attention-based transformers for instance segmentation of cells in microstructures," in Proceedings – 2020 IEEE International Conference on Bioinformatics and Biomedicine, BIBM 2020, 2020, pp. 700–707. doi: 10.1109/BIBM49941.2020.9313305

[101] E. Pahwa, D. Mehta, S. Kapadia, D. Jain, and A. Luthra, "MedSkip: Medical report generation using skip connections and integrated attention," in Proceedings of the IEEE International Conference on Computer Vision, 2021, vol. 2021, October, pp. 3402–3408. doi: 10.1109/ICCVW54120.2021.00380

[102] D. You, F. Liu, S. Ge, X. Xie, J. Zhang, and X. Wu, "AlignTransformer: Hierarchical alignment of visual regions and disease tags for medical report generation," arXiv:2203.1009v1, pp. 72–82, 2021, doi: 10.1007/978-3-030-87199-4_7

[103] D. Demner-Fushman *et al.*, "Preparing a collection of radiology examinations for distribution and retrieval," *J. Am. Med. Informatics Assoc.*, vol. 23, no. 2, pp. 304–310, 2016. doi: 10.1093/jamia/ocv080

[104] M. M. Mohsan, M. U. Akram, G. Rasool, N. S. Alghamdi, M. A. A. Baqai, and M. Abbas, "Vision transformer and Language model based radiology report generation," *IEEE Access*, vol. 11, January, pp. 1814–1824, 2023, doi: 10.1109/ACCESS.2022.3232719

[105] F. Nooralahzadeh, N. P. Gonzalez, T. Frauenfelder, K. Fujimoto, and M. Krauthammer, "Progressive transformer-based generation of radiology reports," *Find. Assoc. Comput. Linguist. Find. ACL EMNLP 2021*, pp. 2824–2832, 2021. doi: 10.18653/v1/2021.findings-emnlp.241

[106] Z. Wang, M. Tang, L. Wang, X. Li, and L. Zhou, "A medical semantic-assisted transformer for radiographic report generation," arXiv:2208.10358v1, pp. 655–664, 2022. doi: 10.1007/978-3-031-16437-8_63

[107] A. E. W. Johnson *et al.*, "MIMIC-CXR-JPG, a large publicly available database of labeled chest radiographs," *Artif. Intell. Rev.*, vol. 14, pp. 1–7, 2019.

[108] M. Li, W. Cai, K. Verspoor, S. Pan, X. Liang, and X. Chang, "Cross-modal clinical graph transformer for ophthalmic report generation," Proc. IEEE Comput. Soc. Conf. Comput. Vis. Pattern Recognit., vol. 2022, June, pp. 20624–20633, 2022. doi: 10.1109/CVPR52688.2022.02000

[109] M. Li *et al.*, "FFA-IR: Towards an explainable and reliable medical report generation benchmark," *Proc. – Neural Information Processing Systems*. *NeurIPS* (Round 2), pp. 1–14, 2021.

[110] O. Alfarghaly, R. Khaled, A. Elkorany, M. Helal, and A. Fahmy, "Automated radiology report generation using conditioned transformers," *Informatics Med. Unlocked.*, vol. 24, pp. 100557, 2021. doi: 10.1016/j.imu.2021.100557

20 Ensemble Embedding and Convolutional Neural Network-Based Big Data Framework for Structure Prediction of Proteins

Leo Dencelin Xavier[1], Ramkumar Thirunavukarasu[2], Rajganesh Nagarajan[3], and Mohamed Uvaze Ahamed Ayoobkhan[4]

[1] Data Science Division, Infosys Ltd, Thiruvananthapuram, India
[2] School of Computer Science Engineering and Information Systems, Vellore Institute of Technology, Vellore, India
[3] Department of Computer Science and Engineering, Sri Venkateswara College of Engineering, Sriperumbudur, Tamilnadu, India
[4] Software Engineering, New Uzbekistan University, Tashkent, Uzbekistan

20.1 INTRODUCTION

Identifying novel patterns from biological data is becoming an interesting research topic in bioinformatics. Particularly, identifying protein structures and their functions from its sequence has been greatly recognized by pharmaceutical organizations during various stages of drug design. Proteins are organic molecules that are diverse in nature in terms of structure and function when compared to other classes of macro-molecules in living organisms. Proteins are made of z sequence of amino acids, which are responsible for catalyzing and regulating biochemical actions and are the basic unit of various structures such as skin, hair, and tendon [1]. The structure of the protein determines the molecular functions and structural similarity is considered as good predictors for functional similarity. The protein function is directly associated with proteins' 3D shape, structure, and the presence of non-protein co-factors [2, 3].

The secondary structure of proteins is referred to as the organization and arrangement of connections or bonds within the amino acid groups. Three types of secondary structures exist that include α helix, β strand, and Coils. Over a period of time, the number of protein sequences deposited in protein data banks have been exceeded and comparatively more than the secondary structures [4, 5]. Hence identification of these secondary structures and motif's (super secondary structures) are helpful to understand different human diseases, which lead to the development of novel enzymes and new drugs [6, 7].

There exists lots of experimental methods such as NMR spectroscopy, and x-ray crystallography techniques to determine the protein structures [8, 9]. However, these experimental methods are highly manual, expensive, and time consuming, which delays the drug discovery process. Traditional methods are very slow and often fail to attain certain accuracy, which demands an efficient computational model for structure prediction [10]. The heterogeneous data with the

DOI: 10.1201/9781003407959-20

provenance information contained in the protein sequences, their increased size, diversity, lack of domain knowledge, absence of standard ontologies for querying data are some of the key challenges in protein structure prediction problems. Hence, an efficient and effective structure prediction mechanism from protein sequences has emerged as an interesting task in the field of proteomics.

With the advent of machine learning-based analytical models, secondary structure prediction is more emphasized among bioinformatics researchers and computational biologists [11]. Agarwal et al. applied support vector machines with the kernel method and the Markov transition encoding scheme for classification of protein structures, which reduces the over-fitting problem that arises during protein structure classification [12, 13]. Prediction of protein structures using a variant of neural network, namely neural network pairwise interaction fields (NNPIF), has been proposed by Mirabello et al. [14]. Their approach considers each residue and extracts contextual information, which is trained to identify native-like conformations from non-native-like conformations. Bouziane et al. [15] implemented an ensemble-based approach that uses most common aggregation rules for decision inference and weighted opinion pooling. The ensemble models which have been used are artificial neural networks and multi-class support vector machines, which showed high efficiency compared to the singleton model.

Dencelin and Ramkumar [16] advocated a neural network-based MLP (multi- layer perceptron) classifier to handle the features of PSSM profile, amino acid sequences, and their properties as input for the perceptron, and implemented in a Spark-based distributed environment. The perceptron trained with different proteins from Protein Data Bank [17]. The split ratio rule of 70:30 was applied for fetching the data for training, validation, and testing phase. In another work, Xavier and Thirunavukarasu [18] proposed an ensemble tree-based approach for the protein secondary structure prediction problem. The proposed approach uses a PSSM profile, amino acid sequence along with the physio-chemical properties of proteins as features which are extracted from the input primary sequence. These features are used by the random forest model for classification of secondary structure of proteins.

A review work on structure prediction methods using a machine learning approach in distributed framework has been carried out [19]. Techniques such as support vector machine, hidden Markov model, ensemble strategies, neural networks are elaborated in the protein structure prediction problem. Also, Hadoop-based map-reduce programming, and Spark-based RDD abstraction are also discussed as part of the distributed programming model in their work. Akbar et al. [20] proposed a PSO-based neural network approach for the protein secondary structure prediction process. They have claimed their novelty by comparing with the ANN-based approaches. However, incorporation of degree of relationship of amino acids using fuzzy sets, modeling uncertainties in the relationship of various parameters of proteins' secondary structure for dynamic systems are needed to be focussed.

The capability of automatic feature extraction found in deep learning-based approaches makes a way to presume all the knowledge without losing any of the information from the input protein sequence [21]. AlGhamdi et al. [22] proposed an ensemble technique named AdaBoost and Bagging-based deep learning models to predict the protein secondary structure. They have developed a deep neural network with bagging and compared their results with the approaches such as DNN, DNN with AdaBoost, and DNN with Bagging and they have concluded a best accuracy with their approach. However, the issue of redundancy is not addressed in their work.

Guo et al. [23] investigated nearly six advanced deep learning architectures with the input features on secondary structure prediction. The deep learning architectures, such as fractal network, convolutional residual memory network are novel for protein secondary structure prediction. However, the performance of the deep learning method has been proved with independent test data sets. Hence, the authors demonstrated the emission/transition probabilities extracted from HMM profiles, which are essential for secondary structure prediction.

A restricted Boltzmann machine (RBM)-based deep learning model for predicting protein structure has been put forth by Spencer et al. [24]. The approach uses PSSM profiles, the Atchley factors (FAC) and the amino acid residues as features. They have used different windows of adjacent residues, which in-turn extract different combinations of these input features. The window sizes and the features are adjusted for attaining the network accuracy. The approach produces an accuracy rate of 80.7 %.

The sequence-based contact prediction required many homologous sequences and an adequate number of correct contacts to achieve correct folds. But it is difficult to assist nonhomologous structure modeling. Hence, Mortuza et al. [25] developed a method named C-QUARK to integrate multiple deep learning and coevolution-based contact maps to guide the replica exchange Monte Carlo fragment assembly simulations. Followed by the success of asymmetric CNN's (ACNNs) and ultra-deep neural networks, another novel method called deep asymmetric long short term memory (deepACLSTM) has been proposed by Guo et al. [26]. DeepACLSTM efficiently applies ACNNs with bidirectional long short term memory (BLSTM) for predicting protein secondary structure. The key idea behind this work is the application of asymmetric convolutional operation for extracting complex local contexts between amino-acid residues. Also, two stacked BLSTM neural networks are used in extracting long-distance interdependencies between amino acid residues.

The above discussed research attempts highlighted the potential implications of machine learning and deep learning models in protein structure prediction task. Still the following associated challenges exist. (i) **Non-account of contextual information in feature extraction:** Most of the discussed feature extraction techniques are unable to handle contextual information, which lead to less accuracy. One of the surprising observations about contextual protein vectors is that they make it easier to disambiguate sequences through various kinds of semantic and syntactic parsing when trained on large protein corpora. It shows that the vector value of a particular amino acid mainly depends on the properties of surrounding amino acids that results in an improved accuracy. Such an aspect of feature extractions has not been focused on in traditional feature extraction techniques. (ii) **Failure of models due to lack of statistical patterns:** Machine learning models are failed due to lack of statistical properties in data, as most of the features extracted might not retain the entire 360-degree view of the amino acid sequence. (iii) **Performance issue due to huge data points:** The volume and variety of data varies, which eventually requires a more sophisticated environment for efficient execution of models.

In this chapter, we have used the capability of the NLP-based vector creation approach for predicting the secondary structures of proteins. The extracted features are then loaded into a convolutional neural network-based deep learning technique for prediction. The rest of the paper has been structured as follows: In Section 20.2, the key highlights and contribution made in the paper has been discussed. The proposed big data framework based on ensemble embedding and convolutional neural network for protein structure prediction problem is introduced in Section 20.3. The performance evaluation of proposed approach based on experimental investigations is elaborated in Section 20.4. The concluding remarks and further research scope are presented in Section 20.5.

20.2   KEY HIGHLIGHTS AND CONTRIBUTIONS

This section elaborates the contribution made in the chapter.

20.2.1   EMBEDDING-BASED VECTORING TECHNIQUES

Treating the primary sequence of protein as a textual string is the idea behind our feature extraction. NLP techniques gain a solid understanding of 'language of life' by encoding the biological sequences using the conceptual analogy – 'biological sequences convey information with-in and between cells'. This is most similar like how humans use languages to communicate with each other. The basic idea behind NLP-based vectorization is that the meaning of any term in a sentence is characterized by its neighboring word. In other words, by its context. Importantly, this data representation model uses pre-built vectors, which is trained only once and can be used for various problems such as relation extraction, identifying similarities between medical terms, clustering, and classification of related context [27]. The same idea can be extended in proteomics wherein protein embeddings is mainly used for protein clustering, identifying homologous proteins, protein side-chain predictions. The vectors inferred from the pre-trained vectors will greatly cover more knowledge, thereby facilitating the researchers to effectively concentrate on the modeling part. Moreover, these vectors when combined with generated vectors of input sequence would give more accuracy.

NLP-based word embedding technique is being widely used to represent the protein sequences. This retains the position information of each amino acid available in primary sequence. Generating vectors using amino acid sequences from an input protein dataset has been initially advocated by Yang et al. in [28]. Two different variations of prediction-based models are used for embeddings' creation. Among these models, **Skip-Gram** predicts the vectors of each context amino acid from its target amino acid whereas, **CBOW** (continuous Bag of Words) calculates the vectors of target amino acid based on context amino acids. Both are unsupervised neural network-based models. However, these methods fail to capture more knowledge, as we are considering only an input protein dataset for creating embedding. In this work, we have used an ensemble-based distributed representation of biological sequence, which combines both types of embeddings. (i) 'Prot2Vec'- Pre-built vectors generated by training huge amount of protein data. (ii) Vectors generated only from the input protein sequence.

20.2.2 CONVOLUTIONAL NEURAL NETWORK-BASED PREDICTION MODEL

While comparing with machine learning techniques, deep learning models are capable of automatically extracting features and patterns from protein sequences. Deep learning consists of cascade sequence of multiple layers with the capability to perform non-linear processing, automatic feature extraction and transformation thereby deriving an additional high level of abstract features from low level input features. Also, deep learning exhibits few properties of representational learning wherein these models accept raw data in vectored form like word embedding or 'n-hot vectors'. Further, deep learning models produce promising results while implementing in a distributed processing framework. Several convolutional neural network architectures are proposed for predicting protein binding [29]. In our proposed framework, convolutional neural network-based deep learning architecture has been used, which is computationally efficient that uses convolution, special pooling operations and performs parameter sharing, and mainly preferred for sequential data. The learning model processes the input vector and classifies the secondary structure labels.

20.2.3 INCLUSION OF HIGH-PERFORMANCE COMPUTING FRAMEWORK

GPU machines are considered as the main pre-requisite for predicting protein secondary structure due to the following: (i) Volume – as accuracy is directly proportional to the volume of protein data. Most of the times, researchers use a high volume of data (~ above 10 TB), which requires efficient processing frameworks. (ii) Velocity – amino acid changes over time due to the physiochemical properties. This dynamic nature of protein data requires a real-time processing framework, which ideally requires high performance computing frameworks. (iii) Variety – protein data is highly unstructured that would appropriately need a big data framework. An open-source framework named Apache Spark [30], which facilitates resilient distributed datasets (RDD) and a distributed file system to cater bioinformatics researchers, a good platform for implementation [31]. In this work, we have leveraged Apache Spark by its capability to handle a large amount of data for structure prediction.

As a summary, this chapter explains an efficient learning pipeline that leverages NLP-based 'word embedding'. Using this, an efficient ensemble-based vectoring technique is applied, followed by a deep learning-based analytical model, which is built in a distributed framework. We have tailored our pipeline by including the entire data science life cycle process for better accuracy. The entire pipeline is implemented in Spark cluster for improved performance, which aids in helping the researchers for real-time predictions. The proposed approach can be easily customized for genomics processing and for other domains as well. In the successive sections, we present in detail on how we are extracting the low-level features using NLP and the consecutive layers of deep learning. Also, we present SPARK-based distributed computing framework, which aids in handling the exponential increasing size of protein data.

20.3 ENSEMBLE EMBEDDING AND CONVOLUTIONAL NEURAL NETWORK-BASED FRAMEWORK

The architectural framework of the proposed approach for secondary structure prediction of proteins based on ensemble embedding and convolutional neural network in a Spark-based distributed environment is shown in Figure 20.1. It provides the entire flow, starting from feature extraction using the NLP technique followed by the deep learning framework implementation in Spark. The various modules involved in our proposed approach are data ingestion, data transformation, data modeling, and evaluation. The initial phase is the data ingestion module, which accepts the input dataset in FASTA format, a text-based format for representing peptide sequences, in which amino acids are represented using single-letter codes. We have used the proteins from UniProt KB [32]. An NLP-based feature representation technique is used by embedding each amino acid sequence in an n-dimensional vector. This representation preserves the native sequential information and patterns from the input dataset by performing an iterative process on the input. Here the subsequences are extracted from the input primary sequence with n-grams, where the size of "n" is 3. The generated vectors are then used to train the convolutional neural network. The CNN used here has two layers implemented with the first layer as the input layer, followed by the convolution layer with the

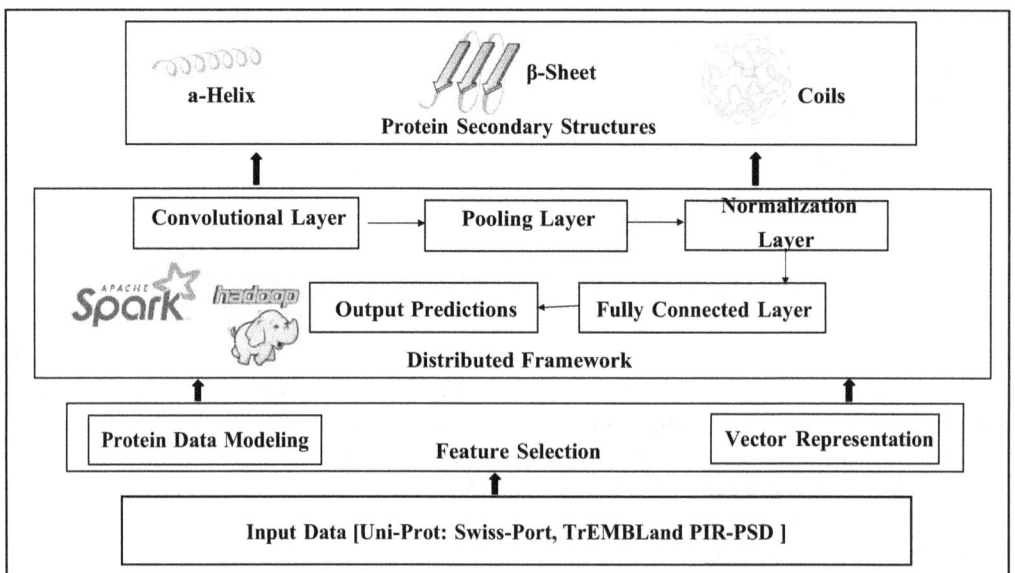

FIGURE 20.1   Proposed Big Data framework for structure prediction of proteins.

max-pooling concepts implemented. This is followed by the second convolutional layer, which in turn ends with the fully connected layer. The various network parameters are modified for better prediction accuracy. The stride and padding values are carefully chosen for performance improvement. The values are then fed into the output layer, which categorizes into α- helix, β-strand, or Coils. Apache SPARK is used to achieve parallelism. Predicted values are evaluated using accuracy metrics. Output of this model is the predicted outcomes of secondary structures.

20.3.1   Input Dataset

We have implemented this approach with all the three databases of Uni-Prot: Swiss-Port, TrEMBL, and PIR-PSD [33] and this approach is efficient in all these three. Uni-Port stands for universal protein resource, which is a central repository by combining all these three databases. This database has annotations, minimal redundancy, and capability to integrate with other DBs, which distinguishes itself from other protein sequence databases. The protein data used for our analysis contains a training data from Swiss-Prot database of 446,790 sequences. For evaluation, we have used 2,45,159 protein sequences from Swiss-Prot from about 5,065 protein families. The format used here is FASTA, which starts with a single-line description followed by lines of sequence data.

20.3.2   NLP-Based Ensemble Embeddings

Proteins are represented as sequence of amino acids. 'Skip gram' and 'CBOW' generate vectors based on input training data. The vector values mainly rely on the amount of training data, and accuracy fails if input does not have enough sequence. This leads to the usage of pre-built embeddings. In this approach, we have encoded input amino acids into an n-D vector and infer the vectors from pre-trained 'Prot2Vec' model. The inferred vectors are then combined with generated vectors to form final vectors. We split the sequence into n-grams using fixed-length splitting approach. Each n-gram is considered as a 'protein term'. A total of 20 amino acids exists, which leads to a maximum of $20*n$ grams. Fixed-length overlapping of n-grams is the easiest and most common technique for sequence data and we used the same here.

- Step 1: The first step involved in this data transformation process is to split the entire FASTA sequences into sub-sequences. Each protein sequence will be represented as three sequences, including sub-sequences 1,2,3 of 3-grams.
  Table 20.1 provides a sample overview on how we are translating the sequence of amino acid into a sequence of words. We have used a window size of 3, region size of 2, and stride length of 3 in the transformation process.
- Step 2: This step transforms the actual subsequences into feature vector. We represent every word as an $R|V|\times1$ vector and the vocabulary size 'V' is equal to the total number of unique word sequence. In our case it's the total number of unique sub-sequences.

TABLE 20.1
3-Grams of Input Protein Sequence

Protein Sequence: TETTSFLITKFSPDQQNLIFQGDGYTTKEKLTLTKA…

$k = 3$

Splitting

Sub-Sequence 1: **TET TSF LIT KFS PDQ….**

Sub-Sequence 2: **ETT SFL ITK FSP….**

Sub-Sequence 3: **TTS FLI TKF SPD….**

TABLE 20.2
n-Gram Representation of the Sub-Sequences Obtained

Sub-Sequence of One Amino Acid	*TET TSF LIT KFS PDQ....*			
	TET TSF	*TSF LIT*	*LIT KFS*	*KFS PDQ*
	.	.	1	.
	.	1	.	.
	.	.	.	.
	.	.	.	1
One hot representation	1	.	.	.
	.	.	.	.
	.	.	1	.
	1	.	.	.
	.	1	.	1
	.	.	.	.

Table 20.2 shows the 'ProtVec' vector representation of the amino acid sequence with a 2D matrix. We have created three sub-sequences from the actual single primary sequence and each of the sub-sequences produces 2D vectors. By doing so, we get a total of 2*3-dimensional vectors as input from the actual primary sequence.

The same procedure is applied on all 673,870 sequences of Swiss-Prot, and a total of 546,790 × 3 = 2,021,610 sequences of 3-grams were obtained. For each biological *n*-grams, we map it to a pre-trained embedding from the dictionary of 9,048 terms, from Swiss-Prot. Finally, the pre-trained vector is combined with generated vectors to form final vectors that cover almost all contextual information. Finally, we transform each sequence into a matrix, on which each row is the embeddings of a protein term. The matrix is then fed into a convolutional neural network [34] to predict the secondary structures. It is worth noting that, instead of creating word embeddings by training the network, which neglects few contextual information, our approach leverages an ensemble way to combine pre-trained and generated vectors to form a final vector, which eventually covers almost all contextual data.

20.3.3 CONVOLUTIONAL NEURAL NETWORK

Convolutional neural network can capture a different level of features by using special convolution, pooling and performs parameter sharing for better performance. We have used CNN architecture for classifying the secondary structure of protein into any one of the three structures: α-helix, β-strand, Coils [35, 36, 37]. The various components and its dimensions are mentioned here.

- Input layer: We have formed an input matrix of size as $M*N*D$, where M is the total number of unique word sequence and in our case we have limited to 64 unique words, 'N' is the max number of words with region size of 2. Considering the total sequence length as 600, we get $N = 100$ and 'D' is input vector dimension of value 6. Matrix input size is 64*100*6 for a single primary sequence.
- Convolutional layer: Convolutional neural networks are different from ordinary neural networks of which neurons in each layer are only sparsely connected to the neurons in the next hidden layer based on the relative location. Reducing the amount of parameters is important to reduce the training complexity. We have used a convolutional layer of size 64*100*6 and have initialized the weight as 3*3*6. We have used two filters of the same

TABLE 20.3
Network Parameters Used and Their Values

Layers Used	Parameters	Values
Input Layer	Input matrix used is $M1*N1*D1$	$M = 64, N = 100, D = 6$
Convolution Layer	Input – 64*100*6	$K = 2, S = 2, F = 3, P = 1$
	Conv parameters	Size of output = 21*31*2
	K – Number of filters	
	P – Zero padding size S-Stride	
	F – Spatial extent	
	Size of Output Volume	
	$M2 = (M1–F+2P)/S+1$	
	$N2 = (N1–F+2P)/S+1$	
	$D2 = K$	

dimensions, and the output of each filter is stacked together to form the depth dimension of the convolved output.

- Stride and padding: A stride factor of size 2 is used to convolute the input and is padded with zeroes (same padding) around the matrix. By doing so, the original values are preserved without any data loss.
- Activation functions: Is a function which is used to transform input value to the output value of neuron. Two important factors determining the quality of activation function that includes ability to handle the diminishing gradient flow to the subsequent layers of the neural network. The most commonly used activation function used in a general ANN is sigmoid and deep learning architectures uses rectified linear units (RELU) and softmax functions. In this study we have used RELU as the activation function. The formula for RELU is $f(x) = \max(x,0)$.
- Pooling: Pooling in general is a form of dimensionality reduction used in convolution neural networks. Its goal is to throw away unnecessary information and only preserve the most critical information. We have used a stride of 2 to get a pooling size of 10.
- DropOut /DropConnect: Each neuron must be able to adapt to the dropped activation values in the neural network. This makes the neurons more adaptive and less restricted to the existing architecture of the neural network. This approach gives more generalized results. Dropout considers only a few neurons using ensemble learning by leveraging more networks at a time, whereas, DropConnect considers only a few weights based on learning results.
- Output layer: We have used the Softmax classification layer as the final layer of the classifier. It uses log probability distribution and is linear in nature. Table 20.3 shows various network parameters and the values used.

20.3.4 Accelerating the Performance

The open-source framework Apache SPARK is used for feature extraction and convolutional neural network. Cluster configuration is here: (1) One EC2 RHEL 7.8 machine(r3.2xlarge). (2) EMR – 1 master node(r3.2xlarge) and 1 core node(r3.2xlarge). Hardware used are as follows: 8 cores with RAM >=128 GB, storage >= 4TB. Complete implementation is using Python 3.6 with all the libraries (Keras, tensorflow, scikit-learn) installed for feature extraction and deep learning model implementation.

20.4   EXPERIMENTAL EVALUATION

The performance of our approach is being measured using various parameters of the network and are modified for better prediction accuracy.

20.4.1   Network Parameters

As performance is tightly coupled with epochs, we have evaluated the performance by varied no. of epochs during training phase. Figure 20.2 depicts that accuracy value increase as the number of epoch increases. We have used different learning rates starting from smaller rates (0.01) and was gradually increased with varied training sets and number of network layers. The model took time to convergence for the smaller learning rates. After that, we have increased the learning rate and found that the loss becomes high. Again, we have gradually decreased the learning rate after each epoch with different training sets thereby facilitating faster training. The learning rates are modified based on two schedules:

1. Drop-based – we have systematically dropped the learning rate based on epochs during the training phase. For this, 0.01 is the initial learning rate and for every 20 epochs it will be dropped. By doing so, first 10 epochs will use 0.1 as the learning rate and the next 10 epochs will use 0.05 and so on.
2. Time-based – the formula used is as below:
   Learning Rate = Learning Rate * 1/(1 + decay * epoch)
   • If decay is zero, there is no effect in learning rate.
   • If decay is specified, there is a decrease in learning rate by the fixed amount from the previous epoch.

20.4.2   Confusion Matrix

A confusion matrix is often used to describe the performance of a classification model on a set of test data for which the true values are known. Confusion matrix entries are shown in Table 20.4.

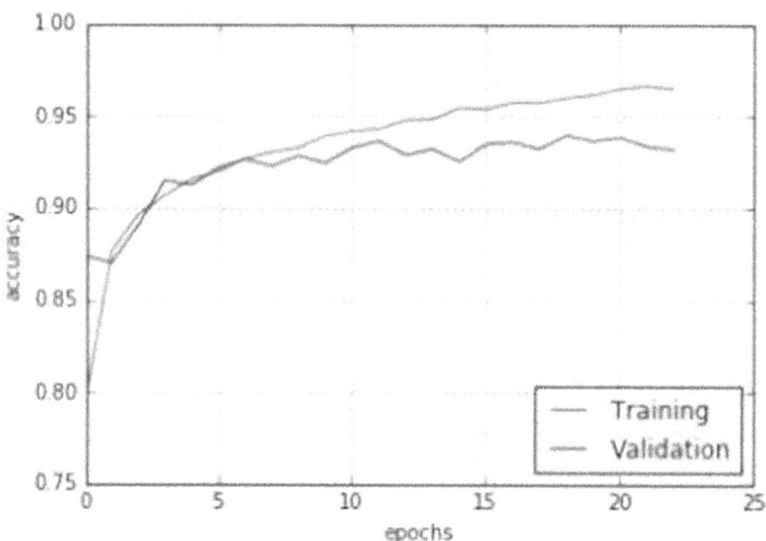

FIGURE 20.2   Plots showing accuracy vs epochs.

TABLE 20.4
Confusion Matrix Obtained for SWISS-PROT Dataset

Secondary Structure – Predicted

	α Helix	β Sheet	Coils
α Helix	79.10%	11.00%	9.90%
β Sheet	10.37%	77.32%	12.31%
Coils	10.53%	5.26%	84.21%

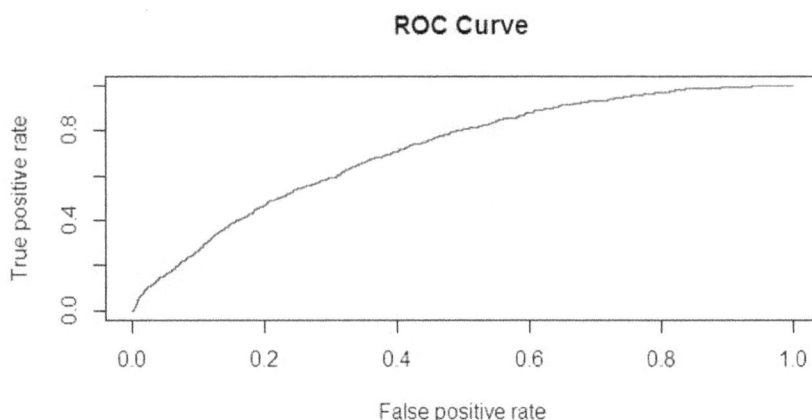

FIGURE 20.3   ROC graph obtained using the confusion matrix values.

ROC is a plot of true positive rate (TPR) against the false positive rate (FPR) for the different possible cut points of the diagnostic test. Figure 20.3 shows the optimal values between sensitivity and specificity (any increase in sensitivity leads to a decrease in specificity).

20.4.3   CLUSTER PARAMETERS

Cluster performance is validated by varying the cluster parameters. We have added a greater number of nodes to attain an optimal performance. Figure 20.4 shows the histogram plotted against the number of instances and running time. It's clearly depicted that, performance increases with an increased number of clusters.

20.4.4   RESULTS AND DISCUSSION

We have evaluated a few approaches for predicting the secondary structures using deep learning architectures. As it is shown in Table 20.5, the ab initio methodology that has been proposed in [24] and achieved (80.7%) Q3 accuracy. This approach provides an extra input profile extraction phase, which extracts the PSSM, Atchley's factors and amino acid (AA) residues from amino acid sequence. In this approach, different window sizes with different combinations of the input features are considered to train the deep learning network. RBNs are used here. Our approach achieved 84.9% accuracy on the test set sequences using a Swiss-Prot protein dataset, which contains annotations of the secondary structures along with the primary sequence. The proposed model suggests that the combination of NLP together with the convolutional neural network is well suited for sensitive

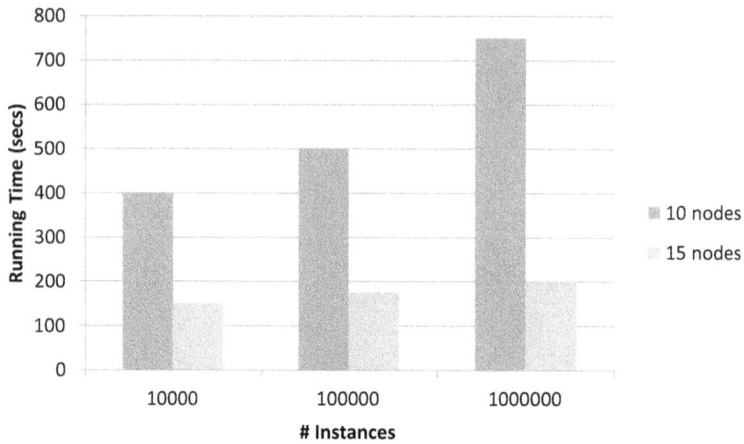

FIGURE 20.4 Protein sequence size and execution time with a varied number of nodes in SPARK cluster.

TABLE 20.5
Comparison of Structure Prediction Techniques: Ab Initio Approach vs NLP-Based Approach

Approaches and Accuracy Comparison

Deep Learning Model	Feature Extraction Method	Accuracy
Deep Neural Network with RBMs [24]	Ab Initio Method Using Input Profile Features using PSSM Profile, Atchley's Factors and AA Residues.	80.7%
Proposed Approch: Word Embeddings using CNN	NLP-Based Feature Representation.	84.9%

TABLE 20.6
Experimental Results of the Three Uni-Prot Dataset

Experimental Results of Three Datasets Used

Dataset	Accuracy	Hidden Layers	Learning Rate ($\dot{\eta}$)	Epochs
Swiss-Prot	84.90%	2	0.3	30
TrEMBL	84.00%	2	0.5	30
PIR-PSD	82.20%	2	0.3	30

data and NLP works best to extract the precise low-level information without any data loss, with improved accuracy compared to traditional methods.

Table 20.6 shows the individual accuracies attained along with learning rates and epochs for Swiss-Prot, TrEMBL, PIR-PSD. Among these results, Swiss-Prot attained good results.

20.5 CONCLUSION AND FUTURE WORK

Convolutional neural networks are capable of learning sophisticated features and play an important role in bioinformatics. This chapter depicts the NLP-based feature representation, together with deep learning approach implemented in a distributed framework. NLP together with deep learning is one of the best approaches to handle protein sequential data. The proposed algorithm achieved an accuracy of 84.9%. Our results clearly confirm that NLP together with deep learning are more accurate than other machine learning techniques. As a final note, natural language processing together with the convolutional neural network in distributed environment improves accuracy and speed, which is required for efficient proteomic study. Application of deep learning in bioinformatics and computational biology is in the research stage, which paves the way for our future investigations on improving the performance and accuracy by investigating new algorithmic methods for identifying the functions, with more parallelized methods in improving the computational performance.

20.5.1 COMPLIANCE WITH ETHICAL STANDARDS AND CONFLICT OF INTEREST

The authors have no conflicts of interest to declare that are relevant to the content of this article.

20.5.2 DATA AVAILABILITY

All datasets used in this study, including training, testing is available in the uniport knowledgebase (www.uniprot.org/).

REFERENCES

1. Ibtehaz N, & Kihara D (2023). Application of sequence embedding in protein sequence-based predictions. In *Machine Learning in Bioinformatics of Protein Sequences: Algorithms, Databases and Resources for Modern Protein Bioinformatics* (pp. 31–55).
2. Wu C. S, & Cheng L (2023). Recent advances towards the reversible chemical modification of proteins. *ChemBioChem.* 24(2): e202200468.
3. Yu Y, Liu X, & Wang J (2021) Protein engineering using unnatural Amino Acids. *Protein Engineering: Tools and Applications*: 243–264.
4. Bittrich S, Bhikadiya C, Bi C, Chao H, Duarte J. M, Dutta S, ... & Rose Y. (2023) RCSB protein data bank: Efficient searching and simultaneous access to one million computed structure models alongside the PDB structures enabled by architectural advances. *J. Mol. Biol.*: 435, 14.
5. Rose Y, Duarte JM, Lowe R, Segura J, Bi C, Bhikadiya C, Westbrook JD (2021) RCSB protein data bank: Architectural advances towards integrated searching and efficient access to macromolecular structure data from the PDB archive. *J. Mol. Biol.* 433(11): 166704.
6. Anand U, Bandyopadhyay A, Jha N. K, Pérez de la Lastra J. M, & Dey A (2023). Translational aspect in peptide drug discovery and development: An emerging therapeutic candidate. *BioFactors*, 49(2): 251–269.
7. Kieffer C, Jourdan JP, Jouanne M, Voisin-Chiret AS (2020). Noncellular screening for the discovery of protein–protein interaction modulators. *Drug Discovery Today*, 25(9), pp.1592–1603.
8. Srivastava A, Nagai T, Srivastava A, Miyashita O, Tama F (2018) Role of computational methods in going beyond x-ray crystallography to explore protein structure and dynamics Int. *J. Mol. Sci.* 19(11): 3401.
9. Lin X, Li X, Lin X (2020) A review on applications of computational methods in drug screening and design. *Molecules.* 25(6): 1375.
10. Wang Y, Mao H, Yi Z (2017) Protein secondary structure prediction by using deep learning method Knowledge-Based Syst. 118: 115–123.
11. Oldfield CJ, Chen K, Kurgan L (2019) Computational prediction of secondary and supersecondary structures from protein sequences. *Methods Mol. Biol.* 1958: 73–100.
12. Shivani Agarwal DM, Pankaj Agarwal (2014) Prediction of protein secondary structure content using support vector machine. *Int. J. Comput. Appl.* 71(5): 2069–2073.

13. Chakraborty A, Mitra S, De D, Pal AJ, Ghaemi F, Ahmadian A, Ferrara M (2021) Determining protein-protein interaction using support vector machine: A review. *IEEE Access*, 9: 12473–12490.

14. Mirabello C, Adelfio A, Pollastri G (2014) Reconstructing protein structures by neural network pairwise interaction fields and iterative decoy set construction. *Biomolecules*. 4(1): 160–180.

15. Bouziane H, Messabih B, Chouarfia A (2015) Effect of simple ensemble methods on protein secondary structure prediction. *Soft Comput*. 19(6): 1663–1678.

16. Dencelin LX, Ramkumar T (2016) Analysis of multilayer perceptron machine learning approach in classifying protein secondary structures. *Biomed. Res*. 2016(2): S166–S173.

17. "RCSB Protein Data Bank – RCSB PDB" *Nucleic Acids Res*. 2000. [Online]. Available: www.rcsb.org/pdb/home/hom.do [Accessed: 03-Sep-2021]

18. Xavier LD, Thirunavukarasu R (2017) A distributed tree-based ensemble learning approach for efficient structure prediction of protein. *Int. J. Intell. Eng. Syst*. 10(3): 226–234.

19. Dencelin L, Ramkumar T (2017) Distributed machine learning algorithms to classify protein secondary structures for drug design – a survey. *Res. J. Pharm. Technol*. 10(9): 3173–3180.

20. Akbar S, Pardasani KR, Khan F (2021) Swarm optimization-based neural network model for secondary structure prediction of proteins. *Netw. Model. Anal. Health Inform. Bioinform*. 10(1): 1–9.

21. Zhang L, Tan J, Han D, Zhu H (2017) From machine learning to deep learning: Progress in machine intelligence for rational drug discovery. *Drug Discov. Today*. 22(11): 1680–1685.

22. AlGhamdi R, Aziz A, Alshehri M, Pardasani KR, Aziz T (2021) Deep learning model with ensemble techniques to compute the secondary structure of proteins. *J. Supercomput*. 77(5): 5104–5119.

23. Guo Z, Hou J, Cheng J (2021) DNSS2: Improved ab initio protein secondary structure prediction using advanced deep learning architectures. *Proteins: Structure, Function, and Bioinformatics* 89(2): 207–217.

24. Matt Spencer JC, Jesse Eickholt (2015) A deep learning network approach to ab initio protein secondary structure prediction. *IEEE/ACM Trans Comput Biol Bioinform*. 12(1): 103–112.

25. Mortuza SM, Zheng W, Zhang C, Li Y, Pearce R, Zhang Y (2021) Improving fragment-based ab initio protein structure assembly using low-accuracy contact- map predictions. *Nat. Commun*. 12(1): 1–12.

26. Guo Y, Li W, Wang B, Liu H, Zhou D (2019) DeepACLSTM: Deep asymmetric convolutional long short-term memory neural models for protein secondary structure prediction. *BMC Bioinform*. 20(1): 1–12.

27. Wang Y, Liu S, Afzal N, Rastegar-Mojarad M, Wang L, Shen F, Kingsbury P, Liu H (2018) A comparison of word embeddings for the biomedical natural language Processing. *J. Biomed. Inform*. 87: 12–20.

28. Yang KK, Wu Z, Bedbrook CN, Arnold FH (2018) Learned protein embeddings for machine learning. *Bioinform*. 34(15): 2642–2648.

29. Zeng H, Edwards MD, Liu G, Gifford GK (2016) Convolutional neural network architectures for predicting DNA-protein binding. *Bioinform* 32(12): i121–i127.

30. Apache Software Foundation. "Apache sparkTM – unified analytics engine for big data," *Apache Spark*, 2018. [Online]. Available: https://spark.apache.org/ [Accessed: 03-Sep-2020]

31. Zaharia M, Xin RS, Wendell P, Das T, Armbrust M, Dave A, Meng X, Rosen J, Venkataraman S, Franklin MJ, Ghodsi A (2016) Apache spark: Engine for big data processing. *Commun. ACM*. 59(11): 56–65.

32. "UniProtKB." [Online]. Available: www.uniprot.org/help/uniprotkb [Accessed: 03-Sep-2020]

33. "UniProt: The universal protein knowledgebase in 2021." *Nucleic Acids Res*. 49, no. D1 (2021): D480–D489.

34. Baek M, DiMaio F, Anishchenko I, Dauparas J, Ovchinnikov S, Lee GR, Baker D (2021) Accurate prediction of protein structures and interactions using a three-track neural network. *Science*. 373(6557): 871–876.

35. Jiang Q, Jin X, Lee SJ, Yao S (2017) Protein secondary structure prediction: A survey of the state of the art. *J. Mol. Graph*. 76: 379– 402.

36. Dougherty DA (2000) Unnatural amino acids as probes of protein structure and function. *Curr Opin Chem Biol*. 4(6): 645–652.

37. Burley SK, Berman HM, Kleywegt GJ, Markley JL, Nakamura H, Velankar S (2017) Protein Data Bank (PDB): The single global macromolecular structure archive. *Methods Mol. Biol*. 1607: 627–641.

21 Deep Learning-Based Automated Diagnosis and Prescription of Plant Diseases

R.K. Kapila Vani, P. Geetha, D. Abhishek, K. Gokul Krishna, and V. Akaash
Department of Computer Science and Engineering, Sri Venkateswara College of Engineering, Pennalur, Sriperumbudur, India

21.1 INTRODUCTION

Research in machine learning (ML) focuses on comprehending and developing "learning" methods. ML has shown to be useful because it can answer problems faster and on a larger scale than human intelligence. It is feasible to train machines to recognize patterns and links between incoming data as well as to automate repetitive tasks using the massive amounts of processing power behind a single activity or a number of related activities. Without being explicitly coded, ML algorithms build a computer model using sampling data, often known as "training data", to make predictions or judgements. These data trends can help businesses make wiser decisions, operate more responsibly, and capture a lot of information

A branch of machine learning called "deep learning" uses algorithms that are modeled after the biological brain's operational principles. The goal of deep learning (DL) is to create a network of nodes that mimics a single human neuron in order to replicate human neurons. The nodes and weights in a typical artificial neural network are dispersed throughout multiple layers. The three sorts of fundamental node layers are input nodes, where input data is given, output nodes, where output prediction is seen, and a hidden layer in between, which looks for patterns in the incoming data. The neural network is trained on the existing dataset using the initial weight biases that have been assigned to each node. Utilizing the required input, weights are modified.

Deep-learning architectures have been applied to a number of fields, including material assessment, drug development, medical imaging, video analysis, and language [1] processing, automatic translation. A multilayer perceptron-like system used by a convolutional neural network (CNN) is used, it has been tuned for quick processing. It features three layers, as depicted in Figure 21.1, including a basis layer, a result layer, and a hidden layer that contains several convolutional, pooling, fully connected, and normalization layers. Images can be transformed into a series of vectors and used as input.Convolution neural networks is commonly used in image recognition. The main distinction between CNN and its predecessors is the automatic recognition of important traits that eliminates the need for human intervention. Two types of data can be combined using the mathematical technique of convolution. Convolution is used in the CNN example [2] to clean up the input and create a feature map. This filter, which is also known to be a kernel, may vary in size from 3×3 to 4×4. The kernel iterates over the input image and multiplies each element's matrices to execute convolution. For each receptive field, the feature map captures the outcome.The creation of a feature map has several applications and one among them is shrinking the size of the input image, and one should be aware that the larger one's strides, the smaller the feature map.

The term "smart agriculture" describes the application of technology like the Internet of Things, sensors, navigational aids, robots, and artificial intelligence to farms. The lack of contemporary

DOI: 10.1201/9781003407959-21

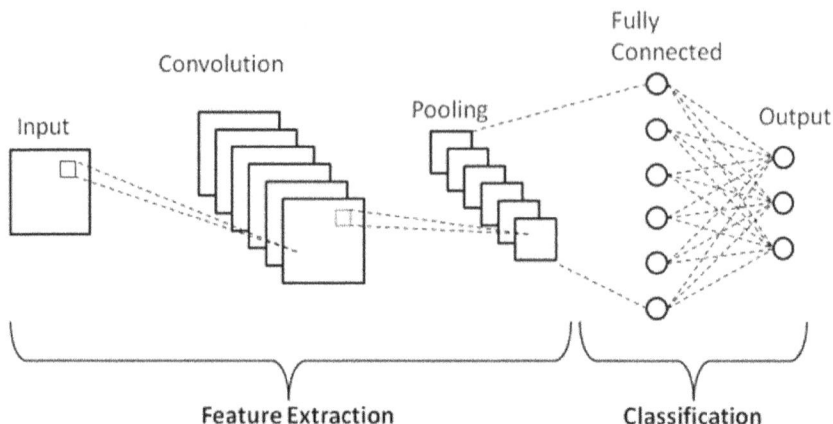

FIGURE 21.1   Layers of the convolutional neural network.

methods for disease identification has mostly hampered crop production. Thus, prompt detection and treatment of numerous crop diseases can help farmers to use effective management techniques. The inability of farmers to recognize diseases in remote areas and the lack of agricultural specialists are two additional problems that reduce the overall harvest productivity. The application of machine learning and deep learning algorithms for identification of plant diseases will serve as a potential solution to the problem.

21.2   LITERATURE REVIEW

Fegade et al. [3] described the development of an automated tool that examines diseased paddy photographs and gives farmers recommendations. The main goal of developing a system for classifying rice diseases is to make it simpler to identify and classify rice diseases using vector assistance and artificial neural networks. Numerous variables are included in the crop forecast, including the amount of precipitation, the lowest and highest temperatures, the type of soil, humidity levels, and the relevance of the soil pH. The agriculture website for Maharashtra provided the data. Nine farming regions were divided up into the data.

Sunil et al. [4] proposed a solution architecture that included a data preprocessing step with background removal. The majority of the time, computer vision algorithms—like OpenCV's image thresholding—removes the background from an image. These methods are useful when the backdrop color contrasts with the fascinating object. In these situations, it is simple to eliminate the background and substituting it with a different color or an image by using green and blue screens. The same technique cannot be used to remove the leaf from a noisy background since there is so much detail to be captured and extracted. They employed the U2Net model to remove the background from the leaf pictures. A mask of the desired area is made from the provided image. Additionally, it performs a binary arithmetic operation on the U2-Net mask and the source image. After eliminating the background, they used the EfficientNet model to make their forecast. After comparing the different EfficientNet versions, they decided to use EfficientNetV2, which was more accurate. Their algorithm had a 98.26% accuracy rate.

More attention should be paid to fuzzy sets, according to Omrani et al. [5]. The brightness level per pixel in the images should be taken into account when estimating the potential quantity of fluidity. When the fuzzy package properly handles picture ambiguity, IFSs are handled. When the satellite can calculate the segmentation's action, the number of uncertain capture photos can be reduced. Depending on the space between the fuzzy set, the clustering technique's classification of the crops' shortcoming will combine an image.

Sabbir Ahmed et al. [6] suggested a light model for mobile apps. It starts with data augmentation in which the dataset training images are flipped and rotated randomly to increase the volume of the dataset. The model they used in their solution is a CNN model called "MobileNet". This model is purely built and designed to run on a mobile system with limited resources. This model has given them an accuracy of 99.3% on the tomato leaf dataset. The main setback in their approach is the lack of data preprocessing is very minimal.

Helong et al. [7] suggested utilizing a method employing K-means clustering and an upgraded deep-learning model to reliably analyze the three prevalent illnesses of leaves from maize such as gray colored spots and rust. Before being included into the revised deep-learning model, sample images were first categorized using K-means to distinguish between these three disorders. In this work, they examined the detection of maize disease in relation to various k values, models like ResNet18, Inception v3, VGG-16, and the improved deep-learning model. The results of their investigation demonstrated that the technique had the best recognition accuracy on 32-means specimens with recall values for the illnesses leaf spot, rust, and gray spot of 89.24%, 100%, and 90.95%, respectively.

According to the researchers Surampalli et al. [8], a study on the detection of tomato leaf diseases employing deep-learning methods should be carried out in 2020. This suggested approach is proposed for the tomato crop in order to detect crop leaf disease utilizing image-preparation techniques. It is dependent on image segmentation and open-source methods with a focus on tomato crops, all of which help to categorize leaf disease in a reliable, secure, and precise manner. A suitable prediction scenario is used to handle the specifics of other plant-assisted illnesses in a broader sense. However, the primary focus is on the tomato crop and the diseases that are related to it.

The most advanced computer vision research to date was presented by Wang and Sun [9] and is based on deep learning. Since it eliminates segmentation based on threshold and labor-intensive feature extraction, it is the most feasible one for a smooth classification of illness similarities. The four levels of disease severity are used as the basis for the botanists' annotation of the images utilizing apple black roots in the dataset of plant villages. Then, deep convolution networks are trained to recognize the disease severity.

A hybrid deep learning model was suggested by Changjian et al. [10] to identify tomato leaf disease by integrating the benefits of dense networks and deep residual networks. They proposed a new model called "restricted residual dense network" (RRDN), which solved the problem of not resizing the images before feeding into the prediction model. It is claimed that this model may decrease the number of training elements to increase computation accuracy while also having improved information flow and gradients. As a consequence, on the dataset of tomato leaves, their model had a 95% accuracy rate.

A method for identifying plant diseases that utilizes CNN-based transfer learning was developed by Varshney et al. [11]. Deep learning was used for extracting features, and support vector machine for categorization. The model's accuracy score was 88.77 when it was tested against the PlantVillage dataset.

Houda et al. [12] conducted an in-depth analysis of advanced transfer learning models and traditional machine learning methods using a dataset collected via the PlantVillage Dataset, which consisted of injured and healthier leaves from plants used for binary classification. They additionally employed several kinds of deep-learning optimization techniques and activation algorithms to further enhance the efficiency of the aforementioned CNN designs. They succeeded to obtain pretty remarkable accuracy in categorizing of leaf illnesses for any model through testing. The InceptionV3 model provides the highest CA (98.01%), whereas NB provides the lowest CA (60.09%).

The best deep-learning (DL) model for increasing detection accuracy with image is the CNN model created by Shoaib et al. [13]. A particular kind of feedforward AI model called deep CNN is made up of a number of hidden layers such as convolutional and pooling. The feature learning and classifying elements make up the first two sections of the CNN model. The feature learning block uses a convolutional layer to extract several types of features, with feature learning taking place at fully connected layers in terms of recognizing plant diseases, the CNN model outperforms all other methods due to its greater accuracy. Studies show that CNNs [14][15] can correctly identify diseases 99% of the time.

A well-liked CNN design for segmenting images is U-Net. The design is referred to as U-Net because to its U-shape and bottleneck between the encoder and decoder components (Shoaib et al. [13]). Through a sequence of convolution and clustering layers, the encoder portion of the network extracts objects from the input image. The feature list from the encoder is linked after these features have been sampled and gone through the bottleneck. As a consequence, the network is capable of making predictions using both surface-level and fundamental image properties. The final segmentation pattern is subsequently produced by the system's decoder element using these linked feature maps. When it comes to tasks involving picture segmentation, the U-Net architecture is very helpful since it can handle class information.

20.3 PROPOSED WORK

Our goal is to use convolution neural networks to detect plant diseases early and suggest proper treatments to the customer. The proposed work uses transfer learning to increase the accuracy of prediction along with faster training time. The model trained after the background removal has a higher accuracy rate with fewer iterations of training. Additionally, to choose the model that would be most effective for the application, we tested various CNN models and employed the approach known as transfer learning. Figure 21.2 depicts the solution's design.

21.3.1 DATA ACQUISITION

The "PlantVillage" dataset [16] that we used in our proposal has 54,309 images of 14 distinct crop varieties and 26 illnesses. Some of the images have noisy backgrounds but some have a solid gray background. This can be removed using the U2Net model [17] in the data preprocessing step. Figures 21.3 and 21.4 show some of the sample images of the dataset.

21.3.2 DATA PREPROCESSING

"U2Net" is employed to successfully eliminate the backgrounds of the images. Salient object detection (SOD) is the main objective of the two-level nested U-structure "U2- Net" system. The architecture enables the network to achieve high resolution without appreciably raising memory and processing costs. Additionally, the U2-Net architecture is demonstrated in Figure 21.5 is entirely constructed using our RSU blocks and does not make use of any pre-trained backbones specifically designed for image categorization, it is adaptable enough to be employed in a wide range of working scenarios with no performance loss. There are two versions of our U2-Net that employs various combinations of filter numbers: the standard version (176.3 MB), and a much smaller version (4.7 MB). In future while moving our solution to a standalone offline app we have to choose a light weight model to run on these small machines.

We are training separate models for each species of plant to achieve better performance. The first step in data preprocessing is data cleaning. Here, we delete the background of the image to cut the leaf using U2-Net. The preprocessing also involves data augmentation, which is converting the bitmap image data to tensor. Tensors are multi-dimensional arrays with a uniform type. The technique of adding additional data points to an existing set of data in an attempt to intentionally enhance the amount of data is known as data augmentation. These consist of arbitrary twists and flips, etc. The CNN model [18] is then trained after receiving the matrix once more. These include random flips, rotations, etc. The tensor is then again fed into the CNN model after which it is trained.

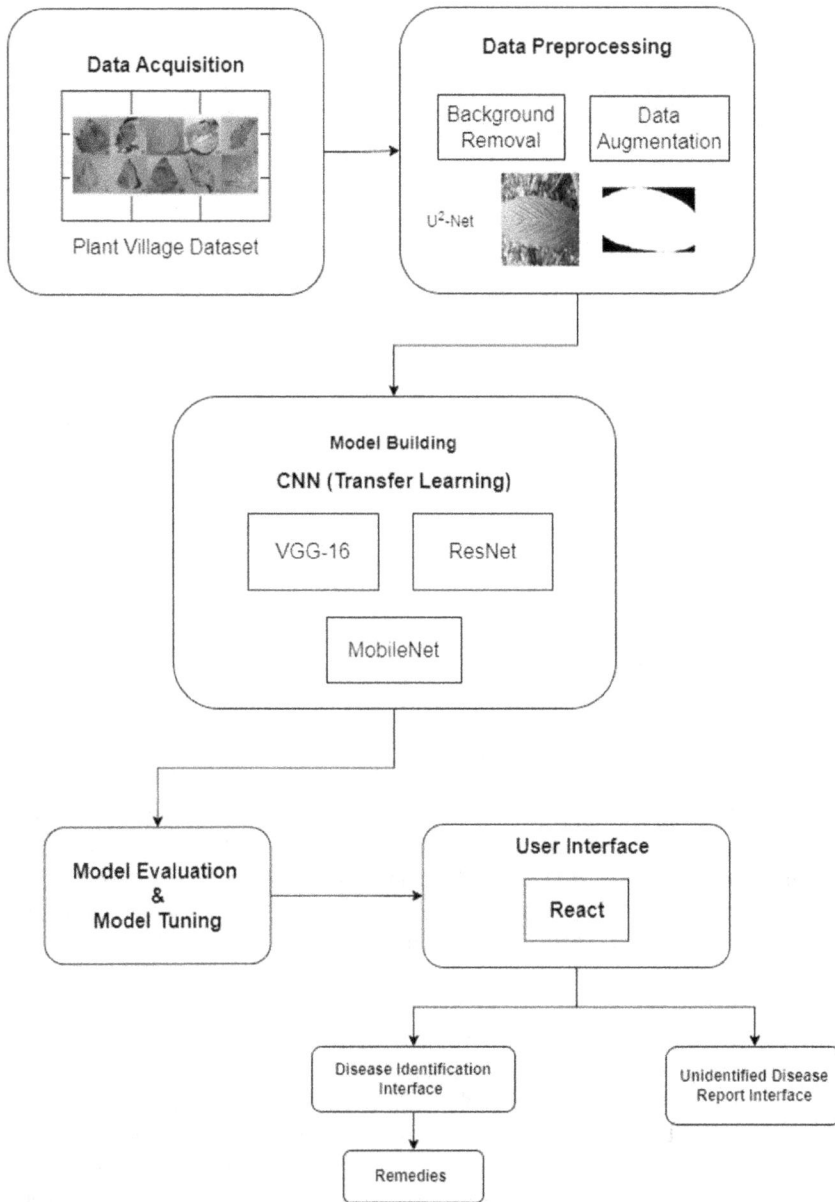

FIGURE 21.2 System architecture.

21.3.3 MODEL BUILDING AND EVALUATION

We have trained each model with single species of a plant, hence we can eliminate the task of classifying the leaf from the model. This will reduce the chance of false predictions. The finest image processing option available is CNN, but choosing a CNN model at random is not the best course of action. So, in order to select the model that is most suitable for the categorization, we compared many models. A pre-trained model is a stored network that has previously undergone extensive training on a large dataset, generally for a more difficult image classification task. Transfer learning [19][20] can employ the pre-trained model exactly as is, or it can be adjusted to match a specific job. Initially, we created our own CNN model with our own layers, but this model took several iterations

▣ README	Pepper,_bell__Bacterial_spot
Tomato__healthy	Peach__healthy
x_Removed_from_Healthy_leaves	Peach__Bacterial_spot
Tomato___Tomato_mosaic_virus	Orange___Haunglongbing_(Citrus_greening)
Tomato___Tomato_Yellow_Leaf_Curl_Virus	Grape__healthy
Tomato___Target_Spot	Grape___Leaf_blight_(Isariopsis_Leaf_Spot)
Tomato___Spider_mites Two-spotted_spider_mite	Grape__Esca_(Black_Measles)
Tomato___Septoria_leaf_spot	Grape___Black_rot
Tomato___Leaf_Mold	Corn_(maize)__healthy
Tomato___Late_blight	Corn_(maize)__Northern_Leaf_Blight
Tomato___Early_blight	Corn_(maize)__Common_rust_
Tomato___Bacterial_spot	Corn_(maize)__Cercospora_leaf_spot Gray_leaf_spot
Strawberry___healthy	Cherry_(including_sour)__healthy
Strawberry___Leaf_scorch	Cherry_(including_sour)__Powdery_mildew
Squash___Powdery_mildew	Blueberry__healthy
Soybean__healthy	Apple__healthy
Raspberry__healthy	Apple___Cedar_apple_rust
Potato__healthy	Apple___Black_rot
Potato___Late_blight	Apple__Apple_scab
Potato___Early_blight	
Pepper,_bell__healthy	

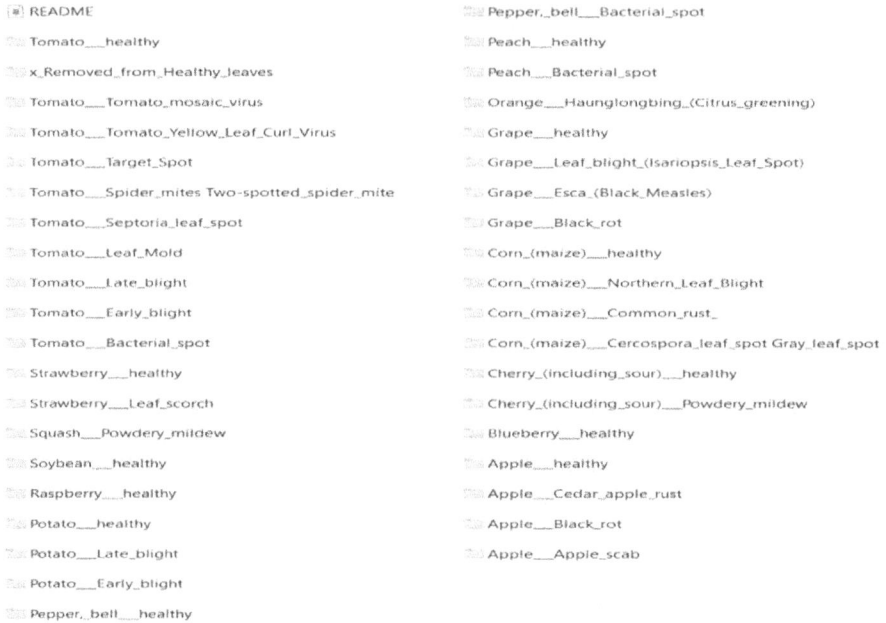

FIGURE 21.3   Dataset file structure.

(epochs) to attain a higher accuracy rate and lesser loss. That is why we have moved to transfer learning.

The concept underlying the application of transfer learning in the classification of images is that a model may effectively serve as a general representation of the visual world provided it is trained on a large enough dataset with adequate generality. Instead of trying to begin from scratch, you can leverage these learnt feature maps to work to your advantage by training a large model on a large dataset. In this instance, three transfer learning models were used: VGG16, ResNet, and MobileNet. VGG16 was recommended in the research work published by Karen Simonyan et al. [21]. This model obtains accuracy of 92.7% on the 14 million images in the ImageNet dataset, which are classified into 1000 classes. Exploding gradients are a problem for VGG16 because of its 138 million parameter count. Exploding gradients are a problem when large weight changes for neural network models during training result from the accumulation of severe error gradients.

The MobileNet model [22] is the first portable image processing framework developed by TensorFlow. MobileNet's convolution may be divided according on depth. It significantly lowers the total number of parameters as compared to a network with traditional convolutions of similar depth in the nets. As a result, compact deep neural networks are created. Our solution is web-based; thus, processing power-intensive tasks are handled by servers rather than users' desktops or mobile devices. Therefore, choosing a lightweight model at the expense of accuracy is invalid.

The diminishing gradient is one of the problems that ResNet attempts to solve. This is due to the fact that at very deep levels of the network, the gradients needed to compute the loss function simply vanish. As a result, learning never takes place as the weight levels are never changed. Also, ResNet achieved higher accuracy and less loss compared to all other mentioned models. Hence, we have used the **ResNet** model for our solution.

21.3.3.1 Custom CNN Model

This is the 16-layer model that was first constructed. Max-pooling layers and six Conv2D layers are present [23]. In Conv2D layers, the "relu" function is used as the activation function. This CNN

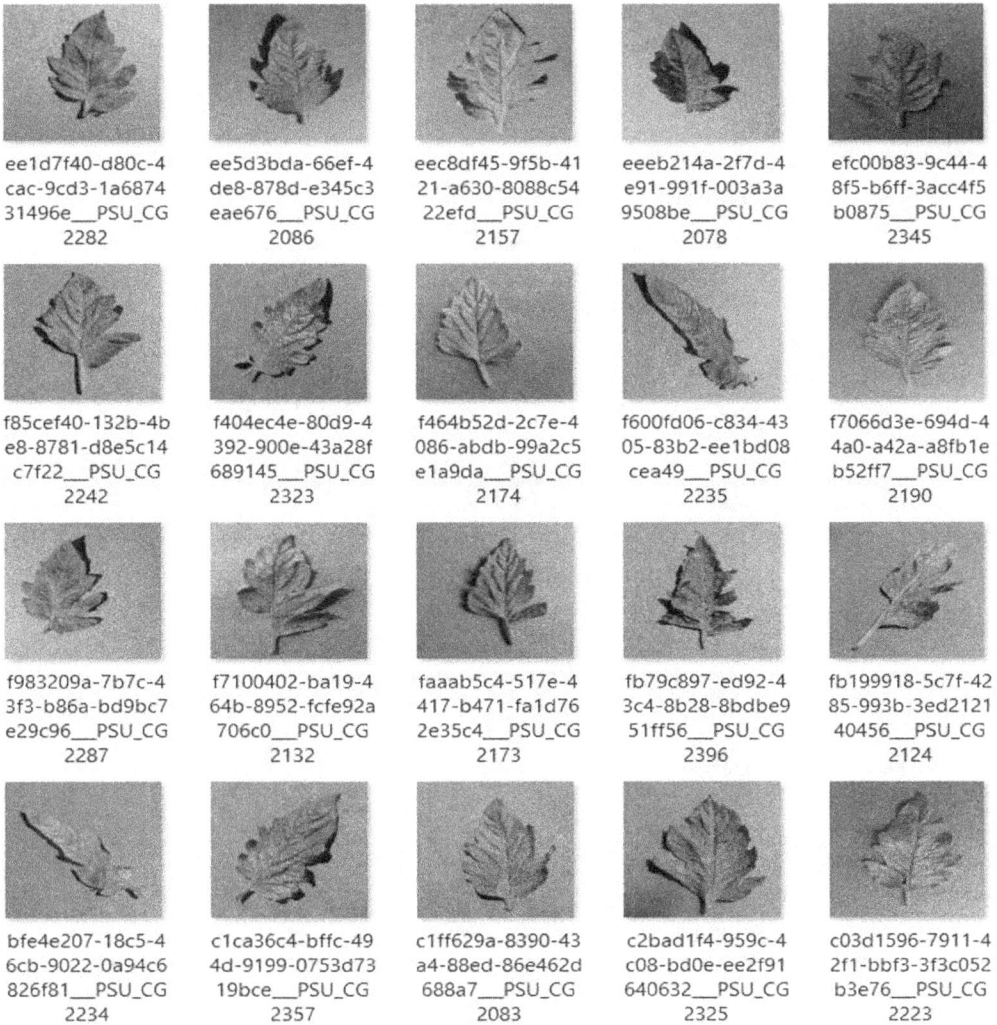

| ee1d7f40-d80c-4
cac-9cd3-1a6874
31496e__PSU_CG
2282 | ee5d3bda-66ef-4
de8-878d-e345c3
eae676__PSU_CG
2086 | eec8df45-9f5b-41
21-a630-8088c54
22efd__PSU_CG
2157 | eeeb214a-2f7d-4
e91-991f-003a3a
9508be__PSU_CG
2078 | efc00b83-9c44-4
8f5-b6ff-3acc4f5
b0875__PSU_CG
2345 |

| f85cef40-132b-4b
e8-8781-d8e5c14
c7f22__PSU_CG
2242 | f404ec4e-80d9-4
392-900e-43a28f
689145__PSU_CG
2323 | f464b52d-2c7e-4
086-abdb-99a2c5
e1a9da__PSU_CG
2174 | f600fd06-c834-43
05-83b2-ee1bd08
cea49__PSU_CG
2235 | f7066d3e-694d-4
4a0-a42a-a8fb1e
b52ff7__PSU_CG
2190 |

| f983209a-7b7c-4
3f3-b86a-bd9bc7
e29c96__PSU_CG
2287 | f7100402-ba19-4
64b-8952-fcfe92a
706c0__PSU_CG
2132 | faaab5c4-517e-4
417-b471-fa1d76
2e35c4__PSU_CG
2173 | fb79c897-ed92-4
3c4-8b28-8bdbe9
51ff56__PSU_CG
2396 | fb199918-5c7f-42
85-993b-3ed2121
40456__PSU_CG
2124 |

| bfe4e207-18c5-4
6cb-9022-0a94c6
826f81__PSU_CG
2234 | c1ca36c4-bffc-49
4d-9199-0753d73
19bce__PSU_CG
2357 | c1ff629a-8390-43
a4-88ed-86e462d
688a7__PSU_CG
2083 | c2bad1f4-959c-4
c08-bd0e-ee2f91
640632__PSU_CG
2325 | c03d1596-7911-4
2f1-bbf3-3f3c052
b3e76__PSU_CG
2223 |

FIGURE 21.4 Sample data images.

algorithm had a 90% accuracy rate and a 31.65% loss rate. Figure 21.11 shows the 20 epochs that were utilized to train this framework.

21.3.3.2 VGG16 Model

One of the best machine vision models at the moment is the VGG16 variant of convolutional neural networks [7]. The model's designers used an architecture with extremely small convolution filters to assess the networks and improve the depth, making it a significant advance over earlier systems. There are thought to be 138 distinct formation traits as a result of increasing the weight of the layer depth to 16–19. In Figure 21.7, the number 16 represents the 16 layers that make up VGG16. Only 16 of VGG16's 21 total layers are weight layers. It comprises 21 layers altogether, 13 of which are convolutional, 5 of which are max pooling, and 3 of which are dense. VGG16 has input tensor dimensions of 3, 224, and 244.

The key feature of VGG16 is the fact that it prioritizes having convolution layers as part of a 3×3 filter in the first step above having many hyper-parameters and that it consistently employs

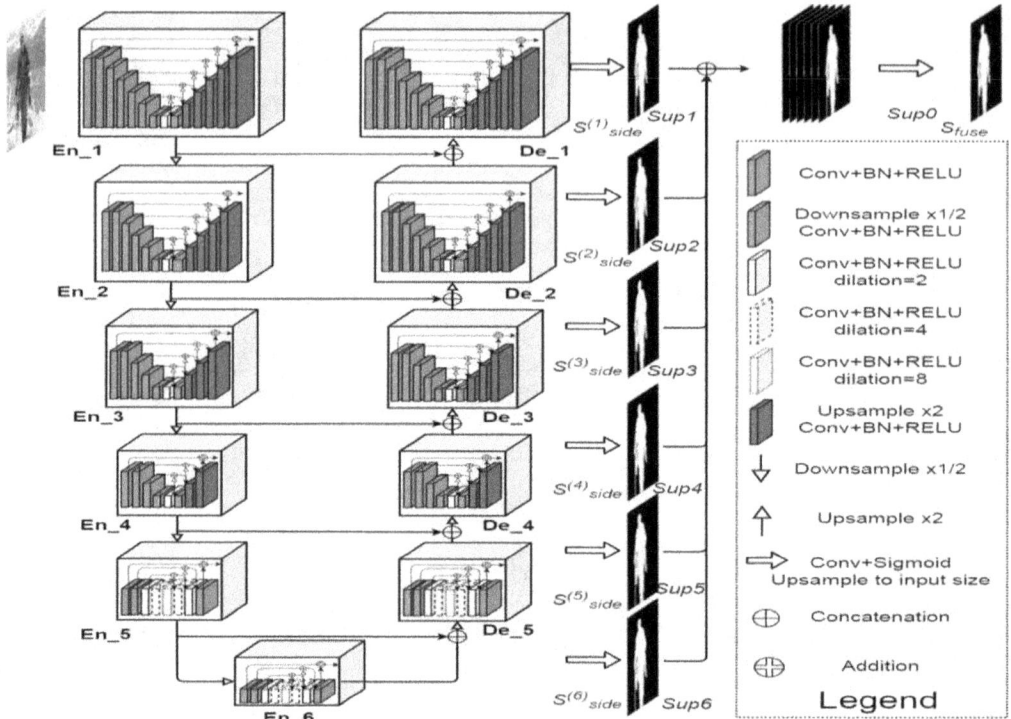

FIGURE 21.5   U2-net architecture.

Source: [17].

comparable storage and a max-pool layer in a 2×2 filter in stage 2. The max pool and the layers of convolution are situated in the same positions throughout the design. 64 filters make up Conv-1 Layer, 128 filters make up Conv-2 Layer, 256 filters make up Conv-3 Layer, 512 filters make up Conv-4 Layer, and 512 filters make up Conv-5 Layer. The third convolutional layer, which performs ILSVRC classification in 1000 distinct ways and contains 1000 channels, is followed by all three fully-connected (FC) layers. The first two FC levels each include 4096 channels, one for each class. The soft-max layer is the base layer. Results from the VGG16 model had an accuracy of 99.69% and a loss of 3.04%.

21.3.3.3   MobileNet Model

Depth-wise segmented convolutions are used by MobileNet [24]. Compared to a system using conventional convolutions, it greatly decreases the overall number of parameters to the same level inside the networks. Compact deep neural networks were as a result produced.

Convolution layers that are depth-wise separable are used to construct MobileNets. A depth-wise convolution and a point-wise convolution make up each layer of each depth-wise separate convolution. The number of layers in a MobileNet is 28, assuming depth-wise and point-wise convolutions are separately considered. By appropriately modifying the width multiplier hyperparameter, the 4.2 million parameters of the standard MobileNet may be further diminished. The supplied picture is 224 by 224 by 3, in size. Table 21.1 provides MobileNet's full architecture. And the comparison of standard convolution layer and depth-wise convolutional layer is given in Figure 21.9. MobileNet model produced an **accuracy of 99.37%** and a **loss of 1.95%** as shown in Figure 21.10.

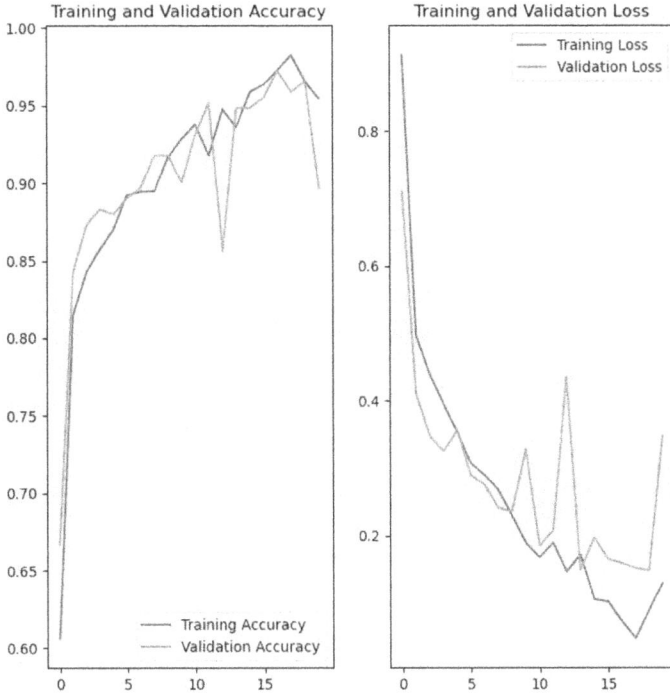

FIGURE 21.6   Accuracy/loss graph for the custom CNN model.

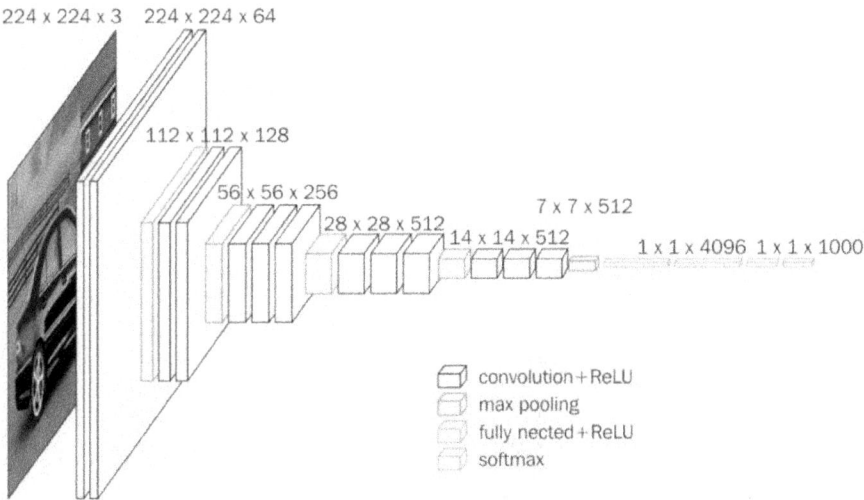

FIGURE 21.7   VGG16 architecture.

Source: [21].

21.3.3.4   ResNet Model

ResNet is a brief for a residual network, is a special type of neural network, which had been proposed in the year 2015 by Kaiming He et al. [25]. ResNet has many variants which utilize the identical concept yet have different values of layers. ResNet50, which is the name, provided to the variant that endorses 50 neural network phases. The ResNet50 architecture is constructed on the ResNet34

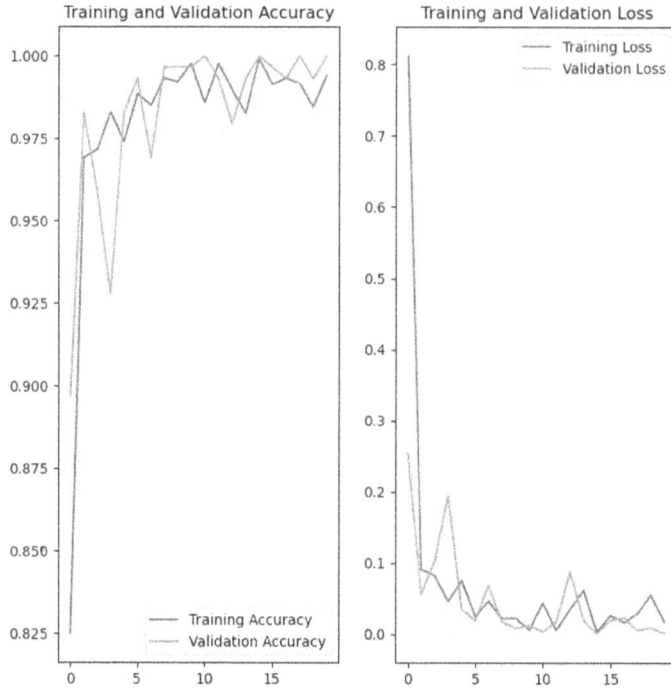

FIGURE 21.8　Accuracy/loss graph for the VGG16 model.

TABLE 21.1
MobileNet Architecture

Type / Stride	Filter Shape	Input Size
Cony / s2	$3 \times 3 \times 3 \times 32$	$224 \times 224 \times 3$
Cony dw / s1	$3 \times 3 \times 32$ dw	$112 \times 112 \times 32$
Cony / s1	$1 \times 1 \times 32 \times 64$	$112 \times 112 \times 32$
Cony dw / s2	$3 \times 3 \times 64$ dw	$112 \times 112 \times 64$
Cony / s1	$1 \times 1 \times 64 \times 128$	$56 \times 56 \times 64$
Cony dw / s1	$3 \times 3 \times 128$ dw	$56 \times 56 \times 128$
Cony / s1	$1 \times 1 \times 128 \times 128$	$56 \times 56 \times 128$
Cony dw / s2	$3 \times 3 \times 128$ dw	$56 \times 56 \times 128$
Cony / s1	$1 \times 1 \times 128 \times 256$	$28 \times 28 \times 128$
Cony dw / s1	$3 \times 3 \times 256$ dw	$28 \times 28 \times 256$
Cony / s1	$1 \times 1 \times 256 \times 256$	$28 \times 28 \times 256$
Cony dw / s2	$3 \times 3 \times 256$ dw	$28 \times 28 \times 256$
Cony / s1	$1 \times 1 \times 256 \times 512$	$14 \times 14 \times 256$
$5 \times {}^{\text{Conv dw/s1}}_{\text{Conv/s1}}$	$3 \times 3 \times 512$ dw	$14 \times 14 \times 512$
	$1 \times 1 \times 512 \times 512$	$14 \times 14 \times 512$
Cony dw / s2	$3 \times 3 \times 512$ dw	$14 \times 14 \times 512$
Conv / s1	$1 \times 1 \times 512 \times 1024$	$7 \times 7 \times 512$
Cony dw / s2	$3 \times 3 \times 1024$ dw	$7 \times 7 \times 1024$
Cony / s1	$1 \times 1 \times 1024 \times 1024$	$7 \times 7 \times 1024$
Avg Pool / s1	Pool 7×7	$7 \times 7 \times 1024$
FC / s1	1024×1000	$1 \times 1 \times 1024$
Softmax / s1	Classifier	$1 \times 1 \times 1000$

Source: [22].

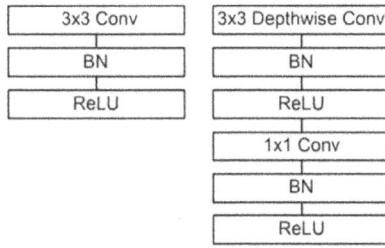

FIGURE 21.9 Comparison of standard and depth-wise convolutional layer.

Source: [22].

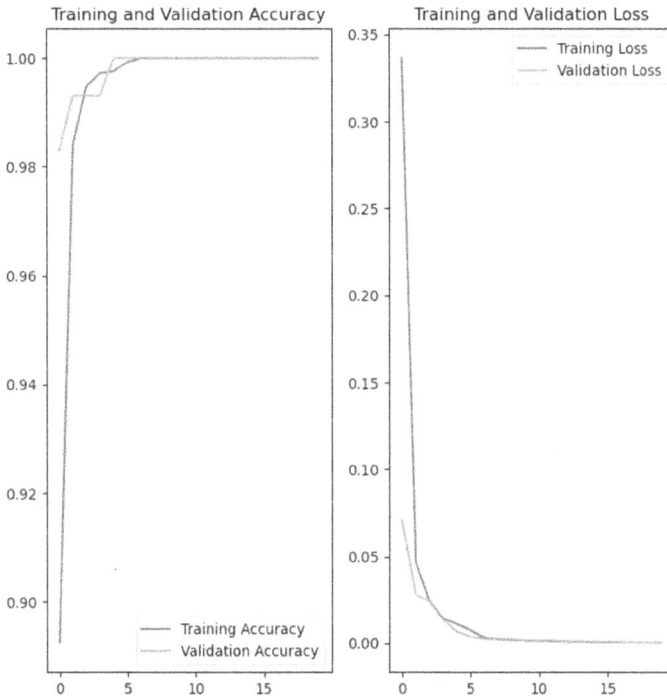

FIGURE 21.10 Accuracy/loss graph for the MobileNet model.

model, but there is one important difference. Concerns about the length of time it would take to teach the layers led to the modification of the building block in this instance into a bottleneck design. The previous two layers were replaced with a stack of three layers in ResNet50. The output of the 50-layer ResNet is 3.8 billion FLOPS. In order to avoid the rising gradients issue that VGG-16 experienced, ResNets were developed. When significant error gradients build up and lead to weight updates for neural network models that are very large during training, this is known as an "exploding gradient" issue.

Special characteristics of ResNet-50:

The 50-layer ResNet employs the bottleneck structure. By using 11 convolutions, a "bottleneck" residual block reduces the variety of factors and matrix multiplications. This significantly speeds up training for each layer. It utilizes a stack of three separate layers as opposed to two.

TABLE 21.2
ResNet Architecture Layers

Layer Name	Output Size	18-Layer	34-Layer	50-Layer	101-Layer
conv 1	112×112	7×7, 64, stride 2			
		3×3 max pool, stride 2			
conv 2_x	56×56	$\begin{bmatrix} 3 \times 3,64 \\ 3 \times 3,64 \end{bmatrix} \times 2$	$\begin{bmatrix} 3 \times 3,64 \\ 3 \times 3,64 \end{bmatrix} \times 3$	$\begin{bmatrix} 1 \times 1,64 \\ 3 \times 3,64 \\ 1 \times 1,256 \end{bmatrix} \times 3$	$\begin{bmatrix} 1 \times 1,64 \\ 3 \times 3,64 \\ 1 \times 1,256 \end{bmatrix} \times 3$
conv 3_x	28×28	$\begin{bmatrix} 3 \times 3,128 \\ 3 \times 3,128 \end{bmatrix} \times 2$	$\begin{bmatrix} 3 \times 3,128 \\ 3 \times 3,128 \end{bmatrix} \times 4$	$\begin{bmatrix} 1 \times 1,128 \\ 3 \times 3,128 \\ 1 \times 1,512 \end{bmatrix} \times 4$	$\begin{bmatrix} 1 \times 1,128 \\ 3 \times 3,128 \\ 1 \times 1,512 \end{bmatrix} \times 4$
conv 4_x	14×14	$\begin{bmatrix} 3 \times 3,256 \\ 3 \times 3,256 \end{bmatrix} \times 2$	$\begin{bmatrix} 3 \times 3,256 \\ 3 \times 3,256 \end{bmatrix} \times 6$	$\begin{bmatrix} 1 \times 1,256 \\ 3 \times 3,256 \\ 1 \times 1,1024 \end{bmatrix} \times 6$	$\begin{bmatrix} 1 \times 1,256 \\ 3 \times 3,256 \\ 1 \times 1,1024 \end{bmatrix} \times 23$
conv 5_x	7×7	$\begin{bmatrix} 3 \times 3,512 \\ 3 \times 3,512 \end{bmatrix} \times 2$	$\begin{bmatrix} 3 \times 3,512 \\ 3 \times 3,512 \end{bmatrix} \times 3$	$\begin{bmatrix} 1 \times 1,512 \\ 3 \times 3,512 \\ 1 \times 1,2048 \end{bmatrix} \times 3$	$\begin{bmatrix} 1 \times 1,512 \\ 3 \times 3,512 \\ 1 \times 1,2048 \end{bmatrix} \times 3$
	1×1	Average pool, 1000-d fc, softmax			
FLOPs		1.8×10^9	3.6×10^9	3.8×10^9	7.6×10^9

Source: [7].

The elements of the 50-layer ResNet architecture are listed in Table 21.2 as follows:

- A convolution is a combination of a 64 extra kernels, a 2-sized step, and a 7×7 kernel.
- A maximum-pooling layer of two stride sizes.
- Nine more layers, including 1×1,64- and 1×1,256-kernel variants of the 3×3,64-kernel convolution. These three stages are repeated three times.
- 12 more layers, each having 4 rounds of 1×1, 128, 3×3, 128, and 512 kernels.
- Nine additional levels, each having three iterations of 1×1512, 3×3,512, and 1, ×1,2048 cores.
- (Up to this point the network has 50 layers.)
- A layer with 1000 completely linked nodes is created using average pooling and the SoftMax activation algorithm.

ResNet model produced an **accuracy of 99.9%** and a negligible loss (4.1097e-04) on apple dataset as shown in Figure 21.12.

21.3.4 Model Training

A workable model is produced during model training, the first phase of machine learning, and it may then be tested, confirmed, and used. The success of a model in training ultimately impacts how well it will function if it is used in an application. The quality of the data used for training and the method choice are the two most crucial variables in the model training phase. Two sets of training data are commonly used: one for training and another for testing and validation.

Machine learning's model training phase produces a functional model that may later be tested, confirmed, and applied. A model's ability to function once it has been integrated into an application is

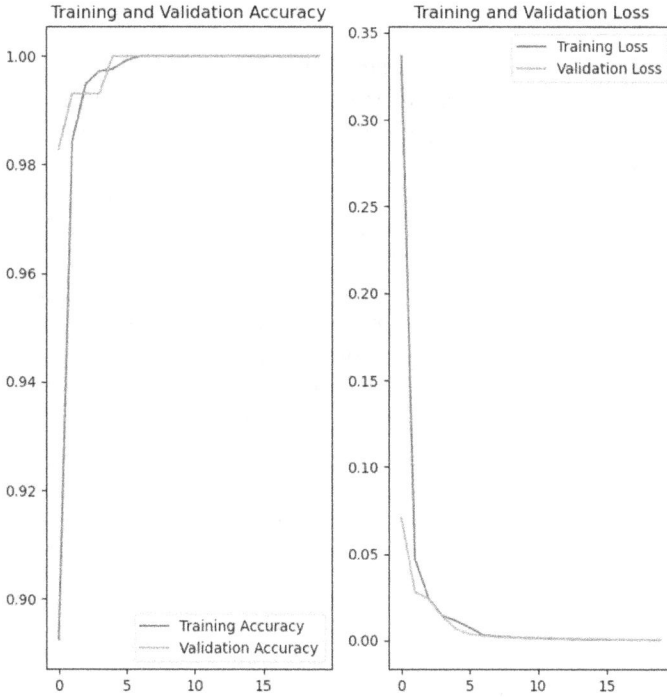

FIGURE 21.11 Accuracy/loss graph for the ResNet model.

Original Leaf Image   **Mask of the Image**   **Final Image**

FIGURE 21.12 Background removal using U2-Net.

ultimately determined by how well it performs during training. The quality of the data used for training and the method choice are the two most important variables in the model training phase. Two sets of information for instruction are commonly used: one for training and another for testing and validation.

The evaluation of a trained model on a validation dataset before validating it on a training dataset is a step in the training process Loss is the cost function's value for the initial training data, whereas loss from validation is the cost function's value for cross-validation data. According to validation data, neurons that use drop out do not exclude random neurons. The cause is that dropping out during training is done to provide some noise in order to prevent over-fitting. The model's accuracy inside the training dataset is detailed by training accuracy, which is similar.

21.3.5 Model Testing

Model testing is a process in which a fully trained model is tested and validated on a test set. This procedure uses data that isn't from the training dataset to assess how well the model performs. Before feeding the image directly into the model, we must do some preprocessing. This includes the background removal of the image.

The image is feed into the U2-Net model, which produces a mask of the given image, and then the mask is used to cut out the leaf from the original image (Figure 21.13).The model is tested with the remaining 10% of the dataset. This gives us the accuracy of the model in real world (Figure 21.14).

User interface is one of the most important parts of any software application. It is the connection between the end user and the actual implementation of the solution. Our aim is to simplify the user interface so that users without out any expert knowledge can utilize the application.

We have used React for the frontend of the application. React is a JavaScript package that is free and open source for creating user interfaces (UI) for online applications. Facebook creates and distributes React. By addressing issues and adding new features, Facebook is constantly improving the React library.

Some advantages of React are as follows:

- It is composable.
- It is declarative.
- Write once, learn anywhere.
- It is simple.
- SEO friendly.
- Fast, efficient, and easy to learn.

React lets us develop application of both mobile and web with slight modification of the underlying code. It enables the developers to develop application a lot quicker and easier for various platforms. React Native the JavaScript library is used to build mobile applications. It follows the same syntax and coding style as ReactJs.

We have used the Flask framework for our solution. We can build web apps using the Flask API for Python. It was developed by Armin Ronacher. Armin Ronacher was the person who created it. Since less code is needed to create a normal web application, Flask's framework is simpler to understand and more comprehensible than Django's. A web framework, also known as a web application framework, is a group of tools and modules that enable programmers to create applications without having to create thread handling procedures and protocols.

The usage of a software infrastructure named Docker facilitates application creation, testing, and deployment. Containers are standardized entities that store all the necessary tools, library resources, and programme code needed for use. They are also how Docker delivers applications.

Docker offers the capability to package and run a programme in a loosely separated environment referred to as a container. Because of the isolation and security, several containers can operate concurrently on a single host. Due to the fact that containers are compact and include all the components necessary to execute apps, you may launch them without relying on whatever already exists on the host. While at work, you may effortlessly share containers with coworkers while ensuring that each one receives a unit that functions identically.

1. It's simple to run an ML model on a computer. It is a difficult challenge to employ that model in other systems during the production stage, though. This task is made simpler, quicker, and more dependable with Docker.
2. We can quickly duplicate the working environment to train and operate the model on many OS systems using Docker.

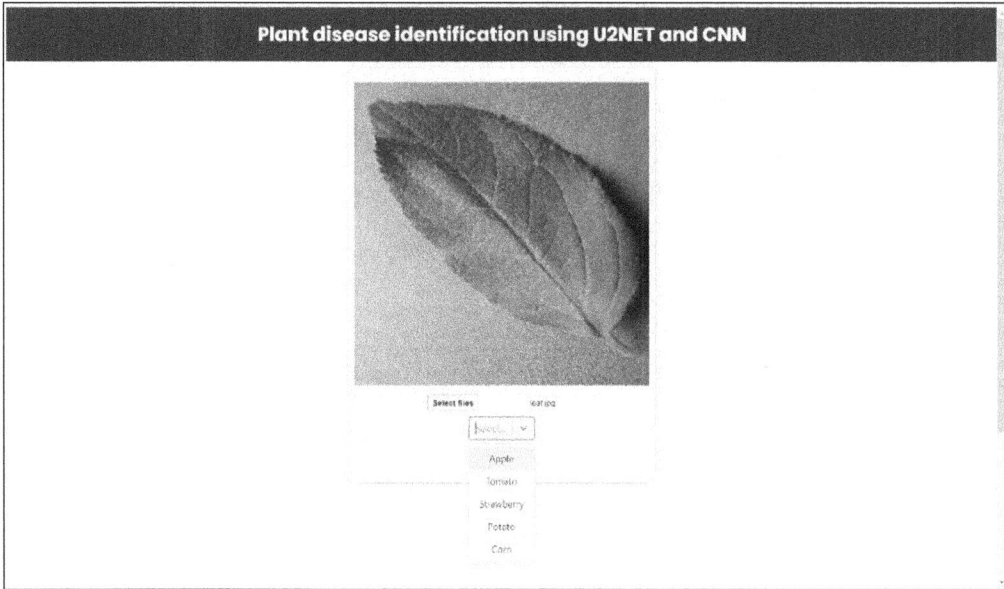

FIGURE 21.13 User interface to select the leaf image.

3. Using tools like OpenShift, a Kubernetes distribution, we can quickly deploy and make your model accessible to the customers.
4. Developers may trace various iterations of a container image, see what was used to build a version, and roll back to earlier iterations.
5. Our machine learning application will continue to function even when it is being repaired, updated, or down for maintenance.

Although the application will almost probably need to communicate with other apps developed in other programming languages, our machine learning model is often written in a single programming language, such as Python. As each microservice can be written in a separate language, allowing for scalability and the quick addition or deletion of independent services, Docker maintains all these interactions.

21.4 RESULTS AND DISCUSSIONS

We have created a website using ReactJs for the frontend and Flask for the backend so that the user can utilize this robust model. The user only chooses the plant species and uploads a photo of a leaf. The image is then forwarded to the backend for disease identification and image processing. To increase the precision of the prediction output and feed it into the model for prediction, the application's backend removes the image's backdrop.

The user gets provided the name and description of the ailment based on the model's results. The user also gets the disease's name, description, and treatment options.

Step 1: The user (farmer) enter the leaf image by clicking the "Select file" button.
Step 2: After the user selects the necessary image, the user selects the species of the leaf and clicks the upload button.
Step 3: The predict result is shown along with the disease description and remedies.
Step 4: The model file generated is served in a docker file to faster access at all time.

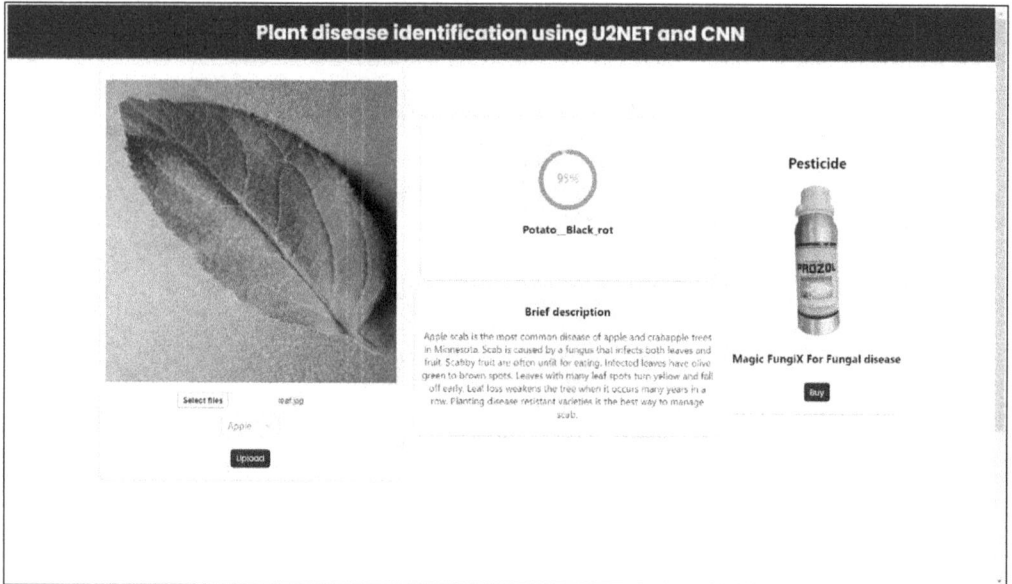

FIGURE 21.14   Predicted result and the remedies for the disease user interface flow.

TABLE 21.3
Comparison of Accuracy of all the Models

Model Name	Accuracy
Custom CNN	90%
VGG16	99.69%
MobileNet	99.37%
ResNet	99.9%

The accuracy the various models are compared and the results are shown in the Table 21.3. After the classifications of diseases are done, appropriate remedies for various diseases are stored in the database and it is shown to the user.

21.5   CONCLUSION AND FUTURE WORK

In conclusion, CNN with transfer learning has proven to be a better fit for this application. The model's accuracy has also been enhanced as a result of the training image's backdrop being removed. The ResNet model is thought to be the best model for the solution based on the comparison, and it has been employed in our solution. The dataset of many species of plants makes it a difficult task to train models for the species of plants or diseases. Hence the farmers can support the community by providing images for the species or disease that is not yet classified. The remedies for various diseases can be collected from various experts and a dataset could be created thereby we can also build an automatic recommendation system.

The future works will include moving the entire cloud-based solution to a standalone app, which will do all the steps such as background removal, prediction, and suggesting remedies. This standalone app will have to be very memory efficient since it will be running on a smart phone. We must also ensure that there is no performance compromise while reducing the size of the models used.

REFERENCES

[1] Hassan Amin, Ashraf Darwish (Member, IEEE), Aboul Ella Hassanien and Mona Soliman, "End-to-End Deep Learning Model for Corn Leaf Disease Classification", *IEEE Access*, vol.10, pp. 31103–31115, 2022.

[2] Lili Li, Shujuan Zhang and Bin Wang, "Plant Disease Detection and Classification by Deep Learning—A Review", IEEE Access, vol.9, pp. 56683–56698,2021.

[3] Tanuja K. Fegade and Bhausaheb Vyankatrao Pawar, (2020). Crop Prediction Using Artificial Neural Network and Support Vector Machine. In: Sharma, N., Chakrabarti, A., Balas, V. (eds) *Data Management, Analytics and Innovation. Advances in Intelligent Systems and Computing*, vol.1016. Springer, Singapore. https://doi.org/10.1007/978-981-13-9364-8_23

[4] C.K. Sunil, C.D. Jaidhar and Nagamma Patil, "Cardamom Plant Disease Detection Approach Using EfficientNetV2", *IEEE Access*, vol.10, pp. 789–804, 2021.

[5] E. Omrani, B. Khoshnevisan, S. Shamshirband, H. Saboohi, N.B. Anuar and M.H.N.M. Nasir, Potential of Radial Basis Function-Based Support Vector Regression for Apple Disease Detection, *Meas.: J. Int. Meas. Confed.* vol.55, pp. 512–519, 2014.

[6] Sabbir Ahmed, Md. Bakhtiar Hasan, Tasnim Ahmed, Md. Redwan Karim Sony and Md. Hasanul Kabir, (Member, IEEE), "Less Is More: Lighter and Faster Deep Neural Architecture for Tomato Leaf Disease Classification", *IEEE Access*, vol.10, pp. 68868–68884, 2022.

[7] Helong Yu, Jiawen Liu, Chengcheng Chen, Ali Asghar Heidari, Qian Zhang, Huiling Chen, Majdi Mafarja and Hamza Turabieh, "Corn Leaf Diseases Diagnosis Based on K-Means Clustering and Deep Learning", *IEEE Access*, vol.9, pp. 143824–143835, 2021.

[8] Surampalli Ashok, Gemini Kishore, Velpula Rajesh and S. Suchitra,"Tomato Leaf Disease Detection Using Deep Learning Techniques", published in 2020 5th International Conference on Communication and Electronics Systems (ICCES). DOI:10.1109/ICCES48766.2020.9137986

[9] G. Wang, Y. Sun and J. Wang, Automatic Image-Based Plant Disease Severity Estimation Using Deep Learning, *Comput. Intell. Neurosci.* vol.2017, pp. 1–8, 2017. https://doi.org/10.1155/2017/2917536 2917536

[10] Changjian Zhou, Sihan Zhou, Jinge Xingi and Jia Song, "Tomato Leaf Disease Identification by Restructured Deep Residual Dense Network", *IEEE Access*, vol.9, pp. 28822–2883, 2021.

[11] D. Varshney, B. Babukhanwala, J. Khan, D. Saxena and A.K. Singh, "Plant Disease Detection Using Machine Learning Techniques," *2022 3rd International Conference for Emerging Technology (INCET)*, Belgaum, India, 2022, pp. 1–5. doi: 10.1109/INCET54531.2022.9824653

[12] H. Orchi, M. Sadik, M. Khaldoun and E. Sabir, Automation of Crop Disease Detection through Conventional Machine Learning and Deep Transfer Learning Approaches. *Agriculture.*vol. 13, pp. 352, 2023.https://doi.org/10.3390/agriculture1302035

[13] M. Shoaib, T. Hussain, B. Shah, I. Ullah, S. M. Shah and F. Ali, et al. (2022a). Deep Learning-Based Segmentation and Classification of Leaf Images for Detection of Tomato Plant Disease. *Front. Plant Sci.* 13. doi: 10.3389/fpls.2022.1031748

[14] Yang Wu, Xian Feng and Guojun Chen, "*Plant Leaf Diseases Fine-Grained Categorization Using Convolutional Neural Networks*", vol.10, pp. 41087–41096, 2022.

[15] Zinon Zinonos, Socratis Gkelios, Ala F. Khalifeh, Diofantos G. Hadjimitsis, Yiannis S. Boutalis and Savvas A. Chatzichristofis, "Grape Leaf Diseases Identification System Using Convolutional Neural Networks and LoRa Technology", *IEEE Access*, vol.10, pp. 122–133, 2021.

[16] "Plant Village Dataset – Dataset of Diseased Plant Leaf Images and Corresponding Labels".https://git hub.com/spMohanty/PlantVillage-Dataset

[17] "U^2-Net: Going Deeper with Nested U-Structure for Salient Object Detection." https://github.com/ xuebinqin/U-2-Net

[18] Stefania Barburiceanu, Serban Meza, Bogdan Orza, Raul Malutan and Romulus Terebes, "Convolutional Neural Networks for Texture Feature Extraction. Applications to Leaf Disease Classification in Precision Agriculture", *IEEE Access*, vol.9, pp. 160085–160103, 2021.

[19] Mobeen Ahmad, Muhammad Abdullah, Hyeonjoon Moon and Dongil Han, "Plant Disease Detection in Imbalanced Datasets Using Efficient Convolutional Neural Networks with Stepwise Transfer Learning", *IEEE Access*, vol.9, pp. 140565–140580, 2021.

[20] Mostafa Mehdipour Ghazi, Berrin Yanikoglu and Erchan Aptoula, Plant Identification Using Deep Neural Networks Via Optimization of Transfer Learning Parameters, *Neurocomputing*. vol.235, pp. 228–235, 2017. ISSN 0925-2312https://doi.org/10.1016/j.neucom.2017.01.018

[21] Karen Simonyan and Andrew Zisserman, "Very Deep Convolutional Networks for Large-Scale Image Recognition", *arXiv: Open access archive*, 2014. https://doi.org/10.48550/arXiv.1409.1556

[22] Andrew G. Howard, Menglong Zhu, Bo Chen, Dmitry Kalenichenko, Weijun Wang, Tobias Weyand, Marco Andreetto and Hartwig Adam, "MobileNets: Efficient Convolutional Neural Networks for Mobile Vision Applications", *arXiv: Open access archive*, 2017.https://doi.org/10.48550/arXiv.1704.04861

[23] S.K. Mahmudul Hassan and Arnab Kumar Maji, "Plant Disease Identification Using a Novel Convolutional Neural Network". *IEEE Access*, vol.10, pp. 5390–5401, 2022.

[24] Elhoucine Elfatimi, Recep Eryigit and Lahcen Elfatimi, "Beans Leaf Diseases Classification Using MobileNet Models", *IEEE Access*, vol.10, pp. 9471–9482,2022.

[25] Kaiming He, Xiangyu Zhang, Shaoqing Ren and Jian Sun, "Deep Residual Learning for Image Recognition", *arXiv: Open access archive*, 2015. https://doi.org/10.48550/arXiv.1512.03385

22 Intelligent Farming Through Weather Forecasting Using Deep Learning Techniques for Enhancing Crop Productivity

V. Ezhilarasi[1], S. Selvamuthukumaran[2], and N. Srinivasan[2]
[1*] Department of Information Technology, A.V.C. College of Engineering, Mayiladuthurai, Tamilnadu, India
[2] Department of Computer Applications, A.V.C. College of Engineering, Mayiladuthurai, Tamilnadu, India

22.1 INTRODUCTION

Crop yield prediction is an essential task for a country in making decisions at national and regional levels. An accurate crop yield prediction will help farmers in deciding on what to grow at the right time. There are many approaches for crop yield prediction. Machine learning is one of the important decision support tools for crop yield prediction, including supporting decisions on what crops to grow and what to do during the growing season of the crops. Several machine learning algorithms have already been proposed to support crop yield prediction. Quick developments in sensing technologies and ML techniques will result in cost-effective solutions in the agricultural sector.

Agriculture is the livelihood for around 70% of the rural people in Tamil Nadu. Figure 22.1 shows the food grain production in Tamilnadu for the past 5 years [1]. Predominance of marginal landholdings, conversion of cultivable lands for purposes other than agriculture, unpredictable climate, paucity of agricultural labor, marketing of agricultural products are the biggest challenges in agriculture. Varieties of various crops based on the prevailing climate, rainfall, and soil fertility along with usage of the latest technologies are now recommended to increase production. Paddy production scheme, nutrient-rich millet mission, pulse production scheme, and oilseed production are in implementation in Tamilnadu with an objective to enhance food grain production.

22.1.1 DEEP LEARNING OVERVIEW

Deep learning has been very successful in many fields including agriculture. As a learning algorithm, deep learning can make better use of datasets for feature selection and extraction. Agricultural technology and modern agriculture has become a new scientific research area that increases agricultural productivity. It also minimizes the impact on the environment by using data-intensive methods. The data produced in modern agricultural processes are provided by various sensors, which can help to understand the operating circumstances including the climatic conditions, soil, and interaction of dynamic crops and the operation itself, thereby improving accuracy and faster decision-making. On implementing smart agriculture, deep learning algorithms can be used to monitor the temperature and water level of the crops. In addition, farmers can also observe their fields from anywhere in the

DOI: 10.1201/9781003407959-22

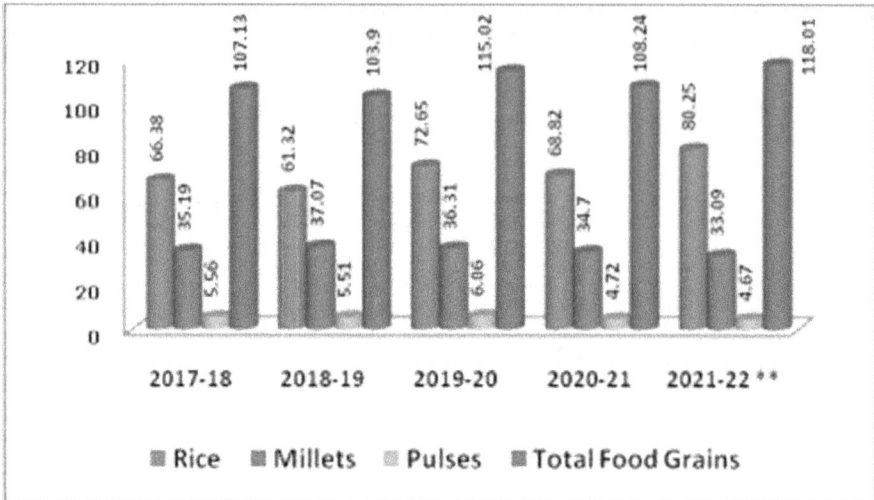

FIGURE 22.1 Food grain production in the past 5 years in Tamilnadu.

world. AI-based smart agriculture seems to be really efficient. Water is the important requirement for proper growth of crop. So far, the most used features in crop prediction are temperature, rainfall, and soil type, and the most applied algorithm is artificial neural networks, convolutional neural networks (CNN), deep neural networks (DNN). long-short term memory (LSTM) is the most widely used deep learning algorithm.

AI plays a very important role in transforming agriculture nowadays, it provides more efficient methods to produce, harvest, and market essential plants, check for defective crops, and enhance healthy crop production. AI is used in weather forecasting, pest or disease identification. It enhances crop management activities and also addresses the challenges that farmers face, such as climate variation, an infestation of pests and weeds that reduces yields.

22.1.2 Factors Affecting Crop Yield Production

Factors such as pH, soil moisture, temperature, humidity, breeze, rainfall, fertility nature of soil, vegetation, water irrigation, pesticides, soil depth or granularity, the salt content, the content of organic mass in soil, heavy metal contamination, the content of nutrients, microbial activity, etc. are all important in cultivation of a crop. Figure 22.2, shows the common factors that are affecting crop yield production. These factors are grouped in three basic categories called technological (agricultural practices, managerial decision, etc.), biological (diseases, insects, pests, weeds), and environmental (climatic condition, soil fertility, topography, water quality, etc.). These factors account for yield differences from one region to another worldwide.

The environmental factors affecting crop yields can be classified into abiotic and biotic constraints. Actually, these factors are more intensified with global warming, which leads to climate change. Abiotic stresses adversely affect growth, productivity, and trigger a series of morphological, physiological, biochemical, and molecular changes in plants. The abiotic constraints include soil properties (soil components, pH, physicochemical, and biological properties), and climatic stresses (drought, cold, flood, heat stress, etc.). On the other hand, biotic factors include beneficial organisms (pollinators, decomposers, and natural enemies), pests (arthropods, pathogens, weeds, vertebrate pests), and anthropogenic evolution.

Some of the applications of AI in agriculture includes crop and soil monitoring, insect and plant disease detection, livestock health monitoring, intelligent spraying, automatic weeding, etc.

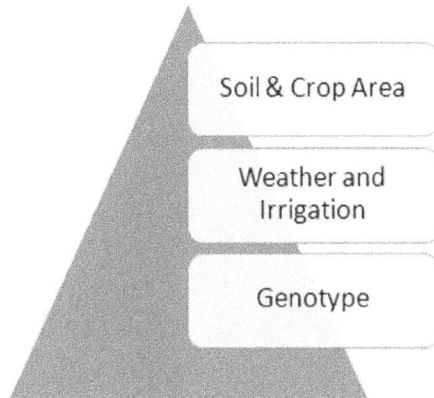

FIGURE 22.2 Factors affecting crop yield.

22.1.3 WEATHER PREDICTION FOR AGRICULTURE

Weather prediction using deep learning helps farmers in knowing about the climatic conditions. Multiple stakeholders in the agriculture ecosystem use weather information. Climatic conditions and seasonal forecasts help farmers in planning for the upcoming season to maximize productivity based on expected weather patterns. The most important decisions made by smallholders are based on the seasonal forecast. Seasonal forecasting helps in deciding which mix of crops and seed varieties can be planted, what seed can be purchased and based on that knowledge farmers prepare their land accordingly. Shorter real-time meteorological information of less than 10 days and daily forecasts helps to determine timing of various activities such as sowing, weeding, spraying, and harvesting. Weather information can be especially impactful if combined with specific advice or tips on the actions that need to be taken by the farmers to address weather patterns.

Real-time daily or 2–3-day forecasts can also help farmers make very practical decisions that can save them time and money or protect them from weather-related damage. For example, with the knowledge that rain is expected farmers can cover or move indoors crops that have been left outdoors for drying, or postpone spraying pesticide on their crops to another day and possibly save both the cost of washed off pesticide as well as the time and labor costs associated with the activity.

22.1.4 WHO CAN USE WEATHER DATA?

Foremost, farmers can use the weather information for cultivation and production of crops, suppliers can use weather forecasts to plan for the upcoming season, commodity buyers and other actors at market level may use weather information to make pricing decisions based on expected crop yield. By predicting harvesting time, availability of supply from different agro-climatic zones can be determined and other quality parameters can be classified.

22.1.5 CHALLENGES OF WEATHER INFORMATION

The accuracy of seasonal and long-term weather forecasts still remains quite low even though access to satellite data and forecasting models are available, the longer the time frame, the higher the possibility of deviance from the forecast. The challenge is to use such probabilistic forecasts to aid decision making for farmers. Weather information needs to be granular and specific to the micro-environment of the farmer. Particularly it should be time sensitive as compared to other information needs of smallholder farmers; especially the short term forecasts may help farmers make day-to-day

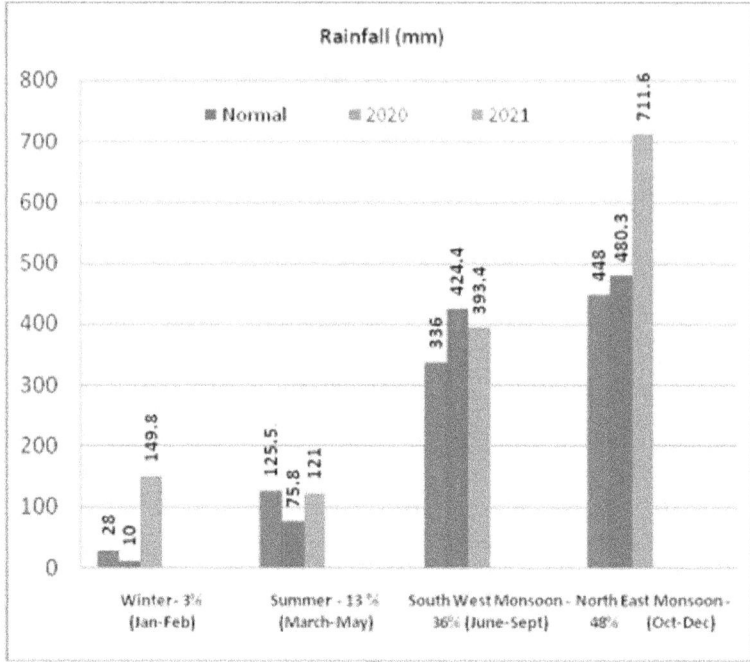

FIGURE 22.3   Seasonal rainfall in Tamilnadu (2021).

decisions. The ability of farmers to understand the weather data as well as use of weather information is limited by the lack of technical knowledge or access to the technology is also yet another
challenge. For all the above reasons, it is needed to minimize the risk or maximize the potential of
expected weather patterns.

Figure 22.3 shows the season-wise rainfall in Tamilnadu for the year 2021 [1]. However, the
current weather forecasting information can be obtained from API for any location. The below is the
code to fetch the weather information of Chennai for 3 days.

22.1.6 RESULTS OF WEATHER INFORMATION FROM WEATHER API

Table 22.1 shows the weather information collected from weather API of Chennai district, Tamilnadu,
India on 21.04.2023. Various parameters such as wind_kph, wind_degree, wind direction, pressure,
precipitation, humidity, temperature are choosen to get the weather information of a particular place.

Since, temperature is an important feature in crop prediction. Figure 22.4 shows the selected minimum, maximum, and average temperature of Tamilnadu in the year 2022.

22.1.7 SUMMARY OF WEATHER DATA

Table 22.2 shows the summary of available weather data used by various researchers in different
prediction models. The availability of publicly accessible datasets and regional datasets helps in
predicting the weather.

22.2 RELATED WORKS

In Khalila et al. [2], the vegetation indices in their crop yield estimations are obtained by satellite
sensors, variables such as weather elements, soil moisture, hydrological conditions, and soil fertility

TABLE 22.1
Weather Information from API

Location:{} 8 keys	current:{} 23 keys	condition:{} 3 keys
me:"Chennai"	last_updated_epoch:1682064900	wind_mph:9.4
region:"Tamil Nadu"	last_updated:"2023-04-21 13:45"	wind_kph:15.1
country:"India"	temp_c:38	wind_degree:60
lat:13.08	temp_f:100.4	wind_dir:"ENE"
lon:80.28	is_day:1	pressure_mb:1004
tz_id:"Asia/Kolkata"		pressure_in:29.65
localtime_epoch:1682065111		precip_mm:0
localtime:"2023-04-21 13:48"		precip_in:0
		humidity:48
		cloud:25
		feelslike_c:52.7
		feelslike_f:126.8
		vis_km:6
		vis_miles:3
		uv:9
		gust_mph:23.7
		gust_kph:38.2

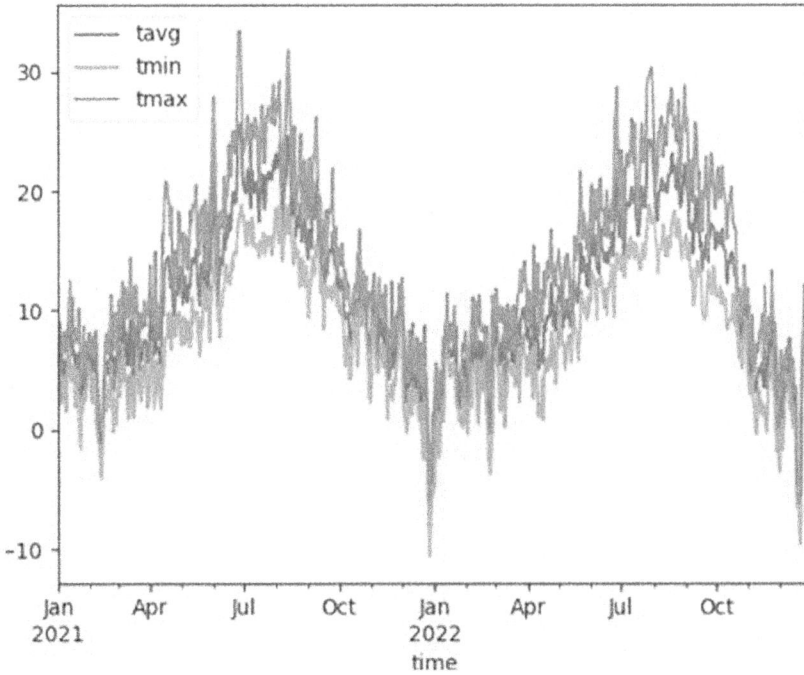

FIGURE 22.4 Selected temperature feature.

TABLE 22.2
Summary of Weather Data

Weather Dataset	Application	Algorithm/Technique Used	Weather Parameters
Synthetic localized weather station data at nearest airport.	For urban building energy modeling and daylight analysis.	Feed forward neural network, RNN -LSTM, GRU, look back.	Temperature, humidity, windspeed, wind direction.
Publicly available Brazil massive weather datasets.	Weather prediction long term	ANN, SVM	Temperature, windspeed, rainfall, precipitation, pressure, dewpoint, ozone level.
Dense weather dataset.	Very-short-term, less than 1-hour, local weather forecasting for efarming, construction	DNN-block type network.	Temperature, precipitation.
Publicly available Total Electron Content (TEC) Maps Europe.	Image sequence prediction	CNN, RNN – LSTM and GRU, Deep Learning.	TEC images, Ionospheric behavior.
Time series SST data for the Korean Peninsula.	Sea surface temperature prediction	LSTM and SST prediction model used	High water temperature

are used. Neural networks and machine learning techniques were employed. The neural network data input varied in the form of normalized histograms of a multispectral image band, normalized vegetation index, absorbed active photosynthetic radiation, canopy surface, and environmental factors. Rapid advancements in satellite technologies and ML techniques provide an affordable and comprehensive solution for accurate grain prediction. Suggesting many remote sensing researches in the field of crop yield prediction.

Kavitha et al. [3] addressed the fact that various machine learning techniques were implemented to estimate the crop yield in Rajasthan state of India on five identified crops. Their results indicate that among all the applied algorithms random forest, SVM, gradient descent, long-short term memory, and Lasso regression techniques [3], the random forest performed better than others with 0.963 R2, 0.035 RMSE, and 0.0251 MAE. The results were validated using R2, root mean squared error, and the mean absolute error to cross-validation techniques. LSTM require a larger quantum of data for a better predictive analysis. Furthermore, based on the observations, most of the models perform better on the specified parameters, whereas models, such as gradient descent and Lasso regression, perform better when applied to the dataset with all of the characteristics. While soil and rainfall quantity are important in crop production and general farming, it is concluded that a deeper investigation of these elements, as well as use of a larger database, it is required for real-life research of such elements using prediction models. Suggesting more deep learning models need to be tested on the dataset to find the best-performing technique.

Rithesh et al.'s [4] research focused on establishing and inter-relating the micronutrients and weather parameters. Classifying the type of crop is based on the micronutrients using a support vector machine and decision tree algorithm. Different types of tools, such as curve fitting and data analysis using Python-3.9.0, have been used for prediction purposes. Out of the different crop production types in this research, three major crops have been considered, such as rice, wheat, and sugarcane. Based on the soil's features and input parameters, the developed model forecasts a suitable crop with a 92% accuracy level (fitness score). SVM with linear Kernel shows better performance compared to other algorithms such as linear SVM and decision tree. The fitness score can be further enhanced by using non-linear curve fittings.

Muyideen AbdulRaheem [5] et al.'s study compares three ML classification techniques for weather prediction. A web-based software application was developed using Flask App to demonstrate weather modeling using three ML models, and the data used for the study was obtained from Kaggle. For the weather prediction, a decision tree (DT), k-nearest neighbor (k-NN), and logistic regression (LR) classifier method were suggested, and comparisons were made between the three classifications techniques. The accuracy results show that with a 100% accuracy rate, the DL classifier outperforms the k-NN with a 78% accuracy rate and LR with a 93% accuracy rate. Variables like humidity and wind speed influence the weather, providing more data leads to better and more accurate models. This predicts weather forecasts using three classifiers using the Seattle weather dataset to test the effectiveness of these three models. These models are validated using Kaggle's meteorological data, including the date, maximum temperature, minimum temperature, precipitation, and rain.

Amin Amani et al. [6] proposed a deep-learning model, which is a viable approach to optimize the information on soil parameters and the effect of agricultural variables in cotton cultivation, even in the case of small datasets. Soil is analyzed to reduce the planting costs by determining the various combinations of soil components and nutrients amounts. Artificial intelligence decreases the expenses in cultivating the crop by predicting at the right time and also increases productivity and profits. Thirteen essential factors are analyzed such as soil parameters and its nutrient content in soil for cotton planting. The proposed DNN model has four hidden layers, which helps solving uncertainty for selecting the factors amount. Other new technologies are recommended, such as IoT can be integrated with artificial intelligence, and different deep learning algorithms and techniques can be considered to increase the performance.

Rishi Gupta et al. [7] collected and analyzed temperature, rainfall, soil, seed, crop production, humidity, and wind speed data in a certain region, these essential parameters help the farmers to improve the production of their crops. The data is pre-processed in Python environment and the Map Reduce framework is applied, which further analyzes and processes the large volume of data. Kmeans clustering is employed on the results gained from Map Reduce, and result is obtained in terms of accuracy. Bar graphs and scatter plots are drawn for results to study the relationship between the crop, rainfall, temperature, soil, and seed type of two regions (Ahmednagar, Maharashtra and, Andaman and Nicobar Islands). Further, a self-designed recommender system has been used to predict the crops and display them on a graphic user interface designed in a Flask environment.

Ju-Young Shin et al. [8] addressed a high-resolution wind speed forecast system for agricultural purposes in South Korea. Logarithmic wind profile, power law, random forests, support vector regression, and extreme learning machine were tested as suitable methods for the downscaling wind speed data. The machine learning-based methods give better performance than traditional methods for downscaling wind speed data. A wind speed forecast system developed provides good performance, particularly in inland areas. Overall, the random forest is considered to be the best downscaling method in this paper. Root mean square error and mean absolute error of wind speed prediction for 48 hours using random forests are approximately 0.8 m/s and 0.5 m/s, respectively.

Liyun Gong et al. [9] addressed a novel deep neural network (DNN)-based methodology to predict the future crop yield based on historical yields and greenhouse environmental parameters (e.g., CO_2 concentration, temperature, humidity, radiation, etc.) information. This method is based on the hierarchical integration of the recurrent neural network (RNN) and temporal convolutional network (TCN), which are both the current state-of-the-art DNN architectures for temporal sequence processing. Evaluations are done through statistical analysis for multiple datasets. Both traditional machine learning and deep learning methods are applied.

Jaseena et al. [10] revealed the available weather datasets around the world, for example: (a) US weather events; (b) historical hourly weather data; (c) rain in Australia; (d) historical daily weather data; (e) weather Istanbul dataset; (f) weather underground; (g) in situ weather data. They suggested that there is a need for using atmospheric images in developing forecasting models, multi-variate models can be employed in prediction so that performance in predicting the weather will be improved, which thereby helps farmers in cultivating the crops, so that the yield can be improved.

Pallab Bharman et al. [11] analyzed works categorized in yield prediction, weed detection, and disease detection. On proper utilization of deep learning and sensor data, farm management systems can be turned into real-time AI-enabled applications. From their survey the LSTM, BBI, DNN, DRL, RNN can be used in yield prediction. CNN, MCNN, AlexNet, LeNET, DNN_JOA, Cafenet, can be used for disease detection. Inception model can be used for detecting weeds in crops. CNN, RFCN, DCNN can be used for detecting weed on broad-leaf, object detection of weeds, weed growth stage estimator, respectively.

22.3   METHODOLOGY

22.3.1   Data Preprocessing

Data preprocessing is a method that is used to convert the raw data into a clean data set. A clean dataset is one without unnecessary, duplicated data. The data are gathered from different sources, it is collected in raw format, which is not feasible for the analysis. By applying different techniques like replacing missing values and null values, we can transform data into an understandable format. The final step on data preprocessing is the splitting of training and testing data. The data usually tend to split unequally because training the model usually requires as much data points as possible. The training dataset is the initial dataset used to train DL algorithms to learn and produce right predictions.

22.3.1.1   Factors Affecting Crop Yield and Production

There are a lot of factors that affects the yield of any crop and its production. These are basically the features that help in predicting the production of any crop over the year. Here factors included are temperature, rainfall, humidity and windspeed, soil conditions, etc.

22.3.1.2   Splitting Dataset into Testing and Training Sets

The final step of data preprocessing is testing and training the data. The division between the training and testing size of the data is not done equally. A larger training set is required as compared to the testing size, since, for prediction, the model needs to be trained over as many data points as possible. For this, the ScikitLearn library and import of the train_test_split module is used. The train_test_split method has been used to split the data, with the test set being 5% of the total dataset. These values give the most accurate possible result, due to the skewness of the dataset as well as its enormity of it. Figure 22.5 shows the design methodology of crop yield prediction. Weather data and soil parameters are taken as input from the dataset. The dataset is cleaned by removing the unnecessary and repeated data. This process is called data preprocessing. The necessary features are selected from the preprocessed data using a backpropagation neural network. For the selected features, the RNN algorithm is applied to predict the crop yield. Transfer learning approach is used to transfer information from one machine learning task to another. A pre-trained model that was trained for one task is re-purposed as the starting point for a new task. Advantage of implementing transfer learning is less time consuming and resources can be saved. The deep transfer learning based on the LSTM RNN model seems to have a high application value.

22.3.2   Feature Selection Using Back Propagation Network For Weather Prediction

It is vital to find important variables and omit the other redundant ones, which may decrease the accuracy of predictive models. Here a guided back propagation method is applied that back propagates the positive gradients to find input variables, which maximize the activation of selected neurons. It

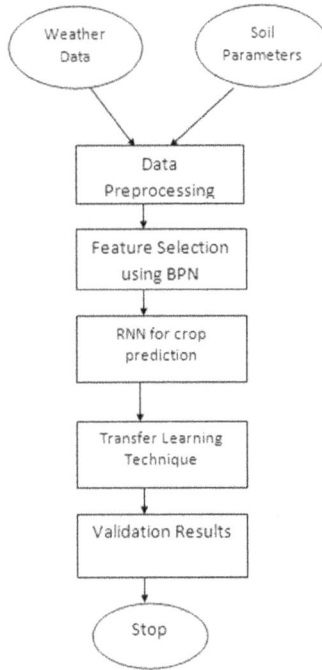

FIGURE 22.5 Design methodology for crop yield prediction.

is less important if an input variable suppresses a neuron with negative gradient somewhere along the path to our selected neurons. First of all, we feed all validation samples to the DNN model and computed the average activation of all neurons in the last hidden layer of the network. We set the gradient of activated neurons to be 1 and the other neurons to be 0. Then, the gradients of the activated neurons were back propagated to the input space to find the associated input variables based on the magnitude of the gradient (the bigger, the more important). The effects of soil conditions and weather components are estimated respectively. The estimated effects indicate the relative importance of each feature compared to the other features. The effects were normalized within each group namely soil conditions, and weather components to make the effects comparable. Solar radiation, temperature, rainfall, precipitation have considerable effects on the variation in crop across different environments. High paddy yield is associated with low temperature and rainfall since lower temperature increases growth duration, thus crops can intercept more radiation. Precipitation is an important factor. High crop yield was associated with less rainfall in the planting period August and September, and above average rainfall throughout October and November, when seed germination and emergence happened. More rainfall with cooler temperatures were also necessary from December, followed by very less rainfall and higher temperatures in January and early February. To evaluate the performance of the feature selection method, we obtained prediction results based on a subset of features. As such, we sorted all features based on their estimated effects, and selected the 10 most important environmental components. The suggested feature selection method successfully finds the important features. Table 22.3 shows the yield prediction performance of deep neural network using selected features.

22.3.2.1 Algorithm for Creating a Neural Network

Step 1: create the network and define its arguments
Step 2: set the number of neurons/nodes for each layer

TABLE 22.3
Yield Prediction Performance of DNN Using Selected Features

Crop	Month	Prediction
Rice	January	Good
Rice	February	Good
Rice	August	Good
Rice	September	Poor
Rice	October	Poor
Rice	November	Poor
Rice	December	Poor

Step 3: initialize the weight matrices.
Step 4: append the hidden layer and output layer
Step 5: define a method to initialize the weight matrices of the neural network
Step 6: train the network
Step 7: run the network with an input vector input_vector
Step 8: test predictions are done.

22.3.3 Recurrent Neural Network in Crop Prediction

It is a type of neural network where the output from the previous step is fed as input to the current step. In traditional neural networks, all the inputs and outputs are independent of each other, but in cases like when it is required to predict the next word of a sentence, the previous words are required and hence there is a need to remember the previous words. Thus, RNN came into existence to solve the issue with the help of a hidden layer. The main and most important feature of RNN is the presence of hidden state, which remembers some information about a sequence. RNN has a "memory", which remembers all information about what has been calculated. It uses the same parameters for each input as it performs the same task on all the inputs or hidden layers to produce the output. This reduces the complexity of parameters, unlike other neural networks.

The crop yield prediction model is a recurrent neural network. Recurrent neural networks are a family of neural networks for processing sequential data. Recurrent networks share parameters in a different way. Each member of the output is a function of the previous members of the output. Each member of the output is produced using the same update rule applied to the previous outputs. The proposed model is based on the LSTM layers, which generate sequence of vectors as their outputs. The model has a dense layer after the LSTM layers. The dense layer is a fully connected layer meaning every neuron of the dense layer is connected to every other neuron. Long-short term memory networks, usually just called LSTMs are a special kind of RNN in the previous layer. The dense layer has only one neuron and outputs a single value. The model is developed to analyze this data as a time series, for better performance. LSTM cells proved to be better performers in processing such time series data. Hence the model is designed with LSTM layers and dense layer. The model is compiled with the loss function, optimizer function, and metric for the performance. This step is significant in the model development as it governs the learning process of the model. The model compilation is followed by fitting. Fitting a model refers to the training process of the model. In the model fit step, the learning process is designed and certain callbacks are utilized for the guidance and extra functionalities during the training. All the above-mentioned design steps are experimented in multiple iterations to arrive at the optimal design specification. The model is designed to predict the yield of one future year in the yield time series, thus only one neuron is used in the dense layer. The output generated from the final dense layer is the final output of the model. This output represents the predicted yield value.

TABLE 22.4
Results of Yield Dataset

Index	Area Code	Element Code	Item Code	Year Code	Value
Count	56717.0	56717.0	56717.0	56717.0	56717.0
Mean	125.65042227198195	5419.0	111.61165082779414	1989.669569970203	62094.66008427808
Std	75.12019495452677	0.0	101.2784353414634	16.133197739658595	67835.93285591144
Min	1.0	5419.0	15.0	1961.0	0.0
25%	58.0	5419.0	56.0	1976.0	15680.0
50%	122.0	5419.0	116.0	1991.0	36744.0
75%	184.0	5419.0	125.0	2004.0	86213.0
Max	351.0	5419.0	489.0	2016.0	1000000.0

Steps to calculate mean, min, max value from the yield dataset are as follows:

Step 1: Import necessary packages like numpy, pandas, etc.
Step 2: Read yield1.csv file.
Step 3: Select the desired columns from the csv file.
Step 4: Rename the fields hg/ha_yield, average rain_fall_mm_per_year,pesticides_tonnes.
Step 5: Drop the unnecessary columns.
Step 6: Calculate mean, min, max from the selected file for yield, rainfall, pesticides used, etc.

Table 22.4 shows the results of yield dataset. The yield1 dataset contains element code, item, item code, area code, value. The df_yield dataset consists of average temperature, pesticides, and average rainfall data for different crops of various years.

22.3.4   DNN in Yield Prediction

A deep neural network is used to build the predictive model. DNN is an artificial neural network algorithm with several hidden layers. The proposed DNN model has four hidden layer. Figure 22.6 shows DNN with three hidden layers, multiple inputs are passed to the hidden layers and different outputs are obtained. The DNN model is created with stochastic gradient descent (SGD). The steps are as follows:

1. The weights are initialized with small values randomly, such as values close to 0, like 0.1, 0.2.
2. Each feature is placed in one input node in the input layer.
3. Forward-propagation operation is applied; the neurons from the input layer to the output layer are activated so that the weights limit each neuron's activation, such operation proceeds until convergence is reached on y prediction.
4. The error is calculated by comparing the prediction and actual value.
5. Then a back propagation operation is applied, the weights are updated based on how much they are relevant for the error, while the learning rate value determines the weight update.
6. Steps 1 to 5 are repeated and the weights are updated after batch learning.
7. When all the process is finished, an epoch is completed; more epochs are done to train a better model.

Each layer has its own bias and weights, each layer's calculation and applying activation function is mentioned in the below equations:

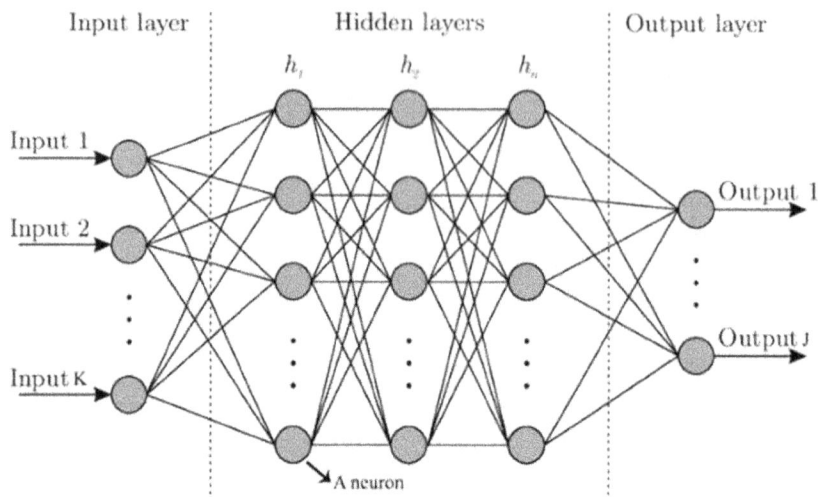

FIGURE 22.6   DNN with three hidden layers.

$$hi^{(1)} = f^{(1)} \left(\Sigma w_{ij}^{(1)} X_j + b_i^{(1)} \right) \tag{22.1}$$

$$hi^{(2)} = f^{(2)} \left(\Sigma w_{ij}^{(2)} X_j^{(1)} + b_i^{(2)} \right) \tag{22.2}$$

$$hi^{(3)} = f^{(3)} \left(\Sigma w_{ij}^{(3)} X_j^{(2)} + b_i^{(3)} \right) \tag{22.3}$$

$$hi^{(4)} = f^{(4)} \left(\Sigma w_{ij}^{(4)} X_j^{(3)} + b_i^{(4)} \right) \tag{22.4}$$

$$yi = f^{(5)} \left(\Sigma w_{ij}^{(5)} X_j^{(4)} + b_i^{(5)} \right) \tag{22.5}$$

where yi is the prediction, w indicates the weight, hi (N) are units in the N-th hidden layer, f is the activation function, Xj represents the input observations.

22.4   EXPERIMENTAL EVALUATION/DISCUSSION

Tensor flow package is employed to implement the deep learning algorithm, and the Scikit-learn package is used for machine learning algorithms. Google colab is used to implement the algorithms. Dataset is obtained from Kaggle. Feature selection is done using back propagation neural network. Among various parameters wind speed, min temperature, max temperature, precipitation is selected for prediction. The selected features are given as input to the DNN, weights are applied and are updated based on the relevancy of error. Both the training and testing phase are carried out on the yield dataset. The yield prediction results are obtained and are compared. The experimental result shows the proposed crop yield prediction performance is quite satisfactory. Table 22.5 shows the yield prediction performance RMSE and correlation co-efficient.

22.5   CONCLUSIONS AND FUTURE ENHANCEMENT

The demand for agriculture food production is increasing rapidly, and without using modern technologies, chances are less to meet the agro demand of the population. Over the past decade, the various industrial applications using deep learning has gained a positive accuracy factor. In this chapter, a DNN-based crop prediction has been done by using back propagation neural network,

TABLE 22.5
Yield Prediction Performance of DNN

Model	Training RMSE	Training Correlation Co-Efficient (%)	Validation RMSE	Validation Correlation Co-Efficient (%)
DNN	12.04	84.04	12.84	81.44

recurrent neural network. DNN-based ensemble methods (back propagation, RNN, and DNN) for crop prediction improves productivity by analyzing weather, which seems to be scalable, simple, and inexpensive in agriculture yield over a large area in a country, with the publicly available sparse multi-source data. This method uses a combination of elements as input and predicts whether the rice growth will be successful or not. The uncertainty problem for choosing the factors have been solved by using back propagation neural network and RNN, consequently the planting costs decrease, and the yield and profits also rise. In future, other new technology, such as IoT, can be integrated with artificial intelligence, and different deep learning algorithms and techniques to increase the yield. More datasets can be collected from different growers on different sites

REFERENCES

[1] *Agricluture and farmers welfare department policy note 2022 – 2023*, Government of Tamlnadu, 2022.

[2] Z.H. Khalil, S.M. Abdullaev, *"Neural network for grain yield predicting based multispectral satellite imagery: Comparitive study"*, Elsevier, 14th International symposium on "Intelligent systems", 2021.

[3] Kavita Jhajharia, Pratistha Mathur, Sanchit Jaina, Sukriti Nijhawan, *"Crop yield prediction using machine learning and deep learning techniques"*, Elsevier, International conference on Machine Learning and Data Engineering, 2023.

[4] Ritesh Dash, Dillip Ku Dash, G.C. Biswa, *"Classification of crop based on macronutrients and weather data using machine learning techniques"*, Elsevier, Results in Engineering, 2021.

[5] Muyideen AbdulRaheem, Joseph Bamidele Awotunde, et al, *"Weather prediction performance evaluation on selected machine learning algorithms"*, IAES International Journal of Artificial Intelligence, 2022.

[6] Amin Amani, Francesco Marinello, " A deep learning-based model to reduce costs and increase productivity in the case of small datasets: A case study in cotton cultivation", *Mohammad, Journal of Agriculture*, MDPI, 2022.

[7] Rishi Gupta, Akhilesh Kumar Sharma, et.al, " *WB-CPI: Weather based crop prediction in India using big data analytics"*, IEEE Access, 2021.

[8] Ju-Young Shin, Byunghoon Min, Kyu Rang Kim, *"High-resolution wind speed forecast system coupling numerical weather prediction and machine learning for agricultural studies – a case study from South Korea"*, Springer, International Journal of Biometeorology, 2022.

[9] Liyun Gong, Miao Yu, Shouyong Jiang, Vassilis Cutsuridis, Simon Pearson, *"Deep learning based prediction on greenhouse crop yield combined TCN and RNN"*, Elsevier, vol. 21, No. 13, 1 July 2021.

[10] K.U. Jaseena, Binsu C. Kovoor, *"Deterministic weather forecasting models based on intelligent predictors: A survey"*, Elsevier, Journal of King Saud University – Computer and Information Sciences, 2020.

[11] Pallab Bharman, Sabbir Ahmad Saad, et. al, "Deep Learning in Agriculture: A Review", *Asian Journal of Computer Science and Information Technology*, February 2022.

23 Plant Disease Detection and Classification Using a Deep Learning Approach for Image-Based Data

D. Tamil Priya and A. Vijayarani
School of Computer Science Engineering and Information Systems,
Vellore Institute of Technology, Vellore, Tamil Nadu, India

23.1 INTRODUCTION

In both the economic and climate change, plants are extremely important. Since, climate change has become a worldwide problem and these plants also play a major role in food industry too. An important problem in our country is to balance the world's food production in a wide range [1]. In order to meet the food demand, which is growing exponentially, various agricultural problems must be addressed by using some techniques. Recent agricultural problem detections are using machine learning techniques widely. In addition to this, deep learning (DL) shows significant developments in the field of agriculture research, because of its automated feature extraction capability [2].

Predicting plant disease in an early state is more vital. It is also a time consuming and difficult task because it should be performed more precisely and efficiently [3]. Previously, prediction of plant disease has been done manually, which takes lot of time and there is inaccuracy in predicting the diseases. A plant disease detector is an automated plant disease diagnostic device that employs machine learning algorithms and computer vision. So that the prediction of disease would be done automatically. For achieving this, a deep learning network such as a convolution neural network (CNN) can be worked well for images.

Proletarians in the agricultural process could benefit greatly from an automated system that detects plant diseases through the appearance of the plant and visual symptoms [4]. This could be a helpful tool for farmers since it has warned them in the appropriate place at the right time before the diseases spread across a large area [5]. In such cases, existing approaches are useful for monitoring vast fields of corps. Automatic disease identification by simply seeing symptoms on plant leaves is both easier and less expensive. This also supports computer vision to provide image-based automatic process control, inspection, and robot guiding.

This study's focus is on the system to detect diseases in plant leaf using deep learning model. The objective is to study the architecture of deep convolutional neural networks to perform a complex analysis of large amounts of plant data by passing it through multiple layers of neurons. Deep convolutional neural networks (DCNN) are used to detect the infected plant leaf by training a large number of data and recognition of the plant disease more accurately. The model can be used on smart phones to differentiate the healthy and infected plants, which suggests fertilizers to the farmer based on types of diseases for different plants.

Integrating data augmentation and training strategies improves the deep learning model's performance. In this chapter, the augmentation techniques such image inversion, image flipping, image rotation, noise injection, gamma correction, deterministic image, zooming image, and resizing

DOI: 10.1201/9781003407959-23

image are used. The proposed deep learning model is tested using validation and training methods with a dataset of plant diseases. The proposed system outperforms accurate classification of diseased plant in leaves and the use of a web application in disease detection in plants.

The remaining part of the paper is organized in the following ways: Section 23.2 represents the review of related works based on various methods, techniques, and models. Section 23.3 describes the image augmentation techniques, proposed detection and classification model and background theories of deep learning techniques. Section 23.4 elaborates the performance of the proposed model, dataset descriptions, and the result analysis. Section 23.5 gives a conclusion followed by future recommendations for achieving more advancements in the visualization, detection, and classification of plant diseases. A method for plant disease identification using deep learning that focuses on individual lesions and spots.

23.2   RELATED WORKS

This section presents the related works for plant disease classification and reviews with various existing models, methods, and datasets.

Saleem et al. [6] discussed the deep learning (DL) approaches used for detecting and classifying the diseases in plant leaves. Various computer vision techniques and deep learning approaches are used for the task of classification. These techniques could be used for plant disease detection, as it is associated with various challenges including the need for labeled data and variability in disease symptoms. The author discussed imaging techniques such as hyperspectral [7], thermal, x-ray, and fluorescence imaging and their respective advantages and limitations used for detecting plant diseases [8]. The new methodology called "ensemble of patches" [1], which overcomes the limitations of deep learning-based approaches by means of dividing the input images into smaller patches and using an ensemble of deep learning models to classify each patch. The authors demonstrated the effectiveness of the proposed method on two datasets of plant diseases and compare it with other methods. Different deep learning models such as CNNs, RNNs, and GANs [2] and their applications for detecting diseases in plants are studied. The author also analyzed the various challenges associated using deep learning techniques for plant disease detection and classification with the possible suggestions for future research directions.

Zhang et al. [7] discussed various methods of hyperspectral data acquisition, including imaging spectroscopy, spectral reflectance, and fluorescence spectroscopy with the importance of data preprocessing, such as calibration, normalization, and feature extraction. Chen et al. explore a range of machine learning and deep learning techniques [9], which have been applied to plant disease detection, including random forests, support vector machines (SVM), convolutional neural networks (CNN) [10], and also using recurrent neural networks (RNN). The chapter also highlights the challenges that remain in this field such as spectral variability, data complexity, and limited availability of labeled data. A novel end-to-end object detection algorithm called DBA_SSD, is proposed by Wang et al. [11] for plant disease detection. The algorithm uses a deep learning technique [6] and a single-shot detection framework for detecting and classifying diseased regions in plant images. The authors, Turkoglu and Hanbay et al. [3], proposed a method for pest detection and for plant disease using a deep learning approach based on plant leaf image features. This proposed method involves training the deep convolution neural network by extracting features from plant images, which are then used to classify the plant as healthy or diseased.

The performance of different convolutional neural network (CNN) models including VGG-16, ResNet-50, and Inception-V3 architecture are applied for plant disease detection using image segmentation techniques [12]. The paper proposed an improved YOLOv5 model [13] for plant disease recognition. The authors make modifications to the YOLOv5 architecture and introduce a multi-scale feature fusion method in order to increase the accuracy rate of the model and it also used the data augmentation method towards improvement of model using large amount of the training dataset. A novel deep learning model "attention embedded residual CNN (ARCNN)" [10], which is used for detecting diseases in tomato leaves [5]. This model integrates attention modules with residual networks in order to increase the accuracy of classification process.

A plant monitoring system with convolutional neural networks (CNNs) [7] in a polyhouse [10] and an adaptive neuro-fuzzy inference system (ANFIS) [12] is studied for plant disease detection. The system uses a Raspberry Pi [16] camera to capture images of the plants and sends them to the cloud for processing. The CNN model is trained on a dataset of images to classify healthy and diseased plants, and it can accurately classify the plants' health status. The proposed system can detect plant diseases in real-time and help farmers to take timely action, which prevents the further spread of diseases. A technique for automatically capturing images of tomato plant leaf disease [14][17] is experimented for detection and recognition of disease by using deep learning approach [5]. This study [18] examined the use of artificial intelligence and image processing techniques to detect plant diseases. It also presents an overview of various methodology for image acquisition, pre-processing, extraction of image features, and classification, including machine learning, deep learning [17, 5], and fuzzy logic.

A method for deploying deep neural networks on edge devices for plant disease detection using dynamic K-means compression is proposed by De et al. [16]. The authors anticipated a pipeline, which trains a deep neural network on a cloud-based server, then compresses it using dynamic K-means clustering to reduce its size, and finally deploys it on a Raspberry Pi [15] device for inference. A method for plant disease identification using deep learning based on individual lesions and spots [4], which involves three stages such as lesion or spot segmentation, extraction of image features, and then classification of disease in plants. The authors, Devi and Nandhini et al. [19], proposed a system which used IoT sensors towards collection of data on plant health, which was then analyzed and determined by using machine learning algorithms for detection of diseases and classification of diseased plants.

To distinguish healthy and unhealthy potatoes, a deep convolutional neural network (CNN) is trained on a dataset of potato tuber images by Oppenheim et al. [20]. Tomato Spotted Wilt Virus (TSWV) detection in capsicum plants utilizing hyperspectral imaging (VNIR and SWIR) [1][7] and by machine learning approaches [21]. The automatic plant disease detection used several processes, including a median filter, thresholding, and the segmentation of segments using several color models [22]. Pattern recognition techniques for disease detection and classification in cotton leaf plants [23][24] is investigated by isolating the diseased spots using an active contour-based segmentation algorithm [25]. The authors, Arsenovic et al. [1] and Zhang et al. [7], examined the use of of hyperspectral imaging and observed normal comparative reflectance in infected leaves. It is possible to perform this with a spectro-radiometer in a lab 32 spectral, features were extracted, then they were evaluated using t-test, correlation analysis, and Fisher linear discriminant analysis [26].

The hyperspectral imaging techniques are utilized to identify chlorophyll and carotenoid content in cucumber plant leaves that have significant angular leaf spot (ALS) infection cases (Wang et al. [27] and Zhang et al. [26]). AgriPest, a domain-specific benchmark dataset is introduced, which is used in finding tiny wild pest recognition and detection system for pest monitoring application [28]. The convolutional neural network-based models are used for insect pest classification, which includes attention, feature pyramid, and fine-grained models [28]. The author, Nanni et al. [29], developed three different methods for image preprocessing and also created three different images for each saliency method in order to detect infected pest images using a convolutional neural network (CNN). The authors investigated CNN architectures for identification of diseases in plants leaf using transfer learning and deep feature extraction approaches. Using the Plant Village dataset, all obtained features could be classified by SVM and KNN [30][31]. A neural network which includes multiple convolution and pooling layers is used for developing a plant disease detection model. To train the model and to detect diseases, the Plant Village dataset is used by Chohan et al. [31].

Harakannanavar et al. [32] used machine learning approaches such as support vector machine (SVM), convolutional neural network (CNN), and K-nearest neighbor (K-NN) for classification of disease on tomato disordered samples. Deboral et al. [33] developed a prototype for plant disease detection using IoT by keeping farmer responsibilities at home simpler and more reachable. This model displays the disease as it exhibits itself on the parts of a plant. By employing images of tomato plant leaves, the hierarchical mixed pooling technique for smoothing to sharpening approach has been applied in the proposed CNN model to detect accurate disease [34]. Using a deep convolutional neural network, Krishna Kewat et al. [35] successfully identify and detect rice diseases and pests, including the healthy plant class. To address the issues of phytopathology in pomegranate plant is employed by k-means clustering to isolate the damaged region of the leaf. The pre-processing method includes locating the region of interest, scaling, color conversion, and filtering. The next step is to extract features and categorize the data using supervised learning models like ResNet and MobileNet [36].

Hassan et al. [37] investigated and identified plant diseases on the Plant Village dataset, on the rice disease dataset and on the cassava dataset by using deep learning approaches. Pandian et al. [38] proposed a novel DCNN model to detect plant leaf diseases. The 14-DCNN model that has been established to train and to identify 42 leaf diseases in 16 different plants using leaf images. To improve the 14-DCNN's performance in this study, the approaches of data augmentation and hyperparameter optimization were also applied. Behera et al. [39] employed the VGG-19-based model to predict nine types of disease in plants before the symptoms developed by using image pre-processing, feature extraction, and classification techniques. Deep learning techniques such as MobileNetV1 and ResNet34 structure were adopted in order to classify the plant diseases more efficiently. In which the ResNet34 model is an efficient approach for disease classification [40]. Shahoveisi et al. [41] employed prospective four different convolutional neural network models such as Xception, Residual Networks (ResNet) 50, EfficientNetB4, and MobileNet, in the detection of rust disease on three commercially significant field crop. The experimental results showed that the EfficientNetB4 model was the most accurate model in the disease detection when compared with by ResNet50.

23.3 METHODOLOGY

The deep convolution neural network (DCNN) model is used in this experiment for detecting diseases in plant leaves using New Plant Diseases Dataset. The classification of diseases required several stages, beginning with data collection and progressing to data preprocessing, feature learning, training and validating the model. The detection and classification model (DCM) is implemented using a deep learning network with multiple layers of convolution and pooling layer. An interface is developed to detect disease at an early stage, to provide solution to the infected plants and also for the benefit of farmers. An example image from the New Plant Diseases Dataset is shown below in Figure 23.1.

22.3.1 PREPROCESSING

Preprocessing the images is an important steps in any image processing applications. In this chapter image augmentation techniques are used to detect plant diseases by identifying the color feature of the image and to speed up the performance of the network.

23.3.1.1 Image Augmentation Techniques

In this experiment, the image augmentation techniques such as inverting image, flipping image, rotating image, noise injection, gamma correction, deterministic image, zooming image, and resizing image are applied on the plant disease dataset. The experiment found that image augmentation

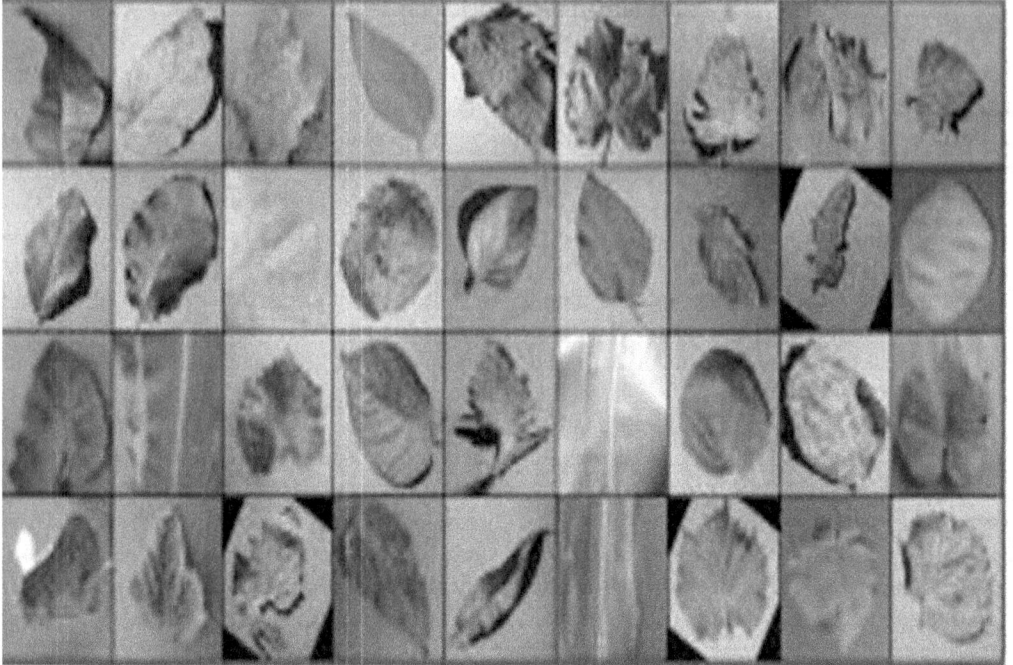

FIGURE 23.1 Sample images from the New Plant Diseases Dataset.

helped to make deep convolution neural network (DCNN) model more accurate and proficient in detecting the diseased leaves in plants. The image augmentation will increase the efficiency of DCNN techniques in terms of accuracy and in recognizing the diseased or healthy plants. The results of image augmentation techniques is shown in Figure 23.2.

23.3.2 DETECTION AND CLASSIFICATION (DCM) MODEL

A typical deep convolution neural network is made up of several layers. Each layer is made up of numerous nodes, each with its own activation function. The first layer is the input layer, which accepts input data, and the last layer is the output layer, which produces output. Between the input and output layers, a random number of hidden layers including, convolutional or convo, pooling, dense or fully connected, and softmax layer. The convolution neural network (CNN) is referred to as deep convolutional neural network (DCNN) if it has two or more hidden layers. The detection and classification model architecture for plant disease detection is represented in Figure 23.3.

23.3.2.1 Loading Dataset

The ImageNet Dataset contains images of fixed size of 224*224 and provided with RGB channels, which have a tensor of 224*224*3 as output. It contains convolution, maxpooling, flatten, fully connected, and softmax layers. For loading dataset, we are directly loading the images as PyTorch tensors by applying image transformation like resizing from 224*224px to 112*112px (for faster processing) and then converting as a new dataset. Then the image is split as the validation and training dataset with 30 percent of the total dataset as validation data. Finally, we create batches of the dataset for faster processing of the dataset.

FIGURE 23.2 Sample images of image augmentation techniques: (a) inverted image; (b) flipped image; (c) rotated image; (d) noise injection; (e) Gamma correction; (f) deterministic image; (g) zoom image; and (h) resized image.

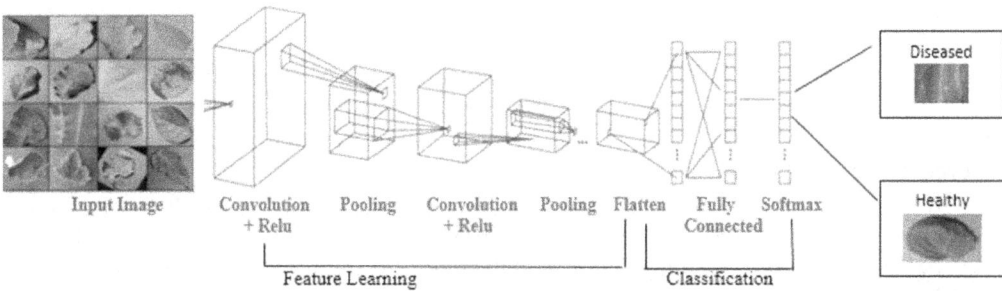

FIGURE 23.3 Representation of detection and classification model (DCM) architecture for plant disease detection.

23.3.2.2 Building DCM Model using CNN

The a convolutional neural network is fed with images. We supply a color image, thus it contains RGB channels. We define a base image classification class with training and validation methods for each batch of data. Using the Image Classification Base class, we define functions, such as training step, validation step, validation epoch end, and epoch end.

The training stage calculates loss for each epoch, which can be used for backward propagation to optimize the model and reduce loss. The validation step is used to calculate the loss and accuracy for the model by verifying the entropy between images and their outputs. The validation epoch end is used to retrieve and validate the batch's loss with accuracy in each epoch. After each epoch, the epoch end is used to print the training loss and to validate the loss and accuracy. The validation

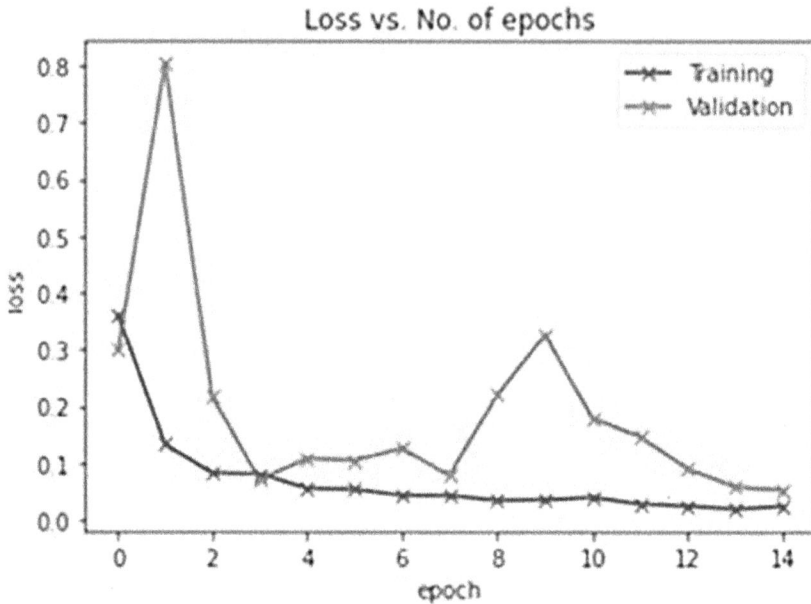

FIGURE 23.4 Diagrammatic representation for each epoch to validate the loss and accuracy.

loss is the metrics used to calculate the performance of deep learning based on validation set. The training loss is the metric used to calculate the performance of the deep learning model based on training set. The loss and accuracy of training and validation data after each epoch is represented in Figure 23.4).

During validation and testing, our loss function only comprises prediction error, resulting in a generally lower loss than the training and we can see how the gap between validation and train loss shrinks after each epoch. This is because as the network learns the data, it also shrinks the regularization loss (model weights), leading to a minor difference between validation and train loss. However, the model is still more accurate on the training set.

For feature learning, we use Conv2D, ReLu, and Maxpool2D layers, and for classification, we use flatten, fully connected, and softmax layers. The softmax layer which classifies the plant leaf as healthy or diseased. Figure 23.5 depicts the block diagram for the proposed plant detection and classification model (DCM) based on deep convolution neural network (DCNN). During training and testing phase, various data augmentation techniques are applied to resize and remove unwanted data. Feature extraction is achieved to learn the image features, which can be done by sequence layers of convolution, ReLu, pooling layer. Then the model classifies the disease as healthy and unhealthy and it generates the loss errors in training and validation data using loss functions.

23.3.2.3 Training and Evaluation

For model evaluation and training, we define two functions: fit and optimizer. The fit function has various parameters for training the model for a specified number of epochs. To train the model, we use the optimizer "Adam" with a learning rate of 0.001. In the fit function, we initially train the model for each epoch and after training each epoch calculate loss and use backward propagation to optimize the model and then finally calculate train loss, val_loss and val_acc.

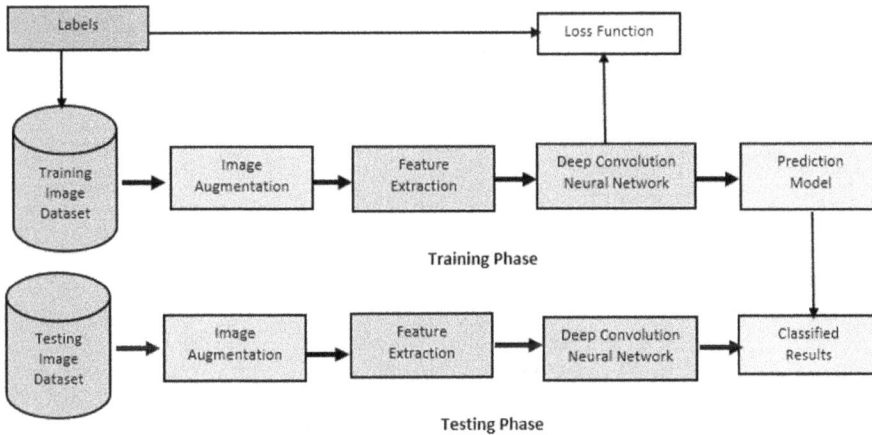

FIGURE 23.5 Block diagram of proposed plant disease detection and classification model using DCNN.

23.4 EXPERIMENTAL ANALYSIS

The proposed DCM model is used to calculate efficiency and performance of networks using a given dataset. A variety of experiments have been conducted with the proposed model. The outcomes of 30% of the Plant Village data, which includes images of both healthy and diseased plants, are evaluated.

23.4.1 DATASET DESCRIPTION

This experiment employs the "New Plant Diseases Dataset" from the Kaggle repository, which contains 70,295 training and 17,572 testing images for plant disease classification. The dataset comprises of given attributes, which could be used to predict plant diseases using deep learning (DL) techniques. After building a robust deep learning model, user friendly web application is developed for the users to know about the infected status of plants. Table 23.1 shows the plant image dataset for each leaf disease class. Unique plants such as pepper, grape, tomato, squash, corn, apple, potato, raspberry, strawberry, cherry, soybean, blueberry, orange, and peach are taken and applied in this work to classify disease as diseased and healthy plants based on various disease types.

23.4.2 EXPERIMENTAL MODEL PERFORMANCE

Plant disease classification and detection model is developed by the deep convolution neural network (DCNN), which is used to detect and classify the disease in leaf of plants. In this work, we built our detection and classification (DCM) model with the TensorFlow deep learning framework and the Keras package. Keras is an open-source deep-learning package that may be used to execute a variety of deep learning applications. We used it to develop DCM architecture, which was inspired by several network layers.

Our proposed model consists of including convolution + Relu, pooling, flatten, fully connected and softmax layers for feature learning and classification. First, resize every image into 224*224, then these images are feed into the convolutional neural network. The convolution layer applies 32 filter size of output channels i.e. $32{\times}224{\times}224$ and converted into $32{\times}222{\times}222$. The ReLU activation function is then used to reduce nonlinearity, and the batch normalization function is used to normalize the neuron weights.

TABLE 23.1
Plant Image Dataset for Leaf Disease Class

S.No	Plant	Disease	Healthy/Diseased
1	Apple	Apple scab	Diseased
2	Apple	Black rot	Diseased
3	Apple	Cedar apple rust	Diseased
4	Apple	No Disease	Healthy
5	Blueberry	No Disease	Healthy
6	Corn	Cercospora leaf spotGray leaf spot	Diseased
7	Corn	Common rust	Diseased
8	Corn	Northern Leaf Blight	Diseased
9	Corn	No Disease	Healthy
10	Cherry	Powdery mildew	Diseased
11	Cherry	No Disease	Healthy
12	Grape	Black rot	Diseased
13	Grape	Esca (Black Measles)	Diseased
14	Grape	Leaf blight (Isariopsis Leaf Spot)	Diseased
15	Grape	No Disease	Healthy
16	Orange	Haunglongbing (Citrus greening)	Diseased
17	Peach	Bacterial spot	Diseased
18	Peach	No Disease	Healthy
19	Pepper bell	Bacterial spot	Diseased
20	Pepper bell	No Disease	Healthy
21	Potato	Early blight	Diseased
22	Potato	Late blight	Diseased
23	Potato	No Disease	Healthy
24	Raspberry	No Disease	Healthy
25	Soybean	No Disease	Healthy
26	Squash	Powdery mildew	Diseased
27	Strawberry	Leaf scorch	Diseased
28	Strawberry	No Disease	Healthy
29	Tomato	Leaf Mold	Diseased
30	Tomato	Septoria leaf spot	Diseased
31	Tomato	Tomato Yellow Leaf Curl Virus	Diseased
32	Tomato	spotted spider mite	Diseased
33	Tomato	No Disease	Healthy

Following that, these images are fed into the maxpooling layer, which selects only the most significant features, yielding an output image with dimensions of $32 \times 112 \times 112$. The same procedure is used for the next convolution neural network. Finally, flatten the output of the final maxpooling layer and feed it to the next linear layer, which is known as a fully connected layer, and then to the softmax layer, which predicts 39 types of plant disease. Finally, we get tensor of 1×39 size as a model output and evaluate the performance of model. The output summary of detection and classification model of layered architecture is shown in Table 23.2 and represents the output of the each layer in DCM architecture based on features and parameters.

23.4.2.1   Evaluation Measures and Results

The efficiency and performance of the proposed neural networks model is measured through precision, recall, accuracy, and the f1-score. Precision denotes correct predictions made from false positives, whereas recall denotes correct predictions made from false negatives. The number of

TABLE 23.2
Summary of DCM Layered Architecture

Layer (type)	Output Shape	Param #
Conv2d-1	[-1, 32, 224, 224]	896
ReLU-2	[-1, 32, 224, 224]	e
BatchNorm2d-3	[-1, 32, 224, 224]	64
Conv2d-4	[-1, 32, 224, 224]	9,248
ReLU-5	[-1, 32, 224, 224]	e
BatchNorm2d-6	[-1, 32, 224, 224]	64
MaxPool2d-7	[-1, 32, 112, 112]	e
Conv2d-8	[-1, 64, 112, 112]	18,496
ReLU-9	[-1, 64, 112, 112]	e
BatchNorm2d-10	[-1, 64, 112, 112]	128
Conv2d-11	[-1, 64, 112, 112]	36,928
ReLU-12	[-1, 64, 112, 112]	e
BatchNorm2d-13	[-1, 64, 112, 112]	128
MaxPool2d-14	[-1, 64, 56, 56]	e
Conv2d-15	[-1, 128, 56, 56]	73,856
ReLU-16	[-1, 128, 56, 56]	e
BatchNorm2d-17	[-1, 128, 56, 56]	256
Conv2d-18	[-1, 128, 56, 56]	147,584
ReLU-19	[-1, 128, 56, 56]	e
BatchNorm2d-20	[-1, 128, 56, 56]	256
MaxPool2d-21	[-1, 128, 28, 28]	e
Conv2d-22	[-1, 256, 28, 28]	295,168
ReLU-23	[-1, 256, 28, 28]	e

correct predictions out of both false positives and false negatives is the accuracy. All of the perform-ance metrics employed in our trained model are calculated using the formulas shown in Eqs (23.1), (23.2), (23.3), and (23.4), which are mentioned below. Finally, we can calculate the model's per-formance and outcomes from the predicted label against true label values, as shown in Figure 23.6.

$$Precision = \frac{TP}{TP + FP} \tag{23.1}$$

$$Recall = \frac{TP}{TP + FP} \tag{23.2}$$

$$Accuracy = \frac{TP + TN}{TP + TN + FN + FP} \tag{23.3}$$

$$F1 - Score = 2 * \frac{precision * recall}{precision + recall} \tag{23.4}$$

Here TP denotes true positives, TN denotes true negatives, FP denotes false positives, and FN denotes false negatives. In this case, the TP and TN are correct predictions, whereas the FP and FN are incorrect predictions produced by the model. The performance of the proposed model is represented in Table 23.3 using several evaluation metrics.

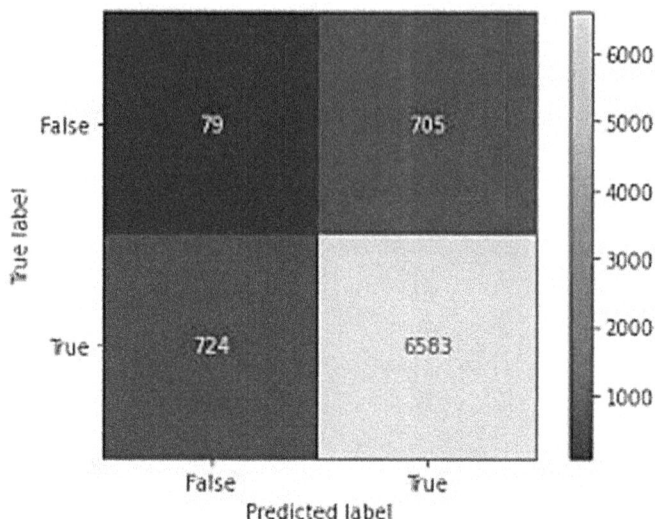

FIGURE 23.6   Performance of our proposed model based on predicted label and true label values.

TABLE 23.3
Performance Metrics Values (in %) by Detection and Classification Model

Evaluation Metrics	Values
Precision	98.42%
Recall	97.99%
F1-Score	97.89%
Accuracy	98.86%

While training and validation, our model generates the loss value over the training data after each epoch of optimization, which represents how the model behaves after each iteration. Accuracy represents the ratio between correct predictions and the total number of predictions in the training data during classification process. Figure 23.7 depicts the accuracy of the detection and classification model after each epoch.

23.4.3   BUILDING INTERFACE MODEL USING DEEP LEARNING

The interface is developed, which provides solution to the farmers for early detection of disease in plant. In order to test plants, the sample leaf can be uploaded in the website. It will identify whether the plant is infected by disease or not. If the plant is infected, the result shows the type of disease and cause of disease. Also it shows the method to prevent the disease in early stage and provides solution to cure the disease. This interface will help the farmer for early detection and prevention of disease. The Figure 23.8 shows the result.

In our proposed method, we need deep learning network for detection and classification of leaf disease using PlantVillage dataset (Figure 23.9).

23.5   DISCUSSION

The performance of the proposed detection and classification model (DCM) using the deep convolution neural network (DCNN) is evaluated. Various data augmentation strategies are used to downsize

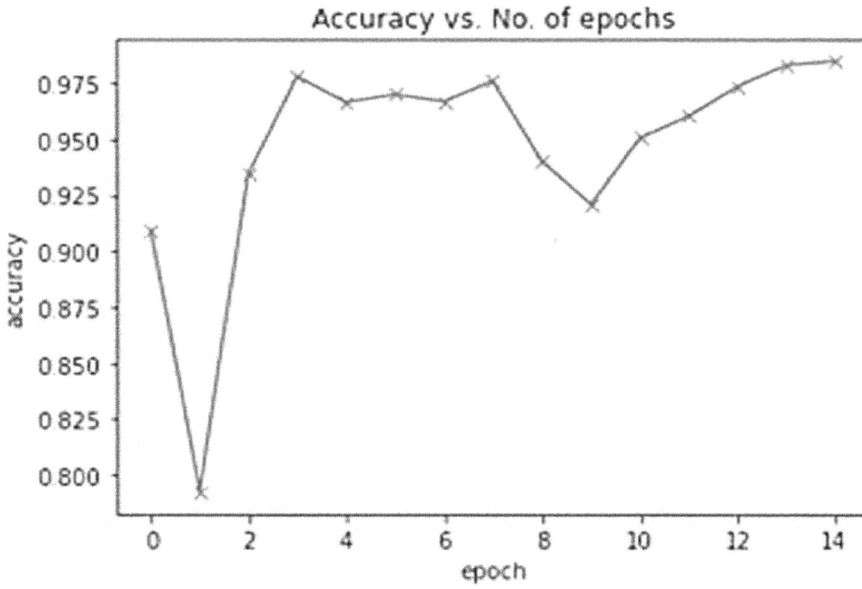

FIGURE 23.7 Performance of the proposed model based on accuracy after each epoch.

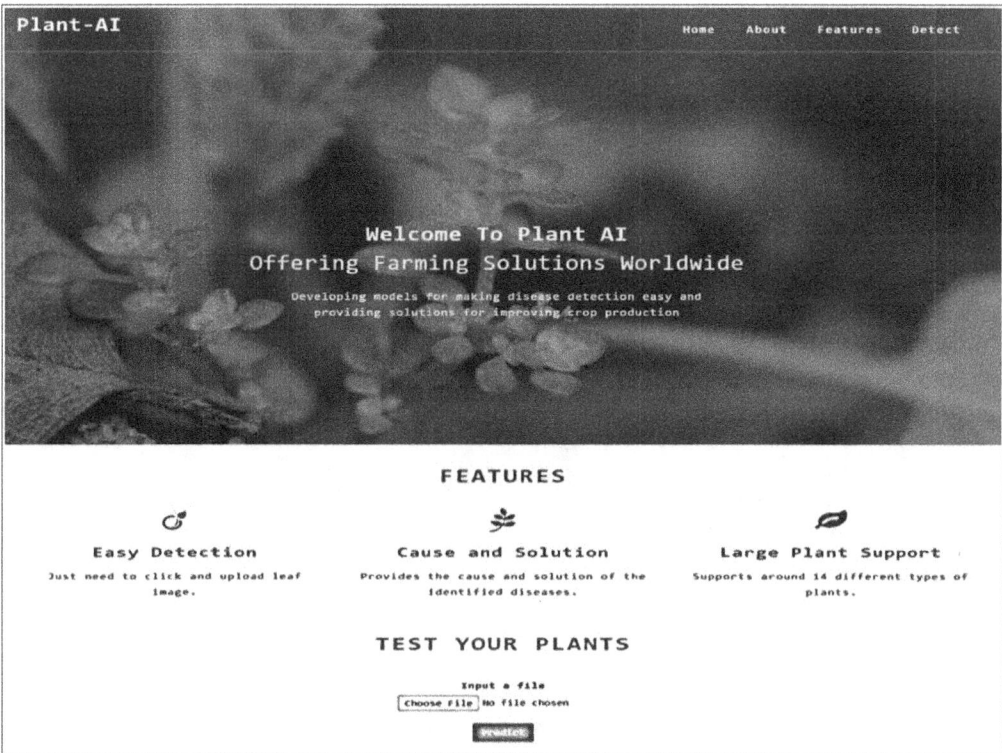

FIGURE 23.8 Interface model to detect and classify the disease of a plant leaf.

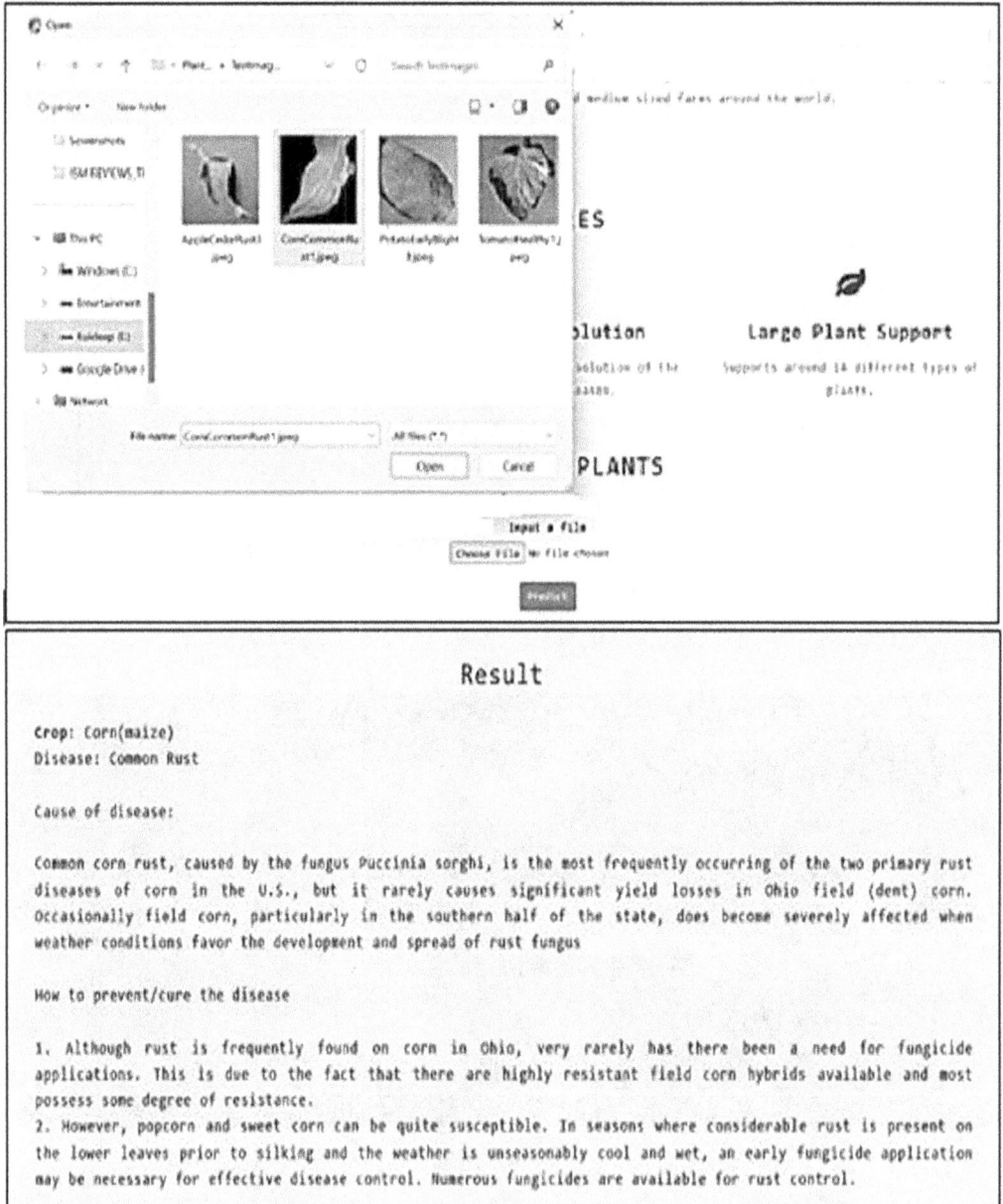

FIGURE 23.9 Result of the plant leaf detection and classification model based on disease type.

and eliminate extraneous data during the training and validation process. Using three different types of layers—convolution, pooling, and activation Relu—feature extraction is accomplished to learn the image features. To identify a plant illness in a leaf image, the fully connected and softmax layers are employed. The performance results obtained by using the deep learning network were evaluated.

Various experiments were conducted using the convolution neural network, deep learning techiques, transfer learning, and tradional methods such as KNN, and SVM, to computer the performance of the conventional and trational methods. The performance of the above said models and classifiers were evaluated using different dataset like Plant Village, cotton plant, insect pest, tomato

TABLE 23.4
Performance of Various Classification Model Based on Accuracy

S.No	Author	Dataset	Classification Model	Accuracy (in %)
1	Barbedo et al. [4]	Plant Village	GoogLeNet	94
2	Turkoglu et al. [3]	Plant Village	CNN	93.67
3	Ramya et al. [24]	Cotton Plant	CNN	90
4	Ung et al. [28]	Insect pest	CNN	98.8
5	Liu et al. [17]	Tomato plant	YoloV3	92.39
6	Nanni et al. [29]	Plant Village	CNN	92.43
7	Chohan et al. [31]	Plant Village	CNN	98.3
8	Harakannanavar et al. [32]	Plant Village	CNN	99.6
9	Kalbande et al. [34]	Tomato plant	CNN	96.46
10	Krishna Kewat et al. [35]	ImageNet	DCNN	96.5
11	Nirmal et al. [36]	Pomegranate leaf image	MobileNet	97.32
12	Hassan et al. [37]	Plant Village	DCNN	99.39
13	Pandian et al. [38]	Open source	DCNN	99.79
14	Behera et al. [39]	Plant Village	VGG19	99.5
15	Divya et al. [40]	Various leaves image	ResNet34	98.91
16	Shahoveisi et al. [41]	Commercially field crops	EfficientNetB4	94.29

plant, ImageNet, pomegranate leaf image, various leaf images, and different commercial field crops. In our proposed method, we used the deep learning network for detection and classification leaf disease using the Plant Village dataset.

Results of the proposed model were compared with different classification models such as GoogLeNet, CNN, YoloV3, DCNN, MobileNet, VGG19, ResNet34, and EfficientNetB4. Table 23.4 represents the performance of various classification model using a different dataset based on accuracy. These comparison results indicates that deep learning model shows good performance when compared with the CNN model using a different dataset. Figure 23.10 illustrates the performance of the various CNN, deep learning, and transfer learning techniques using the Plant Village dataset. The information discussed above suggests that, when compared to existing state-of-the-art methods and models based on network performance, our proposed approach exhibits promising outcomes in terms of accuracy.

23.5 CONCLUSION

This research focuses on developing a deep learning model to detect disease in plant leaves. A deep learning network-based detection and classification model (DCM) has been demonstrated to identify and classify plant diseases in leaf images. First, augmentation techniques such as inverting, flipping, and rotating images, noise injection, gamma correction, deterministic image, zooming image, and resizing image are used to increase the dataset size, reduce noise, and remove unwanted background noise from leaves images. The augmented results also speed up and increase the performance of network. Further, augmented images are given to the deep convolution neural network (DCNN) with multiple convolution and pooling layers are used to detect and classify disease. In this chapter, the Plant Village dataset is used to train and validate the model. After training, the model undergoes extensive testing in order to validate the results. When compared to other models, experimental results show that the proposed model outperforms effectively and has a high accuracy of 98.86%. The interface model also developed for the farmersis used to test the plant and it shows the type and cause of the disease. It also assists farmers in the early detection of diseases in plant leaves and the appropriate steps to be taken to prevent those diseases in future.

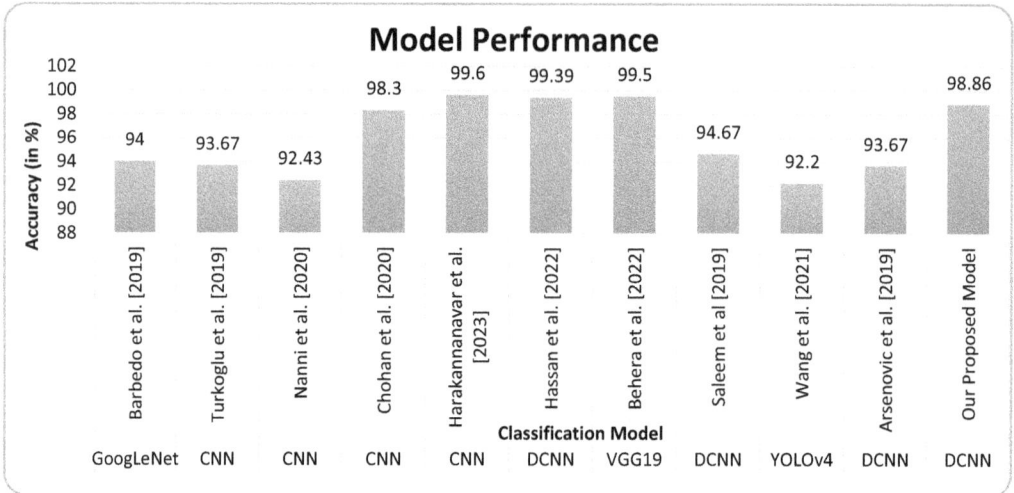

FIGURE 23.10 Performance of various model using the Plant Village dataset.

REFERENCES

1. Arsenovic, M., Karanovic, M., Sladojevic, S., Anderla, A., & Stefanovic, D. (2019). Solving current limitations of deep learning based approaches for plant disease detection. *Symmetry*, 11(7), 939.
2. Li, L., Zhang, S., & Wang, B. (2021). Plant disease detection and classification by deep learning—a review. *IEEE Access*, 9, 56683–56698.
3. Turkoglu, M., & Hanbay, D. (2019). Plant disease and pest detection using deep learning-based features. *Turkish Journal of Electrical Engineering and Computer Sciences*, 27(3), 1636–1651.
4. Barbedo, J. G. A. (2019). Plant disease identification from individual lesions and spots using deep learning. *Biosystems Engineering,* 180, 96–107.
5. De Luna, R. G., Dadios, E. P., & Bandala, A. A. (2018, October). Automated image capturing system for deep learning-based tomato plant leaf disease detection and recognition. In TENCON 2018-2018 IEEE Region 10 Conference (pp. 1414–1419). IEEE.
6. Saleem, M. H., Potgieter, J., & Arif, K. M. (2019). Plant disease detection and classification by deep learning. *Plants,* 8(11), 468. https://doi.org/10.3390/plants8110468
7. Zhang, N., Yang, G., Pan, Y., Yang, X., Chen, L., & Zhao, C. (2020). A review of advanced technologies and development for hyperspectral-based plant disease detection in the past three decades. *Remote Sensing,* 12(19), 3188. https://doi.org/10.3390/rs12193188
8. Singh, V., Sharma, N., & Singh, S. (2020). A review of imaging techniques for plant disease detection. *Artificial Intelligence in Agriculture,* 4, 229–242.
9. Chen, J., Chen, J., Zhang, D., Sun, Y., & Nanehkaran, Y. A. (2020). Using deep transfer learning for image-based plant disease identification. *Computers and Electronics in Agriculture,* 173, 105393.
10. Radha, N., & Swathika, R. (2021, March). A polyhouse: Plant monitoring and diseases detection using CNN. In 2021 International Conference on Artificial Intelligence and Smart Systems (ICAIS) (pp. 966–971). IEEE.
11. Wang, J., Yu, L., Yang, J., & Dong, H. (2021). DBA_SSD: A novel end-to-end object detection algorithm applied to plant disease detection. *Information,* 12(11), 474. https://doi.org/10.3390/info1 2110474
12. Sharma, P., Berwal, Y. P. S., & Ghai, W. (2020). Performance analysis of deep learning CNN models for disease detection in plants using image segmentation. *Information Processing in Agriculture,* 7(4), 566–574.
13. Chen, Z., Wu, R., Lin, Y., Li, C., Chen, S., Yuan, Z., Chen, S., & Zou, X. (2022). Plant disease recognition model based on improved YOLOv5. *Agronomy,* 12(2), 365. https://doi.org/10.3390/agronomy1 2020365

14. Karthik, R., Hariharan, M., Anand, S., Mathikshara, P., Johnson, A., & Menaka, R. (2020). Attention embedded residual CNN for disease detection in tomato leaves. *Applied Soft Computing, 86*, 105933.

15. Sabrol, H., & Kumar, S. (2020). Plant leaf disease detection using adaptive neuro-fuzzy classification. In Advances in Computer Vision: Proceedings of the 2019 Computer Vision Conference (CVC), Vol. 11 (pp. 434–443). Springer International Publishing.

16. De Vita, F., Nocera, G., Bruneo, D., Tomaselli, V., Giacalone, D., & Das, S. K. (2021). Porting deep neural networks on the edge via dynamic K-means compression: A case study of plant disease detection. *Pervasive and Mobile Computing, 75*, 101437.

17. Liu, J., & Wang, X. (2020). Tomato diseases and pests detection based on improved Yolo V3 convolutional neural network. *Frontiers in Plant Science, 11*, 898.

18. Vishnoi, V. K., Kumar, K., & Kumar, B. (2021). Plant disease detection using computational intelligence and image processing. *Journal of Plant Diseases and Protection, 128*, 19–53.

19. Devi, R. D., Nandhini, S. A., Hemalatha, R., & Radha, S. (2019, March). IoT enabled efficient detection and classification of plant diseases for agricultural applications. In 2019 International Conference on Wireless Communications Signal Processing and Networking (WiSPNET) (pp. 447–451). IEEE.

20. Oppenheim, D., Shani, G., Erlich, O., & Tsror, L. (2019). Using deep learning for image-based potato tuber disease detection. *Phytopathology*, 109(6), 1083–1087.

21. Moghadam, P., Ward, D., Goan, E., Jayawardena, S., Sikka, P., & Hernandez, E. (2017, November). Plant disease detection using hyperspectral imaging. In 2017 International Conference on Digital Image Computing: Techniques and Applications (DICTA) (pp. 1–8). IEEE.

22. El Sghair, M., Jovanovic, R., & Tuba, M. (2017). An algorithm for plant diseases detection based on color features. *International Journal of Agricultural Science, 2*, 1–6.

23. Ramesh, K., Sheeba, J., & Priyadarshini, K. (2018). Detection of cotton leaf diseases using image processing techniques. *International Journal of Engineering & Technology*, 7(4.41), 628–632.

24. Ramya, K., & Sharmila, S. (2020). Cotton leaf disease detection using deep learning. *International Journal of Advanced Research in Computer Science and Software Engineering*, 10(9), 87–91.

25. Jyothi, T., & Ramachandra, C. (2018). Cotton plant leaf disease identification and classification using machine learning. *International Journal of Innovative Technology and Exploring Engineering*, 7(5), 281–284.

26. Zhang, Z., Song, Y., & Guo, H. (2021). Deep learning-based cotton leaf disease recognition system. *Journal of Ambient Intelligence and Humanized Computing, 12*(7), 6953–6963.

27. Wang, R., Liu, L., Xie, C., Yang, P., Li, R., & Zhou, M. (2021). Agripest: A large-scale domain-specific benchmark dataset for practical agricultural pest detection in the wild. *Sensors*, 21(5), 1601.

28. Ung, H. T., Ung, H. Q., & Nguyen, B. T. (2021). An efficient insect pest classification using multiple convolutional neural network based models. arXiv preprint arXiv:2107.12189.

29. Nanni, L., Maguolo, G., & Pancino, F. (2020). Insect pest image detection and recognition based on bio-inspired methods. *Ecological Informatics, 57*, 101089.

30. Mohameth, F., Bingcai, C., & Sada, K. A. (2020). Plant disease detection with deep learning and feature extraction using plant village. *Journal of Computer and Communications, 8*(6), 10–22.

31. Chohan, M., Khan, A., Chohan, R., Katpar, S. H., & Mahar, M. S. (2020). Plant disease detection using deep learning. *International Journal of Recent Technology and Engineering, 9*(1), 909–914.

32. Harakannanavar, S. S., Rudagi, J. M., Puranikmath, V. I., Siddiqua, A., & Pramodhini, R. (2022). Plant leaf disease detection using computer vision and machine learning algorithms. *Global Transitions Proceedings, 3*(1), 305–310.

33. Deboral, C. C., Ambhika, C., Nivetha, P., & Sona, D. (2023, March). Prototype of Plant Disease Detection Using IoT. In 2023 International Conference on Innovative Data Communication Technologies and Application (ICIDCA) (pp. 834–839). IEEE.

34. Kalbande, K., & Patil, W. V. (2023). The convolutional neural network for plant disease detection using hierarchical mixed pooling technique with smoothing to sharpening approach. *International Journal of Computing and Digital Systems, 14*(1), 1–1.

35. Kewat, K. (02 March 2023). Plant Disease Classification using Alex Net PREPRINT (Version 1) available at Research Square [https://doi.org/10.21203/rs.3.rs-2612739/v1]

36. Nirmal, M. D., Jadhav, P. P., & Pawar, S. (2023). Pomegranate leaf disease detection using supervised and unsupervised algorithm techniques. *Cybernetics and Systems,* 1–12. doi: 10.1080/01969722.2023.2166192

37. Hassan, S. M., & Maji, A. K. (2022). Plant disease identification using a novel convolutional neural network. *IEEE Access,* 10, 5390–5401.

38. Pandian, J. A., Kumar, V. D., Geman, O., Hnatiuc, M., Arif, M., & Kanchanadevi, K. (2022). Plant disease detection using deep convolutional neural network. *Applied Sciences*, 12(14), 6982.

39. Behera, A., & Goyal, S. (2022, October). Plant disease detection using deep learning techniques. In International Conference on Information Systems and Management Science (pp. 441–451). Cham: Springer International Publishing.

40. Divya Meena, S., Kumar, K.A.Y., Mandava, D., Bhavya Sri, K., Panda, L., Sheela, J. (2023). Plant Diseases Detection Using Transfer Learning. In: Reddy, K.A., Devi, B.R., George, B., Raju, K.S., Sellathurai, M. (eds) Proceedings of Fourth International Conference on Computer and Communication Technologies. Lecture Notes in Networks and Systems, vol 606. Springer, Singapore. https://doi.org/10.1007/978-981-19-8563-8_1.

41. Shahoveisi, F., Taheri Gorji, H., Shahabi, S., Hosseinirad, S., Markell, S., & Vasefi, F. (2023). Application of image processing and transfer learning for the detection of rust disease. *Scientific Reports*, 13(1), 5133. https://doi.org/10.1038/s41598-023-31942-9.

24 Deep Learning-Based Object Detection in Real-Time Video

T. Sukumar
Department of Information Technology, Sri Venkateswara College of
Engineering, Chennai, Tamil Nadu, India

24.1 INTRODUCTION

Object detection is the process of finding and recognizing real-world object instances such as cars, bikes, TV, flowers, and humans out of images or videos. An object detection technique understands the details of a picture or sequence of pictures within an image. Preferably in 3D space, detection and positioning of objects in a real-time scenario with greater accuracy involves a lot of training of datasets of different objects. Object detection is relatively simple if the machine is looking for detecting one particular object. If the machines are unaware of the many object possibilities, solving this challenge will be highly challenging for them.

A lot of applications, particularly those used for video surveillance, heavily rely on object detection in video sequences. It has numerous potential applications in fields like preventing traffic accidents, alerting workers to dangerous products in factories, monitoring military restricted areas, and cutting-edge human-computer interaction. Pre-processing, segmentation, foreground and background extraction, and feature extraction can all be used to find objects in a video stream. The visual system of the human being is quick and precise, and it is capable of carrying out complicated tasks like object identification that require a lot of mental effort. Because of the availability of large amounts of data, faster GPUs, and improved algorithms, computers may be quickly trained to detect and classify multiple elements inside an image with high accuracy.

The majority of 2D object detection research has been done throughout the years. This can be observed using RCNN, fast RCNN, SSD, and masked RCNN. There are 3D items in the physical world. As a result, 3D bounding boxes rather than the frequently employed 2D detections should be used to bound items observed practically. The ability to recognize 3D things is essential because it would allow us to record the dimensions, directions, and locations of items practically.

A key component of enabling a machine to connect with people in an efficient and simple way in object tracking is the capacity of machines to recognize the suspicious object and further identify their behaviors in a certain environment. The existing method for assessing and identifying suspicious objects typically requires special indicators to be connected to the suspicious object, which precludes the extensive use of technology. In this research, the various phases of the prior object tracking method using video sequences are studied and analyzed. The basic flow diagram of object tracking is shown in Figure 24.1.

24.2 LITERATURE REVIEW

In computer vision applications, object detection and classification are crucial for object tracking. Furthermore, tracking is the initial stage in locating or identifying the moving item in a photograph.

FIGURE 24.1   Basic flow diagram of object tracking.

The detected objects could then be classified as moving things such as moving automobiles, humans, birds, and trees. However, tracking the objects using video sequences is a difficult issue when employing the image processing approach.

A unique ordered moving target detection technique based on spatiotemporal saliency was proposed by Shen et al. [1]. Furthermore, they used spatial and temporal saliency data to get the improved detection findings. The experimental findings demonstrate that this method accurately and effectively recognizes moving objects in airborne video. Li et al. proposed an object detection strategy for detecting things in video frames. The simulation results demonstrate that this technique was reliable for detecting objects [2]. A method for text data detection based on a texture in video frames was proposed by Ben Ayed et al. The video frames are divided into multiple fixed-size blocks, and these blocks are then analyzed using the Haar wavelet transform technique [3]. A new methodology to recognize and identify real-time objects regardless of geometrical transformations was proposed by Danyang Cao et al. [4]. The results demonstrate that this technique was quite effective at identifying even minor items with good precision, especially for moving objects. However, item detection is a little chaotic due to angular effect. Ashwani Kumar et al. used a single shot multi-box detector approach to separate the images into various classes based on the labels provided. They were accomplished quite precisely [5].

Weng et al. suggested an algorithm to enhance the effectiveness of natural feature selection in practical settings. They additionally used to accelerate robust features (SURF) for feature extraction from real-time mobile camera images and recognition. These extracted features are calculated utilizing a posture matrix and the homography technique [6]. Zhang et al. [7] developed the frame difference and nonparametric methods for video analysis traceability. Hence, the background noise is eliminated where in applications like product traceability analysis for food and agriculture, which enables to more accurately recognize the moving object. A parameterization-based disparity space was proposed by Houssineau et al. [8] for non-rectified camera networks, expanded to include moving objects, and integrated into a Bayesian multi-object tracking with sensor calibration technique. In order to deal the problem of single-object localization and tracking as well as multi-object tracking, the effectiveness of the obtained framework has been proved for camera calibration on simulated real data. Ye Lyu et al. suggested a method to recognize objects in video using its characteristic. Furthermore, there was a 5% increase in object detector accuracy. However, because some classes aren't included in the dataset, the video object detection task isn't up to par [9].

Fatima et al. proposed an image segmentation method for tracking important objects by specifying the color intensities. On object categorization, a minimal distance classifier strategy is employed. They achieve object tracking by identifying the object centroids in each video frame. The outcomes of the simulation demonstrated that this method was more effective for approximating context [10]. Lecumberry and Pardo [11] introduced an algorithm for probabilistic relaxation-based semi-automatic object tracking in films using a variety of features. The suggested technique performs well for object tracking, in particular for smooth and accurate object borders. Particle filter, mean shift, and Kalman filter have received a lot of attention from researchers. Hence their work improves tracking outcomes. For example, Yang et al. [12] and Zhao et al. [13] previously coupled mean-shift tracking and Kalman filtering, whereas Tang and Zhang [14] developed a combined model of PF with mean-shift tracking. The handling of occlusion is among the most researched occlusion tracking problems in computer vision. Particle filter, Kalman filter, and mean shift tracking algorithms have been claimed to be effective at addressing occlusions in previous studies. Even though the combination of these techniques yields correct results, only specific videos have been

TABLE 24.1
Comparison of Various Techniques

Name of the Technique	Features
R-CNN	This technique can extract image features automatically and reduce errors of positioning and correct the boundary box predictions, however, if there are a lot of features to be extracted then the system can run out of memory and also more complicated than most other existing algorithms at the time. Hence it takes longer time to train the model, which cannot be used for real-time applications due to slow speed.
Fast R-CNN	This technique overcomes the R-CNN model drawbacks and reduces the time taken to train the process of the stage by stage execution with multitasking training and also utilizes ROI pooling to satisfy requests for many scales. However, time-consuming algorithm as the core selective search algorithm is slow itself.
Faster R-CNN	This algorithm produces very good real-time results, better accuracy and map on outputs however this method has overhead due to reshaping the predicted region proposals before predicting the actual offsets for the bounding box and the object proposal requires a lot of time.
YOLO	It is faster than other pre-existing solutions and also it takes less time to detect false positive in images with complex backgrounds. This method is produced high speed detection with 43.5 % AP state-of-the-art results over MS COCO dataset and 65 FPS real-time speed also modified state-of-the-art methods make them more efficient and suitable for single GPU training however the usage of so many layers increases the overhead of deciding the compromise trade off between the mAP and the training and inference speed as to allow the model to run on embedded systems.

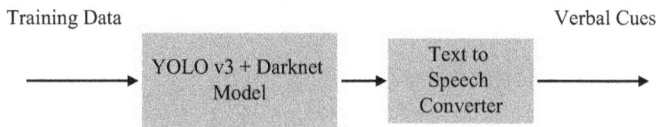

FIGURE 24.2   Proposed combinational YOLOv3 architecture.

evaluated. Additionally, neural network-based face detection techniques are thought to be extremely accurate and computationally demanding. To prepare the models and recognize the objects in videos with low frame rates, Mohd Nazeer et al. used a GPU-enabled platform. That model could be helpful to identify a variety of things [15].

24.3   IMPLEMENTATION

The Darknet-19, a feature extractor utilised by YOLOv3, had 19 convolutional layers. This method's most recent iteration, YOLOv3, uses a brand-new feature extractor called Darknet-53, which makes use of 53 convolutional layers. The overall number of layers used throughout the approach, which consists of 31 additional levels and 75 convolutional layers, is 106. In order to down sample, pooling layers have been eliminated from the design, and another convolutional layer with stride "2" has been inserted in their place. A sizable adjustment was made to avoid attributes from being lost during pooling. The proposed YOLOv3 architecture is shown in Figure 24.2.

The convolutional layers were used before a 2× increase in sampling rate to a 26×26 grid. The upsampling is carried out using the upsample function discussed above until the required feature map is obtained. Outsized items are detected by the first detection layer, whereas smaller areas are detected by the second and third detection layers. Comparing these three results at various scales facilitates the elimination of false positive cases and enhances detection performance. Although the current version of the problem with minor item detection is better than prior versions, this model's limitation still exists. The preservation of the fine-grained qualities and, consequently, the recognition of small objects in the image are supported by the upsampling layers that are related to the preceding layers (shortcut). Using a sample image of 416×416 pixels, the predicted number of boxes will be 10'647 (which is 10 times more boxes than the prediction of the prior model). Another novel improvement came from the loss function used to train the model. The YOLO method uses a modified mAP error to assess how well predictions match reality.

The classification loss, the localization loss, and the confidence loss are the three terms that make up the loss function specifically. The classification loss, which measures the error of detection, is defined by where is equal to 1 if an item exists in cell I is the model's output, and denotes the conditional class probability for class c in cell i. The total of these three contributions determines the final loss. Duplicate detections must be eliminated in order to further improve the detection capabilities. To do this, the YOLO model applies a non-maximal suppression to eliminate duplicates with lower confidence.

24.2.1 IMAGE PREPROCESSING

The photos must be preprocessed in this stage for the model to give a better accuracy rating. Each image underwent the preprocessing processes listed below:

1. Crop the significant area of the picture (which is the most important part of the image).
2. Due to the variety of sizes of the images in the dataset, resize the image to take the shape of (150, 150, 3) = (image width, image height, number of channels).
3. To feed it as an input to the neural network, all photos must therefore have the same shape.
4. Use normalization to scale the pixel values to a range of 0 to 1.

24.2.2 CLASSIFICATION OF OBJECT FEATURE

1. Each input x (picture) that is given to the neural network has the shape of (240, 240, 3). It also traverses the following layers:
2. A layer with zero padding with a pool size of (2, 2).
3. A convolutional layer with 32 filters, a stride of 1, and a filter size of (7, 7).
4. A batch normalization layer to normalize the values of the pixels in order to speed up the calculation.
5. A layer for ReLU activation.
6. A layer of max-pooling with $f=4$ and $s=4$.
7. The same max-pooling layer as before, with $f=4$ and $s=4$.
8. A layer called "flatten" to reduce the three-dimensional matrix to a single-dimensional vector.
9. A dense, completely linked layer with one sigmoid-activated neuron (output unit) (since this is a binary classification task).

The following are the core objectives of this work:

1. To develop a deep neural network image classifier that classifies objects present in the frame and generates class tags.
2. Identifies the RoI (region of interest) and generates the possible image segmentation thereby marking the pixels of the identified image.
3. To implement image classification together with segmentation to generate accurate instance generation.

Once an accurate precision has been achieved for the specified object class, an audio is prompted via the system sound output device.

24.2.3 WORKFLOW DIAGRAM

The workflow of video capture to speech conversion is shown in Figure 24.3.

24.2.4 IMPLEMENTATION

Computer vision uses common deep learning techniques. Convolutional neural network (CNN) layers served as the cornerstone of proposed model architecture. CNNs have a reputation for simulating how the human brain interprets images. An input layer, a few convolutional layers, a few dense layers (also known as fully-connected layers), and an output layer make up the basic structure of a convolutional neural network. These layers are organized in sequence and are linearly stacked.

24.2.4.1 Datasets and Model Formulation

The following sections give an explanation of the dataset and methods used.

The most important aspect to take into account while training an object recognition model is to provide a strong training set. Every photograph in the collection needs to have a unique name and display the sought-after item from various perspectives (ground truth for supervised learning). Because the model has numerous variables that must be altered, a huge volume of images must be used to train a robust classifier. Because of this, the training samples ought to feature a range of backdrops, random objects, and lighting configurations. The uniqueness of the data is certainly one of the most important properties, even though the training set may not just comprise high-quality images. Since some of the available data must be "discarded" and utilized as a test set during the

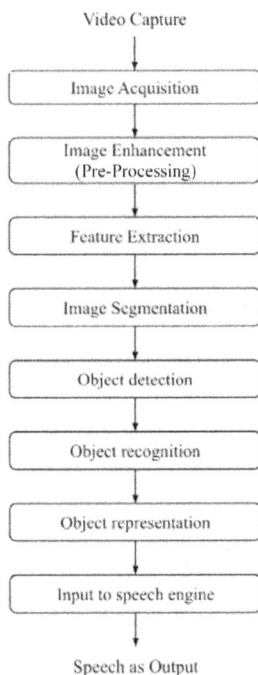

FIGURE 24.3 Workflow – 1.

training phase, the number of samples must be adequate for both methods. It is difficult to provide enough annotated samples to train the YOLO model because it has over 62 million parameters that must be modified. There are many publicly available datasets that can be used to address object detection training challenges.

For numerous deep learning training applications, an expansive open source dataset named COCO is generated. There are likely to be a lot of hand-annotated images available, and these can be very useful for item detection, segmentation, and tagging. The dataset is periodically updated and released in each year. The accessible classes determine the dataset's inherent constraint: Beginning with various animals and moving on to commonplace items and transports, COCO offers 80 different object classes pertaining to general purpose things. Although this restricts the potential uses, it nonetheless remains a highly helpful tool for testing new models 1. More than 200k of the dataset's more than 300k photos has already been tagged. People detection was the main focus of this applications, and as this area is already covered by the available ones, hence the COCO dataset was the best option.

- The object detection problem is considered a hard task for computer vision application but it is a straightforward task for human eyes.
- The increment value chosen is exactly the down-sampling factor performed by the architecture.

The training of the YOLO model requires a lot of time and computational resources. All this work was performed using a cluster machine shared among many users and thus it was impossible to dedicate the full computational resources to a single application.

24.2.4.2 Preprocessing

Images in the dataset are different dimension. Hence the picture had to be rescaled to 150×150 pixels. And then the image is converted to the grayscale using the inbuilt function cvtColor, hence it simplifies the computational requirements. Since, images are usually RGB values, which range from 0 to 255, it is normalized into [$0f$, $1f$] range of arrays.

24.2.4.3 Training and Model Formulation

The YOLO network was trained on these images using different scale dimensions: the images are fed to the network with sizes ranging from 320×320 to 608×608 with increments. This variability helps the sensibility of network (convolutional) filters to the details of the image. Moreover, it helps the detection to identify the object at different scale levels since the filter weights are independent to them. However, the tests highlight that the best results are obtained resizing the image to 608×608.

The data obtained from the Kaggle repository was preprocessed to train the model. For better accuracy, InceptionV3 is used with imagenet weights using transfer learning.

24.3 PROPOSED ALGORITHM

The approach used for implementing the proposed model is discussed below

Step 1: Let the preprocess image be X
- (i) Numpy conversion = Images are converted to Numpy vectors for the ease of transformation and better efficiency.
- (ii) Reshape image (X) to (150, 150, 3) The images that are fed to the AI algorithm vary in size, therefore, the base size for all images fed into the AI algorithms, which here is 224 * 224 pixels.
- (iii) Splitting data = Data is split into a train set and test set in a ratio of 9:1.

Step 2: Apply the image to an input of the model.
Step 3: Retrieve the model's last convolution layer's output.
Step 4: Use GlobalAveragePooling to regularize the output.
Step 5: Use dropout of 50%.
Step 6: Apply dense layer of four units.
Step 7: Apply softmax activation.

24.4   METRICS AND RESULT

The following performance measurement parameters are used to calculate the classification performance of the algorithm:

24.4.1   ACCURACY

It is defined as the number of correct predictions made by the model over the total number of predictions. This is a good measure, especially when the target variable classes are balanced in the data. This can be represented as

$$Number\ of\ correct\ predictions\ (CP) = True\ Positives + True\ Negatives \qquad (24.1)$$

$$\begin{aligned} Number\ of\ total\ predictions\ (TP) &= True\ Positives + True\ Negatives \\ &+ False\ Positives + False\ Negatives \end{aligned} \qquad (24.2)$$

$$Accuracy = \frac{CP}{TP} \qquad (24.3)$$

24.4.2   F1 SCORE

The balanced F-measure is used to measure a test's accuracy. The F1 score is considered to be good if the overall number of false positives and false negatives is low. It is defined as the harmonic mean of precision and recall.

$$F1\ score = \frac{2 \times Precision \times Recall}{\left(Precision + Recall\right)} \qquad (24.4)$$

24.4.3   PRECISION

It is determined by dividing the total number of positive outcomes predicted by the classifier by the number of genuine positive findings.

$$Precision = \frac{TruePositives}{\left(TruePositives + FalsePositives\right)} \qquad (24.5)$$

24.4.4   RECALL

It is determined by dividing the total of true positives and false negatives by the number of true positive outcomes.

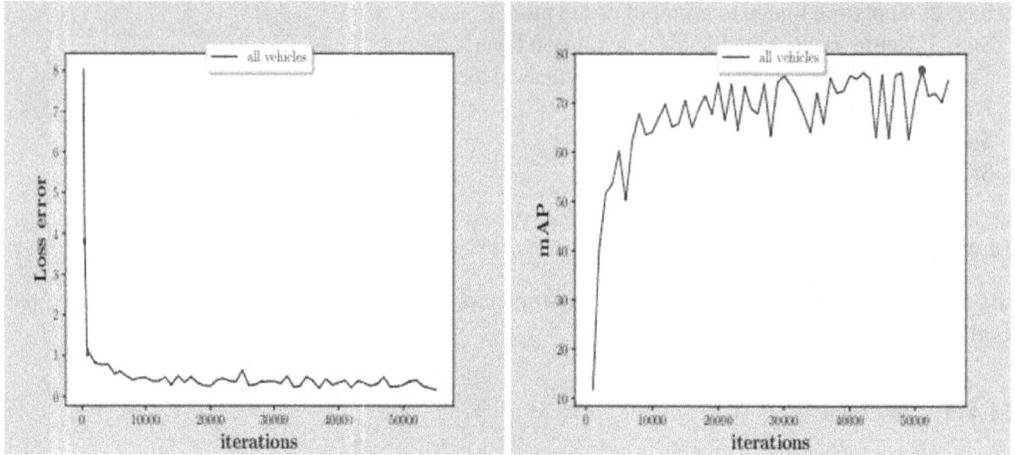

FIGURE 24.4   Loss error and mAP performance of YOLOv3 on the trained dataset.

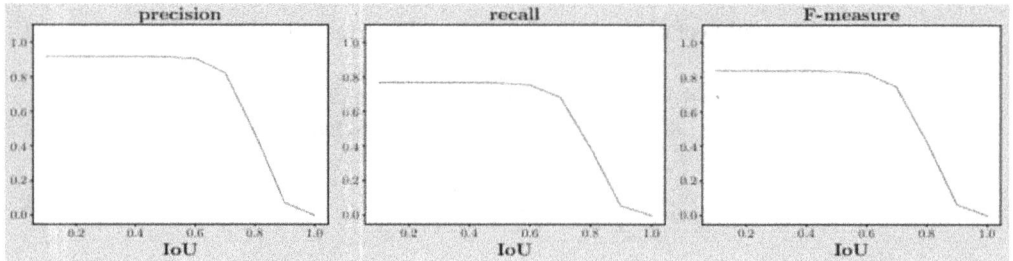

FIGURE 24.5   YOLOv3 metric performance.

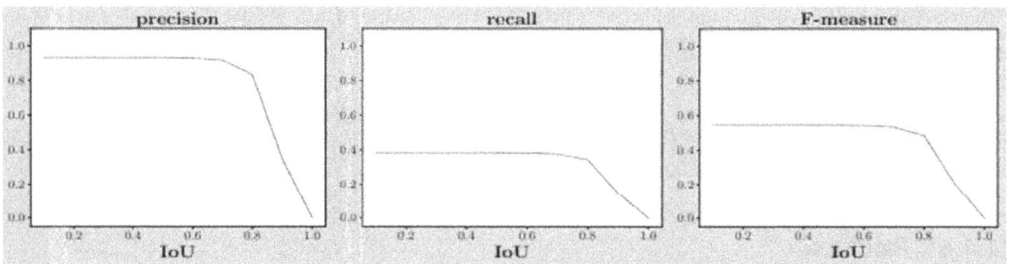

FIGURE 24.6   SSD metric performance.

$$Precision = \frac{TruePositives}{\left(TruePositives + FalsePositives\right)} \qquad (24.6)$$

24.4.5   RESULTS

Figures 24.4, 24.5, and 24.6 are shown performance of loss error, YOLOv3 metric performance and SSD metric performance.

The data source used for the work is a dataset, which contains files that have been trained and tested accurately to further enhance the output prediction of the models of the working work. This

FIGURE 24.7 Image classification.

FIGURE 24.8 Object localization.

is because in real time, the movement of the object can vary to a very high degree leading to misinterpretation and misjudgement of the output prediction.

Hence, the entire predicted process in an observed trained time frame that was carried out to ensure that the output accuracy is high. The demonstration of the various processes such as image classification, object localization, semantic segmentation, semantic instance segmentation are shown in Figures 24.7 to 24.10.

Figure 24.11 shows the command line which contains the detected objects printed on the console. The output is displayed in the command line as well as a speech output is also provided while actualling running the work.

The objects that have been detected in Figures 24.12 and 24.13 are **chairs** with accuracy 97% and 80%, **tv** (actually a desktop monitor) with accuracy 76%. The above GUI output of the model plots the bounding boxes to the segmented objects and helps to identify the accuracy of the bounding boxes and perform fps tests to decide whether to determine if the model has any of the edge failure cases like in the case of over fitting and under fitting scenarios (Figure 24.14).

The accuracy of the predicted object is displayed over the bounding boxes and is formulated by the prediction percentage that is calculated by comparing the feature map of the current image, which is then compared with the pixel data of the trained model. However, the actual output of the

FIGURE 24.9   Semantic segmentation.

FIGURE 24.10   Semantic instance segmentation.

command-line is a voice-based output that is predicted after identifying all the possible weightful objects in the video capture frame. The module gTTS, which has support extended to all consumer operating systems is used. gTTS module also has inbuilt support for translational features to speak the output multilingually.

It can be deduced from the graph represented in the Figure 24.14, which represents the epoch plotted in the x-axis and the classification accuracy along the y-axis. With the increasing number of epochs it can be deduced that the classification accuracy increases along the exponentially increasing plotted line. Hence, the model ensures that epoch results in proper training of the model and doesn't indicate an overfitting criteria. The training accuracy is **99.795%**. Similarly, the validation accuracy is **high (96.60%)** and validation loss is **very low (0.0035).**

24.5   CONCLUSIONS AND FUTURE WORK

Our main goal was to identify different object detection, tracking, recognition techniques, feature descriptors, and segmentation method, which is based on the video frame and various tracking technologies. This approach is used towards increasing object detection with new ideas. Furthermore,

FIGURE 24.11   Command line output prediction.

1	person	41	wine glass
2	bicycle	42	cup
3	car	43	fork
4	motorbike	44	knife
5	aeroplane	45	spoon
6	bus	46	bowl
7	train	47	banana
8	truck	48	apple
9	boat	49	sandwich
10	traffic light	50	orange
11	fire hydrant	51	broccoli
12	stop sign	52	carrot
13	parking meter	53	hot dog
14	bench	54	pizza
15	bird	55	donut
16	cat	56	cake
17	dog	57	chair
18	horse	58	sofa
19	sheep	59	pottedplant
20	cow	60	bed
21	elephant	61	diningtable
22	bear	62	toilet
23	zebra	63	tvmonitor
24	giraffe	64	laptop
25	backpack	65	mouse
26	umbrella	66	remote
27	handbag	67	keyboard
28	tie	68	cell phone
29	suitcase	69	microwave
30	frisbee	70	oven
31	skis	71	toaster
32	snowboard	72	sink
33	sports ball	73	refrigerator
34	kite	74	book
35	baseball bat	75	clock
36	baseball glove	76	vase
37	skateboard	77	scissors
38	surfboard	78	teddy bear
39	tennis racket	79	hair drier
40	bottle	80	toothbrush

FIGURE 24.12   Object classifier names.

FIGURE 24.13   Object detection output in a real-time web capture-1.

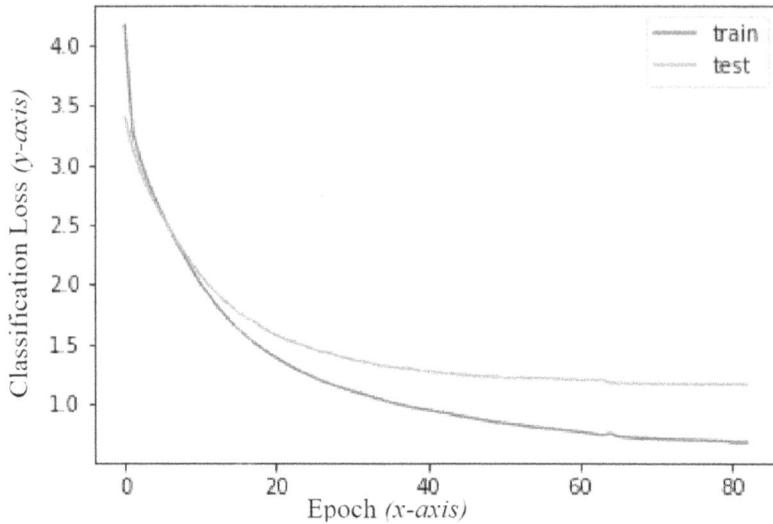

FIGURE 24.14   Graph for epoch vs loss.

tracking the object from the video frames with theoretical explanation is provided in bibliography content. Further, scope of various methods and limitations are discussed. Also, it is observed that some methods give accuracy but have high computational complexity. The statistical techniques, background subtraction, and temporal differencing with the optical flow were specifically covered.

However, these solutions must focus on dealing with abrupt lighting changes, heavier shadows, and object occlusions. The training is only done for the last layer. CNN also extracts raw pixel values with depth, width, and height feature values. Finally, to obtain high precision, the gradient

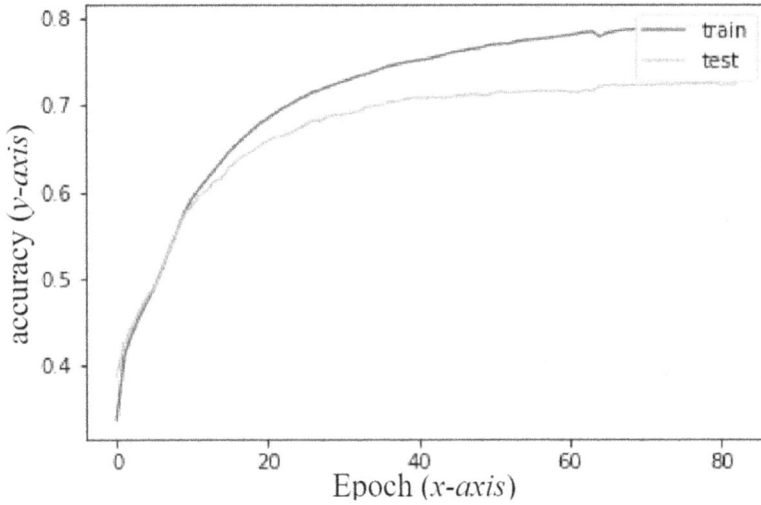

FIGURE 24.15 Graph for epoch vs accuracy.

descent-based loss function is used. The validation accuracy, validation loss, and training accuracy are all computed. The training precision is 99.795%. Similarly, validation accuracy (96.60%) and validation loss (0.0035) are both high.

24.6 FUTURE WORKS

Design and simulation of complex video sequences and test them using the same tracking algorithm. Occlusion is employed in the hypothetical case for an object of the same color as the moving objects, or else employing a larger occlusion with a longer occlusion time. Increasing the number of the object helps to identify the efficiency and functionality of the tracking algorithm. Weight parameters must be added for each pixel's specific intensity level. If an intensity value is allocated as the foreground in an image based on the current frame, there is a lower possibility that the foreground also has comparable pixel coordinates, therefore the BG weightage for the pixel is set to the minimal value rather than the starting value. The advantage of using a weightage that is less than the initial value is that the previous pixel value can be eliminated with the least possibility as opposed to the evolving scene.

24.6.1 Data Availability

The information used to support the study's conclusions is accessible from the first author upon request at any time.

24.6.2 Conflicts of Interest

The authors declare that they have no conflicts of interest.

REFERENCES

1. Hao Shen, Shuxiao Li, Chengfei Zhu and Hongxing Chang, "Moving Object Detection in Aerial Video Based on Spatiotemporal Saliency", Chinese Journal of Aeronautics, October 2013, 26(5), 1211–1217.

2. Shanshan Li, Yuehuan Wang, Wenhui Xie and Zhiguo Cao, "Moving Object Detection and Tracking in Video Surveillance System", Proceeding of SPIE, 7495, MIPPR 2009: Automatic Target Tecognition and Image Anaylsis, 74952D, 30 Oct 2009.
3. Abdelkarim Ben Ayed, Mohamed Ben Halima and Adel M. Alimi, "Map Reduce-Based Text Detection in Big Data Natural Scene Videos", Procedia Computer Science, 2015, 53, 216–223. doi:10.1016/j.procs.2015.07.297.
4. Danyang Cao, Zhixin Chen and Lei Gao, "An Improved Object Detection Algorithm Based on Multi-Scaled and Deformable Convolutional Neural Networks", Human-Centric Computing and Information Sciences, 2020, 10, 14, https://doi.org/10.1186/s13673-020-00219-9.
5. Ashwani Kumar and Sonam Srivastava, "Object Detection System Based on Convolution Neural Networks Using Single Shot Multi-Box Detector", Procedia Computer Science, 2020, 171, 2610–2617.
6. Edmund Ng Giap Weng, Rehman Ullah Khan, Shahren Ahmad Zaidi Adruce and Oon Yin Bee, "Objects Tracking from Natural Features in Mobile Augmented Reality", Procedia – Social and Behavioral Sciences, 2013, 97, 753–760.
7. Jianshu Zhang and Jie Cao Bo Mao, "Moving Object Detection Based on Non-parametric Methods and Frame Difference for Traceability Video Analysis", Procedia Computer Science, 2016, 91, 995–1000.
8. Jeremie Houssineau, Daniel E. Clark, Spela Ivekovic, Chee Sing Lee and Jose Franco, "A Unified Approach for Multi-Object Triangulation, Tracking and Camera Calibration", IEEE Transactions on Signal Processing, June 2016, 64(11), 2934–2948.
9. Ye Lyu, Michael Ying Yang, George Vosselman and Gui-Song Xia, "VideoObject Detection with a Convolutional Regression Tracker", ISPRS Journal of Photogrammetry and Remote Sensing, June 2021, 176, 139–150.
10. Hira Fatima, Syed Irtiza Ali Shah, Muqaddas Jamil, Farheen Mustafa and Ismara Nadir, "Object Recognition, Tracking and Trajectory Generation in Real-Time Video Sequence" International Journal of Information and Electronics Engineering, November 2013, 3(6), 639–642.
11. Federico Lecumberry and Alvaro Pardo, "Semi-Automatic Object Tracking in Video Sequences", Journal of Computer Science and Technology, July 2008, 5(04), 218–224.
12. Xiaodong Yang, Houqiang Li and Xiaobo Zhou, "Nuclei Segmentation Using Marker-Controlled Watershed, Tracking Using Mean-Shift, and Kalman Filter in Time-Lapse Microscopy", IEEE Transactions on Circuits and Systems I, November 2006, 53(11), 2405–2414.
13. Huiyu Zhou, Yuan Yuan and Shi Chunme, "Object Tracking Using SIFT Features and Mean Shift", Computer Vision and Image Understanding, March 2009, 113(3), 345–352.
14. Da Tang and Yu-Jin Zhang, "Combining Mean-Shift and Particle Filter for Object Tracking", 2011 Sixth International Conference on Image and Graphics, 771–776, doi: 10.1109/ICIG.2011.118.
15. Mohd Nazeer, Mohammed Qayyum and Abdul Ahad, "Real Time Object Detection and Recognition in Machine Learning Using Jetson Nano", International Journal from Innovative Engineering and Management Research (IJIEMR), 2022, 11(10), 118–124.

25 Prediction of COVID Stages Using Data Analysis and Machine Learning

Rajalakshmi Gurusamy[1], S. Siva Ranjani[2], and G. Susan Shiny[1]
[1] Department of Information Technology, Sethu Institute of Technology, Madurai, Tamilnadu, India
[2] Department of Computer Science and Engineering, Sethu Institute of Technology, Madurai, Tamilnadu, India

25.1 INTRODUCTION

Coronaviruses are a broad family of viruses that are known to cause illnesses ranging from the common cold to more serious ailments like MERS and SARS. It was initially identified in Wuhan, China in December 2019, and since then, it has spread globally [1]. MERS was originally identified in Saudi Arabia in 2012, while SARS was identified in China in 2002 [2]. An international pandemic brought on by the SARS-CoV-2 coronavirus has resulted in millions of confirmed illnesses and hundreds of thousands of fatalities.

Since then, the high mortality impact has been accompanied by an increase in coronavirus cases. As the role of healthcare epidemiologists has expanded, this has led to an increased need for healthcare epidemiologists to have expertise in data analysis and informatics. They play a crucial role in analyzing and interpreting health data to inform public health policies and interventions [3]. With the rise of electronic health data, healthcare epidemiologists are now able to utilize advanced analytics tools to identify patterns and trends in health data, which can help improve patient outcomes and reduce healthcare costs.

By leveraging their expertise in data analysis and informatics, healthcare epidemiologists are helping to shape the future of healthcare [4]. It is common for health teams to be unable to provide patients with the necessary care due to healthcare shortages of resources and a delay in test findings.

The prediction of coronavirus infection in patients is done using machine learning techniques [5]. By examining enormous amounts of patient data, we can spot trends and risk factors that potentially halt the development of diseases. Machine learning algorithms can be used to find complex patterns and correlations in data, even when they are unknown or difficult to discover by traditional methods [6]. This is particularly useful with respect to risk factors, as revealing hidden correlations can help companies make better decisions and reduce potential pitfalls [7]. Interpretation of medical data related to epilepsy [8, 9], muscle and neurological disorders [10, 11], and heart rhythm [12, 13] has been successfully achieved using machine learning classifiers.

However, machine learning (ML) techniques face several challenges, such as lackluster online databases that limit the scope of research and development. Another difficulty arises because a large amount of data is required to properly train a classifier. Also, when training a classifier with less data, there is the potential for overfitting, which can produce erroneous results. In conclusion, despite the success of machine learning classifiers in medical interpreting, data accessibility and resource demands remain problematic.

DOI: 10.1201/9781003407959-25

Deep learning algorithms are also effective in predicting clinical outcomes in biological research [14], viral diseases [15], and cancer [16]. Deep learning algorithms are used in various industries such as banking, marketing, transportation, and healthcare. These can be used, for example, to predict traffic patterns [17, 18], customer behavior [19], and stock prices [20]. In addition, this approach has also been used in natural language processing tasks [21], machine translation and speech recognition [22]. Deep learning algorithms hold great promise for improving various decision-making processes and solving complex problems. These methods are effective and can be used to predict stages of COVID-19.

To precisely detect the signs and risks of COVID-19, the model will be trained using an enormous data set of patient information as well as medical records. The accuracy and speed of early diagnosis of COVID-19 infection can be improved by using machine learning approaches (SVM, RF, and ANN), which are described and highlighted in our study. Our findings indicate that these approaches may help manage and contain the ongoing pandemic, especially in areas of scarcity health resources.

This study is structured as follows. Section 25.2 describes previous research on COVID-19. Section 25.3 presents various machine learning algorithms to accurately predict the early stages of COVID-19. Section 25.4 presents the results and description of the machine learning algorithm. Study conclusions are presented in Section 25.5.

25.2 RELATED WORKS

A review of the COVID stage prediction literature found that several factors, including age, gender, comorbidities, biomarkers, and imaging findings, are significant predictors of disease severity, mortality, and progression of COVID-19. Models that can forecast the likelihood of a severe illness and the progression of the disease in COVID-19 patients are being created using machine learning techniques. Machine learning algorithms can be trained with this data to produce more accurate disease predictions. Additionally, algorithms that utilize machine learning may be employed to generate customized therapeutic strategies for patients depending on their genetic profile and past medical conditions [23].

Digital medical records were suggested by Estiri et al. [24] as a way to forecast the morbidity and severity of COVID-19 in patients. The authors found that age, sex, and comorbidities were important predictors of COVID-19 severity and mortality. Sujath et al. [25] uses machine learning algorithms to forecast the global spread of COVID-19. The authors anticipate the number of COVID-19 cases in various nations using a variety of machine learning models, including decision trees, ANNs, and SVMs.

A mathematical model is created by Shitharth et al. [26] to forecast the propagation of the COVID-19 infection in India. It takes into account various factors such as population density, temperature, humidity, and age distribution. Dueas et al. [27] estimated the effect of mobility patterns in the transmission of COVID-19 in Colombia using statistical models. The authors use different statistical methods, such as multiple regression analysis and generalized linear models, to analyze the relationship between mobility patterns and COVID-19 transmission.

Pulmonary imaging and clinical features were employed in Feng et al.'s [28] investigation to forecast pneumonia caused by COVID-19 patients' illness development. The authors found that the extent of lung involvement on chest CT scans and the patient's age and comorbidities were important predictors of disease progression. Kim et al. [29] developed an AI model to prioritize and triage COVID-19 chest radiographs in emergency departments based on severity. Xiang et al. [30] review various epidemic models used for COVID-19 prediction, including SEIR (susceptible-exposed-infectious-recovered) and SIR (susceptible-infectious-recovered) models.

In a study by Tsiknakis et al. [31], they proposed an x-ray-based deep learning COVID-19 classification system to improve its performance against state-of-the-art methodologies. The work on COVID-19 classification using transfer learning methods led to improved AUC performance.

A real-time novel coronavirus infection detection and tracking system was suggested by Otoom et al. [32]. This research highlights the potential of machine learning algorithms in conjunction with IoT devices for accurate prediction and tracking. The authors employed seven algorithms for machine learning, ran experiments on each algorithm, and created comparisons. In their study, five machine learning techniques using IoT devices for data gathering and observation over quarantine showed prediction accuracy above 90%.

To test for COVID-19 in the city, Khmaissia and colleagues [33] proposed an unsupervised machine learning strategy to discover commonalities among New York zip codes. They used feature selection and clustering algorithms to compare accessibility, economic, and demographic characteristics to COVID-19 trends. Machine learning approaches have enabled to fully understand the complex interplay between socioeconomic variables and disease prevalence in metropolitan settings. These findings can reduce the impact of COVID-19 on vulnerable groups and ultimately guide targeted measures to reduce infection rates in cities.

To improve COVID-19 detection while preserving privacy, Zhang et al. [34] proposes a dynamic fusion architecture. The proposed fusion framework adds model updates from many medical institutions, thereby improving detection performance. Experimental results show that the fusion framework is more efficient than traditional focused learning techniques. This research provides insight into the potential of collaborative machine learning to solve global health problems and contributes to the growing theme of federated learning for medical applications.

To diagnose the COVID-19 cases, an integrated deep neural network architecture that can process both chest x-rays and CT scans was proposed by Mukherjee et al. [35]. This approach eliminates the need to use different models for each imaging modality, streamlining the diagnostic process and maximizing resource utilization. Experimental results demonstrate that the efficacy and applicability of this method outperform conventional chest radiography and computed tomography. The integrated approach has many advantages such as faster speed, reduced computational requirements, and improved accuracy.

A review by Bhattacharya et al. [36] highlights the potential of deep learning applications in medicine and how important it is to address the unique challenges posed by COVID-19 medical imaging. Monjur et al. [37] provide insight into the importance of predicting COVID-19 and the variables that influence mortality risk. Scientists used cloud-based mobile devices to predict the number of people infected with COVID-19 based on symptoms such as wheezing, high fever, dry cough, and migraines.

Chest CT scan is recognized as a useful imaging modality for detecting lung abnormalities associated with COVID-19 as part of the diagnosis [38, 39]. A deep learning model called COVIDNet-CT is proposed by Gunraj et al. [40]. It is designed to analyze chest CT images and identify possible COVID-19 cases. Kassania et al. [41] provides useful insights and validates the effectiveness of various deep learning-based feature extraction frameworks for COVID-19 identification from chest x-rays. The results show that the DenseNet121 feature extractor outperforms other deep neural frameworks when combined with a bagging tree classifier.

Numerous studies have been done to use machine learning and other statistical techniques to forecast the future phase of COVID-19 patients. This can aid in treatment decisions and resource allocation. However, more research is needed to further refine these models and improve their accuracy.

25.3 PROPOSED METHODOLOGY

To develop models that can more accurately predict the presence of infectious diseases, COVID-19 data can be used to train algorithms involving machine learning that were utilized for recognizing various infectious diseases. The model can be trained to distinguish between mild and severe stages of COVID-19 infection, allowing for more accurate diagnosis and faster treatment. Figure 25.1 shows the COVID-19 prediction architecture. The following steps are involved in the proposed COVID-19 stage prediction:

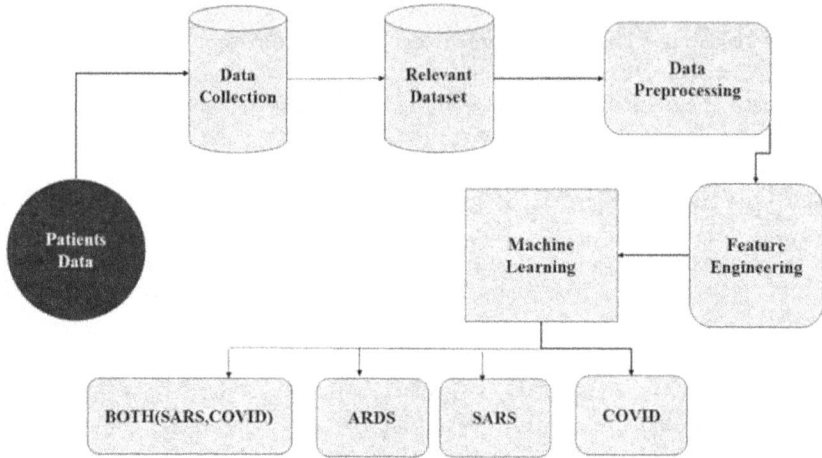

FIGURE 25.1   COVID-19 prediction architecture.

1. Data collection
   Data collection is an important but time-consuming task. Accurate data collection is critical to maintaining consistency, regardless of the research topic. Data acquisition was a rigorous and time-consuming process, as clinical information about patients is not accessible to the public. Due to the current scenario, direct information could not be obtained for hospitals with a high influx of COVID-19 patients. Extensive searches of numerous databases were performed to find open source clinical data on people detected through COVID-19.

2. Relevant dataset
   An open-source dataset provided by Yanyan Xu and stored in the Dataset Y repository was used to train a model to predict COVID-19. The datasets used to train machine learning models contain only the most relevant attributes. All fields considered irrelevant or redundant were removed to ensure that the model could correctly predict the outcome. The remaining fields contain a mixture of numeric and textual information, with the former encoded as an integer value. These features were carefully selected based on their impact on the expected results. All data collection attributes are listed in Table 25.1 for analysis.

3. Data preprocessing
   The accuracy and reliability of machine learning models are greatly affected by data preparation. Frequently collected data is unchecked and contains missing numbers, out-of-range values, etc. Such information can skew experimental results. A mean strategy is used to replace missing values. It uses one-hot encoding techniques to process categorical data.

Additionally, the most pertinent features for the model depicted in Figure 25.2 and Table 25.1 are found using feature importance techniques. After preprocessing, the data can be divided into training and testing sets to assess the model's performance. To further analyze the data and create predictions, we employed machine learning methods like SVM, RF, and ANN.

25.3.1   SUPPORT VECTOR MACHINE (SVM)

Applications for classification or regression might use the SVM-supervised machine learning method. SVM has been used as a classifier for COVID-19 stage prediction cases to determine the severity of a patient's illness. SVM works by identifying optimal boundaries or hyperplanes that can divide different classes of data points depending on the radial basis function kernel (RBF). It has proven effective in tasks that require medical diagnosis and prediction, such as predicting the severity of COVID-19.

To predict COVID-19 stages using SVM, we need to find decision boundaries that separate different classes of COVID-19 stages. The feature vector x_i and the accompanying label y_i are given in the context of a collection of n training examples $\{(x_1, y_1), (x_2, y_2) \ldots, (x_n, y_n)\}$ (i.e. COVID-19 stage), the goal of SVM is to find a hyperplane that separates the various classes with the largest margin.

The hyperplane is defined as

$$\omega^t * x + \beta = 0 \tag{25.1}$$

where, ω^t = weight term, β= bias term.

The SVM algorithm finds the weight vector ω^t and bias term β by solving the following optimization problem.

$$\text{Minimize } \left(\frac{1}{2}\right)\omega^{t2} + C\sum \varepsilon \tag{25.2}$$

$$\text{Subject to } y_i\left(\omega^t * x_i + \beta\right) \geq 1 - \varepsilon_i, \varepsilon_i \geq 0 \text{ for } i = 1, 2,\ldots., n \tag{25.3}$$

where, ε_i= slack variable that allows some classification error, C= hyperparameter controlling the trade-off between maximizing the margin and minimizing the classification error, ω^{t2}= weight vector L2 norm.

The RBF core in SVM for classification problems is described below:

$$K\left(x_i, x_j\right) = e^{(-\gamma x_i \cdot x_j^2)} \tag{25.4}$$

where, γ= hyperparameters that control the shape of the decision boundary.

Based on the new sample feature vector, the resulting weight vector (w) and bias term (b) can be used to predict the stage of COVID-19.

25.3.2 Random Forest (RF)

RF is the powerful machine learning algorithm that combines decision trees, ensemble learning, bagging, and a subset of random functions to produce accurate predictions. Input characteristics for predicting the COVID-19 stage consist of demographic data (such as age and gender), clinical measurements (such as temperature and oxygen saturation), and laboratory test results (such as viral load and inflammatory markers). The COVID-19 prediction stage will be the output variable and can be defined in different ways depending on the exact prediction context (e.g. severity of illness, likelihood of hospitalization, risk of death).

We typically start by splitting the available data between a set for training and a validation set to train a RF model to predict COVID-19 stages. Each decision tree is trained using the training set, the model's performance is evaluated, and its parameters are tuned using the validation set.

The following steps are part of the random forest algorithm for COVID-19 stage prediction:

1. **Create decision trees:** A decision tree approach, on which RF is based, divides the data into subgroups according to the input characteristics. Splits are selected based on the purity of the resulting clusters, evaluated by metrics such as Gini impurity and information gain.

The equation for Gini impurity is

$$Gini_{impurity} = 1 - \left(P_1^2 + P_2^2 + P_3^2 + P_4^2 \right) \tag{25.5}$$

where, $P_1, P_2, P_3,$ and P_4 are the proportions of each COVID-19 stage in the subset. A lower Gini impurity score indicates a purer subset, where all the samples belong to the same COVID-19 stage.

The equation for information gain is

$$Information_{gain} = \Delta_{parent} - \left\{ \begin{array}{c} \left(\sum_{1=1}^{n} P(left\_child_i) log_2 P(left\_child_i) \right) \\ + \left(\sum_{1=1}^{n} P(right\_child_i) log_2 P(right\_child_i) \right) \end{array} \right\} \tag{25.6}$$

where Δ_{parent} = entropy of the parent node, $P(left\_child_i)$ and $P(right\_child_i)$ = weighted average entropies of the left and right child nodes, respectively.

A higher information gain score indicates that the split results in a more informative subset, where the COVID-19 stages are more distinguishable.

2. **Make predictions:** Random forest models combine all the decision tree results to make predictions after they are built. The most common way to predict the stage of COVID-19 is to use the mode (highest prediction) among all outputs of a decision tree. In other words, the outcome is determined by a majority vote across the decision tree.

The equation for the final prediction of COVID-19 stage can be expressed as

$$y = mode(tree_1(x), tree_2(x), \ldots, tree_n(x)) \tag{25.7}$$

where y = predicted COVID-19 stage, x = input features of a patient, $tree_1(x), tree_2(x), \ldots, tree_n(x)$ are the outputs of the decision trees in the random forest.

3. **Probability estimation:** Random forest can also estimate the probability of a particular prediction. The COVID-19 stage prediction probability is computed as the percentage of decision trees in the forest that correctly predict the COVID-19 stage.

The probability of a given prediction can be calculated as

$$P(y|x) = count\left(tree_1(y) = y, tree_2(y) = y, \ldots, tree_n(y) = y \right) \Big/ N \tag{25.8}$$

where, $P(y|x)$ = probability of a patient being classified as having COVID-19 stage y, given their input features x, $count\left(tree_1(y) = y, tree_2(y) = y, \ldots, tree_n(y) = y \right)$ = number of decision trees that predict COVID-19 stage y, and N is the total decision trees in the RF.

25.3.3 Artificial Neural Networks (ANN)

ANNs can learn to predict the COVID-19 stages based on input data, such as demographic information, symptoms, laboratory test results, and comorbidities. The ANN can learn to identify patterns and relationships between the input and output data, allowing it to make accurate predictions.

The following steps are part of the ANN algorithm for COVID-19 stage prediction:

1. Calculation of weighted input:

   The weighted sum of inputs is calculated as follows:

   $$z(i) = \sum_{i=1}^{n} (\omega_1 x_1, \omega_2 x_2, \ldots\ldots, \omega_n x_n) \tag{25.9}$$

   where, $x_1, x_2, \ldots, x_n$ are the input values, $\omega_1, \omega_2, \ldots, \omega_n$ are the weights associated with each input, and $z(i)$ is the output value of the neuron

2. Sigmoid activation function:

   The weighted input z is then passed through a sigmoid activation function, to introduce non-linearity into the network:

   $$\sigma(i) = \frac{1}{\left(1 + e^{-z(i)}\right)} \tag{25.10}$$

   where σ = sigmoid activation function.

3. Calculation of output:

   For each neuron k in the output layer, the weighted input z is calculated as the sum of the product of the hidden layer output values σ and their corresponding weights ω, plus a bias term β:

   $$z(k) = \sum_{i=1} \sigma(i) * \omega(i,k) + \beta(k) \tag{25.11}$$

4. Softmax activation:

   The softmax activation function is used to normalize the output values, ensuring that they sum up to one and represent probabilities of belonging to each class:

   $$y(k) = softmax(z(k)) = \frac{\exp(z(k))}{\sum_j \exp(z(j))} \tag{25.12}$$

5. Calculation of loss function:

   The discrepancy between the output that was predicted and the output that was produced is measured using the loss function. Common loss functions used for multi-class classification tasks, such as predicting COVID-19 stages, include categorical cross-entropy:

$$L = -\sum_{k=1} y(k) * \log\left(y\_pred(k)\right) \tag{25.13}$$

where, y_pred = predicted output.

6. Backpropagationand gradient descent:

The gradients of the loss function with respect to the weights and biases in the network are computed via backpropagation. Gradient descent optimization is then applied to the gradients to update the weights and biases:

$$\omega(i,k) = \omega(i,k) - lr * \partial L / \partial \omega(i,k) \tag{25.14}$$

$$\beta(k) = \beta(k) - lr * \partial L / \partial \beta(k) \tag{25.15}$$

where, lr = learning rate is a hyperparameter that determines the step size of the optimization algorithm.

By iteratively updating the weights and biases using backpropagation and gradient descent, the ANN can learn to predict the COVID-19 stages based on input data.

25.4 RESULTS AND DISCUSSION

25.4.1 DATASET DESCRIPTION

A model that forecasts COVID-19 was trained using an open source dataset provided by Yanyan Xu in the Dataset Y repository. Data collection included details of hospitalized patients who tested positive for COVID-19. This included demographic information, signs and symptoms, old medical records, test results extracted from electronic records, and demographic data. These variables have been removed as they are not needed in the model. A collection of multidimensional data constitutes a dataset.

Some information reveals whether a patient has been diagnosed with certain diseases, such as kidney disease, or gastrointestinal disease, while other information includes specific clinical measurements taken in the past. Some fields have exact values and others have text information. Integer values were used to encrypt text data in the experimental setup. Machine learning model considerations for attributes in data collection are presented in Table 25.1.

25.4.2 PERFORMANCE MEASURES

Measuring the effectiveness of machine learning models is a significant challenge. We have utilized accuracy as an evaluation metric to assess the model's efficacy. In order to produce more accurate predictions in the unknowable test data set, accuracy is a metric that lets us know which model is best at learning patterns in the training set. It is described as

$$Acc = \left(Tr_{Po} + Tr_{Ne}\right) / \left(Tr_{Po} + Tr_{Ne} + Fa_{Po} + Fa_{Ne}\right) \tag{25.16}$$

where, $Tr_{Po} = True_{Positive}$, $Tr_{Ne} = True_{Negative}$, $Fa_{Pa} = False_{Positive}$, $Fa_{Ne} = False_{Neative}$.

TABLE 25.1
Attributes and Feature Importance in the Data Collection

Feature No.	Attribute Names	Feature Value
1.	Respirational system disease	0.162
2.	Chest tightness	0.154
3.	Gastrointestinal disease	0.142
4.	Kidney disease	0.140
5.	Fever	0.133
6.	Malignant tumor	0.098
7.	Lymphocyte count	0.0845
8.	CT findings	0.0673
9.	Cough	0.0621
10.	PCT	0.0591
11.	Renal disease	0.0501
12.	Age	0.0492
13.	WBC count	0.0423
14.	Fatigue	0.0327
15.	Diarrhea Chest	0.0245

25.4.3 EXPERIMENTAL RESULTS

We found that the best classification challenge algorithms for COVID-19 prediction are SVM, RF, ANN, etc. We trained each algorithm using the obtained dataset and analyzed the results. The efficiency of each algorithm was evaluated at several phases on the training set. This project aims to identify the most effective method for COVID-19 prediction.

The data are additionally divided into separate categories so that we may assess which approach functions best with the various datasets. Figure 25.2 illustrates how feature importance is generated in order to assess how a certain characteristic may affect COVID-19 prediction. In order to calculate accuracy at each stage, each set of records is trained using the SVM algorithm. At each stage, the data were split into training and test sets using k-fold cross-validation ($k = 10$). SVM obtains accuracy of 96.36%. Table 25.2 displays the accuracy of each batch of records that the SVM algorithm was able to obtain.

For each set of records, a RF algorithm is similarly trained to determine its accuracy at all levels. Using k-fold cross-validation ($k = 10$), the data were divided into training and test sets at each stage.

The accuracy of RF is 97.89%. For each data set, Table 25.3 displays the RF algorithm's classification accuracy. The ANN algorithm is tested after training on record set data. Using the ANN algorithm improves the classification accuracy to 98.9%. For each data set, Table 25.4 displays the ANN algorithm's classification accuracy.

The accuracy results obtained from the experiments conducted have been compiled and presented in Table 25.5. To provide a visual representation of the performance of each algorithm on different record sets, Figure 25.3 has been included. Table 25.5 and Figure 25.3 offer valuable insights into the effectiveness of each algorithm in handling various record sets. It is clear that RF and ANN algorithms perform better than SVM, depending on the nature of the records being processed. These findings provide a useful reference point for researchers and practitioners working in this field, helping to improve the accuracy and efficiency of record-processing tasks.

This study highlights how machine learning could improve our understanding of and ability to treat viral diseases such as COVID-19. Compared to other algorithms, SVM performed better even with fewer training records. Predictive accuracy did not change significantly as the amount

TABLE 25.2
SVM Accuracy

Number of Patient Records	Accuracy %
125	91.23
175	93.2
220	94.3
370	95.7
420	96.36

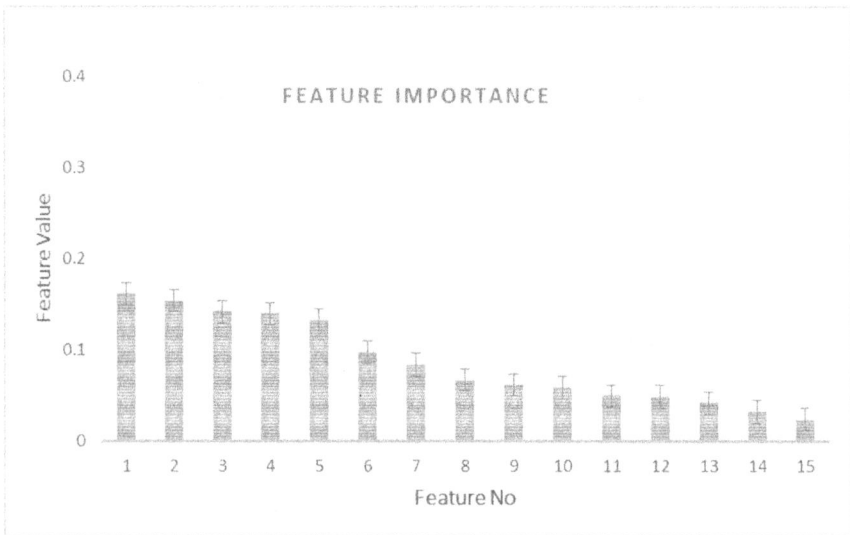

FIGURE 25.2   Feature importance chart.

TABLE 25.3
RF Accuracy

Number of Patient Records	Accuracy %
125	92.75
175	95.32
220	96.78
370	97.23
420	97.89

TABLE 25.4
ANN Accuracy

Number of Patient Records	Accuracy %
125	90.23
175	94.36
220	96.85
370	98.54
420	98.9

TABLE 25.5
Overall Comparisons Accuracy

Number of Patient Records	SVM Accuracy %	RF Accuracy %	ANN Accuracy %
125	0. 9123	0.9275	0.9023
175	0.9320	0.9532	0.9436
220	0.9430	0.9678	0.9685
370	0.9570	0.9723	0.9854
420	0.9636	0.9789	0.9890

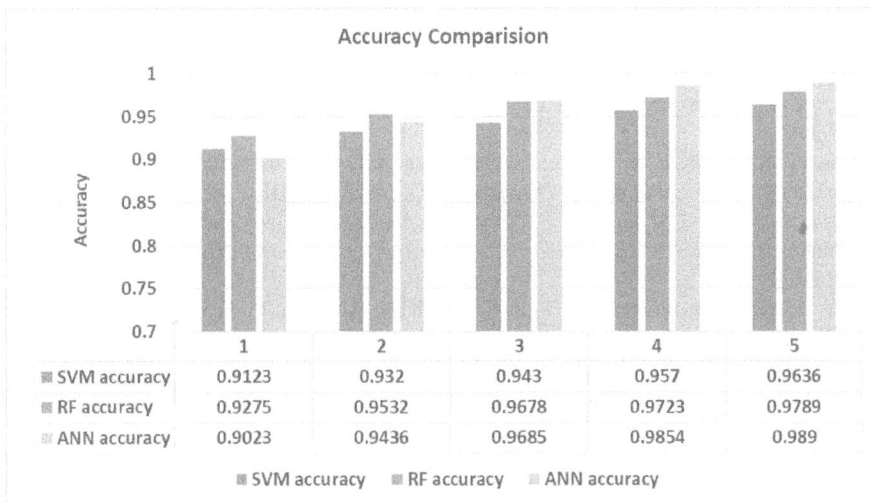

FIGURE 25.3   Comparison of classification accuracy.

of patients data increased. RF performed better in terms of accuracy than the other algorithms. RF continues to improve in accuracy at every stage, but initially it preferred SVM for its performance on smaller datasets. The most advanced algorithm among them is ANN.

ANN showed a steady change in accuracy level as the number of records in the dataset increased, but the accuracy was lowest when the record set was small. Using ANN gives the highest classification accuracy for almost all record sets used. These results demonstrate the importance of using RF and ANN increasing the size of datasets to increase accuracy, which is critical for the development and application of novel coronavirus disease (COVID-19) prediction models.

25.5 CONCLUSIONS

By conducting a bibliographic review, some works were considered relevant to the research subject and field of study. It was decided that no particular algorithm could be considered the best algorithm. Each method has its own advantages. SVM, ANN, and RF were among the algorithms chosen. Clinical patient data were used to train the selected algorithms. Each algorithm is trained on a set of records containing varying numbers of patients to test the accuracy of the machine learning model. The trained algorithms were evaluated using a performance measure of accuracy. After analyzing the results, RF and ANN showed higher prediction accuracy than SVM. Based on research findings, RF and ANN is a suitable machine learning method for predicting COVID-19. Further studies are needed to confirm these results in a larger patient population and to explore the feasibility of combining different algorithms to increase accuracy.

REFERENCES

1. Fanelli, D., and Piazza, F. "Analysis and forecast of COVID-19 spreading in China, Italy and France." *Chaos, Solitons & Fractals*, 134, 109761. 2020.
2. Willman, M., Kobasa, D., and Kindrachuk, J. "A comparative analysis of factors influencing two outbreaks of Middle Eastern respiratory syndrome (MERS) in Saudi Arabia and South Korea." *Viruses*, 11(12), 1119. 2019.
3. Bates, D. W., Saria, S., Ohno-Machado, L., Shah, A., and Escobar, G. "Big data in health care: using analytics to identify and manage high-risk and high-cost patients." *Health Affairs*, 33(7), 1123–1131. 2014.
4. Wiens, J., and Shenoy, E. S. "Machine learning for healthcare: on the verge of a major shift in healthcare epidemiology." *Clinical Infectious Diseases*, 66(1), 149–153. 2018.
5. Chakraborty, C., and Abougreen, A. "Intelligent internet of things and advanced machine learning techniques for covid-19." *EAI Endorsed Transactions on Pervasive Health and Technology*, 7(26), 1–14. 2021.
6. Khan, M. A., Abbas, S., Atta, A., Ditta, A., Alquhayz, H., Khan, M. F., and Naqvi, R. A. "Intelligent cloud based heart disease prediction system empowered with supervised machine learning." *Computers, Materials & Continua,* 65(1), 139–151, 2020.
7. Hossain, B., Morooka, T., Okuno, M., Nii, M., Yoshiya, S., and Kobashi, S. "Surgical outcome prediction in total knee arthroplasty using machine learning." *Intelligent Automation & Soft Computing,* 25(1), 1–14, 2019.
8. Alakus, T. B., and Turkoglu, I. "Detection of pre-epileptic seizure by using wavelet packet decomposition and artifical neural networks." In 2017 10th International Conference on Electrical and Electronics Engineering (ELECO), pp. 511–515. IEEE. 2017.
9. Memarian, N., Kim, S., Dewar, S., Engel, J., and Staba, R. J. "Multimodal data and machine learning for surgery outcome prediction in complicated cases of mesial temporal lobe epilepsy." *Computers in Biology and Medicine,* 64, 67–78. 2015.
10. Yousefi, J., and Hamilton-Wright, A. "Characterizing EMG data using machine-learning tools." *Computers in Biology and Medicine,* 51, 1–13. 2014.

11. Karthick, P. A., Ghosh, D. M., and Ramakrishnan, S. "Surface electromyography based muscle fatigue detection using high-resolution time-frequency methods and machine learning algorithms." *Computer Methods and Programs in Biomedicine,* 154, 45–56. 2018.

12. Alfaras, M., Soriano, M. C., and Ortín, S. "A fast machine learning model for ECG-based heartbeat classification and arrhythmia detection." *Frontiers in Physics*, 103, 1–14. 2019.

13. Ledezma, C. A., Zhou, X., Rodriguez, B., Tan, P. J., and Vanessa D.-Z. "A modeling and machine learning approach to ECG feature engineering for the detection of ischemia using pseudo-ECG." *PloS One,* 14(8), e0220294. 2019.

14. Senior, A. W., Evans, R., Jumper, J., Kirkpatrick, J., Sifre, L., Green, T., Qin, C. et al. "Improved protein structure prediction using potentials from deep learning." *Nature,* 577(7792), 706–710. 2020.

15. Andriasyan, V., Yakimovich, A., Georgi, F., Petkidis, A., Witte, R., Puntener, D., and Greber, U. F. "Deep learning of virus infections reveals mechanics of lytic cells." *BioRxiv*, 798074, 1–18. 2019.

16. Munir, K., Elahi, H., Ayub, A., Frezza, F., and Rizzi, A. "Cancer diagnosis using deep learning: a bibliographic review." *Cancers*, 11(9), 1235. 2019.

17. Gurusamy, R., and Seenivasan, S. R. "DGSLSTM: deep gated stacked long short-term memory neural network for traffic flow forecasting of transportation networks on big data environment." *Big Data, 1–14.* 2022.

18. Gurusamy, R., Shiny, G. S., Prema, A., Dharsini, V., and Latifa Banu, A. "Intelligent vehicle damage assessment and cost estimator insurance." *International Journal of Emerging Technology in Computer Science Electronics (IJETCSE)*, 30(3), 15–22. 2023.

19. Chaudhary, K., Alam, M., Al-Rakhami, M. S., and Gumaei, A. "Machine learning-based mathematical modelling for prediction of social media consumer behavior using big data analytics." *Journal of Big Data,* 8(1), 1–20. 2021.

20. Hu, Z., Zhao, Y., and Khushi, M. "A survey of forex and stock price prediction using deep learning." *Applied System Innovation,* 4(1), 9. 2021.

21. Pandey, B., Pandey, D. K., Mishra, B. P., and Rhmann, W. "A comprehensive survey of deep learning in the field of medical imaging and medical natural language processing: Challenges and research directions." *Journal of King Saud University-Computer and Information Sciences,* 34(8), 5083–5099. 2022.

22. Van, L. T., Thi Le Dao, T., Xuan, T. L., and Castelli, E. "Emotional speech recognition using deep neural networks." *Sensors,* 22(4), 1414. 2022.

23. Sun, N. N., Yang, Y., Tang, L. L., Dai, Y. N., Gao, H. N., Pan, H. Y., and Ju, B. A "Prediction model based on machine learning for diagnosing the early COVID-19 patients." *MedRxiv,* 2020–06, 1–11. 2020.

24. Estiri, H., Strasser, Z. H., Klann, J. G., Naseri, P., Wagholikar, K. B., and Murphy, S. N. "Predicting COVID-19 mortality with electronic medical records." *NPJ Digital Medicine*, 4(1), 15. 2021.

25. Sujath, R. A. A., Chatterjee, J. M., and Hassanien, A. E. "A machine learning forecasting model for COVID-19 pandemic in India." *Stochastic Environmental Research and Risk Assessment*, 34, 959–972. 2020.

26. Shitharth, S., Mohammad, G. B., Ramana, K., and Bhaskar, V. (2021). Prediction of COVID-19 wide spread in India using time series forecasting techniques, Europe PMC plus, 10(1), 1–30. 2021.

27. Dueñas, M., Campi, M., and Olmos, L. E. "Changes in mobility and socioeconomic conditions during the COVID-19 outbreak." *Humanities and Social Sciences Communications*, 8(1), 1–10. 2021.

28. Feng, Z., Yu, Q., Yao, S., Luo, L., Zhou, W., Mao, X., ... & Wang, W. "Early prediction of disease progression in COVID-19 pneumonia patients with chest CT and clinical characteristics." *Nature Communications*, 11(1), 4968. 2020.

29. Kim, C. K., Choi, J. W., Jiao, Z., Wang, D., Wu, J., Yi, T. Y., ... and Bai, H. X. "An automated COVID-19 triage pipeline using artificial intelligence based on chest radiographs and clinical data." *NPJ Digital Medicine*, 5(1), 5. 2022.

30. Xiang, Y., Jia, Y., Chen, L., Guo, L., Shu, B., and Long, E. "COVID-19 epidemic prediction and the impact of public health interventions: A review of COVID-19 epidemic models." *Infectious Disease Modelling*, 6, 324–342. 2021.

31. Tsiknakis, N., Trivizakis, E., Vassalou, E. E., Papadakis, G. Z., Spandidos, D. A., Tsatsakis, A., Sánchez-García, J. et al. "Interpretable artificial intelligence framework for COVID-19 screening on chest X-rays." *Experimental and Therapeutic Medicine*, 20(2), 727–735. 2020.

32. Otoom, M., Otoum, N., Alzubaidi, M. A., Etoom, Y., and Banihani, R. "An IoT-based framework for early identification and monitoring of COVID-19 cases." *Biomedical Signal Processing and Control,* 62, 102149. 2020.

33. Khmaissia, F., Haghighi, P. S., Jayaprakash, A., Wu, Z., Papadopoulos, S., Lai, Y. and Nguyen, F. T. "An unsupervised machine learning approach to assess the zip code level impact of covid-19 in nyc." Proceedings of the Healthcare Systems, Population Health, and the role of Health-Tech (HSYS), 108(1), 1–8. 2020.

34. Zhang, W., Zhou, T., Lu, Q., Wang, X., Zhu, C., Sun, H., Wang, Z., Lo, S. K., and Wang, F.-Y. "Dynamic-fusion-based federated learning for COVID-19 detection." *IEEE Internet of Things Journal,* 8(21), 15884–15891. 2021.

35. Mukherjee, H., Ghosh, S., Dhar, A., Obaidullah, S. M., Santosh, K. C., and Roy, K. "Deep neural network to detect COVID-19: one architecture for both CT Scans and Chest X-rays." *Applied Intelligence,* 51, 2777–2789. 2021.

36. Bhattacharya, S., Maddikunta, P. K. R., Pham, Q.-V., Gadekallu, T. R., Chowdhary, C. L., Alazab, M., and Piran, M. J. "Deep learning and medical image processing for coronavirus (COVID-19) pandemic: a survey." *Sustainable Cities and Society,* 65, 102589. 2021.

37. Monjur, O., Preo, R. B., Shams, A. B., Raihan, M., Sarker, M., and Fairoz, F. "COVID-19 prognosis and mortality risk predictions from symptoms: a loud-based smartphone application." *BioMed,* 1(2), 114–125. 2021.

38. Jin, Q., Cui, H., Sun, C., Meng, Z., Wei, L., and Su, R. "Domain adaptation based self-correction model for COVID-19 infection segmentation in CT images." *Expert Systems with Applications,* 176, 114848. 2021.

39. Yang, D., Xu, Z., Li, W., Myronenko, A., Roth, H. R., Harmon, S., Xu, S. et al. "Federated semi-supervised learning for COVID region segmentation in chest CT using multi-national data from China, Italy, Japan." *Medical Image Analysis,* 70, 101992. 2021.

40. Gunraj, H., Wang, L., and Wong, A. "Covidnet-ct: A tailored deep convolutional neural network design for detection of covid-19 cases from chest ct images." *Frontiers in Medicine,* 7, 608525. 2020.

41. Kassania, S. H., Kassanib, P. H., Wesolowskic, M. J., Schneidera, K. A., and Detersa, R. "Automatic detection of coronavirus disease (COVID-19) in X-ray and CT images: a machine learning based approach." *Biocybernetics and Biomedical Engineering,* 41(3), 867–879. 2021.

26 A Statistical Analysis of Suitable Drugs for Major Drug Resistant Mutations in the HIV-1 Group M Virus

N. Durga Shree, D.A. Steve Mathew,
Ramkumar Thirunavukarasu, and J. Arun Pandian
School of Computer Science Engineering and Information Systems,
Vellore Institute of Technology, Vellore, India

26.1 INTRODUCTION

The human immunodeficiency virus (HIV) attacks the immune system of the body and eventually results in the syndrome AIDS (Acquired Immune Deficiency Syndrome) [1]. Though the medications currently used are effective to control the severity, there is no cure for the syndrome as of now. The mutations in the protein sequences of the HIV-1 Group M virus play a vital role in deciding the effectiveness of the drugs administered. The database used in this study tabulates the resistance from the HIV-1 Group M virus observed in infected patients toward eight different drugs. The resistance of the virus towards these drugs varies based on the mutation patterns observed in the virus. The lesser the resistance offered by the virus to a drug, the better it is and it has a higher probability of being chosen for treatment.

To perform predictions in such an environment by identifying the underlying associations, several methods, such as classification, decision theory, diagnostics, and others can be put into practice. It is interesting to note that all of these depend on the conditional probability as a key parameter. One usually bases a classification, decision, prediction, etc. on some evidence. Hence, using the conditional probability as the base for statistical analysis would be a novel approach for this problem undertaken in this chapter. Conditional probability is of essence to this study because the motive of the study is to identify a suitable drug given a mutation sequence. Statistically, the probability of the outcome given the facts is what one is interested in learning [2]. This can be represented as,

$$P \ (result \mid evidence)$$

The notion behind conditional probability is the probability of a result occurring given evidence (or) a conclusion drawn given a premise. Based on the combinations of mutants in a mutation pattern, various drugs are administered. A drug can be administered for various mutation patterns based on its effectiveness. Conditional probability opens doors to numerous inferences in such a case. This enables us to conduct analysis both ways where we can recommend the best drug for a given mutation pattern and analyze the mutation patterns that can be present in the HIV given that a drug performs well.

The reason to look at this biological problem from a statistical perspective is owing to the potential inter-linkages that can be between the mutations in each tuple of data and the synergistic or

DOI: 10.1201/9781003407959-26

antagonistic effect they might have on the efficiency of each drug. The understanding that can be rendered by mining the associations between the mutations can have powerful mathematical elucidations that support the credibility of the questions in the biological realm using a statistical standpoint. In the Protease Inhibitors Mutation Pattern & Susceptibility dataset used from the Stanford University HIV Drug Resistance Database [3], the common mutation patterns in the virus are tabulated along with the resistance they pose towards each drug. Given the resistance faced by each drug, we can understand which drug is more effective. But, facing the least resistance is not essentially the only criterion for identifying the best drug. Still, this is the criterion we are using in this chapter so it is quantifiable from a statistical perspective. Thus, considering the virtually best drug for each tuple enabled us to convert this problem from being numerically sound to categorical.

Instead of considering the exact numerical value of the resistance for each drug, the best drug is chosen for each mutation pattern. It is defended that this is a valid approach since the resistance values are not constant and are subject to change as we record more sequences. But the predominant nature of the response of a mutation sequence to each drug (more resistant or less resistant) is indicated by the aggregate value of the resistance. Thus, we can conclude that the efficacy of the drug can be quantified in a relativistic manner as compared to an absolute decision.

26.1.1   Contribution Made

* Identifying the probability of different drugs being suitable given that a mutant is observed in the mutation pattern. The confidence with which a drug can be used was very high when certain mutants were observed in the mutation patterns. These exceptional occurrences are recorded.
* Recommending the viable drug for a mutation pattern using the probability of different drugs being suitable given a mutant.

The organization of the chapter is as follows: Section 26.2 surveys and narrates previous studies on the dataset using machine learning models and their corresponding performances. Further, Section 26.3 describes the dataset, and proposes a methodology for analysis and prediction followed by statistical outcomes and prediction. Section 26.4 displays some exceptional cases of mutant–drug relationships and establishes the need and significance to consider conditional probability, and Section 26.5 concludes the chapter.

26.2   LITERATURE SURVEY

Numerous methodologies are being implemented to predict drug resistance and for medical practitioners to suggest a suitable drug for a mutation pattern. Machine learning architectures are at high utilization for this purpose. Artificial neural networks, decision trees, random forests, and logistic regression have been implemented to serve the purpose in better ways.

Neural networks were implemented to predict drug resistance based on the sequence information [4]. Among multilayer perceptron, bidirectional recurrent neural networks, and convolutional neural networks, the comparison proves convolutional neural networks perform better. In line with the idea of our research, model interpretability is the goal of the work done. The results obtained attribute high classification performance. The work [5] reviews and outlines the scope and challenges involved in the computation of drug resistance predictions.

A recurrent neural network is used to analyze sequences in classification problems to predict drug resistance [6]. Considering amino acid energies as features, drug resistances are predicted efficiently with accuracies between 81.3% and 94.7% using bidirectional recurrent neural networks. Single-layered ANNs were also used in similar studies for drug resistance prediction [7]. The HIV resistance to the two protease inhibitors, indinavir and saquinavir, is predicted in a study [8]. In the beginning, a predictor was built using the structural features of the HIV protease-drug inhibitor complex. Then, using the sequence data of multiple drug-resistant mutants, a classifier was built. In both steps, key features were first extracted and patterns were clustered using self-organizing maps. After that, labels were added based on the training set's well-known patterns. Cross-validation was used to evaluate the classifiers' performance in making predictions. With an accuracy of between 60 and 70%, the classifier employed the structure information accurately and identified previously undiscovered mutations.

To find genotypic patterns that are indicative of medication resistance, decision tree classifiers were created [9]. In the study conducted [10,11], random forests were used for the prediction of resistance to HIV reverse transcriptase and protease. Various metrics like sensitivity, specificity, precision, and correlation coefficient were used to come up with a qualitative argument supporting the statistical validity of the model developed. The output of the random forest was directed toward predicting whether a particular mutation sequence belongs to HIV variants that are resistant to a certain drug. From the study, it was proved that this prediction is carried out by the model with an accuracy of 85%.

Regression models are proposed for predicting phenotypic drug resistance from genotypes [12]. Phenotypic resistance testing establishes the optimal dose of the drug such that it prevents viral reproduction to assess the virus's drug susceptibility [13]. But the genotypic resistance testing evaluates the genetic structure of the virus. The patient's blood is drawn, and the HIV is examined for the presence of particular genetic mutations that are known to result in drug resistance [14]. To aid in the understanding of sequence data obtained from genotypic resistance tests, the geno2pheno system, which is based on regression models has been developed in this study, which are based on genotype, may predict the fold-change in drug susceptibility. These models reduce intricate mutational patterns to a single drug resistance factor. In the work conducted [15], they were able to greatly increase overall prediction accuracy for all inhibitors compared to single binary classifiers without any additional information by using multilabel classification models (logistic regression) with cross-resistance data.

According to a study [16], support vector machines perform more accurately when combined with sequence-structure characteristics than random forests classifiers. Additionally, it was claimed that single mutations and the development of multiple mutations have key roles in predicting HIV resistance to specific medication therapies. The discovery of the applicability of these features was also made giving way to estimate resistance and other factors in different complicated ailments.

26.3 METHODS

This section deliberates on the description of the dataset used, the methodology incorporated and its statistical outcomes.

26.3.1 ABOUT THE DATASET

The database source describes major and minor drug resistance mutations (DRMs), which include 36 major DRMs and 22 minor DRMs [3]. The tabulated resistances are reported for mutations in major DRMs. These resistances are an outcome of test results from numerous patients. Based on availability, affordability, and other factors, the number of tests performed for each drug and the number of drugs tested for each mutation pattern vary.

The results of the protease inhibitor (PI) susceptibility database studied [17] are used in this work. The database is the outcome of experimental results on testing drugs against mutation patterns. Some mutation patterns are not tested against all the drugs and a few others do not have experimental results for any drugs. Only those mutation patterns that constitute experimental results for at least one drug are considered for drawing statistical associations. The proposed method is applied to these available results for analysis.

26.3.2 PROPOSED METHODOLOGY

In the previous studies, deep learning techniques such as multilayer perceptron, bidirectional RNN, CNN, machine learning techniques such as random forests, decision trees, support vector machines, and statistical concepts such as density estimation using Gaussian models are implemented to make summarizations, analysis and predictions using the databases available on mutation patterns and corresponding drug resistances [4,10,18,12]. When a more simplistic transparent procedure that gives inferences throughout and that can solve the problem statement can be put to use, it is essential to appreciate its contribution to saving space, time and computational power. To mine the data and propose methods that can serve towards the enhancement of antiretroviral therapy (ART), exploration of various data mining techniques to exhaust all the possibilities of deriving information is essential. Hence, a simplistic data mining method to improve medication by identifying optimal drugs for HIV is proposed here.

The probabilities obtained on application of the concept of conditional probability against mutants in mutation patterns and best drugs are the results from the methodology shown in Figure 26.1. These give us insights into the relationship between a particular mutant and a drug. The detailed step-by-step procedure followed in this study is as follows:

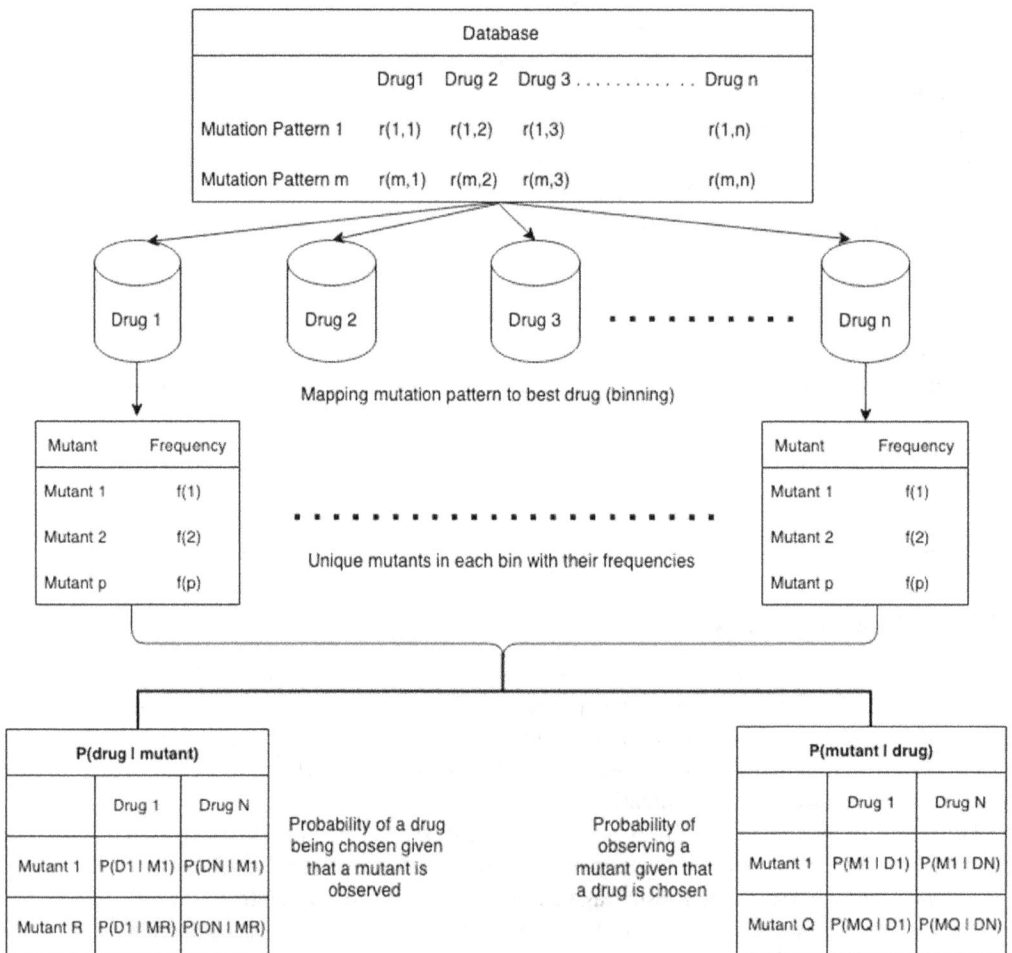

FIGURE 26.1 Proposed methodology.

TABLE 26.1
Frequency of Drugs Acting as the Best Drug

Drug	SQV	NFV	IDV	FPV	ATV	LPV	TPV	DRV
No. of mutation patterns binned	52	1	17	39	7	10	121	39

1. From the database, for each mutation pattern, the drug which witnesses the least resistance from the virus is considered as its best drug in our study. Thus, each mutation pattern is mapped to one optimal drug.
2. All the mutation patterns that have a common best drug are grouped into a bin.
3. Unique mutants corresponding to each bin are identified and their frequency of occurrence in the bin is tabulated. This pictures how often a mutant can be seen while considering a particular drug.
4. Using the available statistics, conditional probabilities can be calculated to devise the probability of:
   1. a drug being chosen given that a mutant observed;
   2. mutant is observed given that a particular drug is chosen.

The lesser the resistance towards a drug, the better the expected performance of the drug on the viruses with the specific mutation pattern. In the study, the best drug is considered as the one which is prone to the least resistance from the mutation pattern of interest. When more than one drug sees the same minimum resistance, the drug which was tested a higher number of times is considered to be the best.

On binning mutation patterns into the corresponding best drugs, as depicted in Table 26.1 and Figure 26.2, the unique set of mutants observed in the mutation patterns can be extracted. The frequency percentage of a mutation is obtained by identifying the ratio between the number of mutation patterns in which the mutation of interest is observed and the number of mutation patterns corresponding to the drug acting as the best. This gives insights into the chances of finding a mutant when a drug is chosen for ART (considering the drug with the least resistance from the virus is always chosen). Statistically, the chances of finding a mutant (frequency of observation) as presented in Tables 26.2–26.9, are based on the concept of conditional probability,

$$P\ (mutation\ |\ drug)$$

For example, Table 26.2 denotes the number of times the mutant is observed when the drug SQV is considered to be the best. Tables 26.3–26.9 tabulate the frequency of observation (%) based on unique mutants observed under best drug classifications. Table 26.10 represents the frequency of mutants observed in the database, and a pictorial representation is given in Figure 26.3.

26.3.3 STATISTICAL OUTCOMES

26.3.3.1   Probability of Observing a Mutant Given that a Drug is Preferred

The heat map generated in Table 26.11 highlights interesting patterns corresponding to mutants. The similarity of behavior between the choice of drug given a mutant and the presence of a mutant given the usage of a drug can be established on those sets that stand out uniquely on the heat map rather than performing computation over all the combinations.

In Table 26.11, for instance, the value corresponding to the mutant 30N and drug SQV is 0.00%. This implies that in no observation of the mutant 30N, the drug SQV has been categorized as the

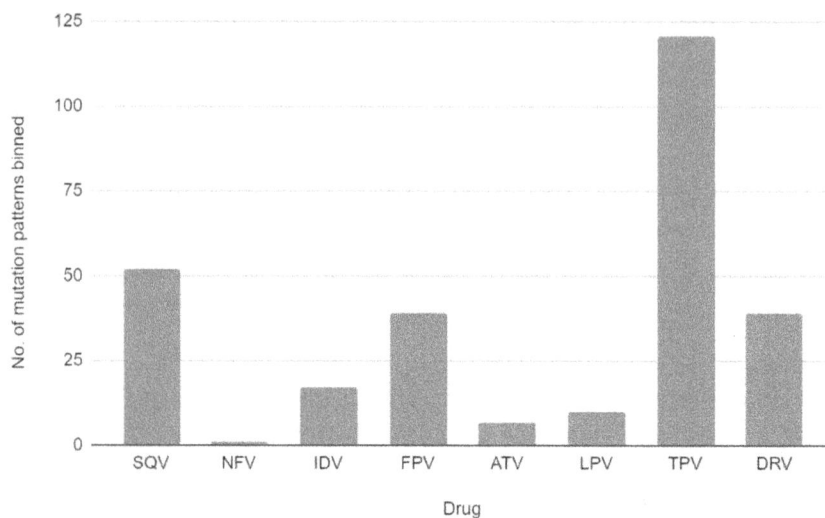

FIGURE 26.2 Frequency of drugs acting as the best drug.

TABLE 26.2
Mutant Composition with SQV as the Best Drug

Mutant	Frequency	Frequency of Observation (%)
46I	32	61.54%
32I	21	40.38%
90M	19	36.54%
82A	16	30.77%
47V	13	25.00%
54M	10	19.23%
76V	8	15.38%
82F	8	15.38%
84V	8	15.38%
54V	8	15.38%
46L	6	11.54%
54L	5	9.62%
82T	4	7.69%
47A	3	5.77%
82L	3	5.77%
50L	2	3.85%
82M	2	3.85%
82S	1	1.92%
50V	1	1.92%
82C	1	1.92%

TABLE 26.3
Mutant Composition with IDV as the Best Drug

Mutant	Frequency	Frequency of Observation (%)
84V	9	52.94%
46I	9	52.94%
90M	9	52.94%
54L	4	23.53%
54V	4	23.53%
82A	3	17.65%
47V	3	17.65%
54M	2	11.76%
46L	2	11.76%
30N	2	11.76%
32I	2	11.76%
82L	1	5.88%
50V	1	5.88%
50L	1	5.88%
48V	1	5.88%
82F	1	5.88%
82C	1	5.88%

TABLE 26.4
Mutant Composition with FPV as the Best Drug

Mutant	Frequency	Frequency of Observation (%)
90M	20	51.28%
54V	19	48.72%
46I	12	30.77%
82A	11	28.21%
46L	10	25.64%
84V	10	25.64%
82T	8	20.51%
88S	7	17.95%
30N	6	15.38%
48V	5	12.82%
32I	3	7.69%
82S	2	5.13%
54T	1	2.56%
48M	1	2.56%
54A	1	2.56%
82C	1	2.56%
47V	1	2.56%
84A	1	2.56%
54L	1	2.56%
48S	1	2.56%

TABLE 26.5
Mutant Composition with ATV as the Best Drug

Mutant	Frequency	Frequency of Observation (%)
46I	5	71.43%
90M	4	57.14%
54V	3	42.86%
76V	3	42.86%
50V	3	42.86%
82A	2	28.57%
46L	1	14.29%
82F	1	14.29%
84V	1	14.29%
47V	1	14.29%
84A	1	14.29%

TABLE 26.6
Mutant Composition with DRV as the Best Drug

Mutant	Frequency	Frequency of Observation (%)
54V	24	61.54%
90M	23	58.97%
82A	16	41.03%
46I	12	30.77%
84V	11	28.21%
46L	9	23.08%
82T	8	20.51%
48V	4	10.26%

best drug. Cumulating the observations, the answers derived from the analysis are driven towards proposing the usage of a drug on observation of mutants in the virus present in a patient.

26.3.3.2 Probability of Using a Drug Given that a Mutant is Present

Given the mutant, the probability of a drug being chosen can be quantified using conditional probabilities,

$$P(drug \mid mutation) = \frac{P(mutation \mid drug) \times P(mutation)}{P(mutation)}$$

$$= \frac{P(drug \cap mutation)}{P(mutation)}$$

$$= \frac{n(drug \cap mutation)}{n(mutation)}$$

TABLE 26.7
Mutant Composition with TPV as the Best Drug

Mutant	Frequency	Frequency of Observation (%)
90M	68	56.20%
82A	62	51.24%
46I	53	43.80%
84V	43	35.54%
54V	35	28.93%
54L	30	24.79%
32I	26	21.49%
46L	25	20.66%
48V	19	15.70%
47V	14	11.57%
50V	12	9.92%
76V	11	9.09%
30N	11	9.09%
54M	11	9.09%
54T	9	7.44%
82T	7	5.79%
54S	6	4.96%
54A	6	4.96%
48M	4	3.31%
82F	3	2.48%
82S	1	0.83%
50L	1	0.83%
47A	1	0.83%
82C	1	0.83%
48L	1	0.83%

TABLE 26.8
Mutant Composition with NFV as the Best Drug

Mutant	Frequency	Frequency of Observation (%)
50L	1	100.00%

The above expression utilizes Tables 26.1–26.10. The values of frequency of Tables 26.2–26.9 serve as the $n(\text{drug} \cap \text{mutation})$ and values of frequency of Table 26.10, also in Figure 26.3 serves as $n(\text{mutation})$.

Table 26.12 gives the probability of choosing a drug for ART given that the mutant is observed in the mutation pattern.

Now, we have the probabilities of choosing a drug given a mutant is observed. But in real-time scenarios, multiple mutants occur simultaneously causing a mutation pattern. If we have to calculate the probability of choosing a particular drug for a mutation pattern, we multiply the individual probabilities (Table 26.12) corresponding to each mutant for the drug of interest. By this, we arrive at the probability that a particular drug is applicable in the case of a particular mutation pattern.

TABLE 26.9
Mutant Composition with LPV as the Best Drug

Mutant	Frequency	Frequency of Observation (%)
46I	6	60.00%
84C	4	40.00%
90M	3	30.00%
84A	3	30.00%
46L	2	20.00%
54M	2	20.00%
48V	1	10.00%
30N	1	10.00%
76V	1	10.00%
32I	1	10.00%
54L	1	10.00%

TABLE 26.10
Frequency of Mutants Observed in the Database

Mutant	Frequency	Mutant	Frequency	Mutant	Frequency
30N	22	48V	30	82F	14
32I	57	50L	7	82L	5
46I	129	50V	17	82M	2
46L	55	54A	7	82T	27
47A	4	54T	10	82S	5
47V	35	54S	8	84A	5
48A	0	54L	41	84C	4
48S	1	54M	26	84V	82
48T	0	54V	93	88S	7
48Q	0	76V	23	88T	0
48L	1	82A	110	88G	0
48M	6	82C	4	90M	146

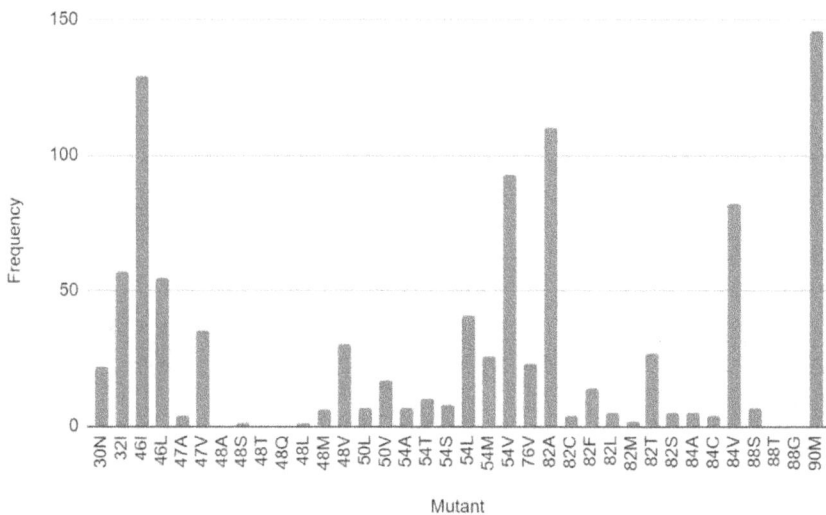

FIGURE 26.3 Frequency of mutants observed in the database.

TABLE 26.11
Heat Map of the Frequency of Observation (%)

	SQV	NFV	IDV	FPV	ATV	LPV	TPV	DRV
30N	0.00%	0.00%	11.76%	15.38%	0.00%	10.00%	9.09%	5.13%
32I	40.38%	0.00%	11.76%	7.69%	0.00%	10.00%	21.49%	10.26%
46I	61.54%	0.00%	52.94	30.77%	71.43%	60.00%	43.80%	30.77%
46L	11.54%	0.00%	11.76%	25.64%	14.29%	20.00%	20.66%	23.08%
47A	5.77%	0.00%	0.00%	0.00%	0.00%	0.00%	0.83%	0.00%
47V	25.00%	0.00%	17.65%	2.56%	14.29%	0.00%	11.57%	7.69%
48A	0.00%	0.00%	0.00%	0.00%	0.00%	0.00%	0.00%	0.00%
48S	0.00%	0.00%	0.00%	2.56%	0.00%	0.00%	0.00%	0.00%
48T	0.00%	0.00%	0.00%	0.0%	0.00%	0.00%	0.00%	0.00%
48Q	0.00%	0.00%	0.00%	0.00%	0.00%	0.00%	0.00%	0.00%
48L	0.00%	0.00%	0.00%	0.00%	0.00%	0.00%	0.83%	0.00%
48M	0.00%	0.00%	0.00%	2.56%	0.00%	0.00%	3.31%	2.56%
48V	0.00%	0.00%	5.88%	12.82%	0.00%	10.00%	15.70%	10.26%
50L	3.85%	100.00%	5.88%	0.00%	0.00%	0.00%	0.83%	5.13%
50V	1.92%	0.00%	5.88%	0.00%	42.86%	0.00%	9.92%	0.00%
54A	0.00%	0.00%	0.00%	2.56%	0.00%	0.00%	4.96%	0.00%
54T	0.00%	0.00%	0.00%	2.56%	0.00%	0.00%	7.44%	0.00%
54S	0.00%	0.00%	0.00%	0.00%	0.00%	0.00%	4.96%	5.13%
54L	9.62%	0.00%	23.53%	2.56%	0.00%	10.00%	24.79%	0.00%
54M	19.23%	0.00%	11.76%	0.00%	0.00%	20.00%	9.09%	2.56%
54V	15.38%	0.00%	23.53%	48.72%	42.86%	0.00%	28.93%	61.54%
76V	15.38%	0.00%	0.00%	0.00%	42.86%	10.00%	9.09%	0.00%
82A	30.77%	0.00%	17.65%	28.21%	28.57%	0.00%	51.24%	41.03%
82C	1.92%	0.00%	5.88%	2.56%	0.00%	0.00%	0.83%	0.00%
82F	15.38%	0.00%	5.88%	0.00%	14.29%	0.00%	2.48%	2.56%
82L	5.77%	0.00%	5.88%	0.00%	0.00%	0.00%	0.00%	2.56%
82M	3.85%	0.00%	0.00%	0.00%	0.00%	0.00%	0.00%	0.00%
82T	7.69%	0.00%	0.00%	20.51%	0.00%	0.00%	5.79%	20.51%
82S	1.92%	0.00%	0.00%	5.13%	0.00%	0.00%	0.83%	2.56%
84A	0.00%	0.00%	0.00%	2.56%	14.29%	30.00%	0.00%	0.00%
84C	0.00%	0.00%	0.00%	0.00%	0.00%	40.00%	0.00%	0.00%
84V	15.38%	0.00%	52.94%	25.64%	14.29%	0.00%	35.54%	28.21%
88S	0.00%	0.00%	0.00%	17.95%	0.00%	0.00%	0.00%	0.00%
88T	0.00%	0.00%	0.00%	0.00%	0.00%	0.00%	0.00%	0.00%
88G	0.00%	0.00%	0.00%	0.00%	0.00%	0.00%	0.00%	0.00%
90M	36.54%	0.00%	52.94%	51.28%	57.14%	30.00%	56.20%	58.97%

The above statistical outcomes are obtained based on binning into the best drugs. This implies that the mutation pattern was tested against at least one drug. In cases where the mutation patterns are observed but the resistances for no drug are available, prediction of the best drug is required.

In Table 26.12, we can observe that there are certain drugs where the probability of choosing them for a particular mutant is 0. This is because these drugs have not acted as the best drugs in even one instance where this particular mutant was observed. When we multiply the probabilities corresponding to each mutant in a mutation pattern, if 0 is one of the probabilities, then the joint probability becomes 0. This completely excludes the drug from being considered for administration. To deal with these exceptional cases, we have to introduce every mutant to every drug.

TABLE 26.12
Probability of Choosing a Drug Given a Mutant is Observed in a Mutation Pattern Set

MUTANTS	SQV	NFV	IDV	FPV	ATV	LPV	TPV	DRV
30N	0.0000	0.0000	0.0909	0.2727	0.0000	0.0455	0.5000	0.0909
32I	0.3684	0.0000	0.0351	0.0526	0.0000	0.0175	0.4561	0.0702
46I	0.2481	0.0000	0.0698	0.0930	0.0388	0.0465	0.4109	0.0930
46L	0.1091	0.0000	0.0364	0.1818	0.0182	0.0364	0.4545	0.1636
47A	0.7500	0.0000	0.0000	0.0000	0.0000	0.0000	0.2500	0.0000
47V	0.3714	0.0000	0.0857	0.0286	0.0286	0.0000	0.4000	0.0857
48A	0.0000	0.0000	0.0000	0.0000	0.0000	0.0000	0.0000	0.0000
48S	0.0000	0.0000	0.0000	1.0000	0.0000	0.0000	0.0000	0.0000
48T	0.0000	0.0000	0.0000	0.0000	0.0000	0.0000	0.0000	0.0000
48Q	0.0000	0.0000	0.0000	0.0000	0.0000	0.0000	0.0000	0.0000
48L	0.0000	0.0000	0.0000	0.0000	0.0000	0.0000	1.0000	0.0000
48M	0.0000	0.0000	0.0000	0.1667	0.0000	0.0000	0.6667	0.1667
48V	0.0000	0.0000	0.0333	0.1667	0.0000	0.0333	0.6333	0.1333
50L	0.2857	0.1429	0.1429	0.0000	0.0000	0.0000	0.1429	0.2857
50V	0.0588	0.0000	0.0588	0.0000	0.1765	0.0000	0.7059	0.0000
54A	0.0000	0.0000	0.0000	0.1429	0.0000	0.0000	0.8571	0.0000
54T	0.0000	0.0000	0.0000	0.1000	0.0000	0.0000	0.9000	0.0000
54S	0.0000	0.0000	0.0000	0.0000	0.0000	0.0000	0.7500	0.2500
54L	0.1220	0.0000	0.0976	0.0244	0.0000	0.0244	0.7317	0.0000
54M	0.3846	0.0000	0.0769	0.0000	0.0000	0.0769	0.4231	0.0385
54V	0.0860	0.0000	0.0430	0.2043	0.0323	0.0000	0.3763	0.2581
76V	0.3478	0.0000	0.0000	0.0000	0.1304	0.0435	0.4783	0.0000
82A	0.1455	0.0000	0.0273	0.1000	0.0182	0.0000	0.5636	0.1455
82C	0.2500	0.0000	0.2500	0.2500	0.0000	0.0000	0.2500	0.0000
82F	0.5714	0.0000	0.0714	0.0000	0.0714	0.0000	0.2143	0.0714
82L	0.6000	0.0000	0.2000	0.0000	0.0000	0.0000	0.0000	0.2000
82M	1.0000	0.0000	0.0000	0.0000	0.0000	0.0000	0.0000	0.0000
82T	0.1481	0.0000	0.0000	0.2963	0.0000	0.0000	0.2593	0.2963
82S	0.2000	0.0000	0.0000	0.4000	0.0000	0.0000	0.2000	0.2000
84A	0.0000	0.0000	0.0000	0.2000	0.2000	0.6000	0.0000	0.0000
84C	0.0000	0.0000	0.0000	0.0000	0.0000	1.0000	0.0000	0.0000
84V	0.0976	0.0000	0.1098	0.1220	0.0122	0.0000	0.5244	0.1341
88S	0.0000	0.0000	0.0000	1.0000	0.0000	0.0000	0.0000	0.0000
88T	0.0000	0.0000	0.0000	0.0000	0.0000	0.0000	0.0000	0.0000
88G	0.0000	0.0000	0.0000	0.0000	0.0000	0.0000	0.0000	0.0000
90M	0.1301	0.0000	0.0616	0.1370	0.0274	0.0205	0.4658	0.1575

To introduce, the minimum frequency count, 1 is given to each mutant with frequency 0 for each bin. To balance the proportion of a mutant's presence in each drug binning, every frequency value is incremented by 1, as shown in Table 26.13.

After the updation, the conditional probability of choosing a drug given a mutant is given in Table 26.14 and Figure 26.4.

26.3.4 RECOMMENDATION

To predict the suitable drug for a mutation pattern, the probabilities for each drug being suitable are computed using joint probability.

TABLE 26.13
Actual Versus Updated Frequency of a Mutant in Each Best Drug Bin

30N	SQV	NFV	IDV	FPV	ATV	LPV	TPV	DRV
Actual	0	0	2	6	0	1	11	2
Updated	1	· 1	3	7	1	2	12	3

TABLE 26.14
Updated Probability of Choosing a Drug Given a Mutant is Observed in a Mutation Pattern Set

Mutants	SQV	NFV	IDV	FPV	ATV	LPV	TPV	DRV
30N	0.0333	0.0333	0.1000	0.2333	0.0333	0.0667	0.4000	0.1000
32I	0.3385	0.0154	0.0462	0.0615	0.0154	0.0308	0.4154	0.0769
46I	0.2409	0.0073	0.0730	0.0949	0.0438	0.0511	0.3942	0.0949
46L	0.1111	0.0159	0.0476	0.1746	0.0317	0.0476	0.4127	0.1587
47A	0.3333	0.0833	0.0833	0.0833	0.0833	0.0833	0.1667	0.0833
47V	0.3256	0.0233	0.0930	0.0465	0.0465	0.0233	0.3488	0.0930
48A	0.1250	0.1250	0.1250	0.1250	0.1250	0.1250	0.1250	0.1250
48S	0.1111	0.1111	0.1111	0.2222	0.1111	0.1111	0.1111	0.1111
48T	0.1250	0.1250	0.1250	0.1250	0.1250	0.1250	0.1250	0.1250
48Q	0.1250	0.1250	0.1250	0.1250	0.1250	0.1250	0.1250	0.1250
48L	0.1111	0.1111	0.1111	0.1111	0.1111	0.1111	0.2222	0.1111
48M	0.0714	0.0714	0.0714	0.1429	0.0714	0.0714	0.3571	0.1429
48V	0.0263	0.0263	0.0526	0.1579	0.0263	0.0526	0.5263	0.1316
50L	0.2000	0.1333	0.1333	0.0667	0.0667	0.0667	0.1333	0.2000
50V	0.0800	0.0400	0.0800	0.0400	0.1600	0.0400	0.5200	0.0400
54A	0.0667	0.0667	0.0667	0.1333	0.0667	0.0667	0.4667	0.0667
54T	0.0556	0.0556	0.0556	0.1111	0.0556	0.0556	0.5556	0.0556
54S	0.0625	0.0625	0.0625	0.0625	0.0625	0.0625	0.4375	0.1875
54L	0.1224	0.0204	0.1020	0.0408	0.0204	0.0408	0.6327	0.0204
54M	0.3235	0.0294	0.0882	0.0294	0.0294	0.0882	0.3529	0.0588
54V	0.0891	0.0099	0.0495	0.1980	0.0396	0.0099	0.3564	0.2475
76V	0.2903	0.0323	0.0323	0.0323	0.1290	0.0645	0.3871	0.0323
82A	0.1441	0.0085	0.0339	0.1017	0.0254	0.0085	0.5339	0.1441
82C	0.1667	0.0833	0.1667	0.1667	0.0833	0.0833	0.1667	0.0833
82F	0.4091	0.0455	0.0909	0.0455	0.0909	0.0455	0.1818	0.0909
82L	0.3077	0.0769	0.1538	0.0769	0.0769	0.0769	0.0769	0.1538
82M	0.3000	0.1000	0.1000	0.1000	0.1000	0.1000	0.1000	0.1000
82T	0.1429	0.0286	0.0286	0.2571	0.0286	0.0286	0.2286	0.2571
82S	0.1538	0.0769	0.0769	0.2308	0.0769	0.0769	0.1538	0.1538
84A	0.0769	0.0769	0.0769	0.1538	0.1538	0.3077	0.0769	0.0769
84C	0.0833	0.0833	0.0833	0.0833	0.0833	0.4167	0.0833	0.0833
84V	0.1000	0.0111	0.1111	0.1222	0.0222	0.0111	0.4889	0.1333
88S	0.0667	0.0667	0.0667	0.5333	0.0667	0.0667	0.0667	0.0667
88T	0.1250	0.1250	0.1250	0.1250	0.1250	0.1250	0.1250	0.1250
88G	0.1250	0.1250	0.1250	0.1250	0.1250	0.1250	0.1250	0.1250
90M	0.1299	0.0065	0.0649	0.1364	0.0325	0.0260	0.4481	0.1558

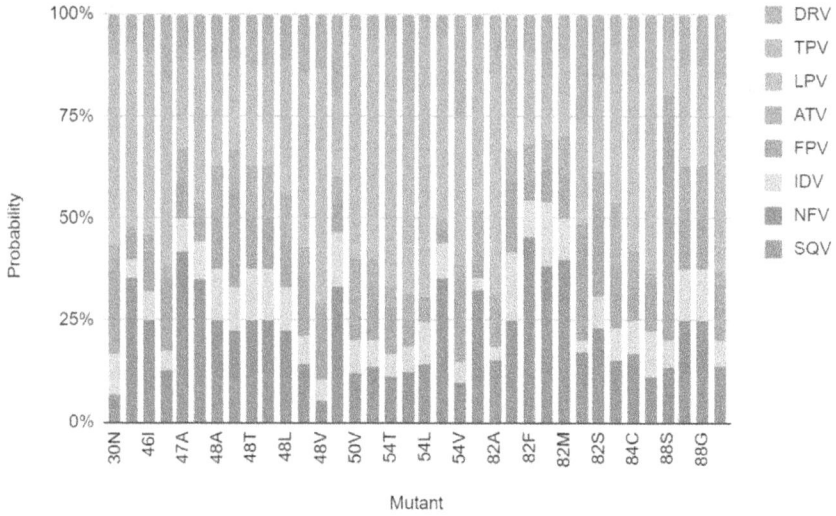

FIGURE 26.4　Probability of choosing a drug given a mutant is observed in a mutation pattern set.

Considering that the mutants in a sequence are independent to each other,

$$P(drug \mid m_1 \cap m_2 \cap .... \cap m_n) = \frac{P\left(drug \cap \left(m_1 \cap m_2 \cap .... \cap m_n\right)\right)}{P\left(m_1 \cap m_2 \cap .... \cap m_n\right)}$$

$$= \frac{P\left(drug \cap m_1\right) \times P\left(drug \cap m_2\right) \times ... \times P\left(drug \cap m_n\right)}{P\left(m_1\right) \times P\left(m_2\right) \times ... \times P\left(m_n\right)}$$

$$= \frac{P\left(drug \cap m_1\right)}{P\left(m_1\right)} \times \frac{P\left(drug \cap m_2\right)}{P\left(m_2\right)} \times ... \times \frac{P\left(drug \cap m_n\right)}{P\left(m_n\right)}$$

$$= P\left(drug \mid m_1\right) \times P\left(drug \mid m_2\right) \times ... \times P\left(drug \mid m_n\right)$$

For a particular mutation pattern, the drug which gets the highest probability is likely to be qualified as the best drug. On comparing the resistances of the predicted best drug and actual best drug (drug corresponding to minimum resistance for a given mutation pattern in the database), the difference between the resistances towards them from the virus can be considered as a measure to quantify the goodness of considering conditional probability for prediction.

26.4　RESULTS AND DISCUSSIONS

26.4.1　Exceptional Relationships Between Drugs and Mutants

When the probabilities of choosing a particular drug given a mutation were considered, certain drugs showed extremely positive results when certain mutants were present in the mutation pattern. In certain cases, certain drugs faced the least resistance given the presence of a mutant irrespective of any other mutant in the pattern. Some of the findings from this analysis are as follows:

- When the mutation sequence has mutant 84C, LPV faces the least resistance at all times. It is also to be noted that in 40% of mutation combinations that are treated with LPV, 84C mutation is found. The probability that the drug LPV faces the least resistance given that the mutant 84C is present in the mutation pattern is

$$P\ (drug=LPV\ |\ mutant=84C)=\frac{P\big(drug=LPV\cap mutant=84C\big)}{P\big(mutant=84C\big)}=\frac{4}{4}=1$$

- When the mutation sequence has mutant **47A**, the drug SQV faces the least resistance in the majority of the cases. The absence of mutant 32I (which is observed in one out of the four cases where 47A is observed), has contributed to lesser resistance against the drug TPV than the drug SQV. In all the patterns, it is observed that mutant 47A always occurs with mutant 46I. The probability that the drug SQV faces the least resistance given that the mutant 47A is present in the mutation pattern is,

$$P\ (drug=SQV\ |\ mutant=47A)=\frac{P\big(drug=SQV\cap mutant=47A\big)}{P\big(mutant=47A\big)}=\frac{3}{4}=0.75$$

- When the mutation sequence has mutant **88S**, the drug FPV faces the least resistance in all the cases. The probability that the drug FPV faces the least resistance given that the mutant 88S is present in the mutation pattern is

$$P\ (drug=FPV\ |\ mutant=88S)=\frac{P\big(drug=FPV\cap mutant=88S\big)}{P\big(mutant=88S\big)}=\frac{7}{7}=1$$

- When the mutation sequence has the mutant 48M, the drug TPV faces the least resistance in most cases. The probability that the drug TPV faces the least resistance given that the mutant 48M is present in the mutation pattern is,

$$P\ (drug=TPV\ |\ mutant=48M)=\frac{P\big(drug=TPV\cap mutant=48M\big)}{P\big(mutant=48M\big)}=\frac{4}{6}=0.66$$

- When the mutation sequence has the mutant 54L, the drug TPV faces the least resistance in most cases. The probability that the drug TPV faces the least resistance given that the mutant 54L is present in the mutation pattern is

$$P\ (drug=TPV\ |\ mutant=54L)=\frac{P\big(drug=TPV\cap mutant=54L\big)}{P\big(mutant=54L\big)}=\frac{30}{41}=0.7371$$

26.4.2 GOODNESS OF CONSIDERING CONDITIONAL PROBABILITY IN THIS STUDY

Deep learning models are considered "black-box" because of their inability to provide explanations for the outcome of the model. When drug prediction is the only expected output, these models perform the best. But when model interpretability is expected by researchers to further explore the mutation sequences and other underlying associations between mutants, mutation patterns, and drugs, a viable methodology as used in this study is necessary. The proposed methodology in this study uses the idea of conditional probability, providing interpretations and reasons for decisions statistically. Though numerous machine learning and deep learning models are used for drug resistance prediction, a simple and yet efficient methodology is crucial for drug prediction, as designed in this study. It is significantly necessary to consider methodologies that give simpler solutions to problems to save resources.

About the recommendation section, one practical hindrance in computing the difference between a few resistances is that the resistance of the predicted best drug is not registered in the database. The average difference in resistance to omitting such cases is 13.76 units. This implies that the resistance towards the drug that is chosen from the algorithm used in this study will on average vary from the best drug by 13.76 units.

In certain cases of mutation patterns, the difference between the resistance towards the predicted best drug and the actual best drug differed by vast amounts. The cases that resulted in these vast amounts were removed using box-plot analysis.

On application of box-plot analysis on the differences, about 39 values out of 286 were categorized as outliers and the average difference in resistance between actual and predicted best drug, in this case, is –0.149. It is also worthwhile to note that since the dominance of zero difference (right prediction) is there, very small values are also considered to be outliers. While applying box-plot analysis, if 0 difference is considered only once, then the number of outliers is reduced to 14 and the average difference in resistance between actual and predicted best drug, in this case, is –0.923. In both cases, it is evident that the resistance of the predicted drug has a very minimum average difference with the resistance of the drug that is the best. This corroborates the prediction of the drug made by the methodology is appreciably effective in identifying the best drug for a given mutation pattern.

26.5   CONCLUSION

The synergistic or antagonistic effect rendered by the combination of mutations is analyzed from the drug resistances to quantitatively comment on the most viable drug(s). The notion of conditional probability has served to be a potential pathway to analyze the relationships between drugs, mutations, and mutation patterns. Though there are many extensive models for suitable drug prediction, conditional probability establishes a way to infer the underlying associations between mutation patterns and the respective potential drug for treatment. Relationships between mutants and drugs have been established in this study using conditional probability. Joint probability has been used to predict the suitable drug for a given mutation pattern using mutant conditional probabilities. This is an attempt to establish concrete statistical associations to enhance antiretroviral therapy for HIV.

REFERENCES

1.   Centers for Disease Control and Prevention, last accessed on November 4, 2022, www.cdc.gov/hiv/basics/whatishiv.html
2.   The Advanced Processor Technologies Research Group, University of Manchester, last accessed on November 26, 2022, https://apt.cs.manchester.ac.uk/ftp/pub/ai/jls/CS2411/prob97/node6.html
3.   Rhee, S.-Y., Gonzales, M. J., Kantor, R., Betts, B. J., Ravela, J., and Shafer, R. W., "Human immunodeficiency virus reverse transcriptase and protease sequence database", *Nucleic Acids Research*, 31(1), 2003, 298–303.
4.   Steiner, M. C., Gibson, K. M., and Crandall, K. A., "Drug resistance prediction using deep learning techniques on HIV-1 sequence data", *Viruses*, 12(5), 2020, 560.
5.   Riemenschneider, M., and Heider, D., "Current approaches in computational drug resistance prediction in HIV", *Currernt HIV Research*, 14(4), 2016, 307–315.
6.   Bonet, I., et al., "Predicting human immunodeficiency virus (HIV) drug resistance using recurrent neural networks", International Work-Conference on the Interplay Between Natural and Artificial Computation, Springer, Berlin, Heidelberg, 234–243, 2007.

7. Amamuddy, O. S., Bishop, N. T., and Bishop, Ö. T., "Improving fold resistance prediction of HIV-1 against protease and reverse transcriptase inhibitors using artificial neural networks". *BMC Bioinformatics* 18(1), 2017, 1–7.

8. Drăghici, S., and Potter, R. B., "Predicting HIV drug resistance with neural networks", *Bioinformatics* 19(1), 2003, 98–107.

9. Beerenwinkel, N., Schmidt, B., Walter, H., Kaiser, R., Lengauer, T., Hoffmann, D., Korn, K., and Selbig, J. "Diversity and complexity of HIV-1 drug resistance: A bioinformatics approach to predicting phenotype from genotype", *Proceeding of the National Academy of Sciences USA*, 99, 2002, 8271–8276.

10. Tarasova, O., et al., "A computational approach for the prediction of HIV resistance based on amino acid and nucleotide descriptors", *Molecules* 23(11), 2018, 2751.

11. Tarasova, O. A., Filimonov, D. A., and Poroikov, V.V., "Computational prediction of human immunodeficiency resistance to reverse transcriptase inhibitors", *Biomed Khim*, 63(5) 2017, 457–460.

12. Beerenwinkel, N., et al., "Geno2pheno: Estimating phenotypic drug resistance from HIV-1 genotypes", *Nucleic Acids Research*, 31(13), 2003, 3850–3855.

13. Andreoni, M., "Phenotypic resistance testing", *Scandinavian Journal of Infectious Diseases Supple*ment, 106, 2003, 35–36.

14. HIV Drug Resistance Testing, Stanford Medicine Health Care, last accessed on December 24, 2022, https://stanfordhealthcare.org/medical-conditions/sexual-and-reproductive-health/hiv-aids/treatments/hiv-drug-resistance-testing.html

15. Heider, D., et al., "Multilabel classification for exploiting cross-resistance information in HIV-1 drug resistance prediction", *Bioinformatics* 29(16), 2013, 1946–1952.

16. Z., Khalid, and Sezerman, O. U. "Prediction of HIV drug resistance by combining sequence and structural properties", in *IEEE/ACM Transactions on Computational Biology and Bioinformatics*, 15(3), 2018, 966–973.

17. Rhee, S.-Y., et al., "HIV-1 protease mutations and protease inhibitor cross-resistance", *Antimicrobial Agents and Chemotherapy*, 54(10), 2010, 4253–4261.

18. Beerenwinkel, N., et al. "Diversity and complexity of HIV-1 drug resistance: A bioinformatics approach to predicting phenotype from genotype", *Proceedings of the National Academy of Sciences* 99(12), 2002, 8271–8276.

Index

A

Activation function, 14–15
Artificial Intelligence, 103–105
Artificial neural network, 321–322
Autoencoders, 27

B

Backward elimination, 6
BERT model, 253–254
  BERT-BiLSTM, 254–255
Bias, 14–15
Big data, 19–20
  rainfall data, 20
Bidirectional LSTM, 194–195
Black-box model, 109
Blockchain, 54–55

C

Century Tech, 218–220
Ciphertext-policy, 144–146
Cognii, 217
Computer Vision, 292–294
Confusion matrix, 316–318
ChatGPT, 216–218
Cloud computing, 86–90, 121–122
  gains, limitation, 89–90
Cloud-based services, 91–93
  SaaS, PaaS, IaaS, 91–92
Cloud deployment models, 90–91
  public, private, hybrid cloud, 90–92
  community cloud, 91–92
Cluster parameter, 317
Convolutional neural networks (CNN), 24–25, 69–70,
  155–156, 207, 314–315
Crop yield production, 339–341
Customer information systems (CIS), 2–3
Cyber-physical systems, 36

D

Darknet-19, 371–372
Data centers, 1–2
Data normalization, 21–22
Data management, 37–38
Data management framework, 39–41
Data processing, 21–22
  data pre-processing steps, 22
Deep learning, 14–17, 61, 154–155, 339–340
Decision tree, 140
Discriminative Modeling, 77–78
Dynamic prediction migration, 7–8

E

Edge computing, 82–84, 121–122
  gains, 84–85
  limitation, 84–85
  challenges, 122–123
Electronic health records (EHRs), 36–37
Electrocardiography, 100–101
Electrooculography (EOG), 100–101
Energy management system (EMS), 2–3

F

Feature extraction, 23–24
  extraction techniques, 23
Fetal monitoring sensors, 270–271
  Ecg sensor, oxygen sensor, 270–271
  Temperature Sensor, 271
  Blood Pressure Sensor, 271
Flood prediction, 16–17
Fog computing, 52–55, 85–86, 121–123
  characteristics, 54–55
  gains, 86
  limitations, 86
F-measure, 209–210

G

Generative modeling, 78
Gated recurrent unit, 26–27
Generative adversarial networks, 26, 72
Gensim, 199
Gradescope, 217
GPU machines, 311–312

H

Healthcare, 35–38, 266–267
  challenges, 38–39
Honeypots, 143
Hybrid text classification model, 163–164
  preprocessing, 163
  text representation, 164
  ensemble models, 164–165
Hydrological models, 20

I

Image augmentation techniques, 355–356
Internet of Things (IoT), 35–37, 52–57
  IoT use cases, 57
  IoT layers, 138–139
  components of The IoT ecosystem, 138–140
Intrusion detection system (IDS), 117–119
  location-based intrusion detection system, 117
  host intrusion detection system (HIDS), 117
  challenges, 119–120
Ivy Chatbot, 218–219

K

Kernel, 141
Keras, 199

L

Long-short term memory network, 26–27, 73–74, 194

M

Machine learning, 140–141
Mean absolute error (MAE), 28, 78–79
Mean squared error (MSE), 78–79
Medical Internet of Things, 38–39
MobileNet model, 328–330
Model interpretability, 105–106
Multi-layer perceptron, 68–69

N

Natural language processing (NLP), 188–189
Network intrusion detection system (NIDS), 119–121
Neural network layers, 14–15

O

Object detection, 369–371
Ontology, 113
Operating systems, 139–140

P

Personalized learning, 214–215
Probabilistic modelling, 78
Precision, 209–210
Polynomial regression, 173–174

R

Random forest (RF), 387–388
Random migration, 7
Recurrent neural network, 25–26, 70–72, 207–208
Regression models, 172–173
Recall, 209–210
ResNet model, 329–330
Restricted Boltzmann machines, 141–142
Ridge regression, 173–175
Root mean squared error (RMSE), 28, 78–79
R-squared (R2), 28, 78–79

S

Signature-based intrusion detection system, 119
Smart grid, 2, 127–128
Smart city, 48–50, 128–130
Sensors, 34–37
Sentimental analysis, 188–190, 202–205
  types, 190–191
  levels, 204–205
Serverless computing, 91–93
Stacked bidirectional LSTM model, 252–253
SPARQL queries, 4–5
Spatiotemporal modelling, 76–78
Support vector machine (SVM), 386–387

T

Transformers, 27
  transformer networks, 297–300
TensorFlow, 198–199

U

Ultrasound imaging, 291

V

Vectoring techniques, 310–311
VGG16 model, 327–329

W

Wearable device, 99–100
Weather prediction, 75

X

X-Ray image modality, 289–291
XAI, 103–112
  challenges, 112–113
XAI models, 108
  data analysis, transformation, 108–109
  data summarization, 109–110
  data squashing, 109–100

Z

Z-score normalization, 23–24

For Product Safety Concerns and Information please contact our EU
representative GPSR@taylorandfrancis.com
Taylor & Francis Verlag GmbH, Kaufingerstraße 24, 80331 München, Germany

www.ingramcontent.com/pod-product-compliance
Lightning Source LLC
Chambersburg PA
CBHW080140220326
41598CB00032B/5126